EINSTEIN SPACES

EINSTEIN SPACES

by

A. Z. PETROV

Professor of Relativity and Gravitation
Kazan University

Translated by

R. F. KELLEHER

Translation edited by

J. WOODROW

THE QUEEN'S AWARD TO INDUSTRY 1966

PERGAMON PRESS

OXFORD · LONDON · EDINBURGH · NEW YORK
TORONTO · SYDNEY · PARIS · BRAUNSCHWEIG

Pergamon Press Ltd., Headington Hill Hall, Oxford
4 & 5 Fitzroy Square, London W.1
Pergamon Press (Scotland) Ltd., 2 & 3 Teviot Place, Edinburgh 1
Pergamon Press Inc., 44–01 21st Street, Long Island City, New York 11101
Pergamon of Canada Ltd., 207 Queen's Quay West, Toronto 1
Pergamon Press (Aust.) Pty. Ltd., 19a Boundary Street, Rushcutters Bay, N.S.W. 2011, Australia
Pergamon Press S.A.R.L., 24 rue des Écoles, Paris 5e
Vieweg & Sohn GmbH, Burgplatz 1, Braunschweig

First English edition 1969

This is a translation of the Russian
Пространства Эйнштейна (Prostranstva Einshteina)
published by Fizmatlit, Moscow, 1961 and contains
amendments and revisions supplied by the author

Library of Congress Catalog Card No. 67-23712

PRINTED IN HUNGARY
08 012315 5

CONTENTS

PREFACE TO THE ENGLISH EDITION

EVERY specialist needs to be fully conversant with the basic facts of his subject and the methods of investigating them. For a physicist specializing in the general theory of relativity, his "subject" comprises the *field* as a physical object and the *four-dimensional Riemann manifold* as the mathematical description of this field. The methods of investigation comprise experiments followed by hypotheses and assembly of data defining four-dimensional Riemannian geometry with signature $(- - - +)$. Naturally, contemporary mathematicians cannot be satisfied with the methods of analysis used by Einstein, Hilbert and the other founders of general relativity; this explains the present tendencies to devise *up-to-date* experiments and to apply new mathematical methods to investigate new topics.

The principal aim of our book is to provide a study of the mathematical basis of the theory of gravitation; invariant methods are used whenever possible, e.g. differential and algebraic invariants, Lie groups. The emphasis is therefore mathematical, the details of the physical theory of fields being scarcely mentioned. The book is really no more than an introduction to modern field theory and probably not complete at that. Nevertheless, we hope that triviality has been avoided and that the book will prove useful.

The scope of the book is considerably broader than implied by the title, which formally means a study of manifolds with a Ricci tensor proportional to the metric tensor. Although these special cases of Riemann manifolds are very important, the problems considered are almost invariably solved in a very general form, limited only by the requirement that the space be four-dimensional with metric signature $---+$; hence by "Einstein spaces" the reader should understand any space-time arising in the general theory of relativity.

None of the so-called "unified theories" are considered in this book, as in our opinion their physical significance is inversely proportional to the number of variants advanced, now more than twenty.

The tendency to introduce problems which the author and his colleagues have recently investigated has resulted in many interesting topics being omitted altogether, apart from references in the bibliography, which we do not claim to be complete. The book has been extensively revised, corrected and modified for the English edition; a new chapter on the geodesic mapping of gravitational fields has been added and many new examples have been provided. The author would be grateful for any criticism or suggestions concerning the problems discussed.

The reader should consider the book as an introduction rather than as a complete survey; if, after reading it, he comes to regard some aspects of relativity in a new light and tries to apply some of the methods and results in his own work, the author's purpose in writing the book will be achieved.

FOREWORD

THIS book is devoted to an investigation of the various spaces which form the basis of the theory of relativity, and the generalization of these to an arbitrary number of dimensions.

As there already exists among publications on relativity theory a series of fundamental monographs such as Landau and Lifshitz's *The Classical Theory of Fields* [156] and Fock's *Theory of Space, Time and Gravitation* [225], we have consciously tried to limit ourselves to a number of questions not considered in these works, which should be of interest to both physicists and mathematicians. This explains why the treatment is given a mathematical emphasis and the special attention paid to 4-spaces with a Lorentz-type signature.

Many new results appearing in a wide range of journals remain unknown to mathematicians and physicists not specifically concerned with aspects of the general theory of relativity. Some of these results are considered in this book; other problems, which for some reason or other cannot be included, are either indicated by references in square brackets [ ] to the bibliography, which we have made as complete as possible, or provide the basis of examples, some requiring simple application of results obtained in the book, and others considerable extension of the theory.

Whilst remaining inside the framework of the general theory of relativity, we do not consider different versions of a single theory, but concentrate the reader's attention on currently important problems—such as the application of Lie groups to the investigation of gravitational fields, the Cauchy problem, and methods for the invariant treatment of Einstein spaces.

A study of the contemporary development of the theory of relativity, leading to the investigation of such problems as gravitational radiation, the behaviour of elementary particles in a gravitational field, and the interaction of fields, necessarily demands a refined approach. In writing this book we have tried to help the reader to acquire such an approach.

We wish to acknowledge with deep appreciation the work done by the editor of the Russian edition, A. F. Lapko; his advice and comments have greatly affected the quality of the book.

NOTATION

CHAPTER 1

BASIC TENSOR ANALYSIS

EVERY Einstein space represents a special case of a Riemann manifold and hence it is essential to start by defining this latter concept.

§ 1. Riemann Manifolds

Let us consider the sets of objects which can be put in one–one correspondence with all the ordered sets of n real or complex numbers $(x^1, \ldots, x^n)$ satisfying the inequalities

$$|x^\alpha - a^\alpha| < \varepsilon^\alpha, \tag{1.1}$$

where a^α $(\alpha = 1, 2, \ldots, n)$ are constants and ε^α are positive numbers. We call such a set *a finite domain of n-dimensional space*, denoted by the symbol X_n, and the objects themselves—*points* in X_n. Depending on the application and the number n, these points can be interpreted in different ways (frequencies in a spectrum, events in a space-time continuum, the state of a dynamic system, straight lines, etc.).

Each one–one correspondence between points and sets of numbers $(x^1, \ldots, x^n)$ is called a *coordinate system*; the numbers x^α are the *coordinates of the corresponding point in* X_n.

If two coordinate systems, $\{x^\alpha\}$ and $\{x^{\alpha'}\}$, are given in some domain, there exists a continuous one–one correspondence between them; this is expressed by the equations

$$x^\alpha = f^\alpha(x^{1'}, \ldots, x^{n'}), \quad x^{\alpha'} = \varphi^{\alpha'}(x^1, \ldots, x^n) \tag{1.2}$$

These equations are said to define a *transformation of coordinates*. We can also consider (1.2) as a *point* transformation, where x^α and $x^{\alpha'}$ are considered as the coordinates of different points in the same coordinate system; point transformations will be indicated by changing the letter of the alphabet or by adding the sign *, whilst keeping the same indices:

$$x^\alpha = x^\alpha(y^1, \ldots, y^n), \quad \overset{*}{x}{}^\alpha = \overset{*}{x}{}^\alpha(x^1, \ldots, x^n). \tag{1.3}$$

A function $f(x^1, \ldots, x^n)$ which has continuous partial derivatives up to order r is called a *function of class* C^r. If f and φ in (1.2) are of class C^r and the Jacobians $\left|\frac{\partial f^\alpha}{\partial x^{\beta'}}\right|$ and $\left|\frac{\partial \varphi^{\alpha'}}{\partial x^\beta}\right|$ differ from zero, for any coordinate transformation in the domain, X_n is said to be a *space of class* C^r.

If $\{x^\alpha\}$ and $\{x^{\alpha'}\}$ are two coordinate systems in X_n, the Jacobian $J = \left|\frac{\partial x^{\alpha'}}{\partial x^\alpha}\right|$ does not change sign throughout X_n; the two coordinate systems are said to *have the same orientation* if the sign of J is positive, and *opposite orientation* if it is negative.

We can now introduce the concept of a *geometric object*, which provides the framework for nearly all the subsequent mathematical development.

Let us consider a definite point P in an X_n in which two systems of coordinates, $\{x^\alpha\}$ and $\{x^{\alpha'}\}$, are defined. Suppose we are given an ordered set of N numbers at P, in addition, an object transformation law showing how to calculate another ordered set of N numbers uniquely, in terms of the first set and partial derivatives of sufficiently high order of x^α with respect to $x^{\alpha'}$ at P. We then say we are given a *geometric object*, and that the first and second sets of N numbers are its components relative to the coordinate systems $\{x^\alpha\}$ and $\{x^{\alpha'}\}$ respectively.

If the N components are functions of x^α throughout X_n, we have a *field* of geometric objects. For example, quantities with a single component $\varphi(x^\alpha)$ and the transformation law

$$'\varphi(x^{\alpha'}) = J^{-k}\varphi(x^\alpha), \tag{1.3a}$$

where $J = \left|\frac{\partial x^{\alpha'}}{\partial x^\alpha}\right|$, and k is some number, are objects, and are known as *scalar field densities of weight k*; in particular when $k = 0$, we have an object called a *scalar field* for which the transformation law is

$$'\psi(x^{\alpha'}) = \psi(x^\alpha). \tag{1.3b}$$

Let us now suppose that in a domain of n-dimensional real space, and associated with the coordinate system $\{x^\alpha\}$, we are given a quadratic, and in general, *indefinite* differential form

$$ds^2 = g_{\alpha\beta}\, dx^\alpha\, dx^\beta \tag{1.4}$$

and we require it to be independent of the system of coordinates, i.e. a scalar. Here, and in what follows, repeated upper or lower indices indicate summation (*Einstein's rule*), and $g_{\alpha\beta} = g_{\beta\alpha}(x^\alpha)$ are functions of a given class C^r, subject to the condition

$$g = |g_{\alpha\beta}| \neq 0. \tag{1.5}$$

The form (1.4) defines the *metric* in the domain of n-dimensional space. A domain with such a metric is called a *Riemann manifold*; the theory of such manifolds is known as *Riemannian geometry* and its foundations were laid in Riemann's classical paper [1], which deals only with positive forms (1.4). These manifolds are denoted by the symbol V_n. A more detailed definition of a V_n involves the concept of covering of topological space or other considerations (Rashevskii [188], pp. 346–66; Yano and Bochner [257], pp. 3–5; cf. also § 43).

Introducing a metric of the form (1.4) means taking ds as the length of the vector with components dx^α. Because of the indefinite nature of (1.4), ds^2 can be greater than, smaller than, or equal to, zero, and in this last case the vector is said to be *isotropic*.

If the coordinates x^α are given as functions of a real parameter t, the quantity

$$s = \int_{t_1}^{t_2} \sqrt{\left| g_{\alpha\beta} \frac{dx^\alpha}{dt} \frac{dx^\beta}{dt} \right|}\, dt \tag{1.6}$$

defines the arc distance between the points t_1 and t_2 along the curve $x^\alpha = x^\alpha(t)$. The possibility that this distance might be zero is not excluded; such *isotropic* curves are later interpreted as *world light lines* in the space manifold of the relativistic theory of gravitation.

Among the possible types of geometric objects, *tensors* are particularly important; in order to define these, it is necessary to give the transformation rule for a transformation of coordinates from the system $\{x^\alpha\}$ to another $\{x^{\alpha'}\}$. To lead up to this rule, we consider the second of formulae (1.2) and, bearing in mind the class of φ^α, we write

$$dx^{\alpha'} = \frac{\partial x^{\alpha'}}{\partial x^\beta} dx^\beta. \tag{1.7}$$

Similarly, let us consider the scalar field $\psi(x)$: by definition

$$\psi(x^{\alpha'}) = \psi(x^\alpha),$$

Denoting the operator $\dfrac{\partial}{\partial x^\alpha}$ by ∂_α, we obtain the partial derivatives

$$\partial_{\alpha'}\psi = \frac{\partial x^\beta}{\partial x^{\alpha'}} \partial_\beta \psi. \tag{1.8}$$

The relations between the old and new components of the objects in formulae (1.7) and (1.8) are both linear but the coefficients are different. We therefore introduce two distinct kinds of indices; *covariant indices* corresponding to (1.8) are written as subscripts and *contravariant indices* corresponding to (1.7) are written as superscripts. Extending this idea, we will consider objects which have both covariant and contravariant indices. If there are r of the former and s of the latter, we refer to tensors of *order* $r+s$. We can now give the following general definition.

A tensor of weight k *and order* $r+s$ is a geometrical object with n^{r+s} components

$$T^{\alpha_1 \dots \alpha_s}{}_{\beta_1 \dots \beta_r} \quad (\alpha_1, \dots \alpha_s, \beta_1, \dots, \beta_r = 1, \dots, n),$$

which under a transformation of coordinates, transform to new components in accordance with the rule

$$T^{\alpha'_1 \dots \alpha'_s}{}_{\beta'_1 \dots \beta'_r} = J^{-k} T^{\gamma_1 \dots \gamma_s}{}_{\delta_1 \dots \delta_r} \frac{\partial x^{\alpha'_1}}{\partial x^{\gamma_1}} \cdots \frac{\partial x^{\alpha'_s}}{\partial x^{\gamma_s}} \frac{\partial x^{\delta_1}}{\partial x^{\beta'_1}} \cdots \frac{\partial x^{\delta_r}}{\partial x^{\beta'_r}}. \tag{1.9}$$

This tensor is said to be *covariant of order* r *and contravariant of order* s. If $k \neq 0$, the tensor is said to be *relative*, and if $k = 0$, we simply refer to a *tensor* omitting any indication of the weight.

If the components of the tensor are functions of x^α throughout X_n, we have a tensor field.

In denoting objects, we use the *kernel-index method*, in which a coordinate transformation is shown by changing the *index fount*, and an *object* transformation by changing the kernel letter ([110], p. 3); only with scalars, which do not have indices, is it necessary to use the "prime" notation as in (1.3a) and even this may be omitted as in (1.8).

Thus the scalar field (1.3b) is a tensor field of order $0+0$ and weight $k = 0$; the scalar (1.4) is a tensor of order $0+0$ and weight $k = 0$. A scalar density has order $0+0$ and weight k.

The next two examples of tensor are vectors: *covariant*,

$$v_{\alpha'} = \frac{\partial x^\beta}{\partial x^{\alpha'}} v_\beta \tag{1.10}$$

and *contravariant*,

$$u^{\alpha'} = \frac{\partial x^{\alpha'}}{\partial x^\beta} u^\beta, \tag{1.11}$$

both with weight equal to zero. In particular (1.7) defines a contravariant vector, and (1.8) a covariant, namely *gradient* field.

By definition $ds' = ds$ for Riemann manifolds, and hence

$$g_{\alpha\beta}\, dx^{\alpha}\, dx^{\beta} = g_{\alpha'\beta'}\, dx^{\alpha'} dx^{\beta'};$$

substituting from (1.7) we obtain

$$\left(g_{\alpha'\beta'} - \frac{\partial x^{\alpha}}{\partial x^{\alpha'}} \frac{\partial x^{\beta}}{\partial x^{\beta'}} g_{\alpha\beta}\right) dx^{\alpha'}\, dx^{\beta'} = 0$$

and in view of the completely arbitrary choice of the vector $dx^{\alpha'}$

$$g_{\alpha'\beta'} = g_{\alpha\beta} \frac{\partial x^{\alpha}}{\partial x^{\alpha'}} \frac{\partial x^{\beta}}{\partial x^{\beta'}},$$

i.e. $g_{\alpha\beta}$ defines a covariant tensor field of order 2. Since

$$\frac{\partial x^{\alpha}}{\partial x_{\beta'}} \frac{\partial x^{\beta'}}{\partial x^{\gamma}} = \delta^{\alpha}_{\gamma}, \quad \frac{\partial x^{\alpha'}}{\partial x^{\beta}} \frac{\partial x^{\beta}}{\partial x^{\gamma'}} = \delta^{\alpha'}_{\gamma'}, \tag{1.12}$$

the matrix $\left[\frac{\partial x^{\alpha'}}{\partial x^{\beta}}\right]$ is the reciprocal of matrix $\left[\frac{\partial x^{\alpha}}{\partial x^{\beta'}}\right]$; δ^{α}_{β} is the *Kronecker symbol*, equal to 1 for $\alpha = \beta$ and zero for $\alpha \neq \beta$.

Let us consider some point in X_n and use the notation

$$\left(\frac{\partial x^{\alpha'}}{\partial x^{\beta}}\right)_P = A^{\alpha'}_{\beta}, \quad \left(\frac{\partial x^{\alpha}}{\partial x^{\beta'}}\right)_P = A^{\alpha}_{\beta'}. \tag{1.13}$$

Then at P the components of a contravariant vector transform according to the law

$$\eta^{\alpha'} = A^{\alpha'}_{\alpha}\eta^{\alpha}, \quad \eta^{\alpha} = A^{\alpha}_{\alpha'}\eta^{\alpha'}, \quad |A^{\alpha'}_{\alpha}| \neq 0, \quad A^{\alpha}_{\alpha'}A^{\alpha'}_{\beta} = \delta^{\alpha}_{\beta}, \tag{1.14}$$

which defines a group of linear homogeneous transformations—the *centro-affine group* corresponding to *Klein geometry*, a generalization to n dimensions of normal affine geometry. Hence each point P in X_n has its own *local* centro-affine space which will be denoted by the symbol E_n. If we identify the origin of the E_n with the point P in the X_n, then the fields of geometric objects in X_n define object fields with constant components in E_n. There are a number of articles and monographs devoted to the axiomatics and detailed study of affine spaces (Shirokov [91]; Norden [170], pp. 11–58; Rashevskii [188], pp. 82–144).

Examples

1. Show that if Kronecker symbols are taken as the components of a mixed tensor of order 1+1 in some coordinate system, then the components of this tensor in a different system will also be Kronecker symbols.

2. Show that if the components of a tensor are zero at some point in one system $\{x\}$, then they are also zero at this point in any other system $\{x'\}$.

3. Show that for any vector field $u^\alpha(x)$ it is possible to find a system of coordinates $\{x'\}$ such that

$$u^{\alpha'} = 0, \quad u^{n'} = 1 \quad (\alpha < n). \tag{1.15}$$

4. Show that $\partial_{\alpha\beta}\varphi$ together with $\partial_\alpha\varphi$ form an object field with $\frac{1}{2}n(n+3)$ components.

§ 2. Tensor Algebra

A *tensor* of any weight is said to be *zero* if all its components are zero in any one system of coordinates; it is then zero in any system (example 2, § 1).

Tensor algebra introduces operations by means of which it becomes possible to associate new tensors with given ones.

Addition and subtraction of tensors. If we add the corresponding components of two tensors of equal weight and with the same number r of covariant and s of contravariant indices, the system of quantities thus obained defines another tensor of the same type. Thus if

$$M^{\alpha'}_{\beta'} = J^{-k}A^{\alpha'}_{\gamma}A^{\delta}_{\beta'}M^{\gamma}_{\delta}, \quad N^{\alpha'}_{\beta'} = J^{-k}A^{\alpha'}_{\gamma}A^{\delta}_{\beta'}N^{\gamma}_{\delta},$$

then

$$M^{\alpha'}_{\beta'}+N^{\alpha'}_{\beta'} = J^{-k}A^{\alpha'}_{\gamma}A^{\delta}_{\beta'}(M^{\gamma}_{\delta}+N^{\gamma}_{\delta}).$$

This proof can clearly be extended to any order.

Multiplication of tensors. Multiplying each component of a tensor of weight k and order $r+s$ by each component of another tensor of weight k' and order $r'+s'$, we obtain the components of a tensor of weight $k+k'$ and order $(r+r')+(s+s')$. To prove this, we multiply the two right- and left-hand sides of the transformation equations (1.9) for the components of the two tensors.

For example, the product of a scalar density and a covariant vector is a relative covariant vector. Similarly, the product of a covariant vector and a contravariant vector is a tensor of order $1+1$ and weight 0.

Contraction is an operation applicable to any tensor of order $r+s$, where $r, s > 0$, and consists in putting a covariant index equal to a contravariant index and then summing over this index. This leads to a tensor of order $(r-1)+(s-1)$. Since the weight of the tensor and the values of r and

s play no part in the proof of this statement, it will be sufficient to consider the tensor

$$M^{\alpha'}{}_{\beta'\gamma'} = A^{\alpha'}_{\alpha} A^{\beta}_{\beta'} A^{\gamma}_{\gamma'} M^{\alpha}{}_{\beta\gamma};$$

contracting this with respect to the indices α' and β' we obtain:

$$M^{\alpha'}{}_{\alpha'\gamma'} = A^{\alpha'}_{\alpha} A^{\beta}_{\alpha'} A^{\gamma}_{\gamma'} M^{\alpha}{}_{\beta\gamma} = \delta^{\beta}_{\alpha} A^{\gamma}_{\gamma'} M^{\alpha}{}_{\beta\gamma} = M^{\alpha}{}_{\alpha\gamma} A^{\gamma}_{\gamma'},$$

i.e. $M^{\alpha}_{\alpha\gamma} = M_{\gamma}$ is a tensor if $M^{\alpha}_{\beta\gamma}$ is a tensor.

The operation of *contracted multiplication of tensors* is a combination of multiplication and contraction and therefore clearly leads once again to a tensor.

Using these operations, we can obtain the *quotient law* often used to investigate tensor equations. It may be stated as follows: suppose that in each coordinate system we are given an ordered set of n^{r+s} numbers $M_{\alpha_1 \ldots \alpha_r \beta_1 \ldots \beta_s}$ such that choosing *any* s contravariant vectors v, the set of numbers

$$N_{\alpha_1 \ldots \alpha_r} = M_{\alpha_1 \ldots \alpha_r \beta_1 \ldots \beta_s} v^{\beta_1} \ldots v^{\beta_s}$$

is a tensor of order $r+0$; then $M_{\alpha_1 \ldots \alpha_r \beta_1 \ldots \beta_s}$ is a tensor of order $(r+s)+0$ It is sufficient to prove the case $s = 1$. We have

$$M_{\alpha'_1 \ldots \alpha'_r \beta'} v^{\beta'} = M_{\alpha_1 \ldots \alpha_r \beta} v^{\beta} A^{\alpha_1}_{\alpha'_1} \ldots A^{\alpha_r}_{\alpha'_r} = v^{\beta'} A^{\beta}_{\beta'} M_{\alpha_1 \ldots \alpha_r \beta} A^{\alpha_1}_{\alpha_1} \ldots A^{\alpha}_{\alpha'_r}.$$

Since the choice of v is completely arbitrary,

$$M_{\alpha'_1 \ldots \alpha'_r \beta'} = M_{\alpha_1 \ldots \alpha_r \beta} A^{\alpha_1}_{\alpha'_1} \ldots A^{\alpha_r}_{\alpha'_r} A^{\beta}_{\beta'}$$

which proves the theorem. This law can, of course, be generalized for the case where the numbers M also have contravariant indices, and the corresponding number of vectors v are covariant. It is frequently used, and in a certain sense defines the converse of tensor multiplication.

If the components of some tensor do not change when two or more indices of the same variance are interchanged, the tensor is said to be *symmetric* with respect to these indices. If, on the other hand, interchanging the indices leads only to a change of sign of the component, the tensor is said to be *antisymmetric* with respect to the indices. It is obvious that symmetry and antisymmetry are invariant under a transformation of coordinates. The most important consequence for any tensor with one of these quite common properties is a reduction in the number of independent components.

Thus a covariant tensor of the second order has in general n^2 independent components; if, however, it is completely symmetric it has $n(n+1)/2$, and if completely antisymmetric only $n(n-1)/2$ independent components.

Conversion to symmetric and antisymmetric forms. If a tensor has several indices of the same variance, it is possible to form from it a new tensor which is symmetric or antisymmetric with respect to all or some of these indices.

Suppose, for example, we are given a tensor of order $m+0$: $B_{\alpha_1 \ldots \alpha_m}$, Taking the $m!$ components produced by all the possible permutations of the indices, adding them, and dividing by $m!$ we obtain a component of the following symmetric tensor

$$B_{(\alpha_1 \ldots \alpha_m)} = \frac{1}{m!}(B_{\alpha_1\alpha_2 \ldots \alpha_m} + B_{\alpha_2\alpha_1 \ldots \alpha_m} + \ldots).$$

This *symmetrizing* process is denoted by the pair of round brackets.

In a similar way the process of *alternation* leads to an antisymmetric tensor; taking the $m!$ components as before, we form a sum in which each component has a plus or minus sign depending on whether the permutation $\begin{pmatrix} \alpha_1\alpha_2 \ldots \alpha_m \\ \alpha_{r_1}\alpha_{r_2} \ldots \alpha_{r_m} \end{pmatrix}$ is even or odd. Dividing by $m!$ leads to the following antisymmetric tensor:

$$B_{[\alpha_1\alpha_2 \ldots \alpha_m]} = \frac{1}{m!}(B_{\alpha_1 \ldots \alpha_m} - B_{\alpha_2\alpha_1 \ldots \alpha_m} + \ldots).$$

The alternating process is denoted by a pair of square brackets.

The division by $m!$ is not essential, but convenient. If the symmetrizing or alternating process is not to apply to some index inside the brackets, this is indicated by isolating the index by two vertical lines, thus

$$B_{(\alpha|\beta|\gamma)} = \tfrac{1}{2}(B_{\alpha\beta\gamma} + B_{\gamma\beta\alpha}).$$

It is easy to show from the definitions that symmetrization and alternation are tensor operations.

As the operations considered above in no way involve the *metric* tensor $g_{\alpha\beta}$, which characterizes Riemann spaces, they apply to any local affine space. Metrizing a space by means of the tensor field $g_{\alpha\beta}$, admits a series of new operations characteristic of Riemann manifolds.

Let $g^{\alpha\beta} = g^{\beta\alpha}$ denote the cofactor of the element $g_{\alpha\beta} = g_{\beta\alpha}$ in the determinant $g = |g_{\alpha\beta}|$ divided by g. The set of numbers $g^{\alpha\beta}$ satisfies the equations

$$g^{\alpha\beta}g_{\beta\gamma} = \delta^{\alpha}_{\gamma} \tag{2.1}$$

at all points in the V_n space.

We shall show that the $g^{\alpha\beta}$ are *components of a symmetric covariant tensor*. Let u^α be the components of an *arbitrary* vector. Then

$$v_\beta = g_{\beta\gamma}u^\gamma$$

is also a vector. Equation (2.1) shows that

$$g^{\alpha\beta}v_\beta = g^{\alpha\beta}g_{\beta\gamma}u^\gamma = \delta^\alpha_\gamma u^\gamma = u^\alpha.$$

As v is an arbitrary vector, we obtain the required result by applying the quotient theorem.

Contracting with respect to α and γ in (2.1) we obtain

$$g^{\alpha\beta}g_{\alpha\beta} = n. \tag{2.2}$$

Denoting the determinant $|g^{\alpha\beta}|$ by $\boldsymbol{g}$, we obtain

$$g\boldsymbol{g} = \begin{vmatrix} 1 & & & \\ & 1 & & \\ & & \ddots & \\ & & & 1 \end{vmatrix} = 1. \tag{2.3}$$

The cofactor of $g^{\alpha\beta}$ in the determinant $\boldsymbol{g}$ divided by $\boldsymbol{g}$ is therefore equal to $g_{\alpha\beta}$.

Raising and lowering of indices. By using the tensors $g_{\alpha\beta}$ and $g^{\alpha\beta}$, it is possible to associate with every tensor another tensor with the same number of indices but of a different variance. Thus if $B_{\alpha\beta\gamma}$ is a tensor, we can associate with it tensors

$$B^\alpha{}_{\beta\gamma} = g^{\alpha\lambda}B_{\lambda\beta\gamma}, \quad B_\alpha{}^\beta{}_\gamma = g^{\beta\lambda}B_{\alpha\lambda\gamma}, \quad B_{\alpha\beta}{}^\gamma = g^{\gamma\lambda}B_{\alpha\beta\lambda}.$$

For each of these tensors the process is reversible; for example,

$$B^\alpha{}_{\beta\gamma}g_{\alpha\nu} = g_{\alpha\nu}g^{\alpha\lambda}B_{\lambda\beta\gamma} = \delta^\lambda_\nu B_{\lambda\beta\gamma} = B_{\nu\beta\gamma},$$

i.e. we return to the original tensor. Thus we can say that to a certain extent, the tensor $g_{\alpha\beta}$ destroys the dualism between covariant and contravariant indices. From this point of view, it is natural to speak of a single tensor, with alternative representations of differing variances. In order to emphasize this, it is convenient to retain the same kernel letter for all the possible representations of a tensor.

We may note that it is possible to use some other tensor, antisymmetric for example, for raising and lowering indices, and in some problems this is in fact more effective ([170], p. 310), but in any case the same tensor must be so used throughout the investigation.

The metric tensor also serves to define the *norm* of any vector:

$$g_{\alpha\beta}u^\alpha u^\beta = u_\alpha u^\alpha = g^{\alpha\beta}u_\alpha u_\beta.$$

Since $g_{\alpha\beta}$ is an indefinite tensor, there is at each point of space a *cone of isotropic directions*

$$g_{\alpha\beta}x^\alpha x^\beta = 0,$$

separating the *+domain (the domain of real vectors with positive norms)* from the *−domain (real vectors with negative norms)*. The norm of any isotropic vector is zero. Vectors with norms equal to ± 1 are called *unit vectors*. The positive square root of the absolute value of the norm is called the *length of the vector*. Incidentally, in relativity theory only vectors in the −domain are said to have a length; those in the +domain are said to have a *duration*.

The angle between two real vectors u^α and v^α lying in the same domain is defined by equation

$$\cos(u, v) = \frac{\pm u_\alpha v^\alpha}{\left|\sqrt{u_\sigma u^\sigma \cdot v_\tau v^\tau}\right|}, \tag{2.4}$$

the sign corresponding to the region.

The angle is real when none of the vectors

$$au^\alpha + bv^\alpha,$$

forming the pencil defined by u^α and v^α are isotropic, since only then is $|\cos(u, v)| \leqslant 1$. Otherwise we obtain either imaginary angles or a meaningless expression. However, we will in all cases call the vectors u^α and v^α *orthogonal* if

$$u_\alpha v^\alpha = 0. \tag{2.5}$$

Isotropic vectors, and they alone, are self-orthogonal.

We may note that from the transformation law for the metric tensor

$$g_{\alpha'\beta'} = A^\alpha_{\alpha'} A^\beta_{\beta'} g_{\alpha\beta}$$

it follows that

$$g' = gJ^{-2}, \tag{2.6}$$

i.e. g is a scalar density of weight 2.

Examples

1. If a tensor is symmetric with respect to the indices α_k and α_l and anti-symmetric with respect to α_l and α_m, it is zero.

2. Show that the following relations are valid for any tensor:

$$S_{\alpha\beta} = S_{(\alpha\beta)} + S_{[\alpha\beta]},$$

$$\Omega_{\alpha\beta\gamma} = \Omega_{(\alpha\beta\gamma)} + \Omega_{[\alpha\beta\gamma]} + \tfrac{2}{3}(\Omega_{[\alpha\beta]\gamma} + \Omega_{[\gamma\beta]\alpha}) + \tfrac{2}{3}(\Omega_{(\alpha\beta)\gamma} - \Omega_{\gamma(\alpha\beta)}).$$

3. If $\varphi = S_{\alpha\beta}u^\alpha u^\beta$ is a scalar for any u^α, $S_{\alpha\beta}$ is a tensor. If $\varphi = 0$, $S_{(\alpha\beta)} = 0$.

4. If the tensor $A_{\alpha\beta\gamma}$ satisfies the relation $A_{\alpha\beta\gamma}u^\alpha u^\beta u^\gamma = 0$ for any u^α and $A_{[\alpha\beta]\gamma} = 0$, then $A_{\alpha\beta\gamma}+A_{\beta\gamma\alpha}+A_{\gamma\alpha\beta} = 0$.

5. If the tensor $T_{\alpha\beta\gamma\delta}$ satisfies the relation $T_{\alpha\beta\gamma\delta}u^\alpha v^\beta u^\gamma v^\delta = 0$ for arbitrary choice of the vectors u^α and v^α, then

$$T_{\alpha\beta\gamma\delta}+T_{\gamma\beta\alpha\delta}+T_{\alpha\delta\gamma\beta}+T_{\gamma\delta\alpha\beta} = 0. \tag{2.7}$$

If in addition T satisfies the conditions

$$T_{\alpha\beta\gamma\delta}+T_{\beta\alpha\gamma\delta} = 0, \quad T_{\alpha\beta\gamma\delta}+T_{\alpha\beta\delta\gamma} = 0,$$
$$T_{\alpha\beta\gamma\delta}+T_{\beta\gamma\alpha\delta}+T_{\gamma\alpha\beta\delta} = 0,$$

it is zero.

6. If $A_{[\alpha\beta]\gamma} = 0$, then

$$A_{(\alpha\beta\gamma)} = \tfrac{1}{3}\,(A_{\alpha\beta\gamma}+A_{\beta\gamma\alpha}+A_{\gamma\alpha\beta}). \tag{2.8}$$

If $B_{(\alpha\beta)\gamma} = 0$, then

$$B_{[\alpha\beta\gamma]} = \tfrac{1}{3}(B_{\alpha\beta\gamma}+B_{\beta\gamma\alpha}+B_{\gamma\alpha\beta}). \tag{2.9}$$

7. If the equation

$$aT_{\alpha\beta}+bT_{\beta\alpha} = 0$$

is valid in any one system of coordinates, it is always valid. If $T_{\alpha\beta} \neq 0$, then $a = \pm b$ ([77], p. 16).

8. If the equation

$$aT_{\alpha\beta\gamma}+bT_{\beta\gamma\alpha}+cT_{\gamma\alpha\beta} = 0$$

is valid in any one system of coordinates, it is always valid. If $T_{\alpha\beta\gamma} \neq 0$, then $a+b+c = 0$ or $a = b = c$.

9. Show that

$$A_{[\alpha\beta\,\dots\,\gamma]}B^{\alpha\beta\,\dots\,\gamma} = A_{\alpha\beta\,\dots\,\gamma}B^{[\alpha\beta\,\dots\,\gamma]} = A_{[\alpha\beta\,\dots\,\gamma]}B^{[\alpha\beta\,\dots\,\gamma]}.$$

10. Prove that

$$\underset{1}{v}^{[\alpha_1}\dots\underset{k}{v}^{\alpha_k]} = \underset{[1}{v}^{[\alpha_1}\dots\underset{k]}{v}^{\alpha_k]} = \underset{[1}{v}^{\alpha_1}\dots\underset{k]}{v}^{\alpha_k}.$$

11. The rank of the matrix formed from the components of the tensor $u_\alpha v_\beta$ is equal to 1, and from the tensor $u_{(\alpha}v_{\beta)}$ is equal to 2 if $u_\alpha \neq \lambda v_\beta$.

12. Prove that

$$T_{\alpha[[\beta\gamma]\delta]} = T_{\alpha[\beta\gamma\delta]}.$$

§ 3. Covariant Differentiation

The basic advantage of tensor calculus lies in the possibility of writing tensor equations in an invariant form, i.e. independent of the system of coordinates. In developing tensor analysis, it is essential first of all to connect differentiation with a tensor-like operation. We assume in this section that the class of functions corresponds to the order of differentiation. Equation (1.8) shows a case where the derivatives of a tensor of order $0+0$ and weight 0 define a tensor of order $1+0$ and weight 0. In general, however, the partial derivatives of a tensor do not define a tensor. A non-unique tensor-type operation can be introduced which generalizes ordinary differentiation.

The tensor nature of *covariant differentiation,* introduced in this section, was first noted by Christoffel ([2], p. 56). A formal definition is given here; its geometrical meaning will be explained in § 5.

The set of functions

$$\Gamma_{\alpha,\,\beta\gamma} = \tfrac{1}{2}(\partial_\beta g_{\alpha\gamma}+\partial_\gamma g_{\alpha\beta}-\partial_\alpha g_{\beta\gamma}), \tag{3.1}$$

where $\partial_\gamma g_{\alpha\beta} = \dfrac{\partial g_{\alpha\beta}}{\partial x^\gamma}$ *are called Christoffel symbols of the first kind.* Clearly

$$\Gamma_{\alpha,\,\beta\gamma}+\Gamma_{\beta,\,\alpha\gamma} = \partial_\gamma g_{\alpha\beta}. \tag{3.2}$$

Hence there is a one–one correspondence between the set of partial derivatives of the metric tensor, and the set of Christoffel symbols of the first kind. The *Christoffel symbols of the second kind* are defined as follows:

$$\Gamma^\alpha_{\beta\gamma} = g^{\alpha\sigma}\Gamma_{\sigma,\,\beta\gamma}. \tag{3.3}$$

It follows from the definitions that

$$\Gamma_{\alpha,\,\beta\gamma} = \Gamma_{\alpha,\,\gamma\beta}, \quad \Gamma^\alpha_{\beta\gamma} = \Gamma^\alpha_{\gamma\beta}. \tag{3.4}$$

The Christoffel symbols of the first and second kinds are not tensors; thus in three-dimensional Euclidean space they are all zero in rectangular Cartesian coordinates, but in cylindrical coordinates some of them differ from zero.

The *Christoffel symbols of the second kind form a geometric object; the Christoffel symbols of the first kind together with the metric tensor also form an object.*

Consider the transformation law for the metric tensor:

$$g_{\alpha'\beta'} = A^\sigma_{\alpha'}A^\tau_{\beta'}g_{\sigma\tau}. \tag{3.5}$$

Since $g_{\sigma\tau} = g_{\sigma\tau}(x)$, $\partial_{\gamma'} g_{\sigma\tau} = A^{\varrho}_{\gamma'} \partial_{\varrho} g_{\sigma\tau}$ and

$$\partial_{\gamma'} g_{\alpha'\beta'} = A^{\sigma}_{\alpha'} A^{\tau}_{\beta'} A^{\varrho}_{\gamma'} \partial_{\varrho} g_{\sigma\tau} + g_{\sigma\tau}(\partial_{\gamma'} A^{\sigma}_{\alpha'} A^{\tau}_{\beta'} + A^{\sigma}_{\alpha'} \partial_{\gamma'} A^{\tau}_{\beta'}). \tag{3.5a}$$

Adding and subtracting the expressions obtained from the above by applying respectively the permutations $\binom{\alpha'\beta'\gamma'}{\gamma'\alpha'\beta'}$ and $\binom{\alpha'\beta'\gamma'}{\beta'\gamma'\alpha'}$, and noting that

$$\partial_{\beta'} A^{\sigma}_{\alpha'} = \frac{\partial^2 x^{\sigma}}{\partial x^{\beta'} \partial x^{\alpha'}} = \partial_{\alpha'} A^{\sigma}_{\beta'},$$

we obtain

$$\Gamma_{\alpha', \beta'\gamma'} = A^{\sigma}_{\alpha'} A^{\tau}_{\beta'} A^{\varrho}_{\gamma'} \Gamma_{\sigma, \varrho\tau} + g_{\sigma\tau} A^{\tau}_{\alpha'} \partial_{\beta'} \cdot A^{\sigma}_{\gamma'}, \tag{3.6}$$

which proves the second part of the theorem. Contracting (3.6) with $g^{\alpha'\delta'}$ and using (1.14), we obtain

$$\Gamma^{\delta'}_{\beta'\gamma'} = A^{\delta'}_{\omega} A^{\tau}_{\beta'} A^{\varrho}_{\gamma'} \Gamma^{\omega}_{\varrho\tau} + A^{\delta'}_{\omega} \partial_{\beta'} A^{\omega}_{\gamma'}, \tag{3.7}$$

which proves the first part.

We can now introduce the concept of the covariant derivative of a vector. The partial derivatives of the covariant components of a vector do not define a tensor since

$$\partial_{\gamma'} \lambda_{\alpha'} = \partial_{\gamma'}(A^{\sigma}_{\alpha'} \lambda_{\sigma}) = \partial_{\gamma'} A^{\sigma}_{\alpha'} \lambda_{\sigma} + A^{\sigma}_{\alpha'} \partial_{\gamma'} \lambda_{\sigma} = \partial_{\gamma'} A^{\sigma}_{\alpha'} \lambda_{\sigma} + A^{\sigma}_{\alpha'} A^{\tau}_{\gamma'} \partial_{\tau} \lambda_{\sigma},$$

but if we subtract

$$\Gamma^{\sigma'}_{\alpha'\gamma'} \lambda_{\sigma'} = (A^{\sigma'}_{\tau} \partial_{\alpha'} A^{\tau}_{\gamma'} + A^{\mu}_{\alpha'} A^{\nu}_{\gamma'} A^{\sigma'}_{\tau} \Gamma^{\tau}_{\mu\nu}) A^{\varrho}_{\sigma'} \lambda_{\varrho},$$

we obtain

$$\partial_{\gamma'} \lambda_{\alpha'} - \Gamma^{\sigma'}_{\alpha'\gamma'} \lambda_{\sigma'} = A^{\tau}_{\gamma'} A^{\sigma}_{\alpha'} (\partial_{\tau} \lambda_{\sigma} - \Gamma^{\nu}_{\sigma\tau} \lambda_{\nu}),$$

indicating a tensor. The expression in brackets, a linear combination of a partial derivative of a vector component and all the components, with coefficients determined by the metric of the space V_n, is called *a covariant derivative of the vector*, and is denoted thus:

$$\lambda_{\alpha, \gamma} \equiv \partial_{\gamma} \lambda_{\alpha} - \Gamma^{\sigma}_{\alpha\gamma} \lambda_{\sigma}. \tag{3.8}$$

The covariant derivative of a scalar is the ordinary derivative; applying this to the scalar $\lambda_{\alpha}\lambda^{\alpha}$ we have

$$(\lambda_{\alpha}\lambda^{\alpha})_{, \gamma} = \partial_{\gamma} (\lambda_{\alpha}\lambda^{\alpha}).$$

Hence

$$\lambda_{\alpha, \gamma}\lambda^{\alpha} + \lambda_{\alpha}\lambda^{\alpha}{}_{, \gamma} = (\partial_{\gamma}\lambda_{\alpha})\lambda^{\alpha} + \lambda_{\alpha}\partial_{\gamma}\lambda^{\alpha}$$

$$(\partial_{\gamma}\lambda_{\alpha})\lambda^{\alpha} - \Gamma^{\sigma}_{\alpha\gamma}\lambda_{\sigma}\lambda^{\alpha} + \lambda_{\alpha}\lambda^{\alpha}{}_{, \gamma} = (\partial_{\gamma}\lambda_{\alpha})\lambda^{\alpha} + \lambda_{\alpha}\partial_{\gamma}\lambda^{\alpha}.$$

Interchanging indices α and σ and cancelling by λ_α, we obtain the following expression for the covariant derivative of a contravariant vector:

$$\lambda^\alpha{}_{,\gamma} \equiv \partial_\gamma \lambda^\alpha + \Gamma^\alpha_{\sigma\gamma}\lambda^\sigma. \tag{3.9}$$

The covariant derivative of a tensor of any order $p+q$ and weight 0 follows directly from (3.8) and (3.9):

$$\begin{aligned} T^{\alpha_1\ldots\alpha_p}{}_{\beta_1\ldots\beta_q,\gamma} = {} & \partial_\gamma T^{\alpha_1\ldots\alpha_p}{}_{\beta_1\ldots\beta_q} + \\ & + \sum_{\sigma=1}^{p} T^{\alpha_1\ldots\alpha_{\sigma-1}\tau\alpha_{\sigma+1}\ldots\alpha}{}_{\beta_1\ldots\beta_q}\Gamma_{\tau\gamma} - \\ & - \sum_{\sigma=1}^{q} T^{\alpha_1\ldots\alpha_p}{}_{\beta_1\ldots\beta_{\sigma-1}\tau\beta_{\sigma+1}\ldots\beta_q}\Gamma^\tau_{\beta_\sigma\gamma}. \end{aligned} \tag{3.10}$$

Thus the covariant derivative is a tensor of covariant order greater by unity than that of the original tensor. As covariant differentiation can be based on any non-degenerate symmetric tensor of order $0+2$, then if two or more such differentiations are needed, they must be distinguished by different notations, for example denoting one by a comma, the second by a semi-colon, and so on. The sign ∇ (nabla) is also used, e.g. $\lambda_{\alpha,\gamma} \equiv \nabla_\gamma \lambda_\alpha$. The term "covariant derivative" was introduced by Ricci [3], although the operation itself applied to Riemannian spaces can be found in Christoffel [2].

The normal rules of differentiation hold for single covariant derivatives of sums, products and contracted products. Thus

$$(\lambda_\alpha \pm \nu_\alpha)_{,\beta} = \lambda_{\alpha,\beta} \pm \nu_{\alpha,\beta}, \tag{3.11}$$

$$(\lambda_\alpha \nu_\beta)_{,\gamma} = \lambda_{\alpha,\gamma}\, \nu_\beta + \lambda_\alpha \nu_{\beta,\gamma}, \tag{3.12}$$

$$(T_\alpha{}^\beta{}_\gamma \nu_\beta \omega^\gamma)_{,\delta} = T_\alpha{}^\beta{}_{\gamma,\delta}\nu_\beta\omega^\gamma + T_\alpha{}^\beta{}_\gamma \nu_{\beta,\delta}\omega^\gamma + T_\alpha{}^\beta{}_\gamma\nu_\beta\omega^\gamma{}_{,\delta}. \tag{3.13}$$

The proof of these formulae follows directly from (3.10). The covariant derivative of a scalar is identical with the partial derivative:

$$\varphi_{,\alpha} = \partial_\alpha \varphi. \tag{3.14}$$

Formulae (3.10)–(3.14) may easily be extended to relative tensors, provided that *the weight remains constant under covariant differentiation*. Thus, for example, for a scalar density of weight k we obtain

$$\theta_{,\alpha} = \partial_\alpha \theta - k\theta\Gamma^\sigma_{\alpha\sigma}. \tag{3.15}$$

For a tensor $\Pi^\alpha_{\beta\gamma}$ of order $1+2$ and weight p the covariant derivative has the form

$$\Pi^\alpha_{\beta\gamma,\delta} = \partial_\delta \Pi^\alpha_{\beta\gamma} + \Gamma^\alpha_{\sigma\delta}\Pi^\sigma_{\beta\gamma} - \Gamma^\sigma_{\beta\delta}\Pi^\alpha_{\sigma\gamma} - \Gamma^\sigma_{\gamma\delta}\Pi^\alpha_{\beta\sigma} - p\Pi^\alpha_{\beta\gamma}\Gamma_\delta, \tag{3.16}$$

where

$$\Gamma_\delta = \Gamma^\sigma_{\delta\sigma}.$$

Covariant derivatives may be introduced into the non-metrical space X_n by means of a special postulate of parallel displacement; the components of the object $\Gamma^\alpha_{\beta\gamma}(x)$, called the *coefficients of connectivity*, are no longer connected with any metric tensor. This theory of spaces of affine connectivity A_n of which the manifolds V_n are a special case, is dealt with by Thomas [92], Norden [170], and Rashevskii [188]. It is not even essential to insist that the objects $\Gamma^\alpha_{\beta\gamma}$ be symmetric with respect to the lower indices—the absence of such symmetry corresponds to spaces with *torsion* L_n.

We will call a tensor covariantly constant, if its covariant derivative is zero. The existence of such tensors in V_n is confirmed by the following theorem; the tensors $g_{\alpha\beta}$, $g^{\alpha\beta}$ and δ^α_β are covariantly constant. For

$$\begin{aligned} g_{\alpha\beta,\,\gamma} &= \partial_\gamma g_{\alpha\beta} - g_{\delta\beta}\Gamma^\delta_{\alpha\gamma} - g_{\alpha\delta}\Gamma^\delta_{\beta\gamma} \quad \text{according to eqn. (3.10)}\\ &= \partial_\gamma g_{\alpha\beta} - \Gamma_{\beta,\,\alpha\gamma} - \Gamma_{\alpha,\,\beta\gamma} \quad \text{according to eqn. (3.3)}\\ &= 0 \quad \text{according to eqn. (3.2).} \end{aligned}$$

The proof of the rest of the theorem is left to the reader. Consequently, the *operations of contraction and of covariant differentiation commute;* for example,

$$\begin{aligned} g^{\alpha\beta}(u_\alpha v_\beta)_{,\,\gamma} &= g^{\alpha\beta}(u_{\alpha,\,\gamma}v_\beta + u_\alpha v_{\beta,\,\gamma}) = u^\beta{}_{,\,\gamma}v_\beta + u^\beta v_{\beta,\,\gamma} = (u^\beta v_\beta)_{,\,\gamma}\\ &= (g^{\alpha\beta}u_\alpha v_\beta)_{,\,\gamma}. \end{aligned}$$

Covariant differentiation enables us to introduce the *curl* of a vector

$$\lambda_{i,\,j} - \lambda_{j,\,i} \equiv \partial_j\lambda_i - \partial_i\lambda_j \quad \left(\partial_j \equiv \frac{\partial}{\partial x^j}\right), \tag{3.17}$$

the *divergence* of a vector or tensor based on $g_{\alpha\beta}$, defined respectively by the expressions

$$\lambda^\alpha{}_{,\alpha}, \quad a^{\alpha\beta}{}_{,\beta}, \quad b^\beta_{\alpha,\,\beta},$$

and the scalar operators

$$\left.\begin{aligned} \Delta_1\varphi &= g^{\alpha\beta}\varphi_{,\alpha}\varphi_{,\beta} \equiv g^{\alpha\beta}\partial_\alpha\varphi\partial_\beta\varphi,\\ \Delta_1(f,\varphi) &= g^{\alpha\beta}f_{,\alpha}\varphi_{,\beta},\\ \Delta_2\varphi &= g^{\alpha\beta}\varphi_{,\,\alpha\beta}, \end{aligned}\right\} \tag{3.18}$$

known as *differential Beltrami parameters of the first and second kinds.*

EXAMPLES

1. In order that any covariant vector λ_α should be a gradient it is necessary and sufficient that its first covariant derivative should be symmetric: $\lambda_{\alpha,\beta} = \lambda_{\beta,\alpha}$.

2. If λ_{ij} is the curl of a vector, then

$\lambda_{ij,k}+\lambda_{jk,i}+\lambda_{ki,j} = \partial_k\lambda_{ij}+\partial_i\lambda_{jk}+\partial_j\lambda_{ki} = 0$ (Eisenhart [77], p. 31).

3. Show that $\partial_\gamma \Delta_1\varphi = 2g^{\alpha\beta}\varphi_{,\alpha}\varphi_{,\beta\gamma}$ and $\lambda^\alpha{}_{,\alpha} = \dfrac{1}{\sqrt{g}}\partial_\alpha(\sqrt{g}\,\lambda^\alpha)$.

4. Show that the divergence of the tensor $a^{\alpha\beta}$ based on $g_{\alpha\beta}$ has the form $a^{\alpha\beta}{}_{,\beta} = \dfrac{1}{\sqrt{g}}\partial_\beta\,(a^{\alpha\beta}\sqrt{g})+a^{\beta\gamma}\Gamma^\alpha_{\beta\gamma}$; if $a^{\alpha\beta}$ is an antisymmetric tensor the second term disappears.

5. Show that $\Gamma_\alpha \equiv \Gamma^\sigma_{\alpha\sigma} = \partial_\alpha \ln\sqrt{g}$.

6. Show that if $\lambda^{[\alpha\beta]} = 0$, $\lambda_{\alpha,\beta}^{\ \beta} = \dfrac{1}{\sqrt{g}}\partial_\beta(\lambda^\beta_\alpha\sqrt{g})+\frac{1}{2}\lambda_{\sigma\tau}\partial_\alpha g^{\sigma\tau}$ (Einstein [36]).

7. Let us consider the quantities $\varepsilon_{i_1,i_2,\ldots,i_n} = \varepsilon^{i_1,i_2,\ldots,i_n}$ which, by definition, are equal to zero if they contain two or more identical indices, and equal to 1 or -1 depending on whether the sequence of the indices $i_1, i_2, \ldots, i_n$ is obtained from the natural series 1, 2, 3,..., n by an even or odd number of permutations. Show that (1) any determinant $A = |a^i_j|\,(i, j = 1, \ldots, n)$ (the index i indicates the number of the column and j the number of the row) can be written in one of the two forms $a = \varepsilon_{i_1,i_2,\ldots,i_n}a^{i_1}_1 a^{i_2}_2 \ldots a^{i_n}_n = \varepsilon^{i_1 i_2\ldots i_n}a^1_{i_1}a^2_{i_2}\ldots a^n_{i_n}$; (2) the quantities $\sqrt{g}\,\varepsilon_{i_1 i_2\ldots i_n}$ define the components of a covariant tensor (the *permutation tensor*), where $g = |g_{\alpha\beta}|$ is the determinant corresponding to the metric tensor; the contravariant components of this tensor are $\dfrac{1}{\sqrt{g}}\varepsilon^{i_1 i_2\ldots i_n}$; (3) the permutation tensor is covariantly constant.

§ 4. Parallel Displacement in a V_n Space

The space V_n is called flat and denoted by the symbol R_n, if it is possible to define a system of coordinates relative to which the components $g_{\alpha\beta}$ of the metric tensor are constant. The invariant nature of flat space will be demonstrated in the following section. If in addition the form $ds^2 = g_{\alpha\beta}\,dx^\alpha\,dx^\beta$ is positive-definite, the R_n is usually called an ordinary space; an example of an ordinary R_n is three-dimensional Euclidean space. V_n is a generalized form of R_n as it is not assumed that a system of coordinates in the domain can be found which reduces the $g_{\alpha\beta}$ to constants.

Parallelism is one of the fundamental concepts of Euclidean geometry. In constructing the geometry of V_n, it is natural to investigate the possibility of parallelism in these manifolds. Parallelism is connected with the need for comparing vectors at two different points in a given domain. If these points are joined by a curve, one can speak of comparing vectors at different points on the curve. We shall see later that it is impossible to separate the idea of parallelism in V_n from the curve along which the vectors are considered, and in view of this fact we now consider parallel displacement along a given curve in V_n.

This concept must be made precise and unique by imposing additional requirements. In order that our definition of parallel displacement of vectors in V_n should appear natural, we introduce considerations not essential to a formal treatment. We therefore consider parallelism in an ordinary R_3 in order to derive properties needed for the definition of parallelism in V_n.

If R_3 is provided with a system of curvilinear coordinates, basic coordinate vectors $\underset{\alpha}{\boldsymbol{l}}$ exist at each point, and the covariant components of any vector $v_\alpha = \boldsymbol{v}\cdot\underset{\alpha}{\boldsymbol{l}}$; $\boldsymbol{v}$ and $\boldsymbol{l}$ are in the notation used in vector analysis. It is easy to show that the v_α are in fact vector components, by verifying their transformation law and that of $\underset{\alpha}{\boldsymbol{l}}$ for a change of coordinates. If some vector $\boldsymbol{u}$ is given a parallel displacement in R_3 along the curve $x^\alpha = x^\alpha(t)$, then $\dfrac{d\boldsymbol{u}}{dt} = 0$. The components of this vector must also be equal to zero, i.e.

$$\frac{d\boldsymbol{u}}{dt}\underset{\alpha}{\boldsymbol{l}} = \frac{d}{dt}(\boldsymbol{u}\underset{\alpha}{\boldsymbol{l}}) - \boldsymbol{u}\frac{d\underset{\alpha}{\boldsymbol{l}}}{dt} = \frac{du_\alpha}{dt} - u^\sigma\overset{\sigma}{\boldsymbol{l}}\frac{d\underset{\alpha}{\boldsymbol{l}}}{dt}$$

$$= \frac{du_\alpha}{dt} - u_\sigma\overset{\sigma}{\boldsymbol{l}}\,\partial_\beta\underset{\alpha}{\boldsymbol{l}}\frac{dx^\beta}{dt} = \frac{du_\alpha}{dt} - P^\sigma_{\alpha\beta}u_\sigma\frac{dx^\beta}{dt} = 0.$$

We thus obtain the following property:

(α) $\dfrac{du_\alpha}{dt}$ depends linearly on u_σ and $\dfrac{dx^\beta}{dt}$.

Here $\overset{\sigma}{\boldsymbol{l}}$ is the reciprocal system of vectors, defined by the equation

$$\overset{\sigma}{\boldsymbol{l}}\underset{\tau}{\boldsymbol{l}} = \delta^\sigma_\tau.$$

As is well known, the reciprocal system always exists if the vectors $\underset{\alpha}{l}$ are independent. It is natural to require that property (α) should also hold in V_n. An analogous argument applies to the contravariant components.

Assuming that property (α) holds for parallel displacement in V_n along the curve $x^\alpha(t)$, we obtain:

$$\frac{du^\alpha}{dt} = -A^\alpha_{\beta\gamma} u^\beta \frac{dx^\gamma}{dt}. \tag{4.1}$$

We also require that, as in R_3,

(β) the norm of a vector remains constant during parallel displacement in V_n:

$$\frac{d}{dt}(g_{\alpha\beta} u^\alpha u^\beta) = 0.$$

Then using (4.1) we obtain

$$\partial_\gamma g_{\alpha\beta} u^\alpha u^\beta \frac{dx^\gamma}{dt} + g_{\alpha\beta}\left(\frac{du^\alpha}{dt} u^\beta + u^\alpha \frac{du^\beta}{dt}\right) = (\partial_\gamma g_{\alpha\beta} - A_{\alpha,\,\beta\gamma} - A_{\beta,\,\alpha\gamma}) u^\alpha u^\beta \frac{dx^\gamma}{dt} = 0,$$

where $A_{\alpha,\,\beta\gamma} = g_{\alpha\sigma} A^\sigma_{\beta\gamma}$. If this is true for any vector u^α on any curve, the expression in brackets symmetrized with respect to α and β must be zero. As this expression is already symmetric with respect to α, β

$$\partial_\gamma g_{\alpha\beta} = A_{\alpha,\,\beta\gamma} + A_{\beta,\,\alpha\gamma} = g_{\alpha\tau} A^\tau_{\beta\gamma} + g_{\beta\tau} A^\tau_{\alpha\gamma}. \tag{4.2}$$

Finally we shall require that

(γ) $$A^\alpha_{[\beta\gamma]} = A_{\alpha,\,[\beta\gamma]} = 0.$$

The geometrical significance of this last condition is as follows. Let us consider some point in the space V_n and two curves passing through this point; suppose that the tangent vectors are dx^α and δx^α. Let us displace the vector δx^α along the curve defined by dx^α and vice versa. If we neglect infinitesimally small quantities of order higher than the second, then after the first displacement the end of the displaced vector will be at the point $x^\alpha + dx^\alpha + \delta x^\alpha - A^\alpha_{\beta\gamma} dx^\beta \delta x^\gamma$, after the second displacement the end of the displaced vector will be at the point $x^\alpha + \delta x^\alpha + dx^\alpha - A^\alpha_{\beta\gamma} \delta x^\beta dx^\gamma$. Hence the "gap" between these points is given by the expression $2A^\alpha_{[\beta\gamma]} dx^\beta \delta x^\gamma$. Condition ($\gamma$) thus reduces to the requirement that infinitesimally small parallelograms should exist in each two-dimensional subspace (to the third order of small quantities). A different interpretation is given by Cartan [59], [83].

If (γ) holds it follows from (3.2) (3.3) and (4.2) that $A^{\alpha}_{\beta\gamma} = \Gamma^{\alpha}_{\beta\gamma}$. Thus under conditions (α), (β) and (γ) parallel displacement of the vector u^{α} along a given curve results in

$$\frac{du^{\alpha}}{dt} = -\Gamma^{\alpha}_{\beta\gamma}u^{\beta}\frac{dx^{\gamma}}{dt}. \tag{4.3}$$

Equations (4.3) are identical in form with the law of parallel displacement in R_3, but while the parallelism is invariant in R_3 (from the very nature of the concept), the invariance of (4.3) in V_n has still to be proved.

Since $u^{\alpha'} = A^{\alpha'}_{\sigma}\, u^{\sigma}$, then by virtue of (4.3)

$$du^{\alpha'} = \partial_{\beta}A^{\alpha'}_{\sigma}u^{\sigma}\, dx^{\beta} + A^{\alpha'}_{\sigma}du^{\sigma} = (\partial_{\sigma}A^{\alpha'}_{\tau} - A^{\alpha'}_{\lambda}\Gamma^{\lambda}_{\tau\sigma})u^{\tau}\, dx^{\sigma}. \tag{4.3a}$$

Substituting from (3.7), the transformation law for the object $\Gamma^{\alpha}_{\beta\gamma}$, we obtain

$$du^{\alpha'} = -\, A^{\lambda'}_{\tau}A^{\nu'}_{\sigma}\Gamma^{\alpha'}_{\lambda'\nu'}u^{\tau}\, dx^{\sigma} = -\Gamma^{\alpha'}_{\lambda'\nu'}u^{\lambda'}\, dx^{\nu'}, \tag{4.3b}$$

which proves the invariance of (4.3) under a transformation of coordinates,

If we are given a curve $x^{\alpha} = x^{\alpha}(t)$ where x^{α} is a continuous function of t, then $\Gamma^{\tau}_{\beta\gamma} = \Gamma^{\tau}_{\beta\gamma}(x^{\alpha}(t))$ and eqns. (4.3) define a set of n ordinary linear differential equations with coefficients continuous within some interval: the existence and uniqueness theorem shows that for each set of initial conditions

$$t = t_0, \quad u^{\alpha} = \overset{0}{u}{}^{\alpha} \qquad (\alpha = 1, \ldots, n), \tag{4.3c}$$

there is a unique solution of the system.

Suppose that two curves defined by the equations $x^{\alpha} = x^{\alpha}(t)$ and $\mathfrak{x}^{\alpha} = \mathfrak{x}^{\alpha}(t)$ pass through two points A and B in V_n and let the parameter be such that on both curves the value t_0 corresponds to the point A and the value t_1 to B. We also suppose that both A and B and the curves connecting them, belong to a domain of V_n where the conditions of the existence and uniqueness theorem are satisfied for the system (4.3), (4.3c). For one and the same vector $\overset{0}{u}{}^{\alpha}$ at A, there are two different sets of equations (4.3) corresponding to the two curves, and two different vectors at B resulting from the parallel diplacement of u^{α}. *Thus parallel displacement according to the law* (4.3) *is unique along a given path; the displacement depends on the path.* These propositions are, of course, of a local nature, since they only hold in a domain defined by the interval within which the existence and uniqueness theorem is valid.

Multiplying (4.3) by dt, taking all terms to the left-hand side and denoting the expression obtained by δu^α, we have

$$\delta u^\alpha = (\partial_\gamma u^\alpha + \Gamma^\alpha_{\beta\gamma} u^\beta)\, dx^\gamma = u^\alpha_{,\gamma}\, dx^\gamma = 0.$$

The expression δu^α is called the *absolute differential* of the field u^α. The absolute differential has a meaning only with respect to a particular direction given by the differentials dx^α, and hence *the field u^α undergoes parallel displacement along the curve $x^\alpha = x^\alpha(t)$ if the absolute differential of u^α for an infinitely small displacement along the curve is zero.*

Generalizing this result for any tensor and introducing the absolute differential of a tensor

$$\delta T^{\alpha_1 \dots \alpha_p}{}_{\beta_1 \dots \beta_q} = T^{\alpha_1 \dots \alpha_p}{}_{\beta_1 \dots \beta_q, \gamma}\, dx^\gamma \tag{4.4}$$

along some curve, we shall say that *a tensor undergoes parallel displacement along the curve $x^\alpha = x^\alpha(t)$ if its absolute differential along this curve is zero.* The parallel displacement of tensors is thus directly connected with covariant differentiation; known as *parallel displacement in Levi-Civita's sense,* after the mathematician who introduced and developed the concept [39], it has recently been generalized in a number of different ways. Other possible definitions of parallel displacement in V_n are equivalent to that derived above (Schouten and Struik [110], pp. 73–89; Rashevskii [188], pp. 429–50).

Examples

1. A sphere and circular cone in Euclidean space touch along a small circle (l). Show that any set of vectors parallel along (l) and lying in the sphere will also define vectors parallel along (l) and lying in the cone; if the cone is unrolled into a plane then these vectors are transformed into vectors which are parallel in the Euclidean sense.

2. In order that the tangents to the curve x^2 = constant in the space V_2 should be parallel along some curve (l), it is necessary and sufficient that this curve (l) should be an integral curve of the equation $\Gamma^2_{1\alpha}\, dx^\alpha = 0$.

3. If the vectors $\underset{1}{u}^\alpha, \dots, \underset{n}{u}$ undergo parallel displacement along some curve, then any linear relations between them remain unaltered.

§ 5. Curvature Tensor of a V_n Space

First order partial derivatives and covariant derivatives obey the same rules, but with second order derivatives a fundamental difference arises. This can be shown most simply by solving a basic problem of equations involving covariant derivatives—that of the conditions for integrability. Whilst repeated partial differentiation is commutative, this is not true for covariant derivatives. Suppose that we are given a vector field $u_\alpha(x)$. The covariant derivative is

$$u_{\alpha,\,\beta} = \partial_\beta u_\alpha - \Gamma^\sigma_{\alpha\beta} u_\sigma. \tag{5.1}$$

Differentiating again, we obtain

$$u_{\alpha,\,\beta\gamma} = \partial_\gamma u_{\alpha,\,\beta} - \Gamma^\lambda_{\alpha\gamma} u_{\lambda,\,\beta} - \Gamma^\lambda_{\beta\gamma} u_{\alpha,\,\lambda}.$$

Since Riemann connectivity has no torsion (cf. § 3), $\Gamma^\sigma_{[\alpha\beta]} = 0$; by alternating this equation with respect to β and γ we obtain

$$u_{\alpha,\,[\beta\gamma]} = (\partial_{[\beta} \Gamma^\sigma_{\gamma]\alpha} + \Gamma^\tau_{\alpha[\gamma} \Gamma^\sigma_{\beta]\tau}) u_\sigma.$$

On the left we have a tensor, u_σ is a vector, and hence in accordance with the quotient law for tensors, the expression in brackets must be a tensor of order 1+3 and weight 0. Introducing the notation

$$\tfrac{1}{2} R^\sigma{}_{\alpha\beta\gamma} = \partial_{[\beta} \Gamma^\sigma_{\gamma]\alpha} + \Gamma^\tau_{\alpha[\gamma} \Gamma^\sigma_{\beta]\tau}, \tag{5.2}$$

we obtain the equation

$$u_{\alpha,\,[\beta\gamma]} = \tfrac{1}{2} R^\sigma{}_{\alpha\beta\gamma} u_\sigma, \tag{5.3}$$

which reveals the difference between partial and covariant derivatives.

The tensor $R^\sigma_{\alpha\beta\gamma}$ is called the *curvature tensor* of the space V_n and is important for its characterization.

Multiplying (5.3) by $g^{\alpha\tau}$ and making use of the fact that the metric tensor is covariantly constant, we obtain

$$u^\tau{}_{,\,[\beta\gamma]} = \tfrac{1}{2} g^{\alpha\tau} R^\sigma{}_{\alpha\beta\gamma} u_\sigma = \tfrac{1}{2} R_\sigma{}^\tau{}_{\beta\gamma} u^\sigma = -\tfrac{1}{2} R^\tau{}_{\sigma\beta\gamma} u^\sigma. \tag{5.4}$$

In the case of a tensor of any order, we obtain

$$S^{\alpha_1 \ldots \alpha_p}{}_{\beta_1 \ldots \beta_q,\,[\gamma\delta]} = \tfrac{1}{2} \sum_{k=1}^{q} S^{\alpha_1 \ldots \alpha_p}{}_{\beta_1 \ldots \beta_{k-1} \sigma \beta_{k+1} \ldots \beta_p} R^\sigma{}_{\beta_k \gamma\delta}$$
$$- \tfrac{1}{2} \sum_{r=1}^{p} S^{\alpha_1 \ldots \alpha_{r-1} \sigma \alpha_{r+1} \ldots \alpha_p}{}_{\beta_1 \ldots \beta_q} R^{\alpha_r}{}_{\sigma\gamma\delta}. \tag{5.5}$$

This relationship, first obtained by Ricci, is called *Ricci's identity* ([13], p. 143). When covariant derivatives are used instead of partial derivatives, this identity is the appropriate condition of integrability. As the curvature tensor is completely determined if the object $\Gamma^\alpha_{\beta\gamma}$ is given, it can also be defined in spaces of affine connectivity with torsion (§ 3), where (5.5) is still valid. Ricci's identity shows that in V_n spaces or spaces of affine connectivity, a tensor field with components of class $C^r (r \geqslant 2)$ cannot be given arbitrarily, since it is connected with the metric of the space or with the coefficients of connectivity, respectively, by eqn. (5.5).

The curvature tensor in V_n can be defined by its covariant components

$$R_{\alpha\beta\gamma\delta} = g_{\alpha\sigma} R^{\sigma}{}_{\beta\gamma\delta}, \tag{5.6}$$

and it is convenient to investigate its properties in this form.

We can write the covariant components of the curvature tensor in an explicit form. As

$$g_{\alpha\sigma} \Gamma^{\sigma}_{\beta\gamma} = g_{\alpha\sigma} g^{\sigma\tau} \Gamma_{\tau,\, \beta\gamma} = \Gamma_{\alpha,\, \beta\gamma}$$

and the $\partial_\gamma g_{\alpha\beta}$ can be expressed in terms of Christoffel symbols by using (3.2), we have

$$g_{\alpha\sigma} \partial_\beta \Gamma^{\sigma}_{\gamma\delta} = \partial_\beta (g_{\alpha\sigma} \Gamma^{\sigma}_{\gamma\beta}) - \Gamma^{\sigma}_{\gamma\delta} \partial_\beta g_{\alpha\sigma} = \partial_\beta \Gamma_{\alpha,\, \gamma\delta} - \Gamma^{\sigma}_{\gamma\delta} (\Gamma_{\alpha,\, \sigma\beta} + \Gamma_{\sigma,\, \alpha\beta})$$

and hence

$$\begin{aligned} R_{\alpha\beta\gamma\delta} = \partial_\gamma \Gamma_{\alpha,\, \beta\delta} - \partial_\delta \Gamma_{\alpha,\, \beta\gamma} + \Gamma^{\sigma}_{\beta\gamma} \Gamma_{\sigma,\, \alpha\delta} - \Gamma^{\sigma}_{\beta\delta} \Gamma_{\sigma,\, \alpha\gamma} \\ = \tfrac{1}{2} (\partial_{\beta\gamma} g_{\alpha\delta} + \partial_{\alpha\delta} g_{\beta\gamma} - \partial_{\beta\delta} g_{\alpha\gamma} - \partial_{\alpha\gamma} g_{\beta\delta}) \\ + g^{\sigma\tau} (\Gamma_{\sigma,\, \alpha\delta} \Gamma_{\tau,\, \beta\gamma} - \Gamma_{\sigma,\, \alpha\gamma} \Gamma_{\tau,\, \beta\delta}). \end{aligned} \tag{5.7}$$

The following identities can be proved directly from (5.7):

$$\left.\begin{aligned} &(\alpha) \qquad R_{\alpha\beta\gamma\delta} = - R_{\beta\alpha\gamma\delta} = - R_{\alpha\beta\delta\gamma}, \\ &(\beta) \qquad R_{\alpha\beta\gamma\delta} = R_{\gamma\delta\alpha\beta}, \\ &(\gamma) \qquad R_{\alpha\beta\gamma\delta} + R_{\alpha\gamma\delta\beta} + R_{\alpha\delta\beta\gamma} = 0. \end{aligned}\right\} \tag{5.8}$$

The set of identities which covers all the algebraic relations between the components of a tensor is called *the complete set of identities*: all other identities satisfied by the components of the tensor can be deduced from the complete set. It can be shown (Thomas [92], p. 131) that in the case of the curvature tensor $R_{\alpha\beta\gamma\delta}$, the complete set of identities is given by (α) and (γ); it is easily seen that (β) is a direct result of (α) and (γ).

Let us calculate the number of independent components of the curvature tensor. The components $R_{\alpha\beta\alpha\beta}$ with only two different indices include only $\binom{n}{2}$ independent ones, because of (α). Among the components with three different indices we find, using (α) and (γ), $3\binom{n}{3}$ independent ones, and finally among the components with four different indices we find $2\binom{n}{4}$ independent components. Hence the total number of independent components is

$$N = \binom{n}{2}+3\binom{n}{3}+2\binom{n}{4} = \tfrac{1}{12}n^2(n^2-1). \tag{5.9}$$

In particular V_2 corresponds to $N = 1$, V_3 to $N = 6$, and V_4 to $N = 20$.

In view of the importance of the curvature tensors of a V_n space defined by a metric tensor $g_{\alpha\beta}$, it is useful to determine the tensors formed from them by contraction. A simple contraction produces, apart from sign, only one tensor

$$R_{\alpha\beta} = R^{\sigma}{}_{\alpha\sigma\beta} = \partial_\sigma \Gamma^{\sigma}_{\alpha\beta} - \partial_{\alpha\beta} \ln \sqrt{g} - \Gamma^{\tau}_{\alpha\sigma}\Gamma^{\sigma}_{\tau\beta} + \Gamma^{\sigma}_{\alpha\beta}\partial_\sigma \ln \sqrt{g}, \tag{5.10}$$

according to (5.2) and example 5, § 3; contraction with respect to indices of an antisymmetric pair naturally gives zero. The tensor $R_{\alpha\beta}$ (which in virtue of (5.10) is symmetric) is called the *Ricci tensor* after the mathematician who first introduced it, and plays a fundamental role in Einstein spaces in general and in the theory of relativity in particular.†

Defining $R^{\alpha}{}_{\beta} = g^{\alpha\sigma}R_{\sigma\beta}$, a second contraction produces the scalar

$$R = R^{\alpha}{}_{\alpha},$$

called *the scalar curvature of the space* V_n.

As a first example of the application of the curvature tensor, we will prove the following theorem: *in order that a space should be flat, it is necessary and sufficient that its curvature tensor should be zero.*

Suppose that we are given a flat space, i.e. in some system of coordinates the components of the metric tensor $g_{\alpha\beta}$ are constant.

Then $\Gamma_{\alpha,\beta\gamma} = \Gamma^{\alpha}_{\beta\gamma} = 0$ and $R_{\alpha\beta\gamma\delta} = 0$ in this system of coordinates and hence in any other system. Conversely, suppose that $R_{\alpha\beta\gamma\delta} = 0$ for some V_n. We will consider the metric tensor $g_{\alpha\beta}$ and show that there exists a non-degenerate coordinate transformation $x^{\alpha} = x^{\alpha}(x^{\beta'})$ such that in the system $\{x^{\alpha'}\}$ the components $g_{\alpha'\beta'}$ are constant. This is equivalent to the requirement that

$$\Gamma_{\alpha',\beta'\gamma'} = \Gamma^{\alpha'}_{\beta'\gamma'} = 0.$$

The transformation law (3.7) for the object $\Gamma^{\alpha}_{\beta\gamma}$ shows that

$$\Gamma^{\alpha}_{\beta\gamma}A^{\alpha'}_{\alpha}A^{\beta}_{\beta'}A^{\gamma}_{\gamma'} + \partial_{\beta'}A^{\omega}_{\gamma'}A^{\alpha'}_{\omega} = 0.$$

Contracting this equation with $A^{\lambda}_{\alpha'}$, we reduce it, by virtue of (1.14), to the equivalent form

$$\partial_{\beta'}A^{\lambda}_{\gamma'} = -\Gamma^{\lambda}_{\beta\gamma}A^{\beta}_{\beta'}A^{\gamma}_{\gamma'}, \quad \partial_{\sigma'}x^{\alpha} = A^{\alpha}_{\sigma'}. \tag{5.11}$$

† The Ricci tensor is sometimes defined by the contraction $R^{\sigma}{}_{\alpha\beta\sigma}$, which changes the signs in (5.10).

The problem has thus been reduced to the compatibility of the system of equations (5.11) for the unknown functions $A^{\alpha}_{\alpha'}$ and x^{α} of the independent variables $x^{\alpha'}$ under the conditions

$$g_{\alpha\beta}A^{\alpha}_{\alpha'}A^{\beta}_{\beta'} = g_{\alpha'\beta'} = \text{const.} \tag{5.12}$$

The theory of integrability (Goursat [53], p. 103) shows that if the conditions for the integrability of (5.11) are satisfied identically, i.e. for all values of $A^{\alpha}_{\alpha'}$, and x^{α}, and if the $A^{\alpha}_{\alpha'}$, are regular in the neighbourhood of $\overset{0}{x}{}^{\alpha}$, $(A^{\alpha}_{\alpha'})_0$, then there exists one and only one set of solutions which are regular in this region and take the values $(A^{\alpha}_{\alpha'})_0$ for $x^{\alpha} = \overset{0}{x}{}^{\alpha}$. When the conditions of integrability are satisfied identically, eqns. (5.12) are required only for the choice of the initial conditions $(A^{\alpha}_{\alpha'})_0$. The integrability conditions for the second group of equations (5.11) are

$$\partial_{[\sigma'\tau']}x^{\alpha} = \partial_{[\tau'}A^{\alpha}_{\sigma']}\,.$$

But according to the first group (cf. example 9, § 2)

$$\partial_{[\tau'}A^{\alpha}_{\sigma']} = -\Gamma^{\alpha}_{\beta\gamma}A^{\beta}_{[\tau'}A^{\gamma}_{\sigma']} = -\Gamma^{\alpha}_{[\beta\gamma]}A^{\beta}_{\tau'}A^{\gamma}_{\sigma'} \equiv 0.$$

The conditions of integrability for the first group of equations (5.11) are

$$\begin{aligned}\partial_{[\beta'\tau']}A^{\lambda}_{\gamma'} &= -\Gamma^{\lambda}_{\beta\gamma}(\partial_{[\tau'}A^{\beta}_{\beta']}A^{\gamma}_{\gamma'}+A^{\beta}_{[\beta'}\partial_{\tau']}A^{\gamma}_{\gamma'})-A^{\gamma}_{\gamma'}A^{\beta}_{[\beta'}A^{\nu}_{\tau']}\partial_{\nu}\Gamma^{\lambda}_{\beta\gamma}\\ &= -A^{\gamma}_{\gamma'}A^{\beta}_{\beta'}A^{\nu}_{\tau'}(\partial_{[}{}^{\nu}\Gamma^{\lambda}_{\beta]\,\gamma}+\Gamma^{\sigma}_{\gamma\beta}\Gamma^{\lambda}_{\sigma\nu}-\Gamma^{\sigma}_{\gamma\nu}\Gamma^{\lambda}_{\sigma\beta})\\ &= -A^{\gamma}_{\gamma'}A^{\beta}_{\beta'}A^{\nu}_{\tau'}R^{\lambda}_{\gamma\nu\beta} \equiv 0\end{aligned}$$

in accordance with the initial assumption. The set of equations (5.11) requires n and n^2 arbitrary constants to define the initial values of x^{α} and $A^{\alpha}_{\beta'}$, respectively; eqn. (5.12) imposes $n(n+1)/2$ conditions, and hence the required coordinate transformation contains $n(n-1)/2$ arbitrary constants. It is possible, by making use of this arbitrariness, to still further simplify the relation $g_{\alpha'\beta'} = \text{const}$, reducing ds^2 for a real path to the form $\sum_{\alpha=1}^{n} l_{\alpha}\, dx^{\alpha^2}$, where $l_{\alpha} = \pm 1$ defines the signature of the metric of the flat space.

Investigations in V_n involve considering not only the curvature tensor, but also its covariant derivative $R_{\alpha\beta\gamma\delta,\,\lambda}$. It is important to determine any algebraic identities connecting the components of this tensor; we will derive the so-called *Bianchi identity* which will partially solve this problem.

Let us consider an arbitrary vector field u_α and the field $u_{\alpha,\beta}$. Applying the Ricci identity (5.5) to these, we obtain

$$u_{\alpha,[\beta\gamma]} = \tfrac{1}{2}R^{\sigma}{}_{\alpha\beta\gamma}u_\sigma, \tag{5.13}$$

$$u_{\alpha,\beta[\gamma\delta]} = \tfrac{1}{2}(R^{\sigma}{}_{\alpha\gamma\delta}u_{\sigma,\beta}+R^{\sigma}{}_{\beta\gamma\delta}u_{\alpha,\sigma}). \tag{5.14}$$

Differentiating (5.13) again shows that

$$u_{\alpha,[\beta\gamma]\delta} = \tfrac{1}{2}(R^{\sigma}{}_{\alpha\beta\gamma,\delta}u_\sigma+R^{\sigma}{}_{\alpha\beta\gamma}u_{\sigma,\delta}). \tag{5.15}$$

Alternating (5.14) and (5.15) with respect to the indices $\beta\gamma\delta$ eliminates the inner brackets (cf. example 12, § 2); hence equating the left-hand sides, we obtain

$$R^{\sigma}{}_{\alpha[\gamma\delta}u_{|\sigma|,\beta]}+R^{\sigma}{}_{[\beta\gamma\delta]}u_{\alpha,\sigma}-R^{\sigma}{}_{\alpha[\beta\gamma,\delta]}u_\sigma-R^{\sigma}_{\alpha[\beta\gamma}u_{|\sigma|,\delta]} = 0.$$

The first and last terms cancel, and the second is zero in virtue of (5.8); hence

$$R^{\sigma}{}_{\alpha[\beta\gamma,\delta]}u_\sigma = 0.$$

As the field u_α is arbitrary

$$R^{\sigma}{}_{\alpha[\beta\gamma,\delta]} = R_{\sigma\alpha[\beta\gamma,\delta]} = 0$$

or, in the expanded form (cf. example 6, § 2),

$$R_{\sigma\alpha\beta\gamma,\delta}+R_{\sigma\alpha\gamma\delta,\beta}+R_{\sigma\alpha\delta\beta,\gamma} = 0. \tag{5.16}$$

This relation is known as *Bianchi's identity* ([14], p. 351).

Differentiating identities (α) and (γ) of (5.8) we also obtain

$$\left.\begin{aligned} R_{\alpha\beta\gamma\delta,\lambda} &= -R_{\beta\alpha\gamma\delta,\lambda} = -R_{\alpha\beta\delta\gamma,\lambda},\\ R_{\alpha\beta\gamma\delta,\lambda}&+R_{\alpha\gamma\delta\beta,\lambda}+R_{\alpha\delta\beta\gamma,\lambda} = 0.\end{aligned}\right\} \tag{5.17}$$

It can be proved that (5.16) and (5.17) are the complete set of algebraic identities for $R_{\alpha\beta\gamma\delta,\lambda}$ (Thomas [92], p. 132). By contraction, we obtain the following corollaries from these identities:

$$R^{\sigma}{}_{\alpha\beta\gamma,\sigma}+2R_{\alpha[\beta,\gamma]} = 0, \tag{5.18}$$

$$R^{\sigma}{}_{\beta,\sigma}-\tfrac{1}{2}R_{,\beta} = 0. \tag{5.19}$$

If $n = 4$, then (5.19) is the equation of the law of conservation of energy-momentum for relativistic gravitational fields.

The curvature tensor plays an exceptional role in the development of the theory of Riemann spaces and in applications where it admits of a physical interpretation (cf. [229], [253], [285], [293], [297], [247]).

Examples

1. In formulae (5.8), obtain (β) from (α) and (γ).

2. Show that if $g_{\alpha\beta}$ and $\overset{*}{g}_{\alpha\beta}$ are the components of two symmetric non-degenerate tensors, and $\Gamma^{\alpha}_{\beta\gamma}$ and $\overset{*}{\Gamma}{}^{\alpha}_{\beta\gamma}$ are the corresponding Christoffel symbols, then $\Omega^{\alpha}_{\beta\gamma} = \overset{*}{\Gamma}{}^{\alpha}_{\beta\gamma} - \Gamma^{\alpha}_{\beta\gamma}$ is a tensor; the curvature tensors corresponding to the Riemann connectivities Γ and $\overset{*}{\Gamma}$ satisfy the relationship

$$\overset{*}{R}{}^{\alpha}{}_{\beta\gamma\delta} = R^{\alpha}{}_{\beta\gamma\delta} + 2\Omega^{\sigma}{}_{\beta[\delta,\,\gamma]} + 2\Omega^{\sigma}{}_{\beta[\delta}\Omega^{\alpha}{}_{|\sigma|\gamma]},$$

where the covariant differentiation is based on the metric $g_{\alpha\beta}$.

3. Show that in V_3

$$R_{\alpha\beta\gamma\delta} = -g_{\alpha\delta}R_{\beta\gamma} + g_{\alpha\gamma}R_{\beta\delta} - g_{\beta\gamma}R_{\alpha\delta} + g_{\beta\delta}R_{\alpha\gamma} + \frac{R}{2}(g_{\alpha\delta}g_{\beta\gamma} - g_{\alpha\gamma}g_{\beta\delta}).$$

4. Prove that a space V_3 with a definite metric referred to a tri-orthogonal system of coordinates has the scalar curvature $R = 0$, only when

$$\sum_{i=1}^{3} \frac{1}{g_{ii}} \Big[\sum_j{}' \frac{1}{\sqrt{g_{jj}}} \partial_{ii} \sqrt{g_{jj}} - \partial_i \ln \sqrt{g_{ii}} \sum_j{}' \partial_i \ln \sqrt{g_{jj}} + \partial_i \ln \sqrt{g_{jj}} \partial_i \ln \sqrt{g_{kk}} \Big] = 0,$$

where $\sum'$ denotes summation over $j \neq i$, and i, j, k are unequal (Eisenhart [292]).

5. Show that by taking $\sqrt{g_{11}} = x_1^{a_1}x_2^{b_1}x_3^{c_1}$, $\sqrt{g_{22}} = x_2^{a_2}x_3^{b_2}x_1^{c_2}$, $\sqrt{g_{33}} = x_3^{a_3}x_1^{b_3}x_2^{c_3}$, where a, b and c are constants, it is possible to satisfy the conditions of example 4 by putting

(1) $a_i = c_i = 0$ $(i = 1, 2, 3)$, $b_i = 1$ when $g_{11} = x_2^2$, $g_{22} = x_3^2$, $g_{33} = x_1^2$;

(2) $a_i = 0$, $c_2 = b_3$, $c_3 = b_1$, $c_1 = b_2$, $c_i = \frac{2}{3}$, $b_i = \frac{2}{3}$ when $g_{11} = (x_2x_3)^{4/3}$, $g_{22} = (x_1x_3)^{4/3}$, $g_{33} = (x_1x_2)^{4/3}$.

§ 6. Geodesics

A straight line in Euclidean space can be defined as (1) a curve whose tangent vectors are all parallel, or (2) the shortest distance between two points, or (3) a line of zero curvature.

In V_n it is possible to define curves which have appropriately generalized but analogous properties. Such curves are called *geodesics of* V_n.

A geodesic of V_n is a curve whose tangent vector can be displaced along it parallel to itself in Levi-Civita's sense. Let $x^{\alpha} = x^{\alpha}(t)$, functions of class

$C^r(r \geqslant 2)$, be the parametric equations of a geodesic. In any continuously differentiable and reversible transformation of the parameter $t = t(\tau)$, the tangent vector $\frac{dx^\alpha}{dt}$ transforms as follows:

$$\frac{dx^\alpha}{d\tau} = \frac{dx^\alpha}{dt}\frac{dt}{d\tau},$$

i.e. at each point it is determined to within a numerical factor. Let λ^α be a vector which can be displaced along the geodesic parallel to itself, then

$$\frac{dx^\alpha}{dt} = \sigma\lambda^\alpha.$$

The absolute differential $\delta\lambda^\alpha$ along the geodesic is zero and hence

$$\frac{\delta}{dt}\left(\frac{dx^\alpha}{dt}\right) = \frac{d\sigma}{dt}\lambda^\alpha = \nu\frac{dx^\alpha}{dt}, \tag{6.1}$$

since $\frac{dx^\alpha}{dt} \neq 0$. Let $\tau = \tau(t)$. Then (6.1) can be written in the form

$$\frac{\delta}{d\tau}\left(\frac{dx^\alpha}{d\tau}\right)\frac{d\tau}{dt} = \left(\nu - \frac{d^2\tau}{dt^2}\Big/\frac{d\tau}{dt}\right)\frac{dx^\alpha}{dt}. \tag{6.1a}$$

If we choose $\tau(t)$ so that

$$\frac{d^2\tau}{dt^2} = \nu\frac{d\tau}{dt},$$

the right-hand side of (6.1a) vanishes, which allows the vector $\frac{dx^\alpha}{d\tau}$ to be displaced parallel to itself along the geodesic. With the *canonical parameter* τ as the independent variable the differential equation of the geodesic is

$$\left(\frac{dx^\alpha}{d\tau}\right)_{,\gamma}\frac{dx^\gamma}{d\tau} = 0 \tag{6.2}$$

or in the expanded form

$$\frac{d^2x^\alpha}{d\tau^2} + \Gamma^\alpha_{\beta\gamma}\frac{dx^\beta}{d\tau}\frac{dx^\gamma}{d\tau} = 0. \tag{6.3}$$

In that domain in V_n where the conditions of the existence and uniqueness theorem are satisfied, eqn. (6.3) has a unique solution for each initial point $(x^\alpha)_0$ and initial direction $\left(\frac{dx^\alpha}{d\tau}\right)_0$; hence *one and only one geodesic passes through a given point in a given direction.* If the metric of V_n is indefinite, the geodesics can include isotropic curves.

Along any geodesic with canonical parametrization we have

$$g_{\alpha\beta}\frac{dx^\alpha}{d\tau}\frac{dx^\beta}{d\tau} = \varepsilon = \text{const}, \tag{6.4}$$

since along the curve (6.2)

$$\partial_\gamma\left(g_{\alpha\beta}\frac{dx^\alpha}{d\tau}\frac{dx^\beta}{d\tau}\right)\frac{dx^\gamma}{d\tau} = \left(g_{\alpha\beta}\frac{dx^\alpha}{d\tau}\frac{dx^\beta}{d\tau}\right)_{,\gamma}\frac{dx^\gamma}{d\tau} = 2g_{\alpha\beta}\frac{dx^\alpha}{d\tau}\left(\frac{dx^\beta}{d\tau}\right)_{,\gamma}\frac{dx^\gamma}{d\tau} = 0.$$

In other words, *the norm of the tangent vector* $\dfrac{dx^\alpha}{d\tau}$ *to the geodesic does not vary along it*; in particular *a geodesic which is isotropic at one point is isotropic everywhere.* Naturally not every isotropic curve is a geodesic.

In an ordinary R_n where curves cannot be isotropic, the vector $\dfrac{d^2x^\alpha}{ds^2} = \mu^\alpha$ is the principal normal vector to the curve in Cartesian coordinates. But if curvilinear coordinates are introduced in R_n, μ^α is equal to the left-hand side of (6.3), provided τ is taken as the arc length s. As a generalization we call

$$\mu^\alpha = \frac{d^2x^\alpha}{d\tau^2} + \Gamma^\alpha_{\beta\gamma}\frac{dx^\beta}{d\tau}\frac{dx^\gamma}{d\tau} \tag{6.5}$$

involving the canonical parameter τ, *the principal normal vector to a curve in* V_n; the fact that μ^α is a vector follows from (6.2). It is then natural to call

$$\frac{1}{\varrho} = \sqrt{|g_{\alpha\beta}\mu^\alpha\mu^\beta|}$$

the curvature of a curve in V_n *space* and to search for curves in V_n with zero curvature, for which

$$g_{\alpha\beta}\mu^\alpha\mu^\beta = 0.$$

If the metric of the space V_n *is definite, only geodesics have zero curvature; if the metric is indefinite these curves can be either geodesics or curves with an isotropic vector as their principal normal.* This result is a generalization of property (3) of straight lines in ordinary R_3.

If $\varepsilon \neq 0$ in (6.4) the geodesic is non-isotropic, and the arc length can be taken as a canonical parameter, implying that $\varepsilon = \pm 1$; other canonical parameters can be defined by a linear transformation containing two arbitrary constants.

A non-isotropic geodesic can also be defined as the shortest distance between two points A and B in V_n. This leads to the following problem in variational calculus; find the path for which the integral

$$\int_A^B ds = \int_{t_0}^{t_1} \sqrt{g_{\alpha\beta}\frac{dx^\alpha}{dt}\frac{dx^\beta}{dt}}\,dt$$

has a stationary value. Denoting the integrand by φ we have

$$\sqrt{g_{\alpha\beta}\frac{dx^\alpha}{dt}\frac{dx^\beta}{dt}} = \varphi = \frac{ds}{dt} \equiv \dot{s},$$

and φ must satisfy the Euler equations (Lavrent'ev and Lyusternik [95], p. 30):

$$\frac{d}{dt}\left(\frac{\partial\varphi}{\partial\dot{x}^\alpha}\right) - \frac{\partial\varphi}{\partial x^\alpha} = 0.$$

Since

$$\frac{\partial\varphi}{\partial\dot{x}^\alpha} = \frac{g_{\alpha\beta}\dot{x}^\beta}{\dot{s}}, \qquad \frac{\partial\varphi}{\partial x^\alpha} = \frac{\partial_\alpha g_{\sigma\beta}\dot{x}^\sigma\dot{x}^\beta}{2\dot{s}},$$

the Euler equations can be written in the form

$$g_{\alpha\beta}\ddot{x}^\beta - g_{\alpha\beta}\dot{x}^\beta\frac{\ddot{s}}{\dot{s}} + \Gamma_{\alpha,\,\sigma\beta}\dot{x}^\sigma\dot{x}^\beta = 0.$$

Whence, contracting with $g^{\alpha\gamma}$, we obtain

$$\ddot{x}^\gamma + \Gamma^\gamma_{\sigma\beta}\dot{x}^\sigma\dot{x}^\beta = \frac{\ddot{s}}{\dot{s}}\dot{x}^\gamma.$$

Introducing the canonical parameter $\tau = s$ we arrive at (6.3). The solution of this problem depends essentially on the assumption that the geodesic is non-isotropic. This result is a generalization of property (2) of straight lines in Euclidean space.

As the concept of parallelism also holds for spaces of affine connectivity, geodesics can be defined in exactly the same way for these manifolds ([170], pp. 148–9); this can be seen directly from eqns. (6.3) which depend only on the coefficients of connectivity.

Examples

1. Show that the object of connectivity $\Gamma^\alpha_{\beta\gamma}$ defines the same geodesics as the object $\Gamma^\alpha_{(\beta\gamma)}$.

2. Show that in Minkowski space defined by the metric $ds^2 = -dx^{1^2} - dx^{2^2} - dx^{3^2} + dx^{4^2}$, every isotropic curve has the form

$$x^1 = \int \varrho \cos\varphi \cos\psi \, ds, \quad x^2 = \int \varrho \cos\varphi \sin\psi \, ds,$$

$$x^3 = \int \varrho \sin\varphi \, ds, \quad x^4 = \int \varrho \, ds,$$

where ϱ, φ, ψ are functions of s. When are these curves geodesics?

§ 7. Special Systems of Coordinates in V_n

In defining different types of Einstein space, special systems of coordinates must be used. In this section we consider some of the systems which can be introduced into any V_n. The possibility of introducing these systems follows from a theorem enunciated by Riemann [1]; this states that: *by means of a transformation of coordinates, n components of the metric tensor can be written in the form of specified functions.* Since the metric of V_n and all permissible transformations are non-degenerate, the choice of these arbitrary functions must not lead to degeneracy of the metric. The coordinate systems considered below represent various possible realizations of this theorem.

Coordinates geodesic at a point. As in general, the curvature tensor in V_n is not zero, the object of Riemann connectivity cannot everywhere be reduced to zero by a transformation of coordinates, but it is always possible to find a special system of coordinates which reduces this object to zero at a specified point P: $(\Gamma^\alpha_{\beta\gamma})_P = 0$. Such a coordinate system is said to be *geodesic at the point P*.

Suppose that V_n is referred to a system of coordinates x^α and that the coordinates of P are $\overset{0}{x}{}^\alpha$. Let us introduce new coordinates given by

$$x^{\alpha'} = C^{\alpha'}_\beta \left(x^\beta - \overset{0}{x}{}^\beta\right) + \tfrac{1}{2} C^{\alpha'}_\lambda \Gamma^\lambda_{\beta\gamma}\left(\overset{0}{x}\right) \left(x^\beta - \overset{0}{x}{}^\beta\right) \left(x^\gamma - \overset{0}{x}{}^\gamma\right), \tag{7.1}$$

where $C^{\alpha'}_\beta$ are constants satisfying the condition $\left|C^{\alpha'}_\beta\right| \neq 0$ but otherwise arbitrary; in particular we may put $C^{\alpha'}_\beta = \delta^{\alpha'}_\beta$. Differentiating (7.1) twice with respect to the x^α and evaluating the derivatives at the point P, we obtain

$$\left(\frac{\partial x^{\alpha'}}{\partial x^\beta}\right)_P = C^{\alpha'}_\beta, \quad \left(\frac{\partial^2 x^{\alpha'}}{\partial x^\beta \partial x^\gamma}\right)_P = C^{\alpha'}_\lambda \Gamma^\lambda_{\beta\gamma}\left(\overset{0}{x}{}^\alpha\right). \tag{7.2}$$

In order that $(\Gamma^{\alpha'}_{\beta'\gamma'})_P = 0$ it is necessary and sufficient that in such a transformation

$$(A^{\alpha'}_{\omega} A^{\tau}_{\beta'} A^{\varrho}_{\gamma'} \Gamma^{\omega}_{\varrho\tau} + A^{\alpha'}_{\omega}\, \partial_{\beta'} A^{\omega}_{\gamma'})_P = 0$$

in accordance with (3.7). Contracting with $A^{\beta'}_{\lambda} A^{\gamma'}_{\nu}$ and noting that

$$A^{\sigma'}_{\beta} A^{\lambda}_{\sigma'} = \delta^{\lambda}_{\beta}, \qquad \frac{\partial}{\partial x^{\lambda}}\left(\frac{\partial x^{\omega}}{\partial x^{\gamma'}}\right)\frac{\partial x^{\alpha'}}{\partial x^{\omega}} = -\frac{\partial}{\partial x^{\lambda}}\left(\frac{\partial x^{\alpha'}}{\partial x^{\omega}}\right)\frac{\partial x^{\omega}}{\partial x^{\gamma'}},$$

we obtain the equivalent condition

$$(A^{\alpha'}_{\omega} \Gamma^{\omega}_{\lambda\nu}) = \left(\frac{\partial^2 x^{\alpha'}}{\partial x^{\lambda}\, \partial x^{\nu}}\right)_P,$$

which is satisfied identically according to (7.2). Instead of (7.1) a polynomial of any degree can be chosen with completely arbitrary coefficients in the terms of the third, fourth, and higher degrees. Hence it is possible to introduce a geodesic system at P in an infinite number of ways. This arbitrariness allows further specialization of the coordinates.

For example, it is possible to introduce a system of coordinates which is *geodesic along a specified curve* (Fermi [60]). Equation (7.2) shows that the subsequent analysis applies to spaces of affine connectivity without torsion (cf. § 3). It is also possible to introduce a system of coordinates in these spaces which is geodesic along a specified curve (Eisenhart [80], p. 64; Rashevskii [188], p. 410). The significance of a system of coordinates geodesic at a point P lies in the fact that in a sufficiently small neighbourhood of P, they approximate to the affine coordinates of an E_n, since the formula for parallel displacement from P reduces to $d\xi^{\alpha} = 0$ in these coordinates; thus the coordinates of the displaced vector are stationary at this point, and hence the increments $\Delta\xi^{\alpha}$ acquired in the displacement are infinitesimals of a higher order. Linear transformations of such coordinates leave them geodesic at P, since the *objects of connectivity behave like tensors* under such transformations (cf. (3.7)).

Riemann and normal coordinates. The possibility of finding special coordinates which are geodesic at a point can be realized in the form of the Riemann coordinates in V_n, first introduced by Riemann [1]. The distinctive feature of this system is that the geodesics passing through the origin, have the same form as the equations of straight lines passing through the origin of a Cartesian system of coordinates in Euclidean geometry.

Let us consider a V_n space referred to an arbitrary system of coordinates with a point P as the origin; we suppose that there are geodesics passing through P in every possible direction. We refer the equation of each geodesic to a canonical parameter, and as any two canonical parameters are connected by the linear relationship $\overset{*}{\tau} = a\tau + b$, we can so choose the parameter τ that the point P corresponds to the value $\tau = 0$ on every geodesic, and the parameter is determined to within a constant factor.

Any geodesic passing through P is completely defined by the tangent vector $\xi^\alpha = \left\{\frac{dx^\alpha}{d\tau}\right\}_P$, and any point A on it by the value of the parameter τ.

These statements are, of course, only valid in that region about P where the conditions of the existence and uniqueness theorem for the differential equation of the geodesics are satisfied; only in such a region is it possible to construct Riemann coordinates. If the original coordinates of A are $x^\alpha (\alpha = 1, \ldots, n)$, the corresponding *Riemann coordinates* are defined as

$$y^\alpha = \xi^\alpha \tau. \tag{7.3}$$

Hence

$$x^\alpha = x^\alpha(y^1, \ldots, y^n). \tag{7.4}$$

y^α is used instead of $x^{\alpha'}$ as an exceptional concession to convention. In the original system of coordinates, the functions x^α on the geodesic depend on the initial conditions and the parameter τ, i.e. $x^\alpha = x^\alpha\left(\tau, \overset{0}{x^\alpha}, \xi^\alpha\right)$; as is well known from the theory of differential equations (Stepanov [101], p. 275) the x^α must be functions of the same class as the components of the Riemann connectivity object $\Gamma^\alpha_{\beta\gamma}$.

Hence the x^α are continuously differentiable functions of the y^α. In order to check the one–one correspondence of x^α and y^α in the neighbourhood of P, we note that

$$\left(\frac{dx^\alpha}{d\tau}\right)_P = \xi^\alpha = \left(\frac{\partial x^\alpha}{\partial y^\beta}\frac{\partial y^\beta}{\partial \tau}\right)_P = \left(\frac{\partial x^\alpha}{\partial y^\beta}\right)_P \xi^\beta.$$

Hence

$$\left\{\frac{\partial x^\alpha}{\partial y^\beta}\right\}_P = \delta^\alpha_\beta,$$

and the Jacobian $\left|\frac{\partial(x)}{\partial(y)}\right|_P \neq 0$; it therefore differs from zero in some region about the origin in which the correspondence between x^α and y^α must be one–one; geometrically this means that in this region, there is only one geodesic between P and A.

When the original system of coordinates is replaced by another, the corresponding Riemann coordinates transform into another set by a linear homo-

geneous transformation, since vectors ξ^α in the flat space tangential to V_n at the point P transform according to this law.

We can indicate necessary and sufficient criteria for a Riemann system of coordinates in V_n; these differ in form but are actually equivalent. For example,

(α) the equations of the geodesics have the form (7.3) if it is assumed that $\xi^\alpha =$ constant;

(β) the components of the Riemann connectivity object satisfy the relation

$$\Gamma^\alpha_{\beta\gamma} y^\beta y^\gamma = 0. \tag{7.5}$$

That (α) is necessary follows from the definition of Riemann coordinates. That it is sufficient follows from the fact that $\dfrac{dy^\alpha}{d\tau} = \xi^\alpha =$ constant along a geodesic and therefore at P, i.e.

$$\left(\frac{dy^\alpha}{d\tau}\right)_0 = \xi^\alpha. \tag{7.6}$$

So that ξ^α is the tangent vector at the origin, and coordinates y^α are Riemannian. Let us suppose that (α) holds, then with ξ^α constant, (7.3) defines the integral curves of eqns. (6.3), i.e.

$$\Gamma^\alpha_{\beta\gamma} \xi^\beta \xi^\gamma = 0,$$

which after multiplication by τ^2 are identical with (7.5). This shows that (β) is a necessary condition. Conversely, if (7.5) is true the curves $y^\alpha = \xi^\alpha t$ with ξ^α constant satisfy eqns. (6.3), and are therefore geodesics with canonical parameter t which pass through the origin, i.e. (α) is true.

Taking the arc length as the canonical parameter for non-isotropic geodesics and any canonical parameter for isotropic geodesics, we obtain

$$\overset{0}{g}_{\alpha\beta} \xi^\alpha \xi^\beta = \varepsilon, \tag{7.7}$$

where $\varepsilon = \pm 1$ for non-isotr pic and 0 for isotropic geodesics and $\overset{0}{g}_{\alpha\beta}$ is the metric tensor at the origin of a Riemann system of coordinates. Multiplying (7.7) by s^2, we obtain

$$\overset{0}{g}_{\alpha\beta} y^\alpha y^\beta = \varepsilon s^2, \tag{7.8}$$

which gives the square of the geodesic distance from the origin for any ε. If we put $s =$ constant and $\varepsilon \neq 0$, we obtain the equations of *concentric hyperspheres with real or imaginary radii.*

The case $\varepsilon = 0$ corresponds to a hypercone formed by isotropic geodesics. For $\varepsilon \neq 0$ it is possible to give another criterion for Riemann coordinates which differs from (α) and (β) ([188], p. 540).

Riemann coordinates are geodesics with respect to their origin.

In fact, dividing (7.5) by s^2 and applying the resultant equation to the origin, we obtain

$$\overset{0}{\Gamma}{}^{\alpha}_{\beta\gamma}\xi^{\beta}\xi^{\gamma} = 0.$$

Since ξ^{α} is an arbitrary vector $\overset{0}{\Gamma}{}^{\alpha}_{(\beta\gamma)} = \overset{0}{\Gamma}{}^{\alpha}_{\beta\gamma} = 0$, which proves the proposition.

The components of the metric tensor $\overset{0}{g}_{\alpha\beta}$ at the origin of a Riemann system of coordinates are constants satisfying the condition $|\overset{0}{g}_{\alpha\beta}| \neq 0$; hence it is always possible to find a non-degenerate real linear homogeneous transformation $y^{\alpha} = C^{\alpha}_{\beta'}y^{\beta'}$ with constant coefficients as a result of which the form $g_{\alpha\beta}\, dy^{\alpha}\, dy^{\beta}$ transforms to the form

$$\sum_{\sigma=1}^{n} l_{\sigma}\, dy^{\sigma 2}. \tag{7.9}$$

The set $l_{\sigma} = \pm 1$ is the signature of the quadratic form $\overset{0}{g}_{\alpha\beta}\, dy^{\alpha}\, dy^{\beta}$. The set of coordinates $y^{\sigma'}$ are another Riemann system with the same origin, since the transformation leaves (α) in force. These specialized Riemann coordinates are called *normal coordinates* for the specified origin. They were first introduced by Riemann in the theory of V_n spaces and subsequently applied to the theory of relativity. Normal coordinates were generalized for A_n spaces with symmetric affine connectivity by Veblen [61]; they are discussed by a number of authors ([66], [67], [78] pp. 42–62, [93]).

If it is assumed that the components of the metric tensor $g_{\alpha\beta}(y)$ are analytic functions in the neighbourhood of the origin in a normal system of coordinates, they can be defined by power series in y^{α}, the coefficients being determined by the curvature tensor and its covariant derivatives at the origin. Let us first consider the tensor field $W_{\alpha_1 \ldots \alpha_p}(x)$, the components of which are analytic functions in the neighbourhood of the origin of a normal system of coordinates. Then

$$W_{\alpha_1 \ldots \alpha_p} = \overset{0}{W}_{\alpha_1 \ldots \alpha_p} + \left(\frac{\partial W_{\alpha_1 \ldots \alpha_p}}{\partial y^{\nu}}\right)_0 y^{\nu} + \frac{1}{2!}\left(\frac{\partial^2 W_{\alpha_1 \ldots \alpha_p}}{\partial y^{\mu}\partial y^{\nu}}\right)_0 y^{\mu}y^{\nu} + \ldots \tag{7.10}$$

It is possible to select a system of coordinates which enables the coefficients of this series to be expressed in terms of the covariant derivatives of W, the components of the curvature tensor and its covariant derivatives, evaluated at the origin. Having determined such a system we will then show that it is normal.

The transformation law (3.7) for the connectivity object and corollaries obtained by successive differentiation show that

$$\left.\begin{aligned} \Gamma^{\alpha'}_{\lambda'\mu'}A^{\nu}_{\alpha'} &= A^{\sigma}_{\lambda'}A^{\tau}_{\mu}\Gamma^{\nu}_{\sigma\tau}+\partial_{\lambda'}A^{\nu}_{\mu'}, \\ \partial_{\omega'}\Gamma^{\alpha'}_{\lambda'\mu'}A^{\nu}_{\alpha'}+\Gamma^{\alpha'}_{\lambda'\mu'}\partial_{\omega'}A^{\nu}_{\alpha'} &= \partial_{\omega'}A^{\sigma}_{\lambda'}A^{\tau}_{\mu'}\Gamma^{\nu}_{\sigma\tau} \\ &\quad +A^{\sigma}_{\lambda'}\partial_{\omega'}A^{\tau}_{\mu'}\Gamma^{\nu}_{\sigma\tau}+A^{\sigma}_{\lambda'}A^{\tau}_{\mu'}A^{\gamma}_{\omega'}\partial_{\gamma}\Gamma^{\gamma}_{\sigma\tau}+\partial_{\lambda'\omega'}A^{\nu}_{\mu'}, \\ &\cdots\cdots\cdots\cdots\cdots\cdots \end{aligned}\right\} \tag{7.11}$$

Let us consider the coordinate transformation defined by

$$x^{\nu} = \overset{0}{x}{}^{\nu}+\sum_{k=1}^{\infty}\frac{1}{k!}\left(\frac{\partial^k x^{\nu}}{\partial y^{\lambda_1}\ldots\partial y^{\lambda_k}}\right)_0 y^{\lambda_1}\ldots y^{\lambda_k}, \tag{7.12}$$

with coefficients as follows:

$$\begin{aligned} &\left(\frac{\partial x^{\nu}}{\partial y^{\lambda_1}}\right)_0 = \delta^{\nu}_{\lambda_1}, \quad \left(\frac{\partial^2 x^{\nu}}{\partial y^{\lambda_1}\partial y^{\lambda_2}}\right)_0 = -\overset{0}{\Gamma}{}^{\nu}_{\lambda_1\lambda_2}, \\ &\qquad\left(\frac{\partial^3 x^{\nu}}{\partial y^{\lambda_1}\partial y^{\lambda_2}\partial y^{\lambda_3}}\right)_0 = 2\overset{0}{\Gamma}{}^{\nu}_{\alpha(\lambda_2}\overset{0}{\Gamma}{}^{\alpha}_{\lambda_1\lambda_3)}-\partial_{(\lambda_3}\overset{0}{\Gamma}{}^{\nu}_{\lambda_1\lambda_2)}, \ \ldots; \end{aligned} \tag{7.13}$$

then in virtue of (7.11)

$$\overset{0}{\Gamma}{}^{\nu'}_{\lambda_1'\lambda_2'} = 0, \quad \partial_{(\lambda_3'}\overset{0}{\Gamma}{}^{\nu'}_{\lambda_1'\lambda_2')} = 0, \quad \ldots, \quad \partial_{(\lambda_3'\ldots\lambda_r'}\overset{0}{\Gamma}{}^{\nu'}_{\lambda_1'\lambda_2')} = 0. \tag{7.14}$$

If we wish to retain coefficients in the series (7.10) up to the rth order, we only need the conditions (7.14). In the new system of coordinates y, eqn. (5.2) for the curvature tensor and corollaries obtained by successive differentiation show that

$$\left.\begin{aligned} \{\partial_{(\beta}\Gamma^{\nu}_{\alpha)\mu}\}_0 &= \tfrac{1}{3}\overset{0}{R}{}^{\nu}_{(\alpha\beta)\mu}, \\ \{\partial_{(\gamma\beta}\Gamma^{\nu}_{\alpha)\mu}\}_0 &= -\tfrac{1}{2}\overset{0}{R}{}_{\mu(\gamma}{}^{\nu}{}_{\beta,\alpha)}, \\ \{\partial_{(\delta\gamma\beta}\Gamma^{\nu}_{\alpha)\mu}\}_0 &= -\tfrac{3}{5}\left(\tfrac{2}{9}\overset{0}{R}{}_{(\alpha}{}^{\varrho}{}_{\beta}{}^{\nu}\overset{0}{R}{}_{\delta}{}^{\omega}{}_{\gamma)\mu}\overset{0}{g}_{\varrho\omega}+\overset{0}{R}{}_{\mu(\delta}{}^{\nu}{}_{\gamma,\alpha\beta)}\right); \\ &\cdots\cdots\cdots\cdots\cdots\cdots \end{aligned}\right\} \tag{7.15}$$

after substituting from (7.14).

The coordinate system thus introduced is geodesic with respect to the origin; when we successively differentiate the tensor $W_{\alpha_1 \ldots \alpha_p}$ and express the partial derivatives at the origin in terms of the corresponding covariant derivatives, we can omit those terms containing the object $\overset{0}{\Gamma}{}^{\alpha}_{\beta\gamma}$. Wherever terms $\partial_{(\gamma \ldots \nu}\Gamma^{\alpha}_{\beta)\delta}$ appear, the y are replaced in accordance with eqns. (7.15). Substituting for the derivatives in (7.10) we obtain the following expansion for the tensor W:

$$W_{\alpha_1 \ldots \alpha_p} = \overset{0}{W}_{\alpha_1 \ldots \alpha_p} + \overset{0}{W}_{\alpha_1 \ldots \alpha_p,\, \mu} y^{\mu}$$
$$+\frac{1}{2!}\left\{\overset{0}{W}_{\alpha_1 \ldots \alpha_p,\, \mu\omega} - \frac{1}{3}\sum_{k=1}^{p} \overset{0}{R}{}^{\nu}{}_{\mu\alpha_k\omega} W_{\alpha_1 \ldots \alpha_{k-1}\nu\alpha_{k+1} \ldots \alpha_p}\right\} y^{\mu} y^{\omega}$$
$$+\frac{1}{3!}\left\{\overset{0}{W}_{\alpha_1 \ldots \alpha_p,\, \mu\omega\sigma} - \sum_{k=1}^{p} \overset{0}{R}{}^{\nu}{}_{\mu\alpha_k\omega} \overset{0}{W}_{\alpha_1 \ldots \alpha_{k-1}\nu\alpha_{k+1} \ldots \alpha_p,\, \sigma}\right.$$
$$\left. - \frac{1}{2}\sum_{k=1}^{p} \overset{0}{R}{}^{\nu}{}_{\mu\alpha_k\omega,\, \sigma} \overset{0}{W}_{\alpha_1 \ldots \alpha_{k-1}\nu\alpha_{k+1} \ldots \alpha_p}\right\} y^{\mu} y^{\omega} y^{\sigma} + \ldots . \tag{7.16}$$

The coefficients of this expansion can be calculated to as many terms as desired. This series, which is interesting in its own right, can be applied to any covariant tensor.

Applying it to the metric tensor $g_{\alpha\beta}$, and noting that all the covariant derivatives are zero, we obtain

$$g_{\alpha\beta} = \overset{0}{g}_{\alpha\beta} - \frac{1}{3}\overset{0}{R}_{\alpha\mu\beta\lambda} y^{\mu} y^{\lambda} - \frac{1}{3!}\overset{0}{R}_{\alpha\gamma\beta\lambda,\, \mu} y^{\lambda} y^{\mu} y^{\gamma}$$
$$+\frac{1}{5!}\left\{-6\overset{0}{R}_{\alpha\delta\beta\gamma,\, \lambda\mu} + \frac{16}{3}\overset{0}{R}_{\lambda\beta\mu}{}^{\varrho}\overset{0}{R}_{\gamma\alpha\delta\varrho}\right\} y^{\lambda} y^{\mu} y^{\gamma} y^{\delta} + \ldots \tag{7.17}$$

We now show that this system of coordinates is normal. The $\overset{0}{g}_{\alpha\beta}$ are first converted to $\delta^{\alpha}_{\beta} l_{\alpha}$, a form of type (7.9), by a non-degenerate linear transformation at the specified origin in V_n. Contracting (7.17) with y^{β} we obtain

$$g_{\alpha\beta} y^{\beta} = \overset{0}{g}_{\alpha\beta} y^{\beta}, \tag{7.18}$$

since in all the remaining terms, the index β is clearly antisymmetric with respect to at least one of the summation indices $\lambda, \mu, \gamma, \ldots$ Differentiating (7.18) with respect to y^{γ}, we obtain

$$\partial_{\gamma} g_{\alpha\beta} y^{\beta} + g_{\gamma\alpha} = \overset{0}{g}_{\gamma\alpha}.$$

Contracting with y^{α} and y^{γ} and using (7.18) we obtain

$$\partial_{\gamma} g_{\alpha\beta} y^{\alpha} y^{\beta} = \partial_{\gamma} g_{\alpha\beta} y^{\gamma} y^{\beta} = 0. \tag{7.19}$$

According to (3.1)

$$\Gamma_{\alpha,\,\beta\gamma}y^\beta y^\gamma = \tfrac{1}{2}(\partial_\beta g_{\alpha\beta}+\partial_\gamma g_{\alpha\beta}-\partial_\alpha g_{\beta\gamma})y^\beta y^\gamma.$$

Substituting from (7.19), interchanging indices where necessary, we have $\Gamma_{\alpha,\,\beta\gamma}y^\beta y^\gamma = 0$; this leads to condition (β) for a Riemann system of coordinates.

Equation (7.17) takes a particularly simple form for the special spaces with zero covariant derivatives of the curvature tensor. Such V_n spaces are called *symmetric* and the following theorem is valid: *for a symmetric V_n in a normal system of coordinates*

$$g_{\alpha\beta} = \overset{0}{g}{}^{\alpha\beta}+\frac{1}{2}\sum_{k=1}^{\infty}\frac{(-1)^k 2^{2k+2}}{(2k+2)!}\,m^{\sigma_1}_{\alpha}m^{\sigma_2}_{\sigma_1}\ldots m_{\sigma_{k-1}\beta}, \tag{7.20}$$

where

$$m^\sigma_\alpha = R_{\alpha\lambda\mu}{}^{\sigma}y^\lambda y^\mu; \qquad m_{\alpha\beta} = \overset{0}{R}_{\alpha\lambda\beta\mu}y^\lambda y^\mu, \tag{7.21}$$

$$\overset{0}{g}_{\alpha\beta} = \begin{cases}\pm 1, & \alpha = \beta\\ 0, & \alpha \neq \beta\end{cases}$$

and $\overset{0}{R}_{\alpha\beta\gamma\delta}$ *are the components of the curvature tensor at the origin.*

The structure of (7.20) follows directly from (7.17). The coefficients of the m's agree as far as the fourth order in the y's. It is easy to verify (7.20) by induction provided (7.21) holds.

Normal coordinates admit of a physical interpretation in the general theory of relativity (Pirani [258]), and are useful in many geometrical and physical problems.

Semi-geodesic coordinates. These coordinates (sometimes called Gaussian coordinates) are a generalization of semi-geodesic coordinates on a surface in three-dimensional Euclidean space. They are defined by the network of a single-parameter family of geodesics and a family of orthogonal trajectories ([155], p 185). In V_n these coordinates are defined by a family of geodesics and a family of hypersurfaces orthogonal to these geodesics; if ds^2 is indefinite, isotropic geodesics with no counterpart in Euclidean geometry are possible.

We first define a surface in V_n; *an m-dimensional surface* in V_n is the set of points given by the equations

$$x^\alpha = x^\alpha(u^i) \qquad (\alpha = 1, \ldots, n;\ i = 1, \ldots, m),$$

where the u^i are *independent* parameters and the rank of the matrix $\left[\dfrac{\partial x^\alpha}{\partial u^i}\right]$

is m. This condition, clearly invariant under any non-degenerate transformation of the coordinates x^α in V_n, prevents the appearance of singular points on the surface and the degeneration of the surface into a geometrical entity of a smaller number of dimensions. This definition of an m-surface is essentially *local*; it defines a "piece" of surface. When $m = 1$ we obtain *a curve* in the space V_n, and when $m = n-1$, a *hypersurface* in the space. If we are given functions $u^i(t)$ of class C^r on an m-surface we obtain *a curve of the corresponding class on this m-surface*. As a curve in the space V_n is obtained in this way, its tangent vector is

$$\frac{dx^\alpha}{dt} = \frac{\partial x^\alpha}{\partial u^i}\frac{du^i}{dt}. \tag{7.22}$$

We refer to $\dfrac{dx^\alpha}{dt}$ as a *space vector* and $\dfrac{du^i}{dt}$ as a *surface vector*; they are connected by relations (7.22). We only consider points on the curve where the $\dfrac{du^i}{dt}$ are not all simultaneously zero, for otherwise the tangent vector becomes indefinable. That the $\dfrac{dx^\alpha}{dt}$ cannot then all be zero follows from the rank of the matrix. For arbitrary curves on an m-surface with arbitrary parametrization, the vectors $\xi^i = \dfrac{du^i}{dt}$ are also arbitrary. Then the vectors

$$\xi^\alpha = \frac{\partial x^\alpha}{\partial u^i}\xi^i \tag{7.23}$$

determine the *linear pencil of the m linearly independent vectors* $\underset{k}{u}^\alpha = \dfrac{\partial x^\alpha}{\partial u^k}$ $(k = 1, \ldots, m)$. The set of vectors defines an m-space A_m tangential at the given point to the m-surface.

A line along which only one of the parameters u^i varies is called a *coordinate line* of the m-surface; the m-linearly independent vectors u^i are the tangent vectors to the coordinate lines of the m-surface. Thus at each point of an m-surface there is a *local m-hedral*.

Let us consider an arbitrary infinitesimal displacement along a curve on the m-surface, i.e. $dx^\alpha = \dfrac{\partial x^\alpha}{\partial u^i}du^i = \underset{i}{u}^\alpha du^i$. The square of the length of the displacement is

$$ds^2 = g_{\alpha\beta}\, dx^\alpha\, dx^\beta = \overset{*}{g}_{ij}\, du^i\, du^j, \tag{7.24}$$

where

$$\overset{*}{g}_{ij}(u) = g_{\alpha\beta}\frac{\partial x^\alpha}{\partial u^i}\frac{\partial x^\beta}{\partial u^j} = g_{\alpha\beta}\underset{i}{u}^\alpha \underset{j}{u}_\beta, \quad \overset{*}{g}_{ij} = \overset{*}{g}_{ji} \quad (i, j = 1, \ldots, m). \quad (7.24a)$$

The scalar products of the coordinate vectors of the surface thus define the fundamental differential quadratic form on the m-surface.

If $|\overset{*}{g}_{ij}| \neq 0$ at all points, the m-surface is a Riemann subspace V_m and is said to be *non-isotropic*; in this case the A_m space is a R_m. If, however, $|\overset{*}{g}_{ij}| = 0$, we have an *isotropic* m-surface which receives an important physical interpretation in investigating wave processes in the theory of relativity ($n = 4$, $m = 3$); these surfaces can only exist in a V_n with an indefinite metric.

We describe a vector n^α of a V_n space as *normal* to an m-surface, if it is orthogonal to all the m coordinate vectors $\underset{k}{u}^\alpha (k = 1, \ldots, m; \alpha = 1, \ldots, n)$:

$$g_{\alpha\beta} n^\alpha \underset{k}{u}^\beta = 0 \qquad (k = 1, \ldots, m). \qquad (7.25)$$

As the rank of the matrices $[g_{\alpha\beta}]$ and $[\underset{k}{u}^\beta]$ are respectively n and m, the rank of the matrix $[g_{\alpha\beta}\underset{k}{u}^\beta]$ is m, and the set of equations (7.25), homogeneous in the unknowns n^α, has $n-m$ linearly independent solutions; *at each point on V_m there exist $n-m$ independent vectors in V_n which are normal to V_m.*

The following theorem is of fundamental importance in the case of isotropic hypersurfaces: *the normal to a hypersurface is isotropic only if the hypersurface is isotropic.*

The definition (7.24a) shows that

$$\overset{*}{g} = |\overset{*}{g}_{ij}| = \sum \underset{n}{\Delta} \underset{n}{\omega},$$

where $\underset{n}{\Delta}$ and $\underset{n}{\omega}$ are the complementary mth order determinants of the two matrices $\left[g_{\alpha\beta}\frac{\partial x^\alpha}{\partial u^i}\right]$ and $\left[\frac{\partial x^\alpha}{\partial u^i}\right]$, as follows from the theory of determinants (Kowalewski [21], p. 77). On the other hand (7.25) and the associated equation $n_\beta \underset{k}{u}^\beta = 0$ show that $\underset{n}{\Delta}$ and $\underset{n}{\omega}$ are proportional to n^α and n_α respectively; hence denoting the two non-zero factors of proportionality by k_1 and k_2 we obtain

$$\overset{*}{g} = k_1 k_2 n_\alpha n^\alpha.$$

If the surface is isotropic then $\overset{\bullet}{g} = 0$ and the normal vector is isotropic; the converse is also true.

We will now introduce a *semi-geodesic system of coordinates*, using as a basis a non-isotropic hypersurface $\varphi(x^\alpha) = 0$; at each point on some region there corresponds the non-isotropic normal

$$\Delta_1\varphi = g^{\alpha\beta}\varphi,_\alpha\varphi,_\beta = n_\alpha n^\alpha \neq 0.$$

A geodesic is drawn through each point of the hypersurface in the direction of the normal, which in virtue of (6.4) is also non-isotropic. We define a new system of coordinates as follows: We arrange that the equation of the hypersurface is of the form $x^1 = 0$, and that the geodesics are the coordinate lines x^1, where $x^1 = s$, the arc length measured in the given direction from the hypersurface if the normal has a norm >0, and $x^1 = is$ if the norm < 0. Applying the fundamental form to an infinitesimal displacement along the geodesic x^1, we obtain $ds^2 = g_{11}\, dx^{1^2}$ and hence $g_{11} = l = \pm 1$, depending on the sign of the norm of the normal vector.

Putting $\tau = x^1 = s$ or is in eqn. (6.3) for the geodesics, we obtain $\partial_1 g_{1\alpha} = 0$, i.e. $g_{1\alpha}$ does not depend on x^1; but $g_{1\alpha} = 0$ for $x^1 = 0$ and hence for any x^1. The *metric in a semi-geodesic system of coordinates* can thus be written in the form

$$ds^2 = l dx^{1^2} + g_{ij} dx^i dx^j \qquad (i, j = 2, \ldots, n) \quad l = \pm 1. \tag{7.26}$$

Putting $x^1 = c$, we obtain hypersurfaces which are also orthogonal to the geodesics x^1; these surfaces are said to be *geodesically parallel* to the hypersurface $x^1 = 0$.

Semi-geodesic coordinates provided with a physical interpretation are used in theory of relativity (Landau and Lifshitz [156], p. 289; Keres [259], pp. 12–25). They can be constructed even if the hypersurface degenerates into an m-surface, $m < n-1$. In this case geodesics are drawn at each point of the m-surface in the direction of the $n-m$ normals to the surface, and hence still depend on $n-1$ parameters. By plotting segments of equal length along them we obtain hypersurfaces geodesically parallel to the m-surface. In particular, by taking $m = 0$, we can construct a semi-geodesic system of coordinates starting from a point; the geodesically parallel hypersurfaces are hyperspheres with centres at the given point; this construction defines a semi-geodesic system of coordinates for a region bounded by a hypercone, and excluding the given point.

We now suppose the initial hypersurface $f(x^1) = 0$ and hence also its normals to be *isotropic*. Then $\Delta_1 f = 0$ and the hypersurfaces $f(x^\alpha) =$ constant and their normals must also be isotropic. Let us consider the set of differential equations $\Delta_1(f, \varphi) = g^{\alpha\beta} f,_\alpha \varphi,_\beta = 0$. This set has $n-1$ independent solutions $\varphi^i(x^\alpha)$ $(i = 2, \ldots, n)$ (Stepanov [101], p. 295). We introduce a new coordinate system by means of the transformation

$$x^{1'} = \psi(x^\alpha), \quad x^{2'} = \varphi^2(x^\alpha), \quad \ldots, \quad x^{n'} = \varphi^n(x^\alpha) \equiv f(x^\alpha),$$

where ψ is some function satisfying the inequality $\dfrac{D(\psi, \varphi^2, \ldots, \varphi^n)}{D(x^1, \ldots, x^n)} \neq 0.$

In this coordinate system

$$g^{2'n'} = g^{3'n'} = \ldots = g^{n'n'} = 0, \quad g^{1'n'} = g^{\alpha\beta}\psi,_\alpha f,_\beta,$$

and $g^{1'n'} \neq 0$ since ψ is functionally independent of f and $\varphi^k (k < n)$ and cannot be a solution of the equation $\Delta_1(f, \psi) = 0$. It is always possible to select $\psi(x^\alpha)$ so that $g^{1'n'} = 1$; then $g_{1'i'} = 0$ $(i = 1, \ldots, n-1)$, $g_{1'n'} = 1$ and the metric of the space takes the form

$$ds^2 = 2dx^1 dx^n + g_{ij}(x^\alpha) dx^i dx^j \qquad (7.27)$$
$$(i, j = 2, \ldots, n; \quad \alpha = 1, \ldots, n).$$

A geometrical interpretation of this system in terms of geodesics is possible but is not so simple as for non-isotropic hypersurfaces. Such *isotropic semi-geodesic systems of coordinates* can only be introduced for a V_n with an indefinite metric.

These coordinate systems are specially convenient for investigating a V_n which admits a group of motions operating on isotropic three-dimensional manifolds (cf. Chapters 4 and 5); they can also be given a physical interpretation in the general theory of relativity.

Harmonic coordinates. This system of coordinates realizes the Riemann theorem in the following manner:

The equation

$$\frac{1}{\sqrt{-g}} \partial_{\beta'} (\sqrt{-g'}\, g^{\alpha'\beta'} \partial_{\alpha'}) \varphi \equiv \Box \varphi = 0 \qquad (\alpha', \beta' = 1, \ldots, n), \quad (7.28)$$

where $\Box$ is the generalized *d'Alembert operator*, has n independent solutions $\varphi^1, \ldots \varphi^n$; harmonic coordinates are defined by the transformation

$$x^{\alpha'} = \varphi^\alpha(x^\sigma).$$

The choice of the n functions φ^α can be further restricted by imposing a number of inequalities; the coordinate system is then defined to within linear Lorentz transformations. This system of coordinates, introduced by de Donder [54] and Lanczos [68], has been given a physical interpretation in the theory of relativity by Fock and his colleagues (cf. Fock [111], [225]; Petrova [164]). A discussion of the physical significance of harmonic coordinates as a privileged system is to be found in references [189], [195], [201], [226], [253], [254], [259], [306]. A Cartesian system in a flat space is an example of a harmonic system. The principal advantage of a harmonic system is that the gravitational field equations take a convenient form for investigation (cf. Chapter 2).

Non-holonomic coordinates. Let us consider some system of coordinates in V_n, for example one of those introduced above. At any point in the space V_n we can define *coordinate hypersurfaces* by equations $x^\alpha =$ constant, and *coordinate lines* as the lines along which all the coordinates x^α except one are constant. Let us consider the coordinate line along which x^α varies, α being a specified index. We can take x^α as a parameter along this curve; the tangent vector to the coordinate line has contravariant components $\underset{\alpha}{l}^\sigma(\sigma = 1, \ldots, n)$ given by

$$\underset{\alpha}{l}^\sigma = \frac{\partial x^\sigma}{\partial x^\alpha} = \delta^\sigma_\alpha. \tag{7.29}$$

Thus n independent basic contravariant vectors $\underset{\alpha}{l}^\sigma$ are defined at each point. For any system of n independent contravariant vectors, there is a unique reciprocal system of independent covariant vectors $\overset{\alpha}{l}_\sigma$; the reciprocal system of (7.29) is

$$\overset{\alpha}{l}_\sigma = \frac{\partial x^\alpha}{\partial x^\sigma} = \delta^\alpha_\sigma,$$

and hence

$$\partial_{[\gamma}\overset{\alpha}{l}_{\sigma]} = 0,$$

i.e. *the system of vectors reciprocal to the basic contravariant vectors is a system of gradient vectors.*

Conversely, if we chose a set of functionally independent scalar fields $\overset{\sigma}{\varphi}$, there is a corresponding set of n gradient vector fields $\overset{\sigma}{\varphi}_{,\alpha}$, which can be taken as the basic covariant vectors in a new coordinate system; the basic contravariant components are defined by the transformation

$$x^{\alpha'} = \overset{\alpha}{\varphi}(x^\sigma). \tag{7.30}$$

If u_α is not a gradient field, i.e. $\partial_{[\gamma} u_{\alpha]} \neq 0$, but *proportional to one*, i.e. $u_\alpha = \omega(x^\alpha) v_\alpha$, where $\omega(x^\alpha)$ is a scalar field and $\partial_{[\gamma} v_{\alpha]} = 0$, then the u_α can also be selected, after dividing by ω, as the set of basic covariant vectors. As is well known ([110], p. 71), the necessary and sufficient condition for the existence of such a field is

$$u_{[\alpha} \partial_\beta u_{\gamma]} = 0. \tag{7.31}$$

If the basic covariant vectors are a field of gradient vectors, the system of coordinates is called *holonomic*. All the coordinate systems we have so far considered are holonomic.

However, it is frequently desirable to select a set of independent vector fields $\overset{\sigma}{v}_\alpha$ of class C^r as basic covariant vectors at each point of a V_n space, which are not gradients, i.e. $\partial_{[\gamma} \overset{\sigma}{v}_{\alpha]} \neq 0$. Such vectors form a *non-holonomic* n-hedral at each point of V_n, and hence define a coordinate system; however, the transformation given by (7.30) cannot be defined over a *domain which cannot be reduced to a point*. This type of system is called *non-holonomic*. The following case provides an example of a non-holonomic system; the eigenvectors of a symmetric tensor field $h_{\alpha\beta}$ in an ordinary V_n are mutually orthogonal; if the eigenvectors are selected as the basic vectors, the $h_{\alpha\beta}$ assume a particularly simple form, but the system of coordinates is non-holonomic over V_n. Here we have an example of a *non-holonomic orthogonal coordinate system* in V_n, whereas *holonomic* orthogonal systems are generally impossible in all V_n with $n > 3$.

Let us suppose that we are given some holonomic system of coordinates $\{x^\alpha\}$. Consider the quantities

$$\Omega^{\alpha'}_{\beta'\gamma'} = A^{\sigma}_{\beta'} A^{\tau}_{\gamma'} \partial_{[\sigma} \overset{\alpha'}{l}_{\tau]}. \tag{7.32}$$

Then the system of coordinates with the basic covariant vector fields $\overset{\alpha'}{l}_\tau$ *is holonomic only if all the components* $\Omega^{\alpha'}_{\beta'\gamma'}$ *vanish*. It is easy to show by verifying the transformation law, that Ω is a geometric object. It is usually called the *object of non-holonomity* (Schouten and van Dantzig [87], p. 646). Every non-holonomic system of coordinates is based on a holonomic system. Whilst a non-zero geometric object can be reduced to zero in some system of reference, a non-zero tensor cannot; when using non-holonomic coordinate systems, the following obvious point is of fundamental importance: if a tensor field vanishes for non-holonomic n-hedrals constructed at each point in a domain of V_n, it vanishes in any system of coordinates in this domain; the converse is also true.

EXAMPLES

1. If, in some system of coordinates, the metric of the space V_n has the form

$$ds^2 = l\,dx^{1^2} + g_{ij}(x^\alpha)dx^i dx^j$$
$$(l = \pm 1;\, i, j = 2, \ldots, n;\, \alpha = 1, \ldots, n)$$

the lines x^1 are geodesics orthogonally intersecting the hypersurfaces $x^1 = c$, which are themselves geodesically parallel to $x^1 = 0$.

2. What is the connection between Riemann coordinates with point P as origin, and semi-geodesic coordinates with geodesically parallel concentric hyperspheres with a common centre at P?

3. Show that in a normal system of coordinates

$$g = |g_{\alpha\beta}| = \pm 1 + \frac{1}{3}\overset{0}{R}_{\alpha\beta}y^\alpha y^\beta + \frac{1}{3!}\overset{0}{R}_{\alpha\beta,\,\gamma}y^\alpha y^\beta y^\gamma$$
$$+\frac{1}{4!}\left(\frac{4}{3}\overset{0}{R}_{\alpha\beta}\overset{0}{R}_{\gamma\delta} + \frac{4}{15}\overset{0}{R}_{\alpha\varrho\beta}{}^{\lambda}\overset{0}{R}_{\gamma\lambda\delta}{}^{\varrho} + \frac{6}{5}\overset{0}{R}_{\gamma\delta,\,\alpha\beta}\right)y^\alpha y^\beta y^\gamma y^\delta + \ldots$$

and that if the Ricci tensor $R_{\alpha\beta} = 0$,

$$g = \pm 1 - \tfrac{1}{90}\overset{0}{R}_{\alpha\lambda\beta}{}^{\mu}\overset{0}{R}_{\gamma\mu\delta}{}^{\lambda}y^\alpha y^\beta y^\gamma y^\delta + \ldots,$$

i.e. in the neighbourhood of the origin of a normal system of coordinates

$$g = \pm 1$$

to the fourth order of small quantities.

4. A normal system of coordinates with a fixed origin is defined to within linear transformations which leave the form $\overset{0}{g}_{\alpha\beta}\, dy^\alpha\, dy^\beta$ invariant (Lorentz transformation with $n = 4$ and the signature $- - - +$).

5. Show that it is always possible to introduce a tri-orthogonal system of coordinates such that $ds^2 = \sum_{\alpha=1}^{3} g_{\alpha\alpha}\,(dx^\alpha)^2$, where $g_{\alpha\alpha} = g_{\alpha\alpha}\,(x^1, x^2, x^3)$ (Cotton [10], p. 410).

6. Show that it is always possible to define a coordinate system in V_n such that $g = \pm 1$ in a domain not reducible to a point. In this system the divergence of a vector has the same form as in vector analysis in ordinary R_3.

7. For a tri-orthogonal system in V_3

$$R_{ij} = \frac{1}{g_{kk}} R_{ikjk}, \quad R_{ii} = \frac{1}{g_{jj}} R_{ijij} + \frac{1}{g_{kk}} R_{ikik} (ij,k \neq),$$

$$R = \sum_{ij} \frac{1}{g_{ii}g_{jj}} R_{ijij}.$$

8. The curvature tensor for any V_3 (cf. example 3, § 5) is

$$R_{\alpha\beta\gamma\delta} = g_{\alpha\gamma}L_{\beta\delta} - g_{\alpha\delta}L_{\beta\gamma} + g_{\beta\delta}L_{\alpha\gamma} - g_{\beta\gamma}L_{\alpha\delta},$$

where $L_{\alpha\beta} = R_{\alpha\beta} + \frac{R}{4} g_{\alpha\beta}$.

9. If $R_{\alpha\beta} = 0$ in V_4 and $R_{\alpha\beta\gamma\delta} = 0$ in V_3, a subspace with a non-isotropic normal,
then $R_{\alpha\beta\gamma\varrho} = 0$ in V_4 [296].

§ 8. Riemannian Curvature of V_n. Spaces of Constant Curvature

The simplest application of the curvature tensor occurs in the case of an *ordinary* V_2 space, a surface embedded in three-dimensional Euclidean space.

If $n = 2$, the curvature tensor has only one independent component, and we put

$$R_{1212} = K(g_{11}g_{22} - g_{12}g_{21}) = K\varphi_1. \tag{8.0}$$

K is called the *Gaussian curvature* of the surface. The intrinsic geometry of the surface V_2 depends on the definite metric. Let the equation of the surface be $z = f(x, y)$ where $z = 0$ is the tangent plane at an origin in the surface; putting $x = x^1$, $y = x^2$, we have

$$ds^2 = (1+p^2)dx^{1^2} + 2pq\, dx^1 dx^2 + (1+q^2)dx^{2^2}$$

and $p_0 = q_0 = 0$. Hence the Christoffel symbols at the origin all vanish and $\underset{0}{R_{1212}} = rt - s^2 = \varphi_2$. Equation (8.0) shows that $K = \varphi_2/\varphi_1$; φ_1, φ_2 are the determinants of the *first and second fundamental forms of the surface* ([155], p. 148).

This interpretation can be extended to any two-dimensional direction in a V_n with $n > 2$. Let us introduce a normal system of coordinates into V_n. We take two vectors $\underset{1}{i}^\alpha$ and $\underset{2}{i}^\alpha$ at the origin, form their linear pencil and draw geodesics in the direction of each vector of this pencil; the set of these geodesics defines a two-dimensional geodesic surface at the origin.

The equation of this surface is

$$y^\alpha = u^1 \underset{1}{i^\alpha} + u^2 \underset{2}{i^\alpha}, \tag{8.1}$$

where $u^\alpha = \underset{\alpha}{k}\tau$, and τ is a canonical parameter which can be taken as the arc length of a non-isotropic geodesic. Equation (8.1) shows that

$$dy^\alpha = \underset{a}{i^\alpha}\, du^\alpha \qquad (a = 1, 2),$$

$$ds^2 = g_{\alpha\beta}\, dy^\alpha\, dy^\beta = \overset{*}{g}_{ab}\, du^a\, du^b, \quad \overset{*}{g}_{ab} = g_{\alpha\beta} \underset{a}{i^\alpha} \underset{b}{i^\beta}, \tag{8.2}$$

where $\overset{*}{g}_{ab}$ is the metric tensor of the locally geodesic surface. Let us consider the case when the surface is non-isotropic $\left(|\overset{*}{g}_{ab}| \neq 0\right)$ and hence forms a V_2. As the coordinate system is normal,

$$(\partial_c \overset{*}{g}_{ab})_0 = \left(\partial_\gamma g_{\alpha\beta} \frac{\partial y^\gamma}{\partial u^c} \underset{a}{i^\alpha} \underset{b}{i^\beta}\right)_0 = 0,$$

and $(\overset{*}{\Gamma}_{abc,})_0 = 0$; the only independent component of the curvature tensor of V_2 is

$$(\overset{*}{R}_{1212})_0 = \{\partial_1 \overset{*}{\Gamma}_{1,\,22} - \partial_2 \overset{*}{\Gamma}_{1,\,12}\}_0 .$$

The Christoffel symbols in the space V_2 are $\Gamma_{\alpha,\,\beta\gamma} \underset{a}{i^\alpha} \underset{b}{i^\beta} \underset{c}{i^\gamma} = \overset{*}{\Gamma}_{a,\,bc}$, and hence the curvature tensor component

$$(\overset{*}{R}_{1212})_0 = R_{\alpha\beta\gamma\delta} \underset{1}{i^\alpha} \underset{2}{i^\beta} \underset{1}{i^\gamma} \underset{2}{i^\delta}.$$

It must also equal

$$K\overset{*}{g} = K(\overset{*}{g}_{11}\overset{*}{g}_{22} - \overset{*}{g}{}^2_{12}) = K(g_{\alpha\gamma} g_{\beta\delta} - g_{\alpha\delta} g_{\beta\gamma}) \underset{1}{i^\alpha} \underset{2}{i^\beta} \underset{1}{i^\gamma} \underset{2}{i^\delta},$$

as in any V_2. Hence

$$K = \frac{R_{\alpha\beta\gamma\delta} \underset{1}{i^\alpha} \underset{2}{i^\beta} \underset{1}{i^\gamma} \underset{2}{i^\delta}}{(g_{\alpha\gamma} g_{\beta\delta} - g_{\alpha\delta} g_{\beta\gamma}) \underset{1}{i^\alpha} \underset{2}{i^\beta} \underset{1}{i^\gamma} \underset{2}{i^\delta}}, \tag{8.3}$$

where $\underset{1}{i^\alpha}$, $\underset{2}{i^\alpha}$ are the two vectors defining the two-dimensional geodesic surface.

The quantity K is called the *Riemannian curvature of the space in the two-dimensional direction* $\left\{\underset{1}{i^\alpha} \underset{2}{i^\alpha}\right\}$; if V_2 is ordinary, K is the *Gaussian curva-*

ture at a given point on the geodesic surface in the V_n space. If the two-surface is isotropic, the denominator in (8.3) vanishes and this expression for K becomes meaningless. However, it is still possible (cf. Chapter 3) to give an interpretation to K by supposing that $K(x^\alpha)$ is a continuous function. The formula (8.3) is always meaningful in an ordinary V_n space.

According to (8.3), K is a homogeneous function of zero degree in the bivector $\underset{[1\,2]}{i^\alpha i^\beta}$ defining the two-dimensional geodesic surface, and it changes when this bivector changes. V_n is called a *space of locally constant curvature* if K is independent of the choice of the two-dimensional geodesic surface. Let

$$S_{\alpha\beta\gamma\delta} = R_{\alpha\beta\gamma\delta} - K\begin{vmatrix} g_{\alpha\gamma} g_{\alpha\delta} \\ g_{\beta\gamma} g_{\beta\delta}\end{vmatrix}.$$

From (8.3) we obtain

$$S_{\alpha\beta\gamma\delta}\underset{1}{i^\alpha}\underset{2}{i^\beta}\underset{1}{i^\gamma}\underset{2}{i^\delta} = 0.$$

As this holds for *all* vectors $\underset{1}{i^\alpha}$ and $\underset{2}{i^\alpha}$,

$$S_{\alpha\beta\gamma\delta} + S_{\gamma\beta\alpha\delta} + S_{\alpha\delta\gamma\beta} + S_{\gamma\delta\alpha\beta} = 0.$$

As, moreover, the curvature tensor properties (5.8) (α), (β) and (γ) also apply to the tensor S, $S_{\alpha\beta\gamma\delta} = 0$ (cf. example 5, § 2); hence the *necessary and sufficient condition that $V_n (n > 2)$ is a space of locally constant curvature is*

$$R_{\alpha\beta\gamma\delta} = K\begin{vmatrix} g_{\alpha\gamma} g_{\alpha\delta} \\ g_{\beta\gamma} g_{\beta\delta}\end{vmatrix}. \tag{8.4}$$

If the Riemannian curvature of V_n is locally constant it is constant throughout V_n (for $n > 2$) (Schur [4], p. 563). It follows from (8.4) that

$$R_{\alpha\beta\gamma\delta,\,\lambda} = K_{,\,\lambda}(g_{\alpha\gamma}g_{\beta\delta} - g_{\alpha\delta}g_{\beta\gamma}).$$

Applying Bianchi's identity and contracting with respect to the indices β and γ, we obtain

$$(n-2)(g_{\alpha\lambda}K_{,\,\delta} - g_{\alpha\delta}K_{,\,\lambda}) = 0.$$

If $n > 2$, then contracting over the indices α and λ, we obtain $(n-1)K_{,\delta} = 0$, i.e. $K =$ constant. If, however, $n = 2$, V_2 is a space of constant curvature only when K is constant; condition (8.4) is always satisfied in this case. It follows from (8.4) that $R_{\alpha\beta\gamma\delta}$ and K vanish simultaneously; hence any flat space is a space of zero Riemannian curvature. When $K < 0$ we have a generalized form of Lobachevskii–Bolyai geometry, whilst $K > 0$ corresponds to a geometry analogous to that of a sphere. *A space of constant curvature* will be denoted by the symbol S_n whatever the signature of the metrical form.

Two n-dimensional Riemann spaces are called *isometric*, if there exists a non-degenerate real transformation which converts the metric of one V_n into that of the other. The following theorem is valid for all S_n: *two S_n with equal Riemannian curvatures and identical signatures are isometric.* Every S_n is a symmetric V_n; hence if we are given a normal system of coordinates in an S_n, then, in accordance with (7.20), its metric is completely determined by the values of the components of the curvature tensor at the origin, $\overset{0}{R}_{\alpha\beta\gamma\delta}$. In the case of an orthogonal n-hedral, eqn. (8.4) shows that the curvature tensor $\overset{0}{R}_{\alpha\beta\gamma\delta}$ depends only on K and the signature of the metric. Hence if the signatures and curvatures K are identical, the spaces S_n are isometric.

A corollary is that if we can find at least one metric with arbitrary signature and curvature K which satisfies (8.4), it is also a general expression for the metric of any S_n in a special system of coordinates. Let us assume that such a metric differs only by a scalar factor from the metric of a flat space *(conformally flat V_n)*:

$$ds^2 = \frac{1}{\sigma^2}\sum_{\alpha=1}^{n} l_\alpha\, dx^{\alpha^2}, \qquad \sigma = \sigma(x^\alpha), \qquad l_\alpha = \pm 1, \qquad n > 2, \tag{8.5}$$

i.e. $g_{\alpha\alpha} = l_\alpha/\sigma^2$, $g_{\alpha\beta} = 0$ $(\alpha \neq \beta)$. Evaluating the components of the curvature tensor from (5.7) and satisfying condition (8.4), we obtain

$$\partial_{\alpha\beta}\sigma = 0, \quad \sigma\{l_\alpha\, \partial_{\beta\beta}\sigma + l_\beta\, \partial_{\alpha\alpha}\sigma\} = l_\alpha l_\beta\{K + \sum_\tau l_\tau(\partial_\tau\sigma)^2\}, \quad \alpha \neq \beta.$$

These equations are satisfied by

$$\sigma = 1 + \frac{K}{4}\sum_\alpha l_\alpha x^{\alpha^2}. \tag{8.6}$$

As the signature of (8.5) is arbitrary, as is the curvature K, (8.5) is the *metric of an arbitrary S_n*. Hence *every S_n is a conformally flat V_n*.

Examples

1. If V_n admits of a vector field φ_α satisfying equations

$$\varphi_{\alpha,\beta} = \varrho g_{\alpha\beta}$$

with $\varrho \neq 0$, a system of coordinates exists in which

$$ds^2 = g_{11}\, dx^{1^2} + \frac{1}{g_{11}} g_{pq}(x^2, \ldots, x^n)\, dx^p\, dx^q,$$

where

$$g_{11} = \frac{1}{2\int \varrho(x^1)\, dx^1 + C},$$

and ϱ, $g_{pq}(p, q \neq 1)$ are arbitrary functions of their arguments (Sinyukov [274]).

2. If the space of example 1 (for $n > 2$) is symmetric ($R_{\alpha\beta\gamma\delta,\lambda} = 0$), it is a space of constant curvature.

3. If a Riemann space V_n with Christoffel symbol Γ^k_{ij} can be mapped on $\tilde{V}_n$ with Christoffel symbol $\boldsymbol{\Gamma}^k_{ij}$ so that geodesics transform into geodesics, then

$$\boldsymbol{\Gamma}^k_{ij} = \Gamma^k_{ij} + \delta^k_i \psi_j + \delta^k_j \psi_i,$$
$$\boldsymbol{R}^p{}_{ijk} = R^p{}_{ijk} + \delta^p_k \psi_{ij} - \delta^p_j \psi_{ik},$$

where ψ_i is a gradient, and $\psi_{ij} = \psi_{i,j} - \psi_i \psi_j$.

4. If V_n (for $n > 2$ and $\psi_i \neq 0$) can be geodesically mapped on a symmetric $\boldsymbol{V}_n$, it is a space of constant curvature (Sinyukov [274]).

§ 9. The Principal Axes Theorem for a Tensor

We have later to solve the problem of reducing a pair of quadratic forms with arbitrary signatures to a canonical form *over the field of real numbers*, assuming that one of these forms is non-degenerate. As a rule, the general problem is avoided in books on linear algebra; the discussion is confined to the reduction of a single form, or of a pair of forms provided that the non-degenerate form is positive-definite, or to the much simpler reduction of a pair of forms *over the field of complex numbers*. The general problem was first solved by Dickson ([27], pp. 357–8). We might remark that the method used was rather cumbersome, and the canonical forms obtained not the simplest; more symmetric canonical forms can be found for any R_n. As, moreover, a purely geometric solution is both possible and more useful to us later on, such a solution will be given (Petrov [165], pp. 37–51).

The *principal axes theorem for a real tensor* a_{ij} in an R_n with a *definite* metric tensor g_{ij}, states ([110], p. 27), that r mutually orthogonal vectors

$\underset{\sigma}{\lambda^i}(\sigma = 1, \ldots, r)$ exist such that $a_{ij} = \sum_{\sigma=1}^{r} \underset{\sigma}{\alpha}\, \underset{\sigma}{\lambda_i} \underset{\sigma}{\lambda_j}$, where the $\underset{\sigma}{\alpha}$ are invariant and r is the rank of $[a_{ij}]$; these vectors are called the principal axes of the tensor a_{ij}. This theorem is a consequence of the theory of elementary divisors (Muth [9], pp. 118–25), which shows that the principal axes are unique only if the non-vanishing roots of the characteristic equation $|a_{ij} - \lambda g_{ij}| = 0$ are simple. Corresponding to a root of multiplicity p, there is a p-dimensional space in which any p mutually orthogonal vectors can be chosen as principal axes of the tensor a_{ij}. The matrices of the tensors g_{ij} and a_{ij} ($|g_{ij}| \neq 0$) are simultaneously reduced to the diagonal form.

This theorem is not valid for an indefinite g_{ij}. This is because the elementary divisors of the λ-matrix $A = [a_{ij} - \lambda g_{ij}]$ need not be simple and the roots of the corresponding characteristic equation may be complex. If the problem is posed over a field of complex numbers, then two pairs of tensors $\{a_{ij}, g_{ij}\}$ and $\{\dot{a}_{ij}, \dot{g}_{ij}\}$ are equivalent only if the elementary divisors of the matrices are identical (Bôcher [90], p. 303). Canonical forms of g_{ij} and a_{ij} can always be obtained from Weierstrass' results in the theory of elementary divisors, but they are not unique. If, however, we demand that the components g_{ij} and a_{ij} and the coefficients of the permissible linear transformations be real, these pairs of matrices can be non-equivalent, even though they have identical elementary divisors. We shall later have to reduce pairs of tensors to the canonical form by using just such *real coefficients*.

Let us consider two forms in R_n: $\varphi = g_{ij}x^i x^j$ and $\psi = a_{ij}x^i x^j$ with real coefficients and $|g_{ij}| \neq 0$. To determine the *eigen* vectors, i.e. the principal axes of a_{ij}, defined by equations $(a_{ij} - \lambda g_{ij})\lambda^j = 0$, one must consider the roots of the equation $| a_{ij} - \lambda g_{ij} | = 0$ and the elementary divisors of the matrix $[a_{ij} - \lambda g_{ij}]$. The roots may be complex for an indefinite g_{ij} and the elementary divisors *non-simple* (Shirokov [91], pp. 43–54).

Provided the matrix $A = G^{-1}T = [a^i_j]$ is real, where $G = [g_{ij}]$ and $T = [a_{kj}]$, the Hamilton–Cayley theorem (Shirokov [91], pp. 174–9; Klein [112], pp. 385–93), shows that real transformations to new coordinates $\dot{x}^i$ can reduce both forms φ and ψ to sums of k forms, i.e. $\varphi = \sum_1^k \varphi_i$, $\psi = \sum_1^k \psi_i$, where the φ_i and ψ_i contain only the variables $\dot{x}^j(j = j_1 \ldots, j_i)$. If the matrices of the forms φ_s and ψ_s are denoted by $\dot{G}_s$

and $\overset{*}{T}_s$ respectively, the characteristic equation of the matrix $\overset{*}{A}_s = \overset{*}{G}{}_s^{-1}\overset{*}{T}_s$ is $(\lambda - \underset{s}{\lambda})^{n_s} = 0$.

In other words, it is possible to select new basic vectors $\underset{p}{\overset{*}{\lambda}}{}^i = \underset{p}{c}{}^q \underset{q}{\lambda}{}^i$ so that

$$\overset{*}{G} = \begin{pmatrix} \overset{*}{G}_{n_1} & & \\ & \ddots & \\ & & \overset{*}{G}_{n_k} \end{pmatrix}, \quad \overset{*}{T} = \begin{pmatrix} \overset{*}{T}_{n_1} & & \\ & \ddots & \\ & & \overset{*}{T}_{n_k} \end{pmatrix}, \quad \overset{*}{A} = \begin{pmatrix} \overset{*}{A}_{n_1} & & \\ & \ddots & \\ & & \overset{*}{A}_{n_k} \end{pmatrix}, \tag{9.0}$$

where $\overset{*}{G}_{n_i}$, $\overset{*}{T}_{n_i}$, $\overset{*}{A}_{n_i}$ are square matrices of order n_i. Thus

$$\overset{*}{a}_{ij} = \overset{*}{g}_{ij} = \overset{*}{a}{}^i_j = 0, \tag{9.1}$$

if the indices i and j correspond to different submatrices along the principal diagonal, and the basic vectors dissociate into k *mutually orthogonal clusters*. Transformations of vectors inside a cluster are called *internal*; conditions (9.1) are valid for any internal transformation.

Let us consider the case when *the latent roots are real*. Using only real internal transformations we reduce the matrices $\overset{*}{A}_{n_i}$ to the Jordan canonical form (Shirokov [91], pp. 180–5; Turnbull and Aitken [89], pp. 60–61).

$$\overset{*}{A}_{n_i} = \begin{pmatrix} B_{s_1} & & \\ & \ddots & \\ & & B_{s_i} \end{pmatrix}, \quad \sum_{l=1}^{i} s_l = n_i, \quad B_{s_l} = \begin{pmatrix} \lambda_i & 1 & & \\ & \lambda_i & & \\ & & \ddots & 1 \\ & & & \lambda_i \end{pmatrix}; \tag{9.2}$$

where B_{s_l} is of order s_l.

We now drop the asterisks.

It is now possible to define the matrices G_{n_i} and T_{n_i}. It may be assumed without loss of generality that A_{n_i} is the leading submatrix in A and contains only two submatrices B_{s_l}. This means that there are two invariant clusters of orders m_1 and m_2, where $m_1 + m_2 = n_1$, $m_1 < m_2$, corresponding to the characteristic equation $(\lambda - \lambda_1)^{n_1} = 0$.

For the matrix A_{n_1} we obtain

$$a^i_j \underset{h}{\lambda}{}^j = \lambda_1 \underset{h}{\lambda}{}^i + \underset{h}{\Delta}\, \underset{h-1}{\lambda}{}^i \qquad (h = 1, \ldots, h_i;\ i, j = 1, \ldots, n_1), \tag{9.3}$$

where $\underset{h}{\Delta} = 0$ for $h = 1, m_1 + 1$ and $\underset{h}{\Delta} = 1$ otherwise. Since $\underset{h}{\lambda}{}^i = \delta^i_h$, then $a_{ij} = \lambda_1 g_{ij} + \underset{j}{\Delta} g_{i,\,j-1}$. As the tensors g_{ij} and a_{ij} are symmetric $\underset{i}{\Delta} g_{j,\,i-1} = \underset{j}{\Delta} g_{i,\,j-1}$. Hence

$$G_{n_i} = \left(\begin{array}{cccc|cccc}
 & & & \alpha_1 & & & & \gamma_1 \\
 & & \cdot^{\cdot^{\cdot}} & \vdots & & & \cdot^{\cdot^{\cdot}} & \vdots \\
 & \cdot^{\cdot^{\cdot}} & & \vdots & & \cdot^{\cdot^{\cdot}} & & \vdots \\
\alpha_1 & \cdots & \cdots & \alpha_{m_1} & \gamma_1 & \cdots & \cdots & \gamma_{m_1} \\ \hline
 & & & \gamma_1 & & & & \beta_1 \\
 & & \cdot^{\cdot^{\cdot}} & \vdots & & & \cdot^{\cdot^{\cdot}} & \vdots \\
 & \cdot^{\cdot^{\cdot}} & & \vdots & & \cdot^{\cdot^{\cdot}} & & \vdots \\
\gamma_1 & \cdots & \cdots & \gamma_{m_1} & \beta_1 & \cdots & \cdots & \beta_{m_2}
\end{array}\right)$$

Within each elementary cluster corresponding to an elementary Jordan submatrix, transformations are possible which leave relations (9.3) invariant. These include non-singular real transformations which replace the coefficients α_i and β_i for $i > 1$ and $\gamma_i (i = 1, \ldots, m)$ by zero (Petrov [165], pp. 41–42). Since $(a_{ij}) = (g_{ik})(a_j^k)$, *the matrices of the forms* $g_{ij}x^i x^j$ *and* $a_{ij}x^i x^j$ *can be reduced to the form*

$$(g_{ij}) = \begin{pmatrix} G_{m_1} & & & \\ & \cdot & & \\ & & \cdot & \\ & & & G_{m_k} \end{pmatrix}, \quad (a_{ij}) = \begin{pmatrix} T_{m_1} & & & \\ & \cdot & & \\ & & \cdot & \\ & & & T_{m_k} \end{pmatrix},$$

$$G_{m_i} = \begin{pmatrix} & & & l_i \\ & & \cdot^{\cdot^{\cdot}} & \\ & \cdot^{\cdot^{\cdot}} & & \\ l_i & & & \end{pmatrix}, \quad T_{m_i} = \begin{pmatrix} & & & l_i\lambda_i \\ & & \cdot^{\cdot^{\cdot}} & l_i \\ & \cdot^{\cdot^{\cdot}} & \cdot^{\cdot^{\cdot}} & \\ l_i\lambda_i & l_i & & \end{pmatrix}, \quad l = \pm 1, \tag{9.4}$$

when the latent roots λ_i *are real, by means of a real transformation.* In (9.4) $\sum m_i = n$ and k the number of invariant clusters equals the number of elementary divisors of the matrix. Though this reduction to the form (9.4) does not depend on the concept of elementary divisors, the latter are completely specified by the submatrices T_{m_i}. It follows from (9.4) that *to any simple elementary divisor, there corresponds a non-isotropic characteristic direction of* a_{ij} *; and to any non-simple divisor—an isotropic direction.*

Let us turn to the case of *complex roots*. If all or some of the roots of the equation $|a_{ij} - \lambda g_{ij}| = 0$ are complex, then using the process described above (but without reducing α and β to ± 1) we again obtain the matrices (9.2) and (9.4); the difference is that the elements are now, in general, complex numbers. But even in this case it is possible to select a real reference system, so that real "canonical" matrices are obtained for g_{ij} and a_{ij}.

As a_{ij} and g_{ij} are real numbers, the complex conjugate of any latent root is also a latent root. If we specify the set of complex basic vectors defining the Jordan form of the matrix (a^i_j) and hence satisfying eqns. (9.3), the conjugate cluster corresponding to the conjugate latent root contains the set of vectors conjugate to the initial vectors, and also satisfies eqns. (9.3), and hence defines a Jordan matrix with the same characteristic.

In other words, if an elementary divisor $(\lambda - \alpha - i\beta)^q$ $(q \leqslant p)$ corresponds to a p-fold root $\alpha + i\beta$, there is also an elementary divisor $(\lambda - \alpha + i\beta)^q$ corresponding to a p-fold root $\alpha - i\beta$. By changing the order of the basic vectors, two such submatrices can be made directly adjacent. It is sufficient to consider two such submatrices G_s and $\bar{G}_s$ in the leading positions in G, conjointly denoted by G_{2s}; we suppose that they correspond to the elementary divisor $(\lambda - \alpha \pm i\beta)^s$. Then

$$G_{2s} = \begin{pmatrix} G_s & 0 \\ 0 & \bar{G}_s \end{pmatrix}, \quad G_s = \begin{pmatrix} & & k \\ & \cdot^{\cdot^{\cdot}} & \\ k & & \end{pmatrix}. \tag{9.5}$$

Let

$$\underset{1}{\lambda^j} = \underset{1}{u^j} + i\underset{1}{v^j}, \quad \ldots, \quad \underset{s}{\lambda^j} = \underset{s}{u} + i\underset{s}{v^j},$$

be the basic vectors of the invariant cluster corresponding to the matrix G_s. Then the basic vectors corresponding to the matrix $\bar{G}_s$ are

$$\underset{s+1}{\lambda^j} = \underset{1}{u^j} - i\underset{1}{v^j}, \quad \ldots, \quad \underset{2s}{\lambda^j} = \underset{s}{u^j} - i\underset{s}{v}.$$

If we put $k = 2(p + iq)$ and $\bar{k} = 2(p - iq)$ where p and q are real numbers, g_{hj} is $(p + iq)$ if $h + j = s + 1$, and $(p - iq)$ if $h + j = 3s + 1$, and 0 otherwise.

Taking the $2s$ real vectors $\underset{1}{u}^j, \ldots, \underset{s}{u}^j, \underset{1}{v}^j, \ldots, \underset{s}{v}^j$, as basic vectors, we renumber them, giving the vectors u the odd numbers and v the even ones. Then the matrices G_{2s} and A_{2s} can be expressed as follows:

$$G_{2s} = \begin{pmatrix} & & & p & q \\ & & & q & -p \\ & & \cdot^{\cdot^{\cdot}} & & \\ p & q & & & \\ q & -p & & & \end{pmatrix}, \quad A_{2s} = \begin{pmatrix} \alpha & \beta & 1 & 0 & & \\ -\beta & \alpha & 0 & 1 & & \\ & & & \ddots & & \\ & & & & \alpha & \beta \\ & & & & -\beta & \alpha \end{pmatrix}.$$

A transformation

$$\begin{aligned} x^1 &= c_1 \overset{*}{x}{}^1 - c_2 \overset{*}{x}{}^2, & x^3 &= c_1 \overset{*}{x}{}^3 - c_2 \overset{*}{x}{}^4, & &\ldots, \\ x^2 &= c_2 \overset{*}{x}{}^1 + c_1 \overset{*}{x}{}^2, & x^4 &= c_2 \overset{*}{x}{}^3 + c_1 \overset{*}{x}{}^4, & &\ldots, \end{aligned}$$

can reduce the numbers p and q to 1, so that the matrix $T_{2s} = G_{2s}A_{2s}$ can be expressed in the form

$$G_{2s} = \begin{pmatrix} & & & 1 & 1 \\ & & & 1 & -1 \\ & & \cdot^{\cdot^{\cdot}} & & \\ 1 & 1 & & & \\ 1 & -1 & & & \end{pmatrix},$$

$$T_{2s} = \begin{pmatrix} & & & & & \alpha-\beta & \alpha+\beta \\ & & & & \cdot^{\cdot^{\cdot}} & & \\ & & & & & \alpha+\beta & -\alpha+\beta \\ & & & \cdot^{\cdot^{\cdot}} & & 1 & 1 \\ & & \cdot^{\cdot^{\cdot}} & & \cdot^{\cdot^{\cdot}} & 1 & -1 \\ & \cdot^{\cdot^{\cdot}} & & \cdot^{\cdot^{\cdot}} & & & \\ \alpha-\beta & \alpha+\beta & 1 & 1 & \cdot^{\cdot^{\cdot}} & & \\ \alpha+\beta & -\alpha+\beta & 1 & -1 & & & \end{pmatrix} \tag{9.6}$$

where α and β have in general different values for every s.

Thus in dealing with the general problem of reducing a pair of quadratic forms, we have shown that *provided* $|g_{ij}| \neq 0$ *the matrices of the real symmetric tensors* g_{ij} *and* a_{ij} *can be reduced by a real linear transformation of coordinates to the form*

$$(g_{ij}) = \begin{pmatrix} G_1 & & \\ & \ddots & \\ & & G_r \end{pmatrix}, \quad (a_{ij}) = \begin{pmatrix} T_1 & & \\ & \ddots & \\ & & T_r \end{pmatrix}, \tag{9.7}$$

where G_i *and* T_i *are given by formulae* (9.4) *in the case of real roots, and by* (9.6) *for each pair of conjugate complex roots.*

The number of invariant clusters r of the tensor a_{ij} is equal to the number of elementary divisors of the λ-matrix $(a_{ij} - \lambda g_{ij})$, if a pair of conjugate complex divisors is counted as one.

The numbers l_i, λ_i, α and β in formulae (9.4) and (9.6) depend in general on the cluster, but may be identical for several matrices T_i when the two or more corresponding elementary divisors have identical characteristics.

In books on linear algebra the treatment is confined to the case where the elementary divisors are simple, i.e. all the Jordan submatrices are of the first order, or else the reduction is given over a complex field when the result lies within the scope of formulae (9.6).

Examples

1. Prove that the matrix of any covariantly constant mixed tensor S^α_β can be reduced to Jordan's canonical form in some holonomic system of coordinates, or in other words that there exists a holonomic system of coordinates relative to which all the components of S^α_β are constant (Shirokov [235]).

2. At each point of V_4 the bivector $v_{\alpha\beta}$ and the non-degenerate tensor $g_{\alpha\beta}(|g_{\alpha\beta}| \neq 0)$ can be simultaneously reduced to canonical form by using a complex linear homogeneous transformation of coordinates. If ξ^α_ϱ are the covariant components of the basic vectors, then in this canonical representation:

$$g_{\alpha\beta} = \sum_\sigma \underset{\sigma}{\xi_\alpha} \underset{\sigma}{\xi_\beta},$$

and the bivector $v_{\alpha\beta}$ is reduced to one of the following forms (to within a scalar factor):

I. $\frac{1}{2} v_{\alpha\beta} = \sqrt{-\lambda}\, \underset{1[1}{\xi_\alpha} \underset{2]}{\xi_\beta} \pm \sqrt{-\lambda}\, \underset{2[3}{\xi_\alpha} \underset{4]}{\xi_\beta}$

(the bivector is isotropic only when $\lambda_1 + \lambda_2 = 0$);

II. $\frac{1}{2} v_{\alpha\beta} = a(\underset{[1\ 2]}{\xi_\alpha \xi_\beta} \pm \underset{[3\ 4]}{\xi_\alpha \xi_\beta}) + b(\underset{[1\ 4]}{\xi_\alpha \xi_\beta} \pm \underset{[2\ 3]}{\xi_\alpha \xi_\beta}) + c(\underset{[2\ 4]}{\xi_\alpha \xi_\beta} \pm \underset{[3\ 1]}{\xi_\alpha \xi_\beta})$

(the bivector is isotropic only if $a^2 + b^2 + c^2 = 0$, when it is also simple);

III. $\frac{1}{2} v_{\alpha\beta} = \underset{[1\ 4]}{\xi_\alpha \xi_\beta} \pm i\underset{[3\ 2]}{\xi_\alpha \xi_\beta}$

the bivector is simple and isotropic);

IV. $\frac{1}{2} v_{\alpha\beta} = i(\underset{[1\ 4]}{\xi_\alpha \xi_\beta} + \underset{[2\ 3]}{\xi_\alpha \xi_\beta}) + \underset{[2\ 4]}{\xi_\alpha \xi_\beta} - \underset{[3\ 1]}{\xi_\alpha \xi_\beta}$

(non-isotropic bivector) (Petrov [171]).

§ 10. Lie Groups in V_n

In Chapter 5, we classify V_n spaces determined by gravitational fields, according to the groups of motions admitted by them; we then require certain results from the theory of continuous groups of transformations (Lie groups). This section contains the basic definitions and facts concerning Lie groups, and several theorems essential for subsequent work.

Given an n-dimensional space X_n referred to coordinates $\{x^\alpha\}$ ($\alpha = 1, \ldots, n$), and an r-dimensional space of parameters $\{a^k\}$ ($k = 1, \ldots, r$) denoted by G_r, then the equations

$$\overset{*}{x}{}^\alpha = f^\alpha(x^1, \ldots, x^n; \quad a^1, \ldots, a^r) \equiv f^\alpha(x, a) \tag{10.1}$$

with fixed values of a^k, define a transformation of the point $P(x)$ in the space X_n into the point $\overset{*}{P}(\overset{*}{x})$. It is assumed that the f^α are functions of some class C^s, and that the a^k are *essential* parameters, i.e. they cannot be expressed in terms of a smaller number of independent parameters; for this to be true (Eisenhart [147], p. 10) it is necessary and sufficient that the functions f^α should not satisfy any equations of the type

$$\sum_s \varphi_s \frac{\partial f^\alpha}{\partial a^s} = 0.$$

The set of transformations (10.1) corresponding to a domain in the parametric space constitutes a continuous group (denoted by G_r) if:

1. functions $\varphi^i(a^s, b^t)$ ($i, s, t = 1, \ldots, r$) exist throughout the domain which are continuously differentiable a sufficient number of times, such

that

$$f^{\alpha}(f(x, a), b) = f^{\alpha}(x, \varphi(a, b)),$$

II. the domain contains a point e^s $(s = 1, \ldots, r)$ such that

$$f^{\alpha}(x, e) = x^{\alpha},$$

III. the functions φ satisfy the following group axioms:

(1) $\varphi^i(a, \varphi(b, c)) = \varphi^i(\varphi(a, b), c)$,

(2) $\varphi^s(a, e) = \varphi^s(e, a) = a^s$

(3) $\varphi^s(a, a^{-1}) = \varphi^s(a^{-1}, a) = e^s$.

Condition I defines the *product* of two elements of the group. Axiom (1) defines the *associative law*, axiom (2) ensures the existence of a *unit of the group* e^s, and axiom (3) implies the existence of a *reciprocal* element for each element a of the group. The independence of the three axioms has been studied by a number of authors from different points of view (Huntingdon [21a]; Dickson [22]; Chebotarëv [120], p. 46).

If only condition I and axiom (1) are satisfied, the set of transformations is called a *semi-group*.

A *one-dimensional subgroup* G_1 of a group G_r corresponds to a curve $a^s = a^s\ (t)$ in the parametric space which passes through the unit e^s of the group, and for which

$$\varphi^s(a(t_1), a(t_2)) = a^s(t_3)$$

for all pairs of points t_1 and t_2. Corresponding to each point on this curve infinitesimally close to e^s, there is an infinitesimal transformation in the space X_n

$$\overset{*}{x}{}^{\alpha} = x^{\alpha} + \xi^{\alpha}(x)\,\delta t, \tag{10.2}$$

with which we can associate the operator

$$X = \xi^{\alpha}(x)\,\partial_{\alpha} \qquad \left(\partial_{\alpha} \equiv \frac{\partial}{\partial x^{\alpha}}\right). \tag{10.3}$$

Equation (10.2) shows that the trajectory of a one-dimensional subgroup in the space X_n that passes through a given point $P\,(\overset{0}{x}{}^{\alpha})$ is the integral curve of the system of ordinary differential equations

$$\frac{dx^{\alpha}}{dt} = \xi^{\alpha}(x) \qquad (\alpha = 1, \ldots, n) \tag{10.4}$$

for the initial conditions $x^\alpha(0) = \overset{0}{x}{}^\alpha$, $a^s = e^s$ ($\alpha = 1, \ldots, n$; $s = 1, \ldots, r$).

A set of operators (10.3) can be put into one–one correspondence (apart from a *constant* factor) with the set of one-dimensional subgroups of the group G_r. Of these group operators only r can be linearly independent; such a set $X_1, \ldots, X_r$ is called a base of the group. Any other operator of the group is a linear combination of the base operators with constant coefficients.

In order that r group operators should constitute a base, it is necessary and sufficient that the Poisson operators or commutators

$$[X_i X_j] \equiv (\underset{i}{\xi}{}^\sigma \partial_\sigma \underset{j}{\xi}{}^\beta - \underset{j}{\xi}{}^\sigma \partial_\sigma \underset{i}{\xi}{}^\beta)\, \partial_\beta \tag{10.5}$$

should be group operators in G_r, as follows from the theory of the compatibility of systems of differential equations. Hence the commutators must have the form

$$[X_i X_j] = C^s_{ij} X_s, \tag{10.6}$$

where the C^s_{ij} are the so-called *structure constants* of the group G_r; this forms the content of the *second principal theorem* of Lie ([147], pp. 53–55). In particular, if all the constants C^s_{ij} are zero, the commutators are also zero and the group is called *Abelian*.

The transformations in the group G_r which leave the point $P(\overset{0}{x}{}^\alpha)$ fixed form a subgroup H_m of G_r; it is known as the *subgroup of stability* of P ([147], p. 65); by definition,

$$\xi^\alpha(\overset{0}{x}) = 0 \tag{10.7}$$

for any operator H_m.

If any two points in X_n can be interchanged by some transformation in the group G_r, the group is called *transitive;* otherwise the group is *intransitive* and X_n can be filled by a set of *surfaces of transitivity* on each of which the group is transitive. Thus, for example, the group of displacements in ordinary R_3 is transitive, while the group of rotations is intransitive, with concentric spherical surfaces of transitivity.

There are some transitive groups for which it is possible to find a set of manifolds, one through each non-singular point in X_n such that if one point of one manifold is transformed into a point of another, each point of the first manifold goes into a point of the second. This is true, for example, of the transitive group of displacements in three-dimensional Euclidean space. These transitive groups are called *imprimitive* and the corresponding manifolds *systems of imprimitivity*. Otherwise the groups are *primitive*. Thus the group of motions in a Euclidean plane is obviously primitive.

A group is transitive only if $r \geqslant n$ and the rank of the matrix $[\underset{s}{\xi}^{\alpha}]$ is n $(\alpha = 1, \ldots, n; s = 1, \ldots, r)$ ([147], p. 72). If $r = n$, the group is said to be *simply* transitive, otherwise *multiply transitive*.

The two groups G_r and $\boldsymbol{G}_r$ defined by the finite equations

$$(\alpha) \quad \overset{*}{x}^{\alpha} = f^{\alpha}(x^{\sigma}, a^s), \qquad (\beta) \quad \overset{*}{\boldsymbol{x}}^{\alpha} = \boldsymbol{f}^{\alpha}(x^{\sigma}, \boldsymbol{a}^s)$$

$$(\alpha, \sigma = 1, \ldots, n; s = 1, \ldots, r),$$

are said to be *similar* if a set of r independent functions $\omega^{\alpha}(a^s)$ exists which permits a non-degenerate transformation of coordinates converting (β) into (α), assuming that $\boldsymbol{a}^{\alpha} = \omega^{\alpha}(a^s)$. The question of whether two groups are similar can be answered by the following theorem proved by Lie (Lie and Engel [5], I, p. 353): two r-parametric groups in the same number of variables and with the same structure constants, for which the ranks of the matrices $(\underset{s}{\xi}^{\alpha})$ and $(\underset{s}{\boldsymbol{\xi}}^{\alpha})$ are less than r, are similar only when these ranks are equal (say q), when each pair of corresponding minors of order q has identical rank, and the corresponding set of equations

$$\boldsymbol{\varphi}_p^h(\boldsymbol{x}) = \varphi_p^h(x) \quad (h = 1, \ldots, q; p = q+1, \ldots, n)$$

where

$$\underset{p}{\boldsymbol{\xi}}^{\alpha} = \boldsymbol{\varphi}_p^h \underset{h}{\boldsymbol{\xi}}^{\alpha}$$

are compatible and do not lead to relations between the variables of either group.

If, however, the matrices $(\underset{s}{\xi}^{\alpha})$ and $(\underset{s}{\boldsymbol{\xi}}^{\alpha})$ have ranks r, two *intransitive* groups in the same number of variables are similar if their structure constants are identical. Two simply transitive groups in the same number of variables and with identical structure constants are similar.

All possible linear combinations of the commutators of a group (*the linear pencil of commutators)* determine a subgroup G_{r_1}, called the *commutant* of the group G_r. In its turn the linear pencil of the commutants defines a commutant $G_{r_2} \subset G_{r_1}$, and so on. Continuing this process we obtain

$$G_r \supset G_{r_1} \supset G_{r_2} \supset \ldots$$

There are two alternatives: $r > r_1 > r_2 \ldots > r_k = 0$ and the group is then *soluble:* $r > r_1 > \ldots > r_k = r_{k+1} = \ldots > 0$ when we have an *insoluble group.*

We also need to consider special types of continuous groups of transformations in Riemann spaces—*groups of motions, which leave the metric of V_n unchanged*; following transformations of a group of motions, the components of the metric tensor remain the same functions of the new independent variables:

$$g_{\alpha\beta}(\dot{x}^1, \ldots, \dot{x}^n)\, d\dot{x}^\alpha\, d\dot{x}^\beta = g_{\alpha\beta}(x^1, \ldots, x^n)\, dx^\alpha\, dx^\beta. \tag{10.8}$$

In order that G_r should be a group of motions in V_n, it is necessary and sufficient that the *Killing equations* ([6], p. 167) be satisfied for *each* operator of the group; in deriving these equations it is convenient to use the concept of a *Lie derivative.* Suppose that we are given a certain object field (the order of the object is of no importance in this discussion and hence the indices are denoted by single letters in brackets). Let us consider some fixed point $P(x^\alpha)$ and the value of the object $\Omega^{(\alpha)}_{(\beta)}\ (x^\alpha)$ at this point. If we displace the system of coordinates by $\xi^\alpha\, dt$, the point

$$x^\alpha + \xi^\alpha\, dt$$

acquires the coordinates x^α. Then the point x^α acquires the coordinates

$$x^\alpha - \xi^\alpha\, dt.$$

We now return to the original system of coordinates. If the original value of the object at the point x^α is $\Omega^{(\alpha)}_{(\beta)}$, its value at this same point after displacement is

$$\Omega^{(\alpha)}_{(\beta)} - \underset{L}{\delta}\Omega^{(\alpha)}_{(\beta)}.$$

As suggested by van Dantzig, the object $\underset{L}{\delta}\Omega^{(\alpha)}_{(\beta)}$ is called the *Lie differential* with respect to $\xi^\alpha\, dt$. The definition shows that the Lie differential does not depend on the coefficients of connectivity of the space, and in a holonomic system of coordinates takes the following forms for covariant and contravariant vectors u with respect to $\xi^\alpha\, dt$:

$$\left.\begin{aligned} \underset{L}{\delta}u^\alpha &= (\xi^\sigma\, \partial_\sigma u^\alpha - u^\sigma\, \partial_\sigma \xi^\alpha)\, dt, \\ \underset{L}{\delta}u_\alpha &= (\xi^\sigma\, \partial_\sigma u_\alpha + u_\sigma\, \partial_\alpha \xi^\sigma)\, dt. \end{aligned}\right\} \tag{10.9}$$

Dividing by dt, we obtain the so-called *Lie derivative with respect to the vector field* ξ^α.

The Lie differential for tensors of a higher order is calculated in exactly the same way, taking each index individually and noting whether it is covariant or contravariant.

The Killing equations provide the necessary and sufficient conditions that a displacement $\xi^\alpha dt$ in V_n is a motion, i.e. a displacement of V_n in itself for which (10.8) is valid. This means that the *displaced* value of the field $g_{\alpha\beta}$ at the point

$$x^\alpha + \xi^\alpha\, dt$$

must equal the natural value of the tensor field $g_{\alpha\beta}(x)$; in other words, the Lie differential of $g_{\alpha\beta}$ with respect to $\xi^\alpha dt$ is zero.

Hence *in order that the vector ξ^α should define a motion in V_n, it is necessary and sufficient that in the direction ξ^α*

$$\underset{L}{\delta} g_{\alpha\beta} = 0. \tag{10.10}$$

Substituting from (10.9) we obtain

$$\xi^\sigma\, \partial_\sigma g_{\alpha\beta} + g_{\alpha\sigma}\, \partial_\beta \xi^\sigma + g_{\beta\sigma}\, \partial_\alpha \xi^\sigma \equiv \xi_{\alpha,\,\beta} + \xi_{\beta,\,\alpha} = 0. \tag{10.11}$$

ξ^α is called a *Killing vector* when it satisfies this equation.

Similarly, the Riemann connectivity must also be conserved in a motion in V_n; hence the Lie differential with respect to $\xi^\alpha\, dt$ must vanish,

$$\underset{L}{\delta} \Gamma^\alpha_{\beta\gamma} = 0, \tag{10.12}$$

which means that

$$\xi^\alpha_{,\beta\gamma} + \xi^\sigma R_{\sigma\gamma}{}^\alpha{}_\beta = 0. \tag{10.13}$$

As, according to (7.17), the metric tensor is completely determined if the curvature tensor and its covariant derivatives are known, the following conditions must also hold with respect to the vector ξ^α defining the motion:

$$\left.\begin{aligned} &\underset{L}{\delta} R_{\alpha\beta\gamma\delta} = 0, \\ &\underset{L}{\delta} R_{\alpha\beta\gamma\delta,\,\lambda} = 0, \\ &\dots\dots\dots\dots\dots \\ &\underset{L}{\delta} R_{\alpha\beta\gamma\delta,\,\lambda_1\dots\lambda_p} = 0, \\ &\dots\dots\dots\dots\dots \end{aligned}\right\} \tag{10.14}$$

These equations are the conditions of integrability for the Killing equations and their corollaries obtained by successive covariant differentiation subject to (10.11) and (10.13). They can be obtained directly as the conditions for compatibility of eqns. (10.11) (cf., for example, [147], pp. 213–14).

A study of eqns. (10.14) plays a decisive role in determining the order r of a group G_r admitted by V_n. As a result of the general theorems of the compatibility of a system of differential equations ([147], § 1) we have:

In order that a group of motions G_r should exist in V_n, it is necessary and sufficient that a positive integer k exists such that the rank of the combined sets of homogeneous algebraic equations (10.14) *up to the* kth *in the unknowns ξ^α and $\xi_{\alpha,\beta}$ satisfying* (10.11) *should equal*

$$\frac{n(n+1)}{2}-r,$$

and that the addition of new equations from the $(k+1)$th *set in* (10.14) *onwards should not alter this rank.*

Any classification of V_n according to groups of motions must necessarily be associated with the structure of the groups considered, i.e. with the structure constants for some canonical base of each group. Knowing the structure of the group, we can determine the corresponding simplest holonomic system of coordinates, and the operators of the group relative to this system. The determination of a V_n with a given group then reduces to the integration of the Killing equations for the unknown functions $g_{\alpha\beta}$.

In applying these ideas to spaces determined by gravitational fields, we are only interested in real groups of motions and real structures. A classification for $r = 2$, 3 by real non-isomorphic structures† was given by Bianchi [44]; $r = 4$ (excluding all G_4 which contain Abelian subgroups G_3) was treated by Kruchkovich ([203], p. 9; [260], p. 209), who based his work on the classification of dissimilar complex structures of the group G_4 given by Lie ([7], § 137). This classification leads to the following possible structures:

$$r = 2$$

$$\left.\begin{array}{l} \text{I. } [X_1X_2] = 0-\text{Abelian } G_2, \\ \text{II. } [X_1X_2] = X_1-\text{non-Abelian } G_2. \end{array}\right\} \qquad (10.15)$$

† Isomorphic groups are defined on p. 140.

$$r = 3$$

$$\left.\begin{array}{lllll}
 & \textit{Soluble groups} & & & \\
\text{I.} & [X_iX_j] = 0, & (i,\ j = 1, 2,\ 3) - \text{Abelian } G_3, & & \\
\text{II.} & [X_1X_2] = 0, & [X_2X_3] = X_1, & [X_3X_1] = 0, & \\
\text{III.} & [X_1X_2] = 0, & [X_2X_3] = 0, & [X_3X_1] = -X_1, & \\
\text{IV.} & [X_1X_2] = 0, & [X_2X_3] = X_1+X_2, & [X_3X_1] = -X_1, & \\
\text{V.} & [X_1X_2] = 0, & [X_2X_3] = X_2, & [X_3X_1] = -X_1, & \\
\text{VI.} & [X_1X_2] = 0, & [X_2X_3] = qX_2, & [X_3X_1] = -X_1 & \\
 & & & (q \neq 0 \text{ or } 1), & \\
\text{VII.} & [X_1X_2] = 0, & [X_2X_3] = -X_1+qX_2, & & \\
 & & [X_3X_1] = -X_2 & (q^2 < 4). & \\
 & \textit{Insoluble groups} & & & \\
\text{VIII.} & [X_1X_2] = X_1, & [X_2X_3] = X_3, & [X_3X_1] = -2X_2, & \\
\text{IX.} & [X_1X_2] = X_3, & [X_2X_3] = X_1, & [X_3X_1] = X_2. &
\end{array}\right\} \quad (10.16)$$

The number of non-isomorphic structures over the field of complex numbers is smaller (types VI and VII are identical, as are VIII and IX).

$$r = 4$$

$$\left.\begin{array}{llll}
 & \multicolumn{3}{l}{\textit{Soluble groups } G_4 \textit{ not containing Abelian subgroups } G_3} \\
\text{I.} & [X_1X_2] = 0, & [X_2X_3] = X_1, & [X_3X_1] = 0, \\
 & [X_1X_4] = cX_1, & [X_2X_4] = X_2, & [X_3X_4] = (c-1)X_3 \\
 & & & (c = \text{any real number}), \\
\text{II.} & [X_1X_2] = 0, & [X_2X_3] = X_1, & [X_3X_1] = 0, \\
 & [X_1X_4] = 2X_1, & [X_2X_4] = X_2, & [X_3X_4] = X_2+X_3, \\
\text{III.} & [X_1X_2] = 0, & [X_2X_3] = X_1, & [X_3X_1] = 0, \\
 & [X_1X_4] = qX_1, & [X_2X_4] = X_3, & [X_3X_4] = \\
 & & & -X_2+qX_3\ (q^2 < 4), \\
\text{IV.} & [X_1X_2] = 0, & [X_2X_3] = X_2, & [X_3X_1] = 0, \\
 & [X_1X_4] = X_1, & [X_2X_4] = 0, & [X_3X_4] = 0, \\
\text{V.} & [X_1X_2] = 0, & [X_2X_3] = X_2, & [X_3X_1] = -X_1, \\
 & [X_1X_4] = X_2, & [X_2X_4] = -X_1, & [X_3X_4] = 0. \\
 & \multicolumn{3}{l}{\textit{Soluble groups } G_4 \textit{ with Abelian subgroups } G_3} \\
\text{VI.} & [X_iX_j] = 0, & [X_iX_4] = C_i^kX_k & \\
 & & & (i, j = 1, 2, 3;\ k = 1, 2, 3, 4). \\
 & \textit{Insoluble groups } G_4 & & \\
\text{VII.} & [X_1X_2] = X_1, & [X_2X_3] = X_3, & [X_3X_1] = -2X_2, \\
 & & [X_iX_4] = 0 & (i = 1, 2, 3), \\
\text{VIII.} & [X_1X_2] = X_3, & [X_2X_3] = X_1, & [X_3X_1] = X_2, \\
 & & [X_iX_4] = 0 & (i = 1, 2, 3)
\end{array}\right\} \quad (10.17)$$

In the complex region the structures I and III are identical, as are IV and VIII.

We will now investigate the structure VI not previously considered in this classification. The operators of the Abelian subgroup G_3 are subjected to the non-degenerate real linear transformation

$$Y_i = A_i^j X_j, \quad Y_4 = X_4, \quad |A_i^j| \neq 0.$$

Then $[Y_i\, Y_j] = 0$ ($i, j = 1, 2, 3$), and

$$[Y_i Y_4] = A_i^j [X_j X_4] = A_i^j C_j^s X_s + C_i^4 X_4 = \overset{*}{C}{}_i^s Y_s + C_i^4 X_4,$$

where $\overset{*}{C}{}_i^s = A_i^j\, C_j^h\, \overset{*}{A}{}_h^s$ $(s, h = 1, 2, 3)$ and $\overset{*}{A}{}_j^i$ is the inverse transformation to A_j^i. The corresponding matrix equation is

$$\overset{*}{C} = ACA^{-1}, \quad |A| \neq 0.$$

We are left with the problem of reducing C_j^i to a canonical form by a non-degenerate real transformation. But only the following four non-isomorphic forms exist (Shirokov [91], pp. 187–201):

$$(1)\quad C = \begin{pmatrix} \lambda_1 & 0 & 0 \\ 0 & \lambda_2 & 0 \\ 0 & 0 & \lambda_3 \end{pmatrix}, \qquad (2)\quad C = \begin{pmatrix} \lambda_1 & 1 & 0 \\ 0 & \lambda_1 & 0 \\ 0 & 0 & \lambda_2 \end{pmatrix},$$

$$(3)\quad C = \begin{pmatrix} \lambda_1 & 1 & 0 \\ 0 & \lambda_1 & 1 \\ 0 & 0 & \lambda_2 \end{pmatrix}, \qquad (4)\quad C = \begin{pmatrix} \alpha & \beta & 0 \\ -\beta & \alpha & 0 \\ 0 & 0 & \lambda \end{pmatrix}, \quad \beta \neq 0,$$

where λ_i, α and β are real numbers.

In order that the C_j^i defined by these matrices and the C_i^4 should be structural constants, it is necessary and sufficient that they satisfy the Jacobi relations ([147], p. 26):

$$\left.\begin{aligned} C_{ab}^l &= -C_{ba}^l, \\ C_{ab}^l C_{lc}^f + C_{bc}^l C_{la}^f + C_{ca}^l C_{lb}^f &= 0. \end{aligned}\right\} \tag{10.18}$$

Imposing these conditions on the C_j^i we obtain the following theorem: *if G_4 contains the Abelian subgroup G_3 then it has one of following non-isomorphic real structures*:

$$\left.\begin{array}{ll}
\mathrm{VI}_1. & [X_iX_j] = 0\ (i, j = 1, 2, 3), \quad [X_1X_4] = aX_1+bX_4, \\
 & [X_2X_4] = cX_2+dX_4, \qquad [X_3X_4] = eX_3+fX_4, \\
 & \text{where essentially the only cases are} \\
 & 1)\ a=b=c=d=e=f=0, \quad 2)\ a=c=e=f=0, \\
 & \quad b=d=1, \quad 3)\ a=c=e=0, \quad b=d=f=1, \\
 & 4)\ c=d=e=f=0,\ a=b=1,\ 5)\ b=d=f=0,\ a=1; \\
\mathrm{VI}_2. & [X_iX_j] = 0\ (i, j=1, 2, 3), \quad [X_1X_4] = kX_1+X_2, \\
 & [X_2X_4] = kX_2, \quad [X_3X_4] = \varepsilon X_3, \\
\mathrm{VI}_3. & [X_iX_j] = 0\ (i, j = 1, 2, 3), \quad [X_1X_4] = kX_1+X_2, \\
 & [X_2X_4] = kX_2+X_3, \quad [X_3X_4] = \varepsilon X_3, \\
\mathrm{VI}_4. & [X_iX_j] = 0\ (i, j = 1, 2, 3), \quad [X_1X_4] = kX_1+X_2, \\
 & [X_2X_4] = -X_1+kX_2, \quad [X_3X_4] = lX_3,
\end{array}\right\} \quad (10.19)$$

where ε is equal to 0 or 1, and k, l are real numbers. This argument can be extended in an obvious way to any r.

Examples

1. Prove that the necessary and sufficient condition for V_n to admit an infinitesimal motion, is that there exists a coordinate system in which none of the coefficients $g_{\alpha\beta}$ contain one of the coordinates x^γ; the coordinate curves x^γ are then the trajectories of the motion ([147], p. 209).

2. In a motion of V_n, geodesics pass over into geodesics.

3. If $\underset{s}{\xi}^\alpha$ are Killing vectors, then $c^s \underset{s}{\xi}^\alpha$ (c = constant) is also a Killing vector.

4. If V_n admits a non-transitive group of motions G_r and if a non-isotropic hypersurface V_{n-1} is a hypersurface of transitivity, the hypersurfaces geodesically parallel to V_{n-1} also possess this same property.

5. If the non-isotropic surfaces of transitivity of a group of motions G_r are of dimension $n-1$, they are geodesically parallel.

6. Motions whose trajectories are geodesics are known as parallel *transfers* or *translations*. The necessary and sufficient condition for the Killing vector $\underset{s}{\xi}^\alpha$ to define a translation is that $g_{\alpha\beta}\, \underset{s}{\xi}^\alpha\, \underset{s}{\xi}^s$ = constant; all points are then displaced equal distances.

7. Prove that vectors of the group of motions G_1 form constant angles with any non-isotropic geodesic.

8. The following theorem is due to Bianchi: the trajectories of two groups G_1 of translations intersect at a constant angle ([44], p. 501). Show that this statement is valid only when the least group containing both groups is transitive (Shirokov [105]).

9. A V_n space of constant curvature admits a G_r with $r = n(n+1)/2$. Every V_n with a group of motions of this order has a constant curvature.

10. V_2 admits a translation only when its curvature is zero.

11. A homogeneous Riemann space $V_n(ds^2 > 0)$ does not admit a group of similarities wider than the group of motions except when V_n is Euclidean (Ku Chao-khao [307]).

CHAPTER 2

EINSTEIN SPACES

IN THIS chapter, we consider the principal facts connected with the special and general theories of relativity, which lead to four-dimensional Einstein spaces and to their extension to an arbitrary number of dimensions.

§ 11. The Basis of the Special Theory of Relativity. Lorentz Transformations

Remaining within the limits set by the scope of the book, we give only a summary of the principles of the special theory of relativity, with the emphasis on the mathematical aspects.

Classical mechanics, with foundations laid down in the works of Galileo and Newton, first obtained general acceptance but failed during the subsequent development of physics to provide a satisfactory explanation of new facts, particularly in the theory of light and electromagnetism.

Two theories of light based on classical mechanics, Newton's corpuscular theory and Huyghens' wave theory, appeared almost simultaneously; preference was given for nearly a century to the first theory. Then under the pressure of the experimental facts of diffraction most physicists turned to the wave hypothesis, which considered light as a *state of* a special medium, the aether, proposed as the *carrier* of this state and endowed with many extraordinary properties which have remained mysterious.

On the other hand, the theory of electromagnetism as described by Maxwell's equations and confirmed by a mass of experimental data, is not invariant from the point of view of classical mechanics. The construction of a theory corresponding more closely to the real nature of things than classical mechanics, had to start from experimentally established facts. Two such facts stated by the following propositions have been so thoroughly confirmed experimentally that they are almost obvious; (1) any phenomenon in a moving system is fundamentally independent of the uniform motion of the system as a whole; (2) *the velocity of light c is independent of the source*, and $c \approx 300{,}000$ km/sec. More exactly, proposition (1) implies the existence of ∞^3 reference systems *in uniform linear* motion relative to each other in which the phenomena are identical. Such reference systems are called *inertial.* When restricted to the mechanical phenomena considered in classical mechanics, proposition (1) is identical with the *Galilean principle of relativity*. Proposition (2) can now be formulated as follows: *the velocity of light in empty space is constant and equals c in any inertial system.* The second proposition does not, of course, hold for *non-inertial* reference systems.

In classical mechanics, the Galilean principle of relativity is indicated by the formulation of the laws not changing when one inertial system is replaced by another. In the simplest case, when y and z remain the same, this change of coordinates can be written as follows:

$$x' = x - vt, \quad y' = y, \quad z' = z, \tag{11.1}$$

and as *the time interval must be independent of the reference system*

$$t' = t. \tag{11.2}$$

Transformations (11.1) and (11.2) are called *Galilean.* The essentials of the problem can be understood by considering this particular case only of the relative motion of inertial systems. If proposition (1) is also to apply to electrodynamics, the Galilean transformation leads to a contradiction.

The problem thus arises of how two inertial systems are related when propositions (1) and (2) are satisfied. This same problem may also be approached by demanding that the Maxwell equations remain invariant under the required transformations from one inertial system to another. Propositions (1) and (2) are, however, insufficient for a unique solution of the problem, and additional assumptions are needed to obtain the appropriate transformations, called *Lorentz transformations.*

The history of the problem is as follows. The formulae were first obtained by Larmor when solving the problem of the longitudinal contraction of bodies during motion, and the change in the time scale ([12], pp. 167–77); this work remained unnoticed. Two papers by Lorentz then followed ([15], [19]); he still assumed the existence of an aether in the first, but nevertheless made a number of valuable suggestions, and in the second obtained the required transformations that ensured that Maxwell's equations remain invariant in charge-free space. The contraction of bodies was again treated as a causally dependent phenomenon. The two subsequent papers by Poincaré fill in the formal gaps left by Lorentz ([23], [25]). The terms "Lorentz transformations" and "Lorentz group" first occur in these works; the principle of relativity is for the first time stated as a general principle. The assumption that the dimensions of a body perpendicular to the direction of motion remain constant is based on the hypothesis that the required group of transformations contains the ordinary group of rotations as a subgroup. Poincaré was able to construct a theory of the electron which was invariant with respect to the given group.

Simultaneously, Einstein's papers appeared, containing in essence all the fundamental results of the above-mentioned investigations, and a profound approach to the problem from new aspects (cf. [24]). Einstein obtained these results from (1) and (2) but also made other assumptions such as the "homogeneity" of space. In this work, the special theory of relativity reached almost its final form. Hypotheses (1) and (2) were introduced as experimental facts, a step later strikingly justified (Vavilov [84]).

The formal axiomatics of the problem continue to attract attention. Thus if the following basic axioms are assumed, the correct transformations are obtained to within the sign and the physical interpretation of the constants; (1) the required transformations form a single-parameter homogeneous linear group; (2) the velocity of the reference system S relative to S' is equal and opposite to that of S' relative to S; (3) the contraction of the scale in S' seen from S is equal to that in S seen from S' (von Ignatowsky [29], [31], Frank and Rothe [30], [32]).

Cattaneo considers a common system of axioms for both classical and relativistic mechanics, and using them in conjunction with additional conditions, obtains either the Galilean or the Lorentz transformations [290]. Alternative conditions to Cattaneo's can be introduced (Fock [225], p. 8; Aleksandrov [190]; Aleksandrov and Ovchinnikov [191]; Matsumoto [227]).

The rigorous formulation of the special theory of relativity developed by Aleksandrov, is based on the following axioms; (1) space-time is the manifold of all events, regarded simply from the structural aspect, and defined by a system of order relations abstracted from all other properties; (2) space-time is a four-dimensional manifold; (3) space-time is maximally homogeneous, i.e. the transformation group maintaining the correct order relations is the largest possible group; (4) both the space-time manifold itself and the transformations indicated in (3) are differentiable [317].

We obtain the Lorentz transformations by assuming that propositions (1) and (2) are valid, and that in addition (3) the transformations are linear, (4) the scales in a plane perpendicular to the direction of motion remain unaltered, (5) the readings of two clocks in the plane $O'y'z'$ agree seen from the system S.

Let the reference system S, consisting of the coordinate axes with some scale and a clock, denote the time and place of some event by the set of numbers $x^1 = x$, $x^2 = y$, $x^3 = z$ and $x^4 = t$, and the system S' correspondingly by the four numbers $x^{1'}$, $x^{2'}$, $x^{3'}$ and $x^{4'}$. Then, assuming that at $x^4 = t = 0$ the systems S and S' coincide, (3) allows us to write the required transformations in the form

$$x^{\alpha'} = c^{\alpha'}_{\beta} x^{\beta} \quad (\alpha, \beta = 1, 2, 3, 4).$$

Suppose that the inertial system S' is moving relative to S with a *constant* velocity v, with the axes Ox and $O'x'$ coinciding. Then, according to (4),

$$x^{2'} = x^2, \quad x^{3'} = x^3.$$

It follows from (5) the $x^{1'}$ and $x^{4'}$ must be independent of x^2 and x^3. Thus the transformations must have the form

$$x^{1'} = \alpha x^1 + \beta x^4, \quad x^{2'} = x^2, \quad x^{3'} = x^3 \quad x^{4'} = \gamma x' + \delta x^4. \quad (11.3)$$

Let us suppose that at the instant $t = x^4 = 0$ when S and S' coincide, a light signal is sent out from the common origin and is propagated with velocity c. The proposition (2) shows that after an interval of time t in the reference system S

$$\sum_1^3 x^{i^2} - c^2 x^{4^2} = 0, \quad (11.4)$$

whereas in the system S',

$$\sum_1^3 x^{i'^2} - c^2 x^{4'^2} = 0. \quad (11.5)$$

Substituting for $x^{\alpha'}$ from (11.3) into (11.5), we should obtain equations equivalent to (11.4) apart from a possible factor. But since $x^{2'} = x^2$

and $x^{3'} = x^3$ this factor must be unity. Hence

$$\alpha^2 - c^2\gamma^2 = 1, \quad \alpha\beta - c^2\gamma\delta = 0, \quad \delta^2 - \frac{\beta^2}{c^2} = 1,$$

i.e.

$$\alpha = \cosh\varphi, \quad \beta = c\sinh\varphi, \quad \gamma = \frac{1}{c}\sinh\varphi, \quad \delta = \cosh\varphi.$$

The motion of the origin of S seen from S' is found by putting $x = 0$, $y = 0$, $z = 0$, so that

$$x^{1'} = ct\sinh\varphi, \quad x^{2'} = x^{3'} = 0, \quad x^{4'} = t\cosh\varphi$$

and hence

$$\frac{x^{1'}}{x^{4'}} = \frac{x'}{t'} = -v = c\tanh\varphi, \quad \tanh\varphi = -\frac{v}{c}.$$

Substituting in (11.3) we obtain

$$x^{1'} = \frac{1}{\sqrt{1-\frac{v^2}{c^2}}}(x^1 - vx^4), \quad x^{2'} = x^2,$$

$$x^{3'} = x^3, \quad x^{4'} = \frac{1}{\sqrt{1-\frac{v^2}{c^2}}}\left(-\frac{v}{c^2}x^1 + x^4\right). \tag{11.6}$$

Taking the formal point of view already suggested by the notation used above, we can refer to a four-dimensional space with coordinates $x^\alpha (\alpha = 1, \ldots, 4)$, with geometry corresponding to the group of transformations of type (11.6); we note that the form

$$ds^2 = -dx^{1^2} - dx^{2^2} - dx^{3^2} + dx^{4^2} \quad (x^4 = ct) \tag{11.7}$$

is invariant under these transformations. If, conversely, we try to find transformations (11.3) which leave (11.7) invariant, we obtain (11.6). The form (11.7) defines the metric of a flat space with a signature $(- - - +)$, known as *Minkowski space* after the mathematician who first introduced this convenient interpretation [26].

Two flat spaces R_n and $\mathring{R}_n$ with the same number of dimensions and signature are said to be *isomorphic* if there exists a real one–one mapping of the points of R_n on to the points of $\dot{R}_n$ such that: (1) if the points A and B in the space R_n correspond to the points $\dot{A}$ and $\dot{B}$ in $\dot{R}_n$, the vector AB in R_n corresponds to the vector $\dot{A}\dot{B}$ in $\dot{R}_n$; (2) if the vector X^α in R_n corresponds to $\dot{X}^\alpha$ in $\dot{R}_n$, the vector aX^α in R_n, where a is a number, corresponds to $a\dot{X}^\alpha$ in $\dot{R}_n$; (3) $g_{\alpha\beta}u^\alpha u^\beta = \dot{g}_{\alpha\beta}\dot{u}^\alpha\dot{u}^\beta$ if u^α corresponds to $\dot{u}^\alpha$.

If two isomorphic spaces R_n and $\dot{R}_n$ are identical, we refer to this as *automorphism* or *motion* in flat space. If (3) is replaced by the condition that (11.7) be invariant under the mapping, we obviously obtain automorphism. *The geometry of Minkowski space is the geometry of the group of Lorentz automorphisms.*

The symmetry of the coordinates x^1, x^2 and x^3, allows two more independent transformations (11.6). Moreover, it is obvious that ordinary rotations and translations, with $x^4 = x^{4\prime}$, do not alter the form (11.7), *and neither does a displacement in time* $x^{4\prime} = x^4 + \overset{0}{x}{}^4$. Hence a group of linear real transformations exists, constituting the group of automorphisms of Minkowski space, which depends on ten arbitrary parameters; it includes three Lorentz rotations, three spatial rotations and four displacements along the axes. It is well known (cf. example 9, § 10) that the maximum number of independent parameters for motions in any V_4 is 10. We neglect mirror reflections of the type $x^{\alpha\prime} = -x^\alpha$, which also leave (11.7) invariant and increase the number of distinct transformations ([188], p. 208).

The essence of the special theory of relativity consists in establishing an interconnection between space and time, formally expressed both by (11.6) and the introduction of four-dimensional Minkowski space.

The motion of a material point can be described if its position is given at each instant of time as follows,

$$x^i = x^i(x^4) \quad (i = 1, 2, 3), \tag{11.8}$$

which defines a curve in Minkowski space. In particular, the functions $x^i(x^4)$ for uniform linear motion are *linear* and define straight lines in four-dimensional space. If in addition, the point moves with the velocity of light c, the form (11.7) equals zero, so that the four-dimensional trajectory of such a point is an *isotropic* straight line.

If the propagation of light in empty space is considered to be similar to the motion of material particles, as in geometrical optics, we again obtain isotropic straight lines in Minkowski space. A light signal at a given point is propagated along the semi-generators of the isotropic hypercone with its apex at the point. Equation (11.6) shows that the velocity of a material particle cannot exceed c, and hence the trajectories of material particles with velocities $< c$ are curves with tangents directed inside the hypercone.

§ 12. Field Equations in the Relativistic Theory of Gravitation

The special theory of relativity establishes an interconnection between space and time, modes of existence of matter, but does not consider the relation between the geometry of space-time and the distribution and motion of matter; the space-time continuum is assumed to be homogeneous. Such a relation, the possibility of which was foreseen by Lobachevskii (cf. [199], pp. 10–18), is established in the relativistic theory of gravitation. This theory assumes that the geometry of the special theory is valid only *locally*, when the gravitational field can be neglected, and that in a region where this is impossible the metric of the space-time depends on the distribution and motion of matter; the general theory of relativity is the theory of this dependence.

In the relativistic theory of gravitation, the properties of the space-time continuum are not consistent with the geometry of a flat space, but require Riemannian geometry for their description. If the metric of such a V_4 is defined by the linear element

$$ds^2 = g_{\alpha\beta}\, dx^\alpha\, dx^\beta, \tag{12.1}$$

the components $g_{\alpha\beta}(x^i, \ldots, x^4)$ of the metric tensor depend on the distribution and motion of matter; for this reason the $g_{\alpha\beta}(x)$ are sometimes called the *potentials* of the gravitational field.

The principal problem in field theory, the determination of the field from the distribution and motion of matter, requires the setting-up of the field equations. These were obtained by Einstein in 1915 in the form

$$R_{\alpha\beta} - \tfrac{1}{2} R g_{\alpha\beta} = -\lambda T_{\alpha\beta}, \tag{12.2}$$

where $T_{\alpha\beta}$ is the energy-momentum tensor, and λ a universal constant [35].

Einstein obtained these equations after investigating many alternative hypotheses; thus in 1914 he still assumed that $R_{\alpha\beta} = \lambda T_{\alpha\beta}$ [34]. Simultaneously, and independently of Einstein, Hilbert obtained eqns. (12.2) using the variational principle [41]; he did not, however, derive them for an arbitrary material system, but based himself on Mie's theory of matter [33]. This method of obtaining the basic field equations was later improved by Klein who used the variation of coordinates which do not vanish at the boundary of the domain of integration [40], [45]; this has frequently been done in classical mechanics since the time of Lagrange, and many relations then take a simpler form (Pauli [149], pp. 68–70, 161; Landau and Lifshitz [156], pp. 296–300; Fock [225], pp. 203–7; Weyl [69], pp. 237–9; Bergmann [152]).

In addition to satisfying eqns. (12.2), the metric tensor of a gravitational field corresponding to a real distribution of matter must be reducible at each point in a region of space time to the form

$$(g_{\alpha\beta})_P = \begin{pmatrix} -1 & 0 & 0 & 0 \\ 0 & -1 & 0 & 0 \\ 0 & 0 & -1 & 0 \\ 0 & 0 & 0 & 1 \end{pmatrix}, \tag{12.3}$$

by non-degenerate real linear transformations, i.e. Minkowski geometry must be valid in the tangent space R_4 at each point P. The left-hand side of (12.2) is completely determined by the geometry of the space-time, and the tensor $T_{\alpha\beta}$ is determined by the distribution and motion of matter. Hence the equation implies that matter determines the geometry of space-time and conversely that the motion of matter is determined by the metric tensor of the space, which is not flat. The physical meaning of condition (12.3) is that the geometry of the special theory of relativity is valid for an infinitesimally small region of space during an infinitesimal interval of time.

The agreement between the consequences derived from eqns. (12.2) and experiment to within the accuracy at present attainable (Tikhov [102]; Bogorodskii [121], p. 3; Vavilov [84]), argues in favour of these equations being correct.

Einstein's basic assumptions in setting up the field equations (12.2) are as follows: (1) in the presence of gravitation, space-time is a V_4: (2) the distribution and motion of energy and momentum are described by a symmetric tensor $T_{\alpha\beta}(x^1, x^2, x^3, x^4)$ satisfying the *conservation law*

$$T^{\alpha\beta}{}_{,\alpha} \equiv 0, \tag{12.4}$$

(3) the field equations must be of a tensor character; (4) they must be linear with respect to the second order partial derivatives of the potentials

$g_{\alpha\beta}(x)$; (5) the energy-momentum tensor must be determined by the geometry of a V_4 space. If certain boundary conditions are satisfied, eqns. (12.2) follow uniquely from these premises. Conversely, if we assume that the gravitational field equations have the form (12.2), all these five assumptions can be deduced.

Deriving (12.2) by the variational principle is based on the following hypotheses: (1) the gravitational field equations can be obtained by varying the sum of the actions of field and matter; (2) the variation of the field action leads to the Euler–Lagrange equations of Hamilton's principle; the variational problem is solved for an integral over a four-dimensional region A:

$$I = \int_A R\sqrt{-g}\, dx^1\, dx^2\, dx^3\, dx^4, \quad \delta I = 0,$$

where R is the scalar curvature of the V_4 space. (3) The action of the matter can be expressed as the integral

$$S = \int \Omega\left(q, \frac{\partial q}{\partial x^\alpha}\right) dx^1\, dx^2\, dx^3\, dx^4$$

where Ω is some function of the generalized coordinates q defining the system and of their derivatives with respect to the space-time coordinates. The energy-momentum tensor is then given by

$$T_{\alpha\beta} = \Omega g_{\alpha\beta} - q_{,\alpha} \frac{\partial \Omega}{\partial q^\beta}$$

under condition (12.4). Additional assumptions about the boundary conditions are necessary.

The derivation of the basic equation (12.2) from a series of such hypotheses is of great heuristic importance. In any case, the field equations are derived by the variational principle in most variants of unified field theory.

The physical validity of (12.2) can only be confirmed experimentally.

§ 13. Einstein Spaces

The energy-momentum tensor $T_{\alpha\beta}$ vanishes in empty space and eqns. (12.2) reduce to

$$R_{\alpha\beta} = 0, \tag{13.1}$$

since contraction over α, β shows that the scalar curvature R is zero; if, for example, matter occurs as point masses, (13.1) applies outside these points. Solutions of the field equations (13.1) are somewhat analogous to solutions of Laplace's equation in classical field theory; they determine the geometry of space-time in regions empty of matter. A great deal of research has been devoted to this type of space. Together with eqns. (13.1) we can consider the field equations

$$R_{\alpha\beta} = \varkappa g_{\alpha\beta}, \tag{13.2}$$

where $\varkappa$ is a constant for $n > 2$, as follows directly from (13.2) and (5.19) These equations can be considered as the field equations when the energy-momentum tensor $T_{\alpha\beta}$ differs from the metric tensor $g_{\alpha\beta}$ only by a constant factor; a physical model would be a homogeneous space with uniform density of matter ([156], p. 332; [225], p. 352). We can refer to the "internal" and "external" problems in integrating the field equations for an "insular" distribution of matter; thus (13.1) defines the solutions of the "external" problem.

V_n spaces satisfying eqns. (13.2) are encountered in many geometrical problems unrelated to the theory of gravitation. Riemann manifolds of this type arise in the theory of Lie groups, and in almost all the present variants of unified field theory. *Riemann manifolds of any dimensionality n and metric signature satisfying condition* (13.2) *are called Einstein spaces* and denoted by the symbol G_n. When n is 4 and the signature of the G_4 is $(- - - +)$ the symbol $\mathring{T}$ is used. If, in addition, $\varkappa = 0$, we use the symbol T, which therefore indicates the gravitational field in free space.

Alternatively, we can define G_n space without using (13.2). We consider any point P in a V_n and the set of directions through P satisfying the relation

$$R_{\alpha\beta}\, dx^\alpha\, dx^\beta = 0;$$

Ricci called the principal directions of this second order cone the *principal directions* of the space V_n at the given point P. If we take a second order surface in the space tangent to V_n at P, with centre at P and equation

$$(R_{\alpha\beta} - \varkappa g_{\alpha\beta})\xi^\alpha \xi^\beta = 1,$$

the principal directions of this surface, sometimes called the *Einstein indicatrix*, are also the Ricci principal directions. Equations (13.2) are then equivalent to requiring that these directions be indeterminate, i.e. *that the Einstein indicatrix at each point be a hypersphere.*

Another interesting characteristic of G_n with a *positive-definite metric* has been given by Thomas [107]. He introduced the scalar invariant

$$Q = g^{\alpha i} g^{\beta j} g^{\gamma k} g^{\delta l} R_{\alpha\beta\gamma\delta} R_{ijkl},$$

naming it the *quadratic curvature* of the V_n, and showed that a V_n with a positive-definite metric is flat only when this scalar vanishes. He further showed that a V_n *with a positive-definite metric is a* G_n *only when the following condition is satisfied:*

$$nR_{\alpha\beta}R^{\alpha\beta} - R^2 = 0. \tag{13.3}$$

The restriction to definite metrics is essential; replacing the $\frac{1}{2}n(n+1)$ conditions (13.2) by one is possible only because an equation of the form $\sum a_i^2 = 0$ is equivalent to the set $a_i = 0$ $(i = 1, 2, \ldots)$, when the a_i are real.

There are also definitions of G_n using a generalized curvature of V_n space based on an m-dimensional instead of a two-dimensional direction (Cartan [98], pp. 30–31; Wrona [150], p. 234).

When n is 4, the following theorem holds:

if one of the conditions

$$\left.\begin{aligned} &(\alpha)\ R_{\alpha\beta} = \varkappa g_{\alpha\beta}, \\ &(\beta)\ 3g_{[\alpha[\nu} R_{\beta\gamma]\lambda\mu]} = 2\varkappa g_{[\lambda[\alpha}\, g_{\mu\beta}\, g_{\nu]\gamma]}, \end{aligned}\right\} \tag{13.4}$$

is satisfied by a V_4, *so is the other.*

A double alternation is implied in condition (β), over $\alpha\beta\gamma$ and $\lambda\mu\nu$ respectively. To prove the theorem, we introduce a possibly imaginary non-holonomic orthonormal tetrad at each point in the V_4, so that

$$g^{\alpha\alpha} = g_{\alpha\alpha} = 1, \quad g_{\alpha\beta} = g^{\alpha\beta} = 0 \qquad (\alpha \neq \beta).$$

In every orthonormal tetrad, (α) and (β) reduce to two sets of equations in the curvature tensor components, which are equivalent to within linear combinations; this proves the theorem.

As the curvature tensor of any V_3 is completely defined in terms of the Ricci and metric tensors

$$\left.\begin{aligned} R_{\alpha\beta\gamma\delta} &= g_{\alpha\gamma} L_{\beta\delta} - g_{\alpha\delta} L_{\beta\gamma} + g_{\beta\delta} L_{\alpha\gamma} - g_{\beta\gamma} L_{\alpha\delta}, \\ L_{\alpha\beta} &= R_{\alpha\beta} + \frac{R}{4} g_{\alpha\beta}, \end{aligned}\right\} \tag{13.5}$$

(cf. § 7, example 8); then if (13.2) holds we obtain expression (8.4) for $R_{\alpha\beta\gamma\delta}$; hence *if* V_3 *is an Einstein space it is a space of constant curvature.*

If we contract (8.4) we obtain (13.2) for any n; hence *every Riemann space of constant curvature is an Einstein space.*

It should be noted that the geometry of four-dimensional space-time in the general theory of relativity has an interesting interpretation in non-holonomic Laguerre geometry, allowing the investigation of this manifold from a new point of view [216].

Examples

1. Every V_2 is a G_2.

2. A G_n, conformal to a flat space, is an S_n (Schouten and Struik [55]).

3. Verify that a V_n with a positive-definite metric is a G_n only when (13.3) holds.

4. In order that every geodesic hypersphere in a V_n should be a G_{n-1}, it is necessary and sufficient that the V_n should be an S_n (Shirokov [72], p. 53).

5. The invariant

$$H = \frac{g_{\alpha\alpha_1\beta\beta_1} R^{\alpha}_{\gamma\delta\lambda} R^{\alpha_1}_{\gamma_1\delta_1\lambda_1} u^{\beta\gamma} u^{\delta\lambda} u^{\beta_1\gamma_1} u^{\delta_1\lambda_1}}{(g_{\alpha\gamma\delta\lambda}\, u^{\alpha\gamma} u^{\delta\lambda})^2}$$

where $g_{\alpha\gamma\delta\lambda} = g_{\alpha[\delta} g_{\lambda]\gamma}$ and $u^{\alpha\beta} = \xi^{[\alpha}\eta^{\beta]}$ is a simple bivector, is called a *quadratic curvature of the V_n space.*† If H is independent of $u^{\alpha\beta}$, the V_n is *a space of constant quadratic curvature.* Show that: (1) the magnitude of the bivector $u^{\alpha\beta}$ displaced parallel to itself along the boundary of a non-singular element of surface $u^{\alpha\beta}\Delta\sigma$ is characterized by the invariant H as far as the fourth order of small quantities [$\Delta\sigma = O(\delta^2)$]; (2) any S_n is a space of constant quadratic curvature; (3) any V_3 of constant quadratic curvature is an S_3; (4) any V_4 of constant quadratic curvature is a G_4 (Petrov [157]).

6. Using Ricci's identity (5.5), show that the constant $\varkappa$ in (13.2) can be expressed in terms of any non-isotropic vector field a^α as follows:

$$\varkappa = a^\sigma (a^{\tau}_{,\tau\sigma} - a^{\tau}_{,\sigma\tau}) / a^\nu a_\nu.$$

When $\varkappa = 0$ any vector field with components of class C^2 satisfies the equation $a^{\tau}_{,\tau\sigma} - a^{\tau}_{,\sigma\tau} = 0$ (Narlikar [245]).

§ 14. Some Solutions of the Gravitational Field Equations

The study of gravitational fields T (or $\dot{T}$) reduces to the investigation of a set of ten second order partial differential equations involving ten unknown functions $g_{\alpha\beta}(x)$ of the four independent variables x^α ($\alpha = 1, 2, 3, 4$). The

† Not to be confused with the quadratic curvature defined by Thomas.

equations are linear in the second order derivatives, with variable coefficients. By choosing special coordinate systems in any V_n, n components of the metric tensor can be converted into predetermined functions (cf. §7). Hence the set of field equations can be reduced to ten equations involving six unknown functions, i.e. an over-determined set.

Because of the complexity of the general problem and the physical requirements of the theory of gravitation, special cases have been solved by imposing additional conditions leading to solutions in closed form. Thus a study has been made of *constant* gravitational fields; a reference system exists in which all quantities, and in particular the metric tensor, are independent of the *time* coordinate x^4. An example is the so-called *static* field, with all bodies at rest relative to the reference system in which $g_{\alpha\beta}$ is independent of the time x^4. As both directions of time are then clearly equivalent, all the metric tensor components $g_{\alpha 4} = 0$ $(\alpha \neq 4)$. A characteristic of all fields of this type is that $\partial_4 g_{\alpha\beta} = 0$ in some system of coordinates. Hence constant fields define a geometry admitting a group of motions of ordner $r \geqslant 1$, since the metric admits the operator $X = \delta_4^\alpha \partial_\alpha$.

The first fields to be investigated were centro-symmetric fields, in which ds^2 is the same at all equidistant points from the centre. Such a field can be set up by a centro-symmetric distribution of matter with velocity everywhere directed along a radius, i.e. centro-symmetric pulsations. Fields possessing *axial* symmetry have also been studied.

Since Einstein's principal work appeared in 1915, many particular solutions of the problem have been given, some generalizing or repeating earlier results; this is due either to unfamiliarity with the literature, or to the fact that, although the problem of the equivalence of quadratic differential forms has long been solved, the actual calculation can be cumbersome (Christoffel [2], p. 60).

We now consider in chronological order some well-known T and $\mathring{T}$ manifolds given in the literature.

1. The T space found by Schwarzschild [37] and based on physical considerations:

$$\left.\begin{aligned} g_{44} &= -g_{11}^{-1} = 1+\frac{c}{x^1}, & g_{22} &= -x^{1^2}, \\ g_{33} &= -x^{1^2}\sin^2 x^2, & g_{ij} &= 0 \ (i \neq j), \end{aligned}\right\} \tag{14.1}$$

where c is an arbitrary constant. The use of polar coordinates enables the quadratic form to be reduced to the orthogonal form. Such a T corresponds to the most general centro-symmetric distribution of matter as was rigorously proved by Hilbert [41] and Birkhoff [66].

In other words, when the metric terms independent of time are spherically symmetric, the integration of (13.1) always gives (14.1).

2. The $\mathring{T}$ space suggested by Kottler is a generalization of the Schwartzschild solution [46]. In polar coordinates, the metric tensor has the following components,

$$\left.\begin{aligned} g_{44} &= -g_{11}^{-1} = 1+\frac{a}{3}x^{1^2}+\frac{c}{x^1}, \quad g_{22} = -x^{1^2}, \\ g_{33} &= -x^{1^2}\sin^2 x^2, \quad g_{ij} = 0 \quad (i \neq j), \end{aligned}\right\} \qquad (14.2)$$

where a and c are arbitrary constants. When $a = 0$ we obtain solution (14.1) and when $c = 0$ space of constant curvature S_4. Both (14.1) and (14.2) belong to the class of V_4's with metrics

$$ds^2 = g_{11}(x^1)\,dx^{1^2} - x^{1^2}\,dx^{2^2} - x^{1^2}\sin^2 x^2\,dx^{3^2} + g_{44}(x^1)\,dx^{4^2},$$

in some coordinate system; substituting in (13.2) shows that this metric leads to (14.2).

3. Weyl [42], [51], Bach and Weyl [62a] and Levi-Civita [47], pp. 183, 220) succeeded in finding a static solution of the field equations (13.1) with axial, but not spherical symmetry, defined by the following metric tensor components:

$$g_{44} = -g_{11}^{-1} = -g_{22}^{-1} = e^{\mu}, \quad g_{33} = -e^{\nu-\mu}, \quad g_{ij} = 0 \quad (i \neq j), \qquad (14.3)$$

where the functions μ and ν depend only on the two variables x^3 and $\varrho = \sqrt{x^{1^2}+x^{2^2}}$, and satisfy the following set of field equations

$$f_1 \equiv \frac{\partial^2 \mu}{\partial \varrho^2} + \frac{1}{\varrho}\frac{\partial \mu}{\partial \varrho} + \partial_{33}\mu = 0,$$

$$f_2 \equiv \frac{\partial \nu}{\partial \varrho} - \frac{\varrho}{2}\left[\left(\frac{\partial \mu}{\partial \varrho}\right)^2 - (\partial_3 \mu)^2\right] = 0,$$

$$f_3 \equiv \partial_3 \nu - \varrho \frac{\partial \mu}{\partial \varrho}\,\partial_3 \mu = 0,$$

$$\partial_3 f_2 - \frac{\partial f_3}{\partial \varrho} = \varrho f_1\,\partial_3 \mu.$$

Clearly, the first two equations result from contracting Bianchi's identity; the problem was solved by assuming that μ and ν vanish at infinity.

4. The T space obtained by Bach [62], assuming a metric with matrix

$$(g_{\alpha\beta}) = \begin{pmatrix} \alpha & \omega & 0 & 0 \\ \omega & \beta & 0 & 0 \\ 0 & 0 & \gamma & 0 \\ 0 & 0 & 0 & \delta \end{pmatrix},$$

where the functions α, β, γ, δ and ω depend only on the variables x^3 and x^4. These solutions are not exact; they are obtained by imposing some additional conditions permitting approximations to within a given order of small quantities. Bach thus obtained three solutions involving rational functions of x^3 and $\sin^2 x^4$. This was one of the first attempts to find an approximate solution of the field equations—a method since developed by Einstein, Infeld and Fock.

5. In a paper [73] on spaces conformally mappable on Einstein spaces, Brinkmann proved the following theorem: an Einstein space can be mapped conformally on another Einstein space, i.e.

$$g_{\alpha\beta} = e^{2\sigma}\overset{*}{g}{}^{\alpha\beta}$$

where

$$\Delta_1\sigma \equiv g^{\alpha\beta}\,\partial_\alpha\sigma\,\partial_\beta\sigma \neq 0$$

only when the metric can be reduced to the form

$$ds^2 = fg_{\alpha\beta}\,dx^\alpha\,dx^\beta + f^{-1}\,dx^{n^2} \quad (\alpha, \beta = 1, \ldots, n-1), \tag{14.4}$$

with

$$f = \frac{1}{n(n-1)}(Rx^{n^2} + 2ax^n + b),$$

where a and b are constants, the $g_{\alpha\beta}$ are independent of x^n, and the form $\varphi = g_{\alpha\beta}\,dx^\alpha\,dx^\beta$ is that of an $(n-1)$-dimensional Einstein space. Thus this result defines a special type of n-dimensional Einstein space. It should be noted that when $n = 4$, the form φ defines an S_3, since every three-dimensional Einstein space is a space of constant curvature; hence assuming that the S_3 is an elliptic space, we obtain

$$ds^2 = \frac{f}{\psi^2}\sum_1^3 dx^{i^2} + \frac{1}{f}\,dx^{4^2},$$

where

$$\psi = 1 + \frac{K}{4}\sum x^{i^2}, \quad f = -\varkappa x^{4^2} + ax^4 + b,$$

after which it is clear that $\dot{T}$ must be an S_4.

6. Kasner [57] searched for a $\overset{*}{T}$ space that can be embedded in a flat $(n+k)$-dimensional space, and moreover assumed that the metric is reducible to the orthogonal form

$$ds^2 = \sum_{i=1}^{n} f_i \, dx^{i^2},$$

where the f_i depend only on one or two variables. In this way he determined the metric

$$ds^2 = x^{1^{2a_1}} dx^{1^2} + x^{1^{2a_2}} dx^{2^2} + x^{1^{2a_3}} dx^{3^2} + x^{1^{2a_4}} dx^{4^2},$$

where the constants a_i satisfy the relations

$$-a_1 + a_2 + a_3 + a_4 = 1, \quad -(a_1+1)^2 + a_2^2 + a_3^2 + a_4^2 = 0$$

and the metric

$$ds^2 = x^{1^{-2}} (dx^{1^2} + dx^{2^2}) + x^{3^{-2}} (dx^{3^2} + dx^{4^2})$$

so that the space is *reducible*.

7. An investigation of *symmetric* Riemann manifolds with a covariantly constant curvature tensor, i.e. $R_{\alpha\beta\gamma\delta,\,\lambda} = 0$ leads to the study of non-reducible V_p's $(p<n)$ which play an essential role in the analysis of so-called *semi-simple* groups [81]. These spaces, among which Einstein spaces are naturally prominent, have been studied by Shirokov ([108], pp. 9–27; [72], p. 111; [204], pp. 71–82), Cartan [166], Rozenfel'd ([169], [158]) and many others. The classification and definition of all symmetric T spaces are given in Chapter 9.

8. Mitter [86] used cylindrical coordinates z, r, θ and t, in investigating V_4's with metrics

$$ds^2 = -e^{\lambda} dz^2 - e^{\mu}(dr^2 + r^2 \, d\theta^2) + e^{\nu} dt^2,$$

satisfying (13.1), where λ, μ and ν are functions of r. This led to a search for a static solution symmetrical about the z-axis. Lewis also investigated analogous solutions of (13.1) ([88]). These investigations lead to Weyl's and Levi-Civita's results.

9. Delsarte considered so-called "binary" T spaces, which have metric forms

$$ds^2 = \varphi^2 \, d\overset{*}{s}{}^2 + f^2 d\sigma^2,$$

where

$$\mathring{ds}^2 = \mathring{ds}^2(u^1, \ldots, u^p), \quad d\sigma^2 = d\sigma^2(v^1, \ldots, v^q),$$
$$f = f(u^1, \ldots, u^p), \quad \varphi = \varphi(v^1, \ldots, v^q), \quad p+q = 4.$$

These spaces are in general non-reducible ([94], [94a]). Assuming that the space has an axis of symmetry and a tetra-orthogonal coordinate system, he obtains a solution of the form

$$ds^2 = \frac{1}{\{\wp(u)-\wp(v)\}^2}\{du^2 - dv^2 + \wp'^2(u)\, dy^2 + \wp'^2(v)\, dz^2\}, \qquad (14.5)$$

where $\wp$ is the Weierstrass elliptic function. In [143] Yung Chow Wong determined when the differential form for T_4 can be reduced to either of the forms

$$ds^2 = \{p(x^k, t)\}^2\, dt^2 + g_{ij}(x^k)\, dx^i\, dx^j, \quad \partial_t p \neq 0 \quad (k = 1, 2, 3)$$
$$ds^2 = \{\sigma(x^k, t)\}^2\, dt^2 + \{p(x^k, t)\}^2 g_{ij}(x^k)\, dx^i\, dx^j,$$

where the form $g_{ij}\, dx^i\, dx^j$ is assumed to be negative-definite. If we assume that $\partial_t p = 0$ in the first form, we obtain another solution previously found by Delsarte. The necessary and sufficient conditions for metrics reducible to these forms are given by the author in differential form.

10. Chou [103] investigated solution of (13.1) for a *static and isotropic* V_4, i.e. assuming that

$$ds^2 = u^2\, dt^2 + g_{ij}\, dx^i\, dx^j \quad (i, j = 1, 2, 3),$$

where u and g_{ij} are independent of t; $g_{ij}\, dx^i\, dx^j$ is the metric of a three-dimensional conformally Euclidean space, the condition for *isotropy* of space in the author's terminology. He showed that

$$\partial_{ij} u = \frac{6u}{c+u^2}\left(\partial_i u\, \partial_j u - \frac{1}{3}\Delta_1 u g_{ij},\right), \quad \Delta_1 u = g^{ij}\, \partial_i u\, \partial_j u$$

where c is an arbitrary constant. If $c = 0$ Kasner's solution is obtained, if $c < 0$ Schwarzschild's solution is a special case and if $c > 0$

$$ds^2 = \cot^2\left(\frac{x^1}{2}\right) dt^2 - \frac{\sin^4\left(\frac{x^1}{2}\right)}{4K^2 x^{2^2}}\, [x^{2^2}\, dx^{1^2} + dx^{2^2} + dx^{3^2}] \qquad (14.6)$$

as a special case.

11. If T is a *reducible* space, i.e. has a metric represented by the sum of two or more quadratic forms with neither variables nor differentials in

common, clearly these quadratic forms also define G_q spaces ($q<4$). We therefore obtain, apart from the trivial case S_4, only the following metrics by selecting convenient initial conditions:

$$\left.\begin{aligned} ds^2 &= -\cosh^2(\sqrt{\varkappa}x^2)\,dx^{1^2} - dx^{2^2} - \cos^2(\sqrt{\varkappa}x^4)\,dx^{3^2} + dx^{4^2} \\ ds^2 &= -\cosh^2(\sqrt{\varkappa}x^2)\,dx^{1^2} - dx^{2^2} + \cosh^2(\sqrt{\varkappa}x^4)\,dx^{3^2} - dx^{4^2} \end{aligned}\right\} \varkappa > 0, \quad (14.7)$$

$$\left.\begin{aligned} ds^2 &= -\cos^2(\sqrt{-\varkappa}x^2)\,dx^{1^2} - dx^{2^2} + \cosh^2(\sqrt{-\varkappa}x^4)\,dx^{3^2} - dx^{4^2} \\ ds^2 &= -\cos^2(\sqrt{-\varkappa}x^2)\,dx^{1^2} - dx^{2^2} + \cos^2(\sqrt{-\varkappa}x^4)\,dx^{3^2} - dx^{4^2} \end{aligned}\right\} \varkappa < 0, \quad (14.8)$$

where $\varkappa$ is the Gaussian curvature of corresponding surfaces of constant curvature. The second metrics in (14.7) and (14.8) are *static* by virtue of the signature, whilst the first metrics in (14.7) and (14.8) are *nonstatic*. The solutions are fundamentally different since they cannot be reduced one to another by *real* coordinate transformations; *complex* transformations exist which convert one metric to another.

12. Takeno [127] has shown that the spherically symmetric solutions of the equations $R_{\alpha\beta,\,\gamma} = 0$ lead to the Schwarzschild or Kottler solutions.

13. We have also the following theorem due to Petrov ([146], pp. 8–12); the metric

$$ds^2 = \sum_{i=1}^{n-1} g_i(x^n)\,dx^{i^2} + dx^{n^2}$$

defines an Einstein space with signature $(- - \ldots - +)$ if the g_i are chosen as follows: if $\varkappa > 0$ in (13.2)

$$g_i = -\sin^{\frac{2}{n-1}}(\alpha x^n)\tan^{\frac{2\alpha_i}{\alpha}}\left(\frac{\alpha x^n}{2}\right), \quad g_n = 1 \quad (i = 1, \ldots, n-1); \qquad (14.9)$$

if $\varkappa < 0$,

$$g_i = -\sin^{\frac{2}{n-1}}(\beta x^n)\tanh^{\frac{2\beta_i}{\beta}}\left(\frac{\beta x^n}{2}\right), \quad g_n = 1; \qquad (14.10)$$

and if $\varkappa = 0$ we obtain either an S_n, or

$$g_i = -(x^n)^{c_i}, \quad g_n = 1, \quad \sum c_i = \sum c_i^2 = 1. \qquad (14.11)$$

If $\varkappa = 0$ in solutions (14.9) and (14.10), the constants α_i and β also vanish and $\dot{T}_n$ degenerates into a flat space. Lobachevskii space is obtained as a particular solution where $\varkappa < 0$. The problem of which groups of motions are admitted by these $\dot{T}_n$ can be solved by means of the following theorem: if $g_i \neq g_j$, the space admits only the trivial system of operators ξ^i = constant, $\xi^n = 0$, $X_i f = p_i$ $(i<n)$; if, however, $g_i = g_j$, a rotation operator exists for each such pair.

14. Narlikar and Karmarkar [144] obtained solutions (14.11) in a different system of coordinates in the form

$$ds^2 = dt^2 - \nu^p\, dx^2 - \nu^q\, dy^2 - \nu^r\, dz^2, \quad \nu = 1+kt, \tag{14.12}$$

where k is constant, and p, q and r are constants connected by the relations $p+q+r = 2$, $pq+qr+rp = 0$.

15. Reducible spaces appear in the investigation of the problem of embedding Einstein spaces in flat spaces and spaces of constant curvature (Fialkow [109]) (cf. Chapter 5).

16. Einstein and Rosen [104], seeking a solution of (13.1) for cylindrical waves, obtained a metric admitting the orthogonal coordinate system

$$\left.\begin{array}{l} -g_{11} = g_{44} = A, \quad -g_{22} = B, \quad -g_{33} = C, \\ g_{\alpha\beta} = 0 \quad (\alpha \neq \beta), \quad A,\, B,\, C\, (x^1, x^4), \end{array}\right\} \tag{14.13}$$

where A, B and C are defined by a set of differential equations, which are reduced by the substitutions

$$\alpha = \ln A, \quad \beta = \frac{1}{2}\ln\left(\frac{B}{C}\right), \quad \gamma = \frac{1}{2}\ln(BC)$$

to the form

$$2\partial_{44}\gamma + \tfrac{1}{2}[(\partial_4\beta)^2 + 3(\partial_4\gamma)^2 + (\delta_1\beta)^2 - (\partial_1\gamma)^2 - 2\partial_1\alpha\,\partial_1\gamma - 2\partial_4\alpha\,\partial_4\gamma] = 0,$$

$$2(\partial_{11}\alpha - \partial_{44}\gamma) + 2(\partial_{11}\gamma - \partial_{44}\alpha) + [(\partial_1\beta)^2 + (\partial_1\gamma)^2 - (\partial_4\beta)^2 - (\partial_4\gamma)^2] = 0,$$

$$\partial_{11}\beta - \partial_{44}\beta + \partial_1\beta\,\partial_1\gamma - \partial_4\beta\,\partial_4\gamma = 0,$$

$$2\partial_{11}\gamma + \tfrac{1}{2}[(\partial_1\beta)^2 + 3(\partial_1\gamma)^2 + (\partial_4\beta)^2 - (\partial_4\gamma)^2 - 2(\partial_1\alpha\,\partial_1\gamma + \partial_4\alpha\,\partial_4\gamma)] = 0,$$

$$2\partial_{14}\gamma + \partial_1\beta\,\partial_4\beta + \partial_1\gamma\,\partial_4\gamma - 2(\partial_1\alpha\,\partial_4\gamma + \partial_4\alpha\,\partial_1\gamma) = 0.$$

This solution, based on formal considerations, has a physical interpretation.

Gravitational waves transmit energy, and the energy density becomes the source of a gravitational field deforming the metric; hence the waves must be considered in a space with its special metric. In 1954 Rosen considered those cases with

$$A = e^{2(\gamma-\psi)}, \quad B = e^{-2\psi}\varrho^2, \quad C = e^{2\psi},$$

where

$$\psi = \psi(\varrho, t), \quad \gamma = \gamma(\varrho, t)$$

with independent variables ϱ, φ, z and t ([206]). The problem of determining wave solutions was also attacked by Takeno, using Einstein's idea of obtaining plane waves, but not assuming a weak field [242]. Only solutions obtained for wave coordinates which cannot be eliminated by transformations are regarded as real.

17. Buchdahl investigated the static solutions of (13.1) and introduced the concept of solutions *reciprocal* to a given static solution ([215], [248]). Thus the solution reciprocal to Schwarzschild's is

$$ds^2 = -\gamma^2(\gamma^{-1}\,dr^2 + r^2\,d\theta^2 + r^2\sin^2\theta\,d\varphi^2) + p^{-1}\,dt^2, \tag{14.14}$$

where $p = 1-(2m/r)$. It can be shown that (14.14) differs only trivially from (14.1) (cf. Chapter 4).

18. At different times the author has investigated several solutions of (13.1), which will be given later ([157], [176], [178], [181], [182], [228], [243], [244], [261]).

19. Verma and Roy [246] indicated the following solution of eqns. (13.2):

$$ds^2 = -A(dx^2+dy^2+dz^2)+dt^2, \tag{14.15}$$

where

$$A^{\frac{1}{2}} = \frac{F(t)}{c_1(x^2+y^2+z^2)+c_2x+c_3y+c_4z+c_5},$$

all $\quad c_i = \text{const}, \quad \dfrac{d}{dt}(F^2) = \dfrac{1}{3\varkappa F^2}+\alpha, \quad \alpha = -4c_1c_5+c_2^2+c_3^2+c_4^2.$

Recently, a number of new solutions have been published, e.g. wave solutions by Bondi, Robinson, Trautman and Pirani ([353], [354], [413], [465]).

We have hitherto considered solutions in closed form of the field equations in free space or of eqns. (13.2). If, however, we consider the general field equations, we must determine the components of the energy-momentum tensor, as well as the unknown potentials $g_{\alpha\beta}$; the system becomes *undetermined.* It is, as already stated, non-linear with respect to the $g_{\alpha\beta}$. As, however, Newton's linear theory gives a very accurate description of some gravitational processes, e.g. celestial mechanics, it is natural to assume that this linear theory must be obtained from Einstein's theory as a first but close approximation, and that *at least for those problems where Newton's theory gives good agreement* with experiment, the more precise non-linear gravitational fields must only lead to additional second order effects. These considerations form the basis for the method of "linearization" of the general field equations (12.2) developed by Einstein and his co-workers on the one hand [167], and by Fock [111] and Papapetrou [238] on the other.

References [225] and [238] are devoted to a study of this problem, which is not considered in this book.

CHAPTER 3

GENERAL CLASSIFICATION OF GRAVITATIONAL FIELDS

THE historical summary in § 14 suggests the need for a general classification of all possible gravitational fields and in particular, Einstein spaces. Such a classification must be invariant in nature. In this chapter, a method of classifying all possible gravitational fields is used, which is based on a study of the algebraic structure of the space-curvature and energy-momentum tensors. The special cases of empty spaces and Einstein spaces are considered first, as this leads to a simple solution in the general arbitrary gravitational field.

§ 15. **Bivector Spaces**

We base the classification of Einstein spaces on a study of the algebraic structure of the curvature tensor; as the components of this tensor satisfy the identities (5.8)

$$R_{\alpha\beta\gamma\delta} = -R_{\beta\alpha\gamma\delta} = -R_{\alpha\beta\delta\gamma}, \quad R_{\alpha[\beta\gamma\delta]} = 0, \tag{15.1}$$

constituting the *complete* series of algebraic identities for $R_{\alpha\beta\gamma\delta}$, the classification must be based on (15.1). The *algebraic* study of $R_{\alpha\beta\gamma\delta}$ can be undertaken at any point of the V_n space, and we are only interested in certain special tensors at the point. We first show that the ensemble of all such tensors can be mapped on the set of all the tensors of another space, and we then carry out the investigation in this second space.

We single out all tensors satisfying the following two conditions: (1) the covariant and contravariant orders are even, (2) the covariant and contravariant indices can be divided into separate *pairs*, the tensor being antisymmetric for each pair. Examples of such tensors are *bivectors*, i.e. antisymmetric tensors of the second order specified by the components

$V^{\alpha\beta}$ or $V_{\alpha\beta}$, and the curvature tensor whether defined by the components $R_{\alpha\beta\gamma\delta}$, $R^{\alpha\beta}{}_{\gamma\delta}$ or $R^{\alpha\beta\gamma\delta}$. Such tensors are henceforward called *bitensors*.

The set of all such tensor fields in V_n is called the *bivector* set, and its representation at a given point—the *local bivector set*. We now consider any tensor belonging to the local bivector set and treat each antisymmetrical pair of indices $\alpha\beta$ as one *collective* index. We fix on one of the two possible pairs $\alpha\beta$ and $\beta\alpha$, $\alpha\beta$ for example, and ignore components with indices $\beta\alpha$, which differ only in sign. We then renumber all the collective indices: this can be done arbitrarily but the total number of indices $N = n(n-1)/2$.

Let us consider the transformation of component $\Pi^{\alpha\beta}$ of a bivector in a V_n space; it takes the form

$$\Pi^{\alpha'\beta'} = A^{\alpha'}_{\alpha} A^{\beta'}_{\beta} \Pi^{\alpha\beta} = 2A^{\alpha'}_{[\alpha} A^{\beta'}_{\beta]} \Pi^{\alpha\beta}$$

or in collective indices,

$$\Pi^{a'} = A^{a'}_{a} \Pi^{a}, \tag{15.2}$$

where $a = 1, \ldots, N$, and

$$A^{a'}_{a} \to 2A^{\alpha'}_{[\alpha} A^{\beta'}_{\beta]}. \tag{15.3}$$

Thus a bivector transformation in V_n has the same form as a vector transformation in E_N; however, not all linear non-singular transformations are permissible but only some subgroup of a centro-affine group, defined by conditions (15.3); this agrees with a theorem proved by Gurevich [172].

The ensemble of bivectors with contravariant components at a given point of a V_n space, defines in collective indices an ensemble of vectors, each with N contravariant components. We identify these vectors with the points of an N-dimensional manifold, which is an *affine manifold* E_N only if a Klein geometry with the group

$$\left.\begin{aligned} \eta^{a'} &= A^{a'}_{a}\eta^{a}, \quad \eta^{a} = A^{a}_{a'}\eta^{a'}, \\ |A^{a'}_{a}| &\neq 0, \quad A^{a}_{b'}A^{b'}_{c} = \delta^{a}_{c}, \end{aligned}\right\} \tag{15.4}$$

is valid within it; the $A^{a'}_{a}$ have the specific form (15.3) so that E_N, if such it be, has a special form.

The first two conditions (15.4) follow directly from (15.2); the third follows from the fact that the determinant

$$C = |A^{\alpha'}_{\beta} \cdot A^{\gamma'}_{\delta}| = |A^{\alpha'}_{\alpha}|^{2n} \quad (\alpha, \alpha' = 1, \ldots, n)$$

(Kagan [64], p. 410) and that, on the other hand, by interchange and linear combination of rows and columns, C can be reduced to the form

$$C = p\,|A_a^{\alpha'}| \quad (a, a' = 1, \ldots, N),$$

i.e. $|A_a^{a'}| \neq 0$, if $|A_\alpha^{\alpha'}| \neq 0$. The fourth condition (15.4) can easily be verified from the first two. Hence (15.4) applies and E_N exists; thus *every local bivector set in* V_n *can be mapped on a centro-affine* E_N $[N = \frac{1}{2}n(n-1)]$. The definition of this mapping shows that it is *isomorphic* relative to the operations of tensor addition, subtraction and multiplication (without contraction); contracted multiplication must be excluded because it can lead to tensors which do not admit such a mapping.

Hence, a local E_N with group (15.4) can be associated with every point in a V_n; such an E_N is called a *bivector space*. A tensor at a given point in V_n with the above mentioned properties defines a tensor of one-half the original order in E_N.

It is now possible to *metrize* the bivector space. Suppose that V_n contains a tensor field $T_{\alpha\beta\gamma\delta}$ such that

$$T_{\alpha\beta\gamma\delta} = T_{\gamma\delta\alpha\beta} = -T_{\beta\alpha\gamma\delta} = -T_{\alpha\beta\delta\gamma};$$

then all the E_N spaces corresponding to the points in V_n where the tensor field is defined can be provided with a metric tensor

$$T_{ab} \rightarrow T_{\alpha\beta\gamma\delta},$$

providing only that $|T_{\alpha\beta}| \neq 0$. This tensor is otherwise arbitrary, but it is most appropriate to link the metric introduced into the E_N with that of the original V_n. We therefore introduce the metric tensor

$$g_{ab} \rightarrow g_{\alpha\beta\gamma\delta} \equiv g_{\alpha\gamma}g_{\beta\delta} - g_{\alpha\delta}g_{\beta\gamma}, \tag{15.5}$$

where $g_{\alpha\beta}$ is the metric tensor of the V_n, and the collective indices are the antisymmetric pairs

$$\alpha\beta \rightarrow a, \quad \gamma\delta \rightarrow b.$$

The tensor $g_{ab}(a, b = 1, \ldots, N)$ is *symmetric* and *non-singular*, i.e. $|g_{ab}| \neq 0$; this is easily proved by reducing $g_{\alpha\beta}$ to the diagonal form at a given point in V_n and evaluating g_{ab} from formula (15.5). If $g_{\alpha\beta}$ is a positive-definite or negative-definite tensor, g_{ab} is *positive-definite*; if $g_{\alpha\beta}$ is indefinite, the metric g_{ab} may be indefinite, with signature depending on that of $g_{\alpha\beta}$. The symbol in (15.5) denoting mapping, could have been replaced by *connecting* quantities as follows,

$$g_{ab} = B_a^{\alpha\beta}B_b^{\gamma\delta}(g_{\alpha\gamma}g_{\beta\delta} - g_{\alpha\delta}g_{\beta\gamma}),$$

where the connecting quantities B represent a covariant vector in E_N and a contravariant bivector in the R_n corresponding to the given point in V_n.

We note that the matrix (g_{ab}) is the second compound matrix of $(g_{\alpha\beta})$. More generally, if we regard the operation of contraction $g_{\alpha\beta}V^\alpha$ in affine space as a linear vector function associating a hyperplane harmonically conjugate with respect to the hyperquadric $g_{\alpha\beta}x^\alpha x^\beta = 1$ with the point V^α, and apply this vector function to each vector V^k in the alternating product defining the *simple* multivector

$$V^{\alpha_1\ldots\alpha_p} \equiv \underset{[1}{V}{}^{\alpha_1}\cdots \underset{p]}{V}{}^{\alpha_p},$$

the multivector transforms into

$$V_{\beta_1\ldots\beta_p} = g_{\beta_1\ldots\beta_p\alpha_1\ldots\alpha_p}V^{\alpha_1\ldots\alpha_p},$$

where the tensor

$$g_{\beta_1\ldots\beta_p\alpha_1\ldots\alpha_p} = p!g_{[\alpha_1[\beta_1}g_{\alpha_2\beta_2}\cdots g_{\alpha_p]\beta_p]}$$

defines the pth compound form. When $p = 2$ this tensor is identical with g_{ab}. After introducing the tensor g_{ab} into the bivector affine space, E_N becomes a *metric* space R_N.

§16. Classification of Einstein Spaces

The concepts introduced in § 15 enable us to investigate the algebraic structure of the curvature tensor of any Riemann space (Petrov [171], [207]). Mapping the curvature tensor $R_{\alpha\beta\gamma\delta}$ of a V_n space in the metrized bivector space R_N, and noting (15.1), we obtain a symmetric tensor R_{ab} $(a, b = 1, \ldots, N)$ in R_N which can be associated with the λ-matrix

$$(R_{ab} - \lambda g_{ab}). \tag{16.1}$$

We can formulate a classification of V_n valid for any n by reducing this λ-matrix to the canonical form *using real coefficients* (cf. § 9). The *type* of space is determined by the *Segre characteristic* of the λ-matrix, and is the same throughout any domain in which the characteristic does not vary.

Suppose that the λ-matrix (16.1) corresponding to the point P of a V_n space has the set of elementary divisors

$$\begin{array}{llll}(\lambda-\lambda_1)^{m_1}, & (\lambda-\lambda_2)^{m_1'}, & (\lambda-\lambda_3)^{m_1''}, & \ldots, \\ (\lambda-\lambda_1)^{m_2}, & (\lambda-\lambda_2)^{m_2'}, & (\lambda-\lambda_3)^{m_2''}, & \ldots, \\ \cdots & \cdots & \cdots & \cdots\end{array}$$

Then the *Segre characteristic* corresponding to this λ-matrix

$$[(m_1, m_2, \ldots)(m_1', m_2', \ldots)(m_1'', m_2'', \ldots)\ldots] \tag{16.2}$$

defines the type of the space in a domain A with a common characteristic.

The invariants λ_i, the bases of the elementary divisors, are also the roots of the characteristic equation

$$|R_{ab} - \lambda g_{ab}| = 0. \tag{16.3}$$

Bases $\lambda_i(x)$ of the elementary divisors in the domain A can, however, be identical on certain k-dimensional manifolds. These k-surface manifolds are defined by equations

$$\lambda_i(x) - \lambda_j(x) = 0.$$

Thus if we take the hypersurface defined by

$$\lambda_1(x) - \lambda_2(x) = 0$$

in a region A where the elementary divisors are in general $(\lambda - \lambda_1)^p$ $(\lambda - \lambda_2)^q$, then the Segre characteristic is in general $[p, q, \ldots]$ in A, and $[(p, q), \ldots]$ on the given hypersurface; we state that this coincidence of bases *does not alter the type* of the V_n in the given domain. In determining the type of the λ-matrix of a V_n space, the tensor R_{ab} is limited only by the relations (15.1).

If V_n is an Einstein space, we must take into account the relations

$$R_{\alpha\beta} = \varkappa g_{\alpha\beta} \tag{16.4}$$

as well as (15.1). We now introduce a non-holonomic orthonormal polyad $\underset{\sigma}{\xi}{}^{\alpha}$ ($\sigma, \alpha = 1, 2, \ldots, n$) at a given point P in V_n such that

$$(g_{\alpha\alpha})_P = e_\alpha = \pm 1, \quad (g_{\alpha\beta}) = 0 \quad (\alpha \neq \beta), \tag{16.5}$$

where e_α is defined by the signature of the form $g_{\alpha\beta}\, dx^\alpha\, dx^\beta$. Then

$$g_{aa} = e_\alpha e_\beta, \quad g_{ab} = 0 \quad (a, b = 1, \ldots, N;\ a \neq b), \tag{16.6}$$

and the curvature tensor at the point P in the G_n space with *orthogonal* components connected by the relations

$$\left.\begin{aligned} &R_{\alpha\beta\gamma\delta} = -R_{\beta\alpha\gamma\delta} = -R_{\alpha\beta\delta\gamma}, \quad R_{\alpha[\beta\gamma\delta]} = 0, \\ &\sum_\sigma e_\sigma R_{\sigma\alpha\sigma\alpha} = \varkappa e_\alpha, \quad \sum_\sigma e_\sigma R_{\sigma\alpha\sigma\beta} = 0 \quad (\alpha \neq \beta) \end{aligned}\right\} \tag{16.7}$$

maps into the symmetric tensor R_{ab}. Expressing (16.7) in collective indices gives the *complete series* of algebraic conditions imposed on R_{ab} in R_N;

the type of space is then defined by the Segre characteristic of the λ-matrix (16.3). The number of possible types of V_n increases rapidly with n and depends on the possible signatures, i.e. on the choice of e_α. A detailed classification for $n = 4$ will be given later, the most interesting case as far as physical applications are concerned.

We now illustrate the above concepts by the simple example of a space of constant curvature S_n. If V_n is an S_n,

$$R_{\alpha\beta\gamma\delta} = K(g_{\alpha\gamma}g_{\beta\delta} - g_{\alpha\delta}g_{\beta\gamma}) \tag{16.8}$$

and hence the λ-matrix is

$$((K-\lambda)g_{ab}).$$

Introducing the orthonormal poly-ad (16.5) at the point P reduces this matrix to the diagonal form; hence all the elementary divisors are linear with bases equal to K and the Segre characteristic is [(1, ..., 1)]. Conversely, if the characteristic has this form, then (cf. § 9)

$$g_{ab} = \sum_\sigma e_\sigma \underset{\sigma}{\xi}_a \underset{\sigma}{\xi}_b, \quad R_{ab} = \sum_\sigma e_\sigma \underset{\sigma}{\lambda} \underset{\sigma}{\xi}_a \underset{\sigma}{\xi}_b = K g_{ab} \qquad (\underset{\sigma}{\lambda} = K),$$

so that (16.8) holds and the space has constant Riemann curvature for $n > 2$. Thus a V_n *with* $n > 2$ *is an* S_n *only when the Segre characteristic of the* λ*-matrix is* [(1, ..., 1)]. In particular, V_n is flat if the bases of the elementary divisors are zero.

Example

1. If the number of elementary divisors of the λ-matrix is p, and the rank of the matrix (R_{ab}) r, then $r \geq [n(n-1)/2] - p$ [261].

§ 17. Principal Curvatures

A simple geometrical interpretation of the bases λ_i of the elementary divisors of a λ-matrix can be given for any V_n ([261], pp. 36–40).

Let us consider the Riemann curvature of a V_n space in the two-dimensional direction defined by the simple bivector $V^{\alpha\beta} = \underset{[1}{V^\alpha} \underset{2]}{V^\beta}$:

$$K = \frac{R_{\alpha\beta\gamma\delta}V^{\alpha\beta}V^{\gamma\delta}}{g_{\alpha\beta\gamma\delta}V^{\alpha\beta}V^{\gamma\delta}}, \tag{17.1}$$

where $g_{\alpha\beta\gamma\delta}$ has the form (15.5). We now interpret the invariant K in (17.1) in a generalized sense by dropping the condition that *the bivector* $V^{\alpha\beta}$ *be simple*. This more general K defined by (17.1) is known as *the bivector curvature in the direction of the given bivector*. Mapping on to the bivector space R_N, we obtain

$$K = \frac{R_{ab}V^aV^b}{g_{ab}V^aV^b} \qquad (a, b = 1, \ldots, N). \tag{17.2}$$

We now try to determine the critical values of this invariant K, considered as a homogeneous function of zero order in V^a; this is equivalent to finding those vectors V^a for which K takes the critical values. We call the critical values of K the *principal curvatures* of the V_n space at a given point, and the vectors V^a corresponding to these values, *principal vectors* of R_N (or *principal non-simple bivectors* in V_n at the given point). If, in particular, a vector V^a defining the simple bivector $V^{\alpha\beta}$ in V_n, corresponds to a certain principal curvature, this principal curvature is identical with the Riemann curvature of V_n in the two-dimensional direction $V^{\alpha\beta}$.

There are two different problems depending on whether or not the bivector has to be simple. If we interpret K in the Riemannian sense, we have the problem of determining *the restricted principal directions*. If, however, we have the *bivector* curvature in mind, the problem reduces to determining the *unrestricted principal* V^a, in which case the necessary and sufficient conditions are

$$\frac{\partial}{\partial V^\alpha} K = 0. \tag{17.3}$$

Thus, it is assumed that K is a continuously differentiable function of V^a. It should be noted that if V_n has an *indefinite* metric, the principal directions may be *isotropic*,

$$g_{ab}V^aV^b = 0. \tag{17.4}$$

We postpone consideration of this case. Substituting from (17.2) into (17.3) we obtain

$$(R_{ab} - Kg_{ab})V^b = 0, \tag{17.5}$$

which are the equations determining the eigen directions of the tensor R_{ab} in R_N. The principal curvatures are then identical with the latent roots of (16.1); the λ are the *principal curvatures* of V_n. If $n > 3$, then in general a non-simple bivector corresponds to a principal direction in V_n.

We now suppose that (17.4) holds. We are interested only in those directions V^a for which K varies continuously; as (17.4) holds for an *isotropic* principle V^a, so must

$$R_{ab}V^aV^b = 0. \tag{17.6}$$

As K is continuously differentiable with respect to V^a we can calculate K for an isotropic principal direction by using the relation

$$K(V^a) = \lim_{dV^a \to 0} K(V^a + dV^a).$$

Let

$$\varphi = g_{ab} V^a V^b, \quad \psi = R_{ab} V^a V^b,$$

so that

$$K(V^a) = \lim_{dV^a \to 0} \frac{\psi(V_a + dV^a) - \psi(V^a)}{\varphi(V^a + dV^a) - \varphi(V^a)} = \lim \frac{\left(\sum \frac{\partial}{\partial V^c} \psi\right) dV^c + \dots}{\left(\sum \frac{\partial}{\partial V^c} \varphi\right) dV^c + \dots}.$$

As this limit cannot depend on the way in which dV^a varies,

$$K(V^a) = \frac{\frac{\partial}{\partial V^c} \psi}{\frac{\partial}{\partial V^c} \varphi} = \frac{R_{cb} V^b}{g_{cb} V^b} (b, c = 1, \dots, N),$$

so that (17.5) still holds; an isotropic principal bivector also determines a principal direction of R_{ab}, and the corresponding principal curvature is identical with the base of a non-linear elementary divisor (cf. § 9). Thus *the principal bivector directions of* V_n *are identical with the principal directions* R_{ab} *in* R_N, *and the principal curvatures with the bases of the elementary divisors; hence their number is* $\leqslant N = \frac{1}{2}n(n-1)$. When the metric is indefinite, the principal curvatures and the corresponding principal bivectors in V_n are, in general, *complex*.

§18. The Classification of Einstein Spaces for $n = 4$

Four-dimensional Einstein spaces over a field of real numbers can have three fundamentally different signatures, apart from overall multiplication by -1:

$$\left.\begin{array}{ll} (\alpha) & + + + + , \\ (\beta) & - - - + , \\ (\gamma) & - - + + . \end{array}\right\} \qquad (18.1)$$

We consider all three possibilities. In case (α), *a positive-definite metric*, we can introduce a non-holonomic orthonormal tetrad at a given point, relative to which $g_{\alpha\alpha} = 1$, $g_{\alpha\beta} = 0$ and, hence,

$$g_{aa} = 1, \quad g_{ab} = 0 \qquad (a \neq b).$$

The λ-matrix has then only simple elementary divisors, and the principal curvatures are real (cf. § 9); the Segre characteristic must be [111, 111]. *There is therefore only one type of Einstein space with the signature* (++++).

We now consider (β), the *Minkowski* signature. We can introduce a *real* non-holonomic orthonormal tetrad at any given point P of a T (or $\dot{T}$) space defined by a gravitational field, relative to which

$$(g_{\alpha\beta})_P = \begin{pmatrix} -1 & 0 & 0 & 0 \\ 0 & -1 & 0 & 0 \\ 0 & 0 & -1 & 0 \\ 0 & 0 & 0 & 1 \end{pmatrix}. \tag{18.2}$$

Substituting from (18.2) into (15.5), we obtain

$$(g_{ab}) = \left\{ \begin{matrix} -1 & 0 & 0 & 0 & 0 & 0 \\ 0 & -1 & 0 & 0 & 0 & 0 \\ 0 & 0 & -1 & 0 & 0 & 0 \\ 0 & 0 & 0 & 1 & 0 & 0 \\ 0 & 0 & 0 & 0 & 1 & 0 \\ 0 & 0 & 0 & 0 & 0 & 1 \end{matrix} \right\} \tag{18.3}$$

in the corresponding R_6 space, i.e. g_{ab} defines an indefinite metric. The following system of numbering the collective indices in R_6 is used throughout the book:

$$14 \to 1,\ \ 24 \to 2,\ \ 34 \to 3,\ \ 23 \to 4,\ \ 31 \to 5,\ \ 12 \to 6. \tag{18.4}$$

We now prove the following theorem: *The matrix* (R_{ab}) *for the orthonormal tetrad* (18.2) *is doubly symmetric.*

As the contravariant and covariant components of the metric tensor are identical in the orthonormal tetrad (18.2), setting down the field equations $R_{\alpha\beta} = \varkappa g_{\alpha\beta}$, gives the two sets

$$\text{(I)} \ \sum_{\sigma} \underset{\sigma}{e}\, R_{\alpha\sigma\beta\sigma} = 0 \qquad (\alpha \neq \beta),$$

$$\text{(II)} \ \sum_{\sigma} \underset{\sigma}{e}\, R_{\alpha\sigma\alpha\sigma} = \varkappa e_{\alpha}, \qquad e_{\alpha} = \pm 1.$$

Putting $\alpha, \beta = 1, 2, 3, 4$ in equation (I) and changing over to the collective indices (18.4), we obtain

$$\begin{aligned} R_{12}+R_{45} &= R_{13}+R_{46} = R_{23}+R_{56} = R_{15}-R_{24} \\ &= R_{16}-R_{34} = R_{26}-R_{35} = 0. \end{aligned} \tag{18.5}$$

Equation (II) in *collective indices* reduces to

$$R_{11}+R_{44} = R_{22}+R_{55} = R_{33}+R_{66} = 0,$$
$$R_{11}+R_{22}+R_{33} = -\varkappa. \tag{18.6}$$

We must also take into account the identity $R_{\alpha[\beta\gamma\delta]}=0$, containing *all the restraints* imposed on the tensor R_{ab}; *in collective indices, this gives*

$$R_{14}+R_{25}+R_{36} = 0. \tag{18.7}$$

Let

$$R_{ab} = m_{ab}, \quad R_{a,\,b+3} = n_{ab}, \quad a, b \leqslant 3. \tag{18.8}$$

Equations (18.5) and (18.6) show that

$$(R_{ab}) = \left(\begin{array}{c|c} M & N \\ \hline N & -M \end{array}\right), \tag{18.9}$$

where M and N are two *symmetric* 3×3 matrices,

$$M = (m_{ab}),\ N = (n_{ab}),\ m_{ab} = m_{ba},\ n_{ab} = n_{ba} \quad (a, b = 1, 2, 3), \tag{18.10}$$

which proves the theorem.

The following relations also hold:

$$\sum_{s=1}^{3} m_{ss} = -\varkappa, \quad \sum_{s=1}^{3} n_{ss} = 0; \tag{18.11}$$

Kagan [79] obtained doubly symmetric matrices in studying the group of Lorentz transformations, with, however, the *additional condition that they should be orthogonal.* This type of matrix, with the same assumption of orthogonality, has also been studied by Dubnov [82] and Lopshits [85].

Differentiating eqns. (13.2) k times covariantly, we obtain

$$R_{\alpha\beta,\,\lambda_1 \dots \lambda_k} = 0.$$

Introducing collective indices as above in the orthonormal tetrad (18.2) for antisymmetric pairs before the comma, we can repeat the previous analysis except that $\varkappa$ must now be put equal to zero, and the components m_{ab} and n_{ab} given additional indices $\lambda_1, \dots, \lambda_k$. Hence, *relative to the orthonormal tetrad* (18.2), *the matrix* $(R_{ab,\,\lambda_1,\dots,\lambda_k})$ *is doubly symmetric for any fixed combination of indices* $\lambda_1, \dots, \lambda_k$ $(\lambda_1, \dots, \lambda_k = 1, 2, 3, 4)$; i.e.

$$(R_{ab,\,\lambda_1 \dots \lambda_k}) = \left(\begin{array}{c|c} M,_{\lambda_1 \dots \lambda_k} & N,_{\lambda_1 \dots \lambda_k} \\ \hline N,_{\lambda_1 \dots \lambda_k} & -M,_{\lambda_1 \dots \lambda_k} \end{array}\right), \tag{18.12}$$

where

$$\sum_{a=1}^{3} m_{aa,\,\lambda_1\ldots\lambda_k} = \sum_{a=1}^{3} n_{aa,\,\lambda_1\ldots\lambda_k} = 0 \tag{18.13}$$

and

$$m_{ab,\,\lambda_1\ldots\lambda_k} = m_{ba,\,\lambda_1\ldots\lambda_k} \qquad n_{ab,\,\lambda_1\ldots\lambda_k} = n_{ba,\,\lambda_1\ldots\lambda_k}$$
$$(a, b = 1, \ldots, 6; \quad \lambda_i = 1, 2, 3, 4; \quad i = 1, 2, \ldots).$$

We now prove a fundamental theorem which solves the problem of the general classification of four-dimensional Einstein spaces with signature $(---+)$ based on the algebraic struction of the curvature tensor.

THEOREM. *There are only three types of space defined by gravitational fields* ($R_{\alpha\beta} = \varkappa g_{\alpha\beta}$) *with signature* $(---+)$.

As (g_{ab}) for (18.2) can be written as

$$(g_{ab}) = \begin{pmatrix} -\varepsilon & 0 \\ 0 & \varepsilon \end{pmatrix},$$

where ε is the three-dimensional unit matrix, the λ-matrix is of the form

$$(R_{ab} - \lambda g_{ab}) = \left(\begin{array}{c|c} M+\lambda\varepsilon & N \\ \hline N & -M-\lambda\varepsilon \end{array}\right).$$

We now apply *elementary transformations* which leave the elementary divisors and the Segre characteristic of the λ-matrix unchanged [91], p. 34). Multiply the submatrices in the second column by i and add to those in the first column; we obtain

$$\left(\begin{array}{c|c} M+iN+\lambda\varepsilon & N \\ \hline -i(M+iN+\lambda\varepsilon) & -M-\lambda\varepsilon \end{array}\right).$$

Multiply the submatrices in the first row by i and add to the corresponding submatrices of the second row; the matrix is reduced to the form

$$\left(\begin{array}{c|c} M+iN+\lambda\varepsilon & N \\ \hline 0 & -M+iN-\lambda\varepsilon \end{array}\right).$$

Multiply the submatrices of the first column by $i/2$ and add to the corresponding submatrices of the second column, and then multiply the submatrices in the second row by $i/2$ and add to the corresponding submatrices in the first row; we obtain the following matrix, equivalent to the original λ-matrix to within elementary transformations

$$\left(\begin{array}{c|c} M+iN+\lambda\varepsilon & 0 \\ \hline 0 & M-iN+\lambda\varepsilon \end{array}\right) \equiv \begin{pmatrix} Q(\lambda) & 0 \\ 0 & \bar{Q}(\lambda) \end{pmatrix}. \tag{18.14}$$

The problem has been reduced to the investigation of two 3×3 matrices $Q(\lambda)$ and $\bar{Q}(\lambda)$, the corresponding elements of which are complex conjugates. The elementary divisors of these matrices are therefore also complex conjugates, and their Segre characteristics identical; the Segre characteristic of our λ-matrix thus divides into two *identical parts*.

The three-by-three matrix $Q(\lambda)$ can have one of the only three possible Segre characteristics: 1°. [111], 2°. [21], 3°. [3]; we disregard special cases of elementary divisors with a common base, when numbers in the square brackets must be enclosed in round brackets, e.g. [(11)1], [(21)]. As the matrix $\bar{Q}(\lambda)$ has the same Segre characteristic, the original λ-matrix can have one of the following Segre characteristics only,

$$1°.\ [1\ 1\ 1,\ \overline{1\ 1\ 1}],\qquad 2°.\ [2\ 1,\ \overline{2\ 1}],\qquad 3°.\ [3,\ 3].$$

The bar over the second half of a Segre characteristic denotes elementary divisors with bases which are complex conjugates of those of the elementary divisors corresponding to the first half. No bar is needed for the third type, as the elementary divisors have real bases in this case, as will be shown in the next section. In the following chapter, examples of spaces of all three types will be given. This proves the theorem, derived by the author in 1949 and first published in 1950 [171], [176], [178]. In 1954, the author provided the second proof reproduced above [207]; simultaneously, Norden gave a third based on his investigations of biplanar spaces [208]. A fourth version (for $\varkappa = 0$) was given by Géhéniau in 1957 [262]. The physical applications of this theorem have been discussed by Pirani [258] and others [308], [321], [322], [324].

From this point on, $T(R_{\alpha\beta} = 0)$ and $\mathring{T}(R_{\alpha\beta} = \varkappa g_{\alpha\beta})$ spaces are given an index denoting their type; T_i, $\mathring{T}_i (i = 1, 2, 3)$.

We now consider case (γ), with metric signature $(- - + +)$.

For such metrics, the matrix $(g_{\alpha\beta})$ at a point can be reduced to

$$(g_{\alpha\beta}) = \begin{pmatrix} -1 & 0 & 0 & 0 \\ 0 & -1 & 0 & 0 \\ 0 & 0 & 1 & 0 \\ 0 & 0 & 0 & 1 \end{pmatrix} = (g^{\alpha\beta}), \tag{18.15}$$

relative to a no n-holonomic orthonormal tetrad, so that

$$(g_{ab}) = \begin{pmatrix} -1 & 0 & 0 & 0 & 0 & 0 \\ 0 & -1 & 0 & 0 & 0 & 0 \\ 0 & 0 & 1 & 0 & 0 & 0 \\ 0 & 0 & 0 & -1 & 0 & 0 \\ 0 & 0 & 0 & 0 & -1 & 0 \\ 0 & 0 & 0 & 0 & 0 & 1 \end{pmatrix}. \tag{18.16}$$

Setting down the equations $R_{\alpha\beta} = \varkappa g_{\alpha\beta}$ for the metric (18.15), and using the notation of (18.8), we construct the matrix (R_{ab}) as follows:

$$(R_{ab}) = \left(\begin{array}{ccc|ccc} m_{11} & m_{12} & m_{13} & n_{11} & n_{12} & n_{13} \\ m_{21} & m_{22} & m_{23} & n_{21} & n_{22} & n_{23} \\ m_{31} & m_{32} & m_{33} & -n_{31} & -n_{32} & n_{33} \\ \hline n_{11} & n_{12} & -n_{13} & m_{11} & m_{12} & -m_{13} \\ n_{21} & n_{22} & -n_{23} & m_{21} & m_{22} & -m_{23} \\ n_{31} & n_{32} & n_{33} & -m_{31} & -m_{32} & m_{33} \end{array}\right), \tag{18.17}$$

where

$$m_{ab} = m_{ba}, \quad n_{ab} = n_{ba},$$

and

$$\left.\begin{aligned} -m_{11}-m_{22}+m_{33} &= \varkappa, \\ n_{11}+n_{22}+n_{33} &= 0. \end{aligned}\right\} \tag{18.18}$$

Hence the λ-matrix defining the structure of the space is, by virtue of (18.16),

$$(R_{ab} \quad \lambda g_{ab}) = \left(\begin{array}{ccc|ccc} m_{11}+\lambda & m_{12} & m_{13} & n_{11} & n_{12} & n_{13} \\ m_{21} & m_{22}+\lambda & m_{23} & n_{21} & n_{22} & n_{23} \\ m_{31} & m_{32} & m_{33}-\lambda & -n_{31} & -n_{32} & n_{33} \\ \hline n_{11} & n_{12} & -n_{13} & m_{11}+\lambda & m_{12} & -m_{13} \\ n_{21} & n_{22} & -n_{23} & m_{21} & m_{22}+\lambda & -m_{23} \\ n_{31} & n_{32} & n_{33} & -m_{31} & -m_{32} & m_{33}-\lambda \end{array}\right). \tag{18.19}$$

We now carry out the following successive operations on the λ-matrix. (1) Multiply the sixth row and sixth column by -1; (2) multiply the last three columns by e and add them to the corresponding first three columns; (3) multiply the first three rows by $-e$ and add to the corresponding last three rows, (4) multiply the first three columns by $-e/2$ and add to the corresponding last three columns; (5) multiply the last three rows by $e/2$, and add to the corresponding first three rows.

If we now impose the condition that

$$e^2 = 1,$$

and introduce the so-called *duplex numbers* $a + eb$ (Rozenfel'd [229], pp. 441–4), where a and b are real numbers, then

$$(R_{ab} - \lambda g_{ab}) \approx \begin{pmatrix} P(\lambda) & 0 \\ 0 & \bar{P}(\lambda) \end{pmatrix},$$

where

$$P(\lambda) = \begin{pmatrix} m_{11}+en_{11}+\lambda & m_{12}+en_{12} & m_{13}-en_{13} \\ m_{21}+en_{21} & m_{22}+en_{22}+\lambda & m_{23}-en_{23} \\ m_{31}-en_{31} & m_{32}-en_{32} & m_{33}-en_{33}-\lambda \end{pmatrix},$$

and

$$m_{\alpha\beta} = m_{\beta\alpha}, \quad n_{\alpha\beta} = n_{\beta\alpha},$$

and $\bar{P}(\lambda)$ is the *duplex conjugate* matrix, obtained by replacing e by $-e$ in all the elements of $P(\lambda)$. A peculiarity of the algebra of duplex numbers is the existence of non-zero elements with "norm" $z\bar{z} = 0$; they have no reciprocal elements and are *divisors of zero*. Dual numbers with $e^2 = 0$ were first considered by Clifford [65], and were systematically applied to geometry by Kotel'nikov [8], [11].

The algebra of duplex numbers [leaving aside (18.19)], is analogous to that of complex numbers. In particular, if $f(z_1, z_2, \ldots, z_n)$ is a rational function, then $f(\bar{z}_1, \bar{z}_2, \ldots, \bar{z}_n) = \bar{f}(z_1, z_2, \ldots, z_n)$. Hence the elementary divisors, giving the principal curvatures of $P(\lambda)$ and $\bar{P}(\lambda)$, are duplex conjugate and their Segre characteristics are therefore identical. Hence the following theorem holds; *as far as the structure of the curvature tensor is concerned there are only three types of Einstein space with a signature* $(- - + +)$; these types are defined by the Segre characteristics of the λ-matrix $(R_{ab} - \lambda g_{ab})$, as follows:

$$1°.\ [1\ 1\ 1,\ \overline{1\ 1\ 1}], \quad 2°.\ [2\ 1,\ \overline{2\ 1}], \quad 3°.\ [3,\ 3].$$

Four-dimensional Einstein spaces with signatures $(- - - +)$ and $(- - + +)$ both have the same number of different types, but whereas the λ-matrix for the Minkowski signature is mapped onto a *three-dimensional complex metric space* by means of (18.14), the mapping for the second signature takes place onto a *three-dimensional duplex metric space.*

The matrix (18.19) is not doubly symmetric over the field of complex numbers.

§19. The Canonical Form of the Matrix (R_{ab}) for T_i and $\dot{T}_i$ Spaces

As the number of independent components of the curvature tensor of a V_n space is $n^2(n^2-1)/12$, there are twenty such components when $n=4$. The ten field equations for $\dot{T}_i$ spaces reduce this number to 10. We can, however, still choose six parameters to define an orthonormal tetrad. These parameters may be interpreted geometrically as the three spatial and three Lorentz rotations possible in choosing the orthonormal tetrad. Thus, *the curvature tensor at each point of a $\dot{T}$ space is completely defined if we are given not more than four parameters* (not counting the parameter $\varkappa$ for $\dot{T}$ spaces). We must link the way the components depend on these parameters with the type of the space. We consider $\dot{T}_i$ spaces, as the results for T_i can be obtained by putting $\varkappa=0$.

1. *$\dot{T}_1$ spaces with Segre characteristic* $[111, \overline{111}]$. As the Segre characteristic is simple, the tensor $R_{ab}(a, b = 1, \ldots, 6)$ in the bivector space has six non-isotropic mutually orthogonal eigen directions (cf. § 9). These directions in R_6 for a given point in $\dot{T}_1$ determine bivectors with a specific structure.

We denote the coordinate vectors of a real orthonormal tetrad at a point of $\dot{T}_1$ by $\underset{\alpha}{\xi}^\sigma(\alpha, \sigma = 1, 2, 3, 4)$, and the *simple* bivectors $\underset{[\alpha}{\xi}^\sigma\underset{\beta]}{\xi}^\tau$ by $\underset{\alpha\beta}{\xi}^{\sigma\tau}$. In R_6, these *simple* bivectors define six independent real non-isotropic mutually orthogonal coordinate vectors $\underset{c}{\xi}^a = \delta^a_c$, and any vector in R_6, including the eigen vectors of R_{ab}, can be expressed in terms of the vectors $\underset{c}{\xi}^a$. We now show that the eigenvectors, which are uniquely determined only if the roots of the equation $|R_{ab}-\lambda g_{ab}|=0$ are distinct, are of the form

$$w^a = \sigma(\underset{1}{\xi}^a \pm i\underset{4}{\xi}^a)+\mu(\underset{2}{\xi}^a \pm i\underset{5}{\xi}^a)+\nu(\underset{3}{\xi}^a \pm i\underset{6}{\xi}^a) \qquad (a=1, \ldots, 6). \quad (19.1)$$

For w^a is an eigenvector of R_{ab} provided that

$$(R_{ab}-\lambda g_{ab})w^b = 0. \quad (19.2)$$

Equations (18.8) and (18.9), indicating the double symmetry of the λ-matrix, reduce this set of six equations to the three following,

$$\sigma(m_{a1} \pm in_{a1}+\lambda\delta_{a1})+\mu(m_{a2} \pm in_{a2}+\lambda\delta_{a2})+\nu(m_{a3} \pm in_{a3}+\lambda\delta_{a3}) = 0$$
$$(a = 1, 2, 3).$$

To ensure that σ, μ and ν are non-zero solutions of this set, it is necessary and sufficient that λ be a root of either equation $Q(\lambda)=0$ or $\bar{Q}(\lambda)=0$, i.e. a latent root, which proves the original statement.

To a vector W^a in R_6 for a given point of $\overset{*}{T}_1$ there corresponds a non-singular bivector in $\overset{*}{T}_1$, as follows:

$$W^{\alpha\beta} = \sigma(\underset{14}{\xi^{\alpha\beta}} \pm i\underset{23}{\xi^{\alpha\beta}}) + \mu(\underset{24}{\xi^{\alpha\beta}} \pm i\underset{31}{\xi^{\alpha\beta}}) + \nu(\underset{34}{\xi^{\alpha\beta}} \pm i\underset{12}{\xi^{\alpha\beta}}). \tag{19.3}$$

It is clear that any orthogonal transformation converts $W^{\alpha\beta}$ into a bivector with the same structure, and replaces σ, μ and ν by $\overset{*}{\sigma}$, $\overset{*}{\mu}$ and $\overset{*}{\nu}$ respectively, the norm of the bivector remaining invariant, i.e.

$$\sigma^2 + \mu^2 + \nu^2 = \overset{*}{\sigma}{}^2 + \overset{*}{\mu}{}^2 + \overset{*}{\nu}{}^2.$$

Suppose that the roots $\underset{s}{\lambda}(s = 1, 2, 3)$ of the equation $| R_{ab} - \lambda g_{ab} | = 0$ correspond to the eigenvectors $\underset{s}{W^a}$; then the roots $\underset{s+3}{\lambda}$ must correspond to the complex conjugate eigenvectors $\underset{s}{\overline{W^a}}$.

We represent the bivector $\underset{1}{W^{\alpha\beta}}$ as a combination of two real bivectors

$$\underset{1}{W^{\alpha\beta}} = \underset{1}{V^{\alpha\beta}} + i\underset{1}{\overset{*}{V}{}^{\alpha\beta}}.$$

Hence

$$\underset{4}{W^{\alpha\beta}} = \underset{1}{V^{\alpha\beta}} - i\underset{1}{\overset{*}{V}{}^{\alpha\beta}}.$$

Putting $\sigma = \underset{1}{a} + i\underset{1}{b}$, $\mu = \underset{2}{a} + i\underset{2}{b}$, $\nu = \underset{3}{a} + i\underset{3}{b}$, where the $\underset{s}{a}$ and $\underset{s}{b}$ are real $(s = 1, 2, 3)$, we find that

$$\underset{1}{V^{\alpha\beta}} = \underset{1}{a}\underset{14}{\xi^{\alpha\beta}} + \underset{2}{a}\underset{24}{\xi^{\alpha\beta}} + \underset{3}{a}\underset{34}{\xi^{\alpha\beta}} - \underset{1}{b}\underset{23}{\xi^{\alpha\beta}} - \underset{2}{b}\underset{31}{\xi^{\alpha\beta}} - \underset{3}{b}\underset{12}{\xi^{\alpha\beta}},$$

$$\underset{1}{\overset{*}{V}{}^{\alpha\beta}} = \underset{1}{b}\underset{14}{\xi^{\alpha\beta}} + \underset{2}{b}\underset{24}{\xi^{\alpha\beta}} + \underset{3}{b}\underset{34}{\xi^{\alpha\beta}} + \underset{1}{a}\underset{23}{\xi^{\alpha\beta}} + \underset{2}{a}\underset{31}{\xi^{\alpha\beta}} + \underset{3}{a}\underset{12}{\xi^{\alpha\beta}}.$$

As $\underset{1}{W^a}$ is a non-isotropic vector in R_6 (cf. § 9), it can always be taken as a unit vector, i.e. $g_{ab}\underset{1}{W^a}\underset{1}{W^b} = 1$ $(a, b = 1, \ldots, 6)$; hence

$$\sum_{s=1}^{3} \underset{s}{a}\underset{s}{b} = 0, \quad \sum_{s=1}^{3} (\underset{s}{b}{}^2 - \underset{s}{a}{}^2) > 0. \tag{19.4}$$

We now introduce the concepts of *degree of parallelism* and *degree of orthogonality* of bivectors, and the corresponding symbols. Suppose we are given two generally *non-simple* bivectors which therefore lie in flat clusters ε_p and ε_q with dimensions $p \leqslant q \leqslant 4$, when $n = 4$. If the clusters containing these bivectors have k common directions, the degree of parallelism of these bivectors is said to be k/p. Similarly, if ε_p contains s independent directions which are orthogonal to ε_q, i.e. to all the directions in ε_q, these bivectors are said to be *s/p-orthogonal*.

We now show that the bivectors $\underset{1}{V}^{\alpha\beta}$ and $\underset{1}{\overset{*}{V}}{}^{\alpha\beta}$ are *simple*, i.e. lie in two-dimensional flat clusters. To ensure that the bivector, i.e. antisymmetric tensor $u^{\alpha\beta}$, is simple, i.e. equal to the alternated product of two vectors $\eta^{[\alpha}\gamma^{\beta]}$, the following condition is necessary and sufficient when $n = 4$:

$$u^{14}u^{23}+u^{24}u^{31}+u^{34}u^{12} = 0. \tag{19.5}$$

Expanding the determinant

$$\begin{vmatrix} \underset{1}{u}^1 & \underset{1}{u}^2 & \underset{1}{u}^3 & \underset{1}{u}^4 \\ \underset{2}{u}^1 & \underset{2}{u}^2 & \underset{2}{u}^3 & \underset{2}{u}^4 \\ \underset{1}{u}^1 & \underset{1}{u}_2 & \underset{1}{u}^3 & \underset{1}{u}_4 \\ \underset{2}{u}^1 & \underset{2}{u}^2 & \underset{2}{u}^3 & \underset{2}{u}^4 \end{vmatrix} \equiv 0$$

in minors of the first two rows, we obtain (19.5). Conversely, if (19.5) holds, it is always possible to find two vectors η^α and γ^α such that $u^{\alpha\beta} = \eta^{[\alpha}\gamma^{\beta]}$; when $n = 4$ this can be verified directly. Condition (19.5) is satisfied for $\underset{1}{V}^{\alpha\beta}$ and $\underset{1}{\overset{*}{V}}{}^{\alpha\beta}$ because of the first condition (19.4); both bivectors $\underset{1}{V}^{\alpha\beta}$ and $\underset{1}{\overset{*}{V}}{}^{\alpha\beta}$ lie in two-dimensional flat clusters.

We now show that these bivectors are 0/2-*parallel.* They cannot be 2/2-parallel, as their components would then be proportional and thus reduce to zero. Neither can they be 1/2-parallel, since then $\underset{1}{W}^{\alpha\beta}$ would be a simple bivector, making condition (19.5) contradict (19.4). Hence there remains only one possibility: 0/2-parallel.

These bivectors are 2/2-orthogonal, i.e. any vector lying in the plane of $\underset{1}{V}^{\alpha\beta}$ is orthogonal to any vector lying in the plane of the bivector $\underset{1}{\overset{*}{V}}{}^{\alpha\beta}$. For this to be true, it is necessary and sufficient that the following equation should hold for any α and β (cf. § 19, example 6):

$$\underset{1}{V}_{\alpha\beta}\underset{1}{\overset{*}{V}}{}^{\beta\delta} = 0.$$

Clearly this condition follows from (19.4).

Let us consider the simple bivector $\underset{1}{V}^{\alpha\beta}$. Because of (19.4), its norm is

$$g_{ab}\underset{1}{V}^a\underset{1}{V}^b = \sum_1^3 (b_s^2-a_s^2) > 0.$$

Two real orthogonal and non-isotropic vectors η^α and γ^α can always be selected in the plane of this real bivector. Then the norm of the bivector is $2\eta_\sigma\eta^\sigma\gamma_\tau\gamma^\tau$. Hence these two vectors have norms of the same sign. The signature of the form $(---+)$ and the *law of inertia* for real transformations, in conjunction, show that the norms of these two vectors are negative. After renumbering the two vectors, we can take them as the orthonormal tetrad vectors $\underset{2}{\xi}^{\alpha}$ and $\underset{3}{\xi}^{\alpha}$.

Similarly, we determine two orthogonal real and non-isotropic vectors in the plane $\underset{1}{\overset{*}{V}}{}^{\alpha\beta}$, but with norms of opposite sign since $g_{ab}\underset{1}{\overset{*}{V}}{}^{a}\underset{1}{\overset{*}{V}}{}^{b} < 0$; we denote these vectors by $\underset{1}{\xi}^{\alpha}$ and $\underset{4}{\xi}^{\alpha}$. In this non-holonomic reference system

$$\underset{1}{W}^{\alpha\beta} = \underset{14}{\xi}^{\alpha\beta} + i\underset{23}{\xi}^{\alpha\beta}, \qquad \underset{4}{W}^{\alpha\beta} = \underset{14}{\xi}^{\alpha\beta} - i\underset{23}{\xi}^{\alpha\beta}.$$

We note that the orthonormal tetrad $\{\underset{\sigma}{\xi}^{\alpha}\}$ is chosen to within a rotation in the plane $\{\underset{2}{\xi}\,\underset{3}{\xi}\}$ and a Lorentz rotation in the plane $\{\underset{1}{\xi}\,\underset{4}{\xi}\}$. As $\underset{2}{W}^{\alpha\beta}$ must be orthogonal to $\underset{1}{W}^{\alpha\beta}$ (cf. § 9),

$$\underset{2}{W}^{\alpha\beta} = \mu_2(\underset{24}{\xi}^{\alpha\beta} + i\underset{31}{\xi}^{\alpha\beta}) + \nu_2(\underset{34}{\xi}^{\alpha\beta} + i\underset{12}{\xi}^{\alpha\beta}).$$

This arbitrariness can always be used to convert ν_2 to zero, and μ_2 to 1, without altering the form of $\underset{1}{W}^{\alpha\beta}$ and therefore of $\underset{2}{W}^{\alpha\beta}$ (cf. example 1 in this section). Moreover, $\underset{3}{W}^{\alpha\beta}$ is orthogonal to the first two bivectors (in R_6), so that

$$\underset{1}{W}^{\alpha\beta} = \underset{14}{\xi}^{\alpha\beta} + i\,\underset{23}{\xi}^{\alpha\beta}, \qquad \underset{2}{W}^{\alpha\beta} = \underset{24}{\xi}^{\alpha\beta} + i\,\underset{31}{\xi}^{\alpha\beta}, \qquad \underset{3}{W}^{\alpha\beta} = \underset{34}{\xi}^{\alpha\beta} + i\,\underset{12}{\xi}^{\alpha\beta}.$$

Because of complex conjugacy

$$\underset{4}{W}^{\alpha\beta} = \overline{\underset{1}{W}}{}^{\alpha\beta}, \qquad \underset{5}{W}^{\alpha\beta} = \overline{\underset{2}{W}}{}^{\alpha\beta}, \qquad \underset{6}{W}^{\alpha\beta} = \overline{\underset{3}{W}}{}^{\alpha\beta}.$$

If conditions (19.2) are set down for all these bivectors, whilst noting that $\underset{p}{\xi}^{a} = \delta^{a}_{p}$ $(a, p = 1, \ldots, 6)$, we easily find that

$$m_{ss} = \alpha_s, \qquad m_{st} = n_{st} = 0 \qquad (s \neq t), \qquad n_{ss} = \beta_s, \qquad (s = 1, 2, 3),$$

where α_s and β_s are the real and imaginary parts of the principle curvatures of $\overset{*}{T}_1$ (i.e. the roots of the characteristic equation). Hence *for a space of type* $\overset{*}{T}_1$, *the matrix* (R_{ab}) $(a, b, = 1, \ldots, 6)$ *in some non-holonomic orthonormal tetrad has the following canonical form:*

$$(R_{ab}) = \begin{pmatrix} M & N \\ N & -M \end{pmatrix}, \quad M = \begin{pmatrix} \alpha_1 & 0 & 0 \\ 0 & \alpha_2 & 0 \\ 0 & 0 & \alpha_3 \end{pmatrix}, \quad N = \begin{pmatrix} \beta_1 & 0 & 0 \\ 0 & \beta_2 & 0 \\ 0 & 0 & \beta_3 \end{pmatrix}, \tag{19.6}$$

where

$$\sum_1^3 \alpha_s = -\varkappa, \quad \sum_1^3 \beta_s = 0. \tag{19.7}$$

There are thus four independent parameters defining the structure of the space *in an invariant manner*; this number is the maximum and is smaller for $\overset{*}{T}_2$ and $\overset{*}{T}_3$.

2. $\overset{*}{T}_2$ *spaces; λ-matrix with Segre characteristic* $[21, \overline{21}]$. As was shown above, the eigenvectors and invariant clusters of the λ-matrix must be those of the matrices $Q(\lambda)$ and $\bar{Q}(\lambda)$. We need therefore only consider the λ-matrix $Q(\lambda)$, for example, which has complex elements and Segre characteristic [21]. In this case the tensor

$$p_{ij} = m_{ij} + in_{ij}$$

in a three-dimensional space (cf. § 9) has one non-isotropic eigen direction $\underset{1}{W}^i$,

$$(p_{ij} - \underset{1}{\lambda} g_{ij})\underset{1}{W} = 0, \tag{19.8}$$

one *isotropic* eigen direction $\underset{2}{W}^i$, *orthogonal* to $\underset{1}{W}^j$,

$$(p_{ij} - \underset{2}{\lambda} g_{ij})\underset{2}{W}^j = 0, \tag{19.9}$$

and in addition an *isotropic* vector $\underset{3}{W}^j$ orthogonal to $\underset{1}{W}^j$ but not to $\underset{2}{W}^j$, which with W^j_2 forms the *invariant* cluster $\{\underset{2}{W}, \underset{3}{W}\}$ of the tensor p_{ij}; this is expressed by the relation

$$(p_{ij} - \underset{2}{\lambda} g_{ij})\underset{3}{W}^j = \varrho \underset{2}{W}_i, \tag{19.10}$$

where ϱ is an arbitrary non-zero scalar, for otherwise $\underset{3}{W}$ would also be an eigen direction. This arbitrariness is a consequence of the fact that $\underset{2}{W}^i$ and $\underset{3}{W}^i$, being isotropic, can be multiplied by any non-zero number without change of norm.

Every eigen direction or cluster of the tensor p_{ij} determines a principal direction and invariant clusters of R_{ab} in R_6; all these vectors have the structures (19.3), as was shown at the beginning of this section.

Suppose that the root $\underset{1}{\lambda}$ corresponds to the linear elementary divisor $(\lambda - \underset{1}{\lambda})$ of the λ-matrix and a principal direction defined by the bivector $\underset{1}{W}^{\alpha\beta}$. As this bivector is *non-isotropic*, all the arguments used above in connection with $\overset{*}{T}_1$ spaces can still be applied. Hence a real orthonormal tetrad can be chosen to within a rotation in the two-dimensional plane

$\{\underset{2}{\xi}\,\underset{3}{\xi}\}$ and a Lorentz rotation in $\{\underset{1}{\xi}\,\underset{4}{\xi}\}$, relative to which

$$\underset{1}{W}^{\alpha\beta} = \underset{14}{\xi}^{\alpha\beta} + i\underset{2}{\xi}^{\alpha\beta}.$$

Because of the orthogonality of $\underset{1}{W}$ to $\underset{2}{W}$ and $\underset{3}{W}$, we obtain

$$\underset{s}{W}^{\alpha\beta} = \mu_s(\underset{24}{\xi}^{\alpha\beta} + i\underset{31}{\xi}^{\alpha\beta}) + \nu_s\underset{34}{\xi}^{\alpha\beta} + i\underset{12}{\xi}^{\alpha\beta}) \qquad (s = 2, 3).$$

As bivectors W_2 and W_3 are isotropic

$$\nu_2 = e_2 i\mu_2, \qquad \nu_3 = e_3 i\mu_3,$$

where e_2 and e_3 are either ± 1. As $\underset{2}{W}$ and $\underset{3}{W}$ are not orthogonal, we must have $e_3 = -e_2$. We can therefore take

$$\underset{2}{W}^{\alpha\beta} = \underset{24}{\xi}^{\alpha\beta} + i\underset{31}{\xi}^{\alpha\beta} + i\underset{34}{\xi}^{\alpha\beta} - \underset{12}{\xi}^{\alpha\beta},$$

$$\underset{3}{W}^{\alpha\beta} = \sigma\{\underset{24}{\xi}^{\alpha\beta} + i\underset{31}{\xi}^{\alpha\beta} - i\underset{34}{\xi}^{\alpha\beta} + \underset{12}{\xi}^{\alpha\beta}\},$$

where σ is some arbitrary scalar $\neq 0$. Writing conditions analogous to (19.8) (19.9) and (19.10), for the tensor R_{ab} in R_6, we obtain

$$\left.\begin{aligned} (R_{ab} - \underset{1}{\lambda} g_{ab})\underset{1}{W}^b = 0, \quad (R_{ab} - \underset{2}{\lambda} g_{ab})\underset{2}{W}^b = 0,\\ (R_{ab} - \underset{2}{\lambda} g_{ab})\underset{3}{W}^b = \sigma g_{ab}\underset{2}{W}^b. \end{aligned}\right\} \tag{19.11}$$

The isotropic vectors can be normalized by taking $\sigma = 1$, and then the *canonical* orthonormal tetrad is determined without any arbitrariness. As $\underset{\varrho\sigma}{\xi}^{\alpha\beta} \to \underset{p}{\xi}^a = \delta^a_p$, and the tensor g_{ab} is given by (18.3), putting $a = 1, \ldots, 6$ in (19.11) provides a set of equations which can easily be solved for all the components R_{ab}. These can be expressed in terms of the real and imaginary parts of the principal curvatures:

$$\underset{1}{\lambda} = \alpha_1 + i\beta_1, \quad \underset{2}{\lambda} = \alpha_2 + i\beta_2 \qquad (\underset{2}{\lambda} \equiv \underset{3}{\lambda}).$$

Hence *the canonical form of the matrix* (R_{ab}) *for* $\dot{T}_2$ *spaces can be taken as*

$$\left.\begin{aligned} (R_{ab}) = \begin{pmatrix} M & N \\ N & -M \end{pmatrix},\\ M = \begin{pmatrix} \alpha_1 & 0 & 0 \\ 0 & \alpha_2+1 & 0 \\ 0 & 0 & \alpha_2-1 \end{pmatrix}, \quad N = \begin{pmatrix} \beta_1 & 0 & 0 \\ 0 & \beta_2 & 1 \\ 0 & 1 & \beta_2 \end{pmatrix}, \end{aligned}\right\} \tag{19.12}$$

where

$$\alpha_1 + 2\alpha_2 = -\varkappa, \quad \beta_1 + 2\beta_2 = 0. \tag{19.13}$$

3. $\dot{T}_3$ *spaces; Segre characteristic* [3, 3]. For this Segre characteristic, the tensor p_{ij} has one eigen direction $\underset{1}{W}^j$ and, in addition, a three-dimensional invariant cluster defined by the vectors $\{\underset{1}{W}, \underset{2}{W}, \underset{3}{W}\}$; these vectors must satisfy the conditions (cf. § 9)

$$\left.\begin{aligned} (p_{ij}-\underset{1}{\lambda}g_{ij})\underset{1}{W}^j = 0, \quad (p_{ij}-\underset{1}{\lambda}g_{ij})\underset{2}{W}^j = \sigma g_{ij}\underset{1}{W}^j, \\ (p_{ij}-\underset{1}{\lambda}g_{ij})\underset{3}{W}^j = \tau g_{ij}\underset{2}{W}^j, \end{aligned}\right\} \tag{19.14}$$

where σ and τ are arbitrary non-zero numbers and $i, j = 1, 2, 3$. The vector $\underset{2}{W}^j$ is non-isotropic while $\underset{3}{W}^i$ is isotropic. In addition, $\underset{1}{W}^j$ is orthogonal to $\underset{2}{W}^j$, but not to $\underset{3}{W}^j$, and $\underset{2}{W}^j$ is orthogonal to $\underset{3}{W}^j$.

As $\underset{2}{W}^{\alpha\beta}$ is a non-isotropic bivector, we can, as in the two previous cases, so select an orthonormal tetrad with two degrees of freedom that this bivector takes the form

$$\underset{2}{W}^{\alpha\beta} = \underset{24}{\xi}^{\alpha\beta} + i\underset{31}{\xi}^{\alpha\beta}.$$

Applying the conditions for orthogonality and isotropy involving the bivectors $\underset{1}{W}^{\alpha\beta}$ and $\underset{3}{W}^{\alpha\beta}$, we obtain

$$\underset{1}{W}^{\alpha\beta} = \underset{14}{\xi}^{\alpha\beta} + i\underset{23}{\xi}^{\alpha\beta} + i\underset{34}{\xi}^{\alpha\beta} - \underset{12}{\xi}^{\alpha\beta}, \qquad \underset{3}{W}^{\alpha\beta} = \varrho\{\underset{14}{\xi}^{\alpha\beta} + i\underset{23}{\xi}^{\alpha\beta} - i\underset{34}{\xi}^{\alpha\beta} + \underset{12}{\xi}^{\alpha\beta}\},$$

where ϱ is any non-zero number. The investigation then proceeds as for the previous types of space $\dot{T}_1$ and $\dot{T}_2$. We now write down the analogous conditions to (19.14) but in the real R_6 space, stating that $\underset{1}{W}^\alpha$ is an eigenvector of R_{ab}, and that the vectors $\underset{2}{W}^a$ and $\underset{3}{W}^a$ together with $\underset{1}{W}^a$ define an invariant cluster of the tensor R_{ab}. These conditions are

$$\left.\begin{aligned} (R_{ab}-\underset{1}{\lambda}g_{ab})\underset{1}{W}^b = 0, \quad (R_{ab}-\underset{1}{\lambda}g_{ab})\underset{2}{W}^b = \sigma g_{ab}\underset{1}{W}^b, \\ (R_{ab}-\underset{1}{\lambda}g_{ab})\underset{3}{W}^b = \tau g_{ab}\underset{2}{W}^b, \end{aligned}\right\} \tag{19.15}$$

where σ and τ are non-zero numbers.

As the vector $\underset{p}{W}^a$ in R_6 corresponds to the bivector $\underset{\sigma\tau}{W}^{\alpha\beta}$ at a given point of $\dot{T}_3$, and $\underset{\gamma\lambda}{\xi}^{\alpha\beta} \leftrightarrow \underset{p}{\xi}^a = \delta^{\mathrm{a}}_{\mathrm{p}}$ for an orthonormal hexad in R_6, writing down (19.15) for $a, b = 1, 2, \ldots, 6$ provides a set of only nine independent equations; if the R_{ab} are expressed in terms of m_{ij} and n_{ij}, these reduce to

$$\left.\begin{aligned}
m_{11}+in_{11}+im_{13}-n_{13} &= -\lambda,\\
m_{12}+in_{12}+im_{23}-n_{23} &= 0,\\
m_{13}+in_{13}+im_{33}-n_{33} &= -i\lambda,\\
m_{12}+in_{12} &= -\sigma,\\
m_{22}+in_{22} &= -\lambda,\\
m_{23}+in_{23} &= -i\sigma,\\
m_{11}+in_{11}-im_{13}+n_{13} &= -\lambda,\\
m_{12}+in_{12}-im_{23}+n_{23} &= -\tau,\\
m_{13}+in_{13}-im_{33}+n_{33} &= i\lambda,
\end{aligned}\right\} \tag{19.16}$$

where $\lambda = \alpha+i\beta$ is one of the two possible complex conjugate roots of the equation $|R_{ab}-\lambda g_{ab}| = 0$, and σ and τ are arbitrary non-zero numbers. This arbitrariness follows that of the number ϱ itself, and is a consequence of the isotropy of the vectors $\underset{1}{W}{}^a$ and $\underset{3}{W}{}^a$. The orthonormal hexad can be uniquely defined putting $\sigma = \pm 1$.

Solving the set of equations (19.16) subject to the conditions

$$\sum_{s=1}^{3} m_{ss} = -\varkappa, \quad \sum_{1}^{3} n_{ss} = 0,$$

we obtain

$$\tau = 2, \quad \beta = 0, \quad \alpha = \frac{\varkappa}{3}.$$

Hence *the canonical form of the matrix* (R_{ab}) *for* $\dot{T}_3$ *spaces can be taken as*

$$\left.\begin{aligned}
&(R_{ab}) = \begin{pmatrix} M & N \\ N & -M \end{pmatrix},\\
&\text{where}\\
&M = \begin{pmatrix} -\frac{\varkappa}{3} & 1 & 0 \\ 1 & -\frac{\varkappa}{3} & 0 \\ 0 & 0 & -\frac{\varkappa}{3} \end{pmatrix}, \quad N = \begin{pmatrix} 0 & 0 & 0 \\ 0 & 0 & -1 \\ 0 & -1 & 0 \end{pmatrix}.
\end{aligned}\right\} \tag{19.17}$$

The single, and *real*, root of the equation $|R_{ab}-\lambda g_{ab}| = 0$ is $\varkappa/3$, justifying the statement made in the previous section about the reality of the roots and bases of the elementary divisors in $\dot{T}_3$ spaces.

The following theorem summarizes the results:

there is a real orthonormal tetrad in all three possible types of $\dot{T}_i$ *space* ($i = 1, 2, 3$), *relative to which the orthogonal components of the curvature tensor are defined by canonical forms of the matrix* (R_{ab}); *for* $\dot{T}_1$ *the canonical form is* (19.6) *subject to* (19.7) *for* $\dot{T}_2$ (19.12) *subject to* (19.13) *and for* $\dot{T}_3$, (19.17); *in all three cases the orthonormal tetrad is uniquely defined.*

As this theorem has been proved without assuming that $\varkappa \neq 0$, we can obtain the analogous results for spaces defined by the equation $R_{\alpha\beta} = 0$ by putting $\varkappa = 0$. We thus obtain the following theorem: *the canonical form of the matrix* (R_{ab}) *defining the orthogonal components of the curvature tensor in* T_i *defined by gravitational fields in empty space, is*

$$(R_{ab}) = \begin{pmatrix} M & N \\ N & -M \end{pmatrix},$$

where for T_1

$$M = \begin{pmatrix} \alpha_1 & 0 & 0 \\ 0 & \alpha_2 & 0 \\ 0 & 0 & \alpha_3 \end{pmatrix}, \quad N = \begin{pmatrix} \beta_1 & 0 & 0 \\ 0 & \beta_2 & 0 \\ 0 & 0 & \beta_3 \end{pmatrix}, \tag{19.18}$$

$$\sum \alpha_i = \sum \beta_i = 0;$$

for T_2

$$M = \begin{pmatrix} \alpha_1 & 0 & 0 \\ 0 & \alpha_2+1 & 0 \\ 0 & 0 & \alpha_2-1 \end{pmatrix}, \quad N = \begin{pmatrix} \beta_1 & 0 & 0 \\ 0 & \beta_2 & 1 \\ 0 & 1 & \beta_2 \end{pmatrix}, \tag{19.19}$$

$$\alpha_1+2\alpha_2 = \beta_1+2\beta_2 = 0;$$

and for T_3

$$M = \begin{pmatrix} 0 & 1 & 0 \\ 1 & 0 & 0 \\ 0 & 0 & 0 \end{pmatrix}, \quad N = \begin{pmatrix} 0 & 0 & 0 \\ 0 & 0 & -1 \\ 0 & -1 & 0 \end{pmatrix}. \tag{19.20}$$

These canonical forms embody the basic theorem proved in § 18, and play a fundamental role in the *invariant* characterization of four-dimensional Einstein spaces with signature $(- - - +)$. The three types of T_i discussed clearly admit of a more detailed classification, e.g. the cases of repeated roots or real and imaginary roots; this will be given later as the need arises.

The following comment may be made as to the method of proof. As the Segre characteristic of $\dot{T}_i$ is known, it might seem possible at first sight to write down the canonical form of the matrices (R_{ab}) directly by using the general theorem of § 9. This, however, is impossible for the only permissible transformations are those defined by the matrices

$$(A_a^{a'}) \to 2(A_{[\alpha}^{\alpha'} A_{\beta]}^{\beta'}),$$

where the $A_a^{a'}$ are the coefficients of some real orthogonal transformation at the given point P in the $\dot{T}_i$ space. Hence we can only use transformations in the subgroup of the total group of orthogonal real transformations in six-dimensional space, which leave a metric with the signature $(---+++)$ invariant.

An analysis of the well-known examples of Einstein spaces shows that the overwhelming number belong to type T_1 (cf. Chapter 4). Only recently, in investigating T_i admitting a group of motions (Chapter 4), have T_2 and T_3 spaces been obtained. Some interesting physical theories, e.g. the problem of gravitational radiation, etc., are connected with these spaces (cf. [177], [263], [265], [271], [291], [323], [293], [294], [296], [297], [300], [321], [324], [342], [343], [344], [353], [354], [362]).

Examples

1. Prove that T_2 spaces ($R_{\alpha\beta} = 0$, $\alpha_i = \beta_i = 0$) contain a vector field l^α, such that

$$l^\alpha R_{\alpha\beta\gamma\delta} = 0, \qquad l_\alpha R_{\beta\gamma\lambda\mu} + l_\beta R_{\gamma\alpha\lambda\mu} + l_\gamma R_{\alpha\beta\lambda\mu} = 0$$

(Bel [291]).

2. Prove that T_3 spaces ($R_{\alpha\beta} = 0$) contain a vector field l^α, satisfying the conditions

$$R_{\alpha\beta\gamma\delta} l^\beta l^\delta = e_{\alpha\beta\sigma\tau} R^{\sigma\tau}_{\gamma\delta} l^\beta l^\delta = 0,$$

where $e_{\alpha\beta\sigma\gamma}$ is the permutation tensor of the space; the field l^α defines the isotropic geodesics of the space

$$l^\alpha_{,\sigma} l^\sigma = \varphi l^\alpha$$

(Bel [299]).

3. Prove that T_3 spaces ($R_{\alpha\beta} = 0$) contain vector fields h^α, k^α and r^α as well as the field l^α defined in example 2, which together with l^α form a system of independent vectors satisfying the relations

$$l^\alpha l_\alpha = r^\alpha r_\alpha = l^\alpha h_\alpha = l^\alpha k_\alpha = h^\alpha k_\alpha = h^\alpha r_\alpha = k^\alpha r_\alpha = 0,$$
$$h^\alpha h_\alpha = k^\alpha k_\alpha = -\, l^\alpha r_\alpha = -1.$$

These fields satisfy the conditions

$$R_{\alpha\beta\gamma\delta}l^{\beta}h^{\delta} = l_{\alpha}l_{\gamma}, \quad R_{\alpha\beta\gamma\delta}l^{\beta}l^{\delta} = 0, \quad R_{\alpha\beta\gamma\delta}l^{\beta}r^{\delta} = l_{\alpha}h_{\gamma},$$
$$R_{\alpha\beta\gamma\delta}h^{\beta}h^{\delta} = 0, \quad R_{\alpha\beta\gamma\delta}h^{\beta}k^{\delta} = -k_{\alpha}l_{\gamma},$$
$$R_{\alpha\beta\gamma\delta}h^{\beta}r^{\delta} = -(l_{\alpha}r_{\gamma}+k_{\alpha}k_{\gamma}), \quad R_{\alpha\beta\gamma\delta}k^{\beta}k^{\delta} = l_{\alpha}h_{\gamma}+h_{\alpha}l_{\gamma},$$
$$R_{\alpha\beta\gamma\delta}k^{\beta}r^{\delta} = h_{\alpha}k_{\gamma}, \quad R_{\alpha\beta\gamma\delta}r^{\beta}r^{\delta} = -(h_{\alpha}r_{\gamma}+h_{\gamma}r_{\alpha}).$$

4. Show that in the case of T_3 ($R_{\alpha\beta} = 0$) the vectors l, $\bar{h}$, $\bar{k}$ and $\bar{r}$ referred to the orthonormal tetrad leading to (19.20) have components

$$l^{\alpha}(0, 1, 0, 1), \quad h^{\alpha}(1, 0, 0, 0), \quad k^{\alpha}(0, 0, 1, 0),$$
$$r^{\alpha}(0, -\tfrac{1}{2}, \ 0, \tfrac{1}{2}),$$

and that therefore

$$R_{\alpha\beta\gamma\delta} = -4(l_{[\alpha}h_{\beta]}l_{[\gamma}r_{\delta]}+l_{[\alpha}r_{\beta]}l_{[\gamma}h_{\delta]}+l_{[\alpha}k_{\beta]}k_{[\gamma}h_{\delta]}+k_{[\alpha}h_{\beta]}l_{[\gamma}k_{\delta]}).$$

5. When the conditions of examples 3 and 4 are satisfied, the divergence of the isotropic-geodesic field l^{α} is zero;

$$l^{\sigma}_{,\sigma} = 0.$$

6. Prove that simple bivectors are 2/2-orthogonal only when

$$\underset{1}{V_{\alpha\beta}}\underset{2}{V^{\beta\delta}} = 0$$

([91], p. 391).

7. Prove that in T_2 spaces with $\alpha_i = \beta_i = 0$, the biparametric group $x^{\alpha} = C^{\alpha}_{\alpha'}x^{\alpha'}$ where

$$[C^{\alpha}_{\alpha'}] = \begin{bmatrix} \tfrac{1}{2}(\alpha^2+\beta^2-2), & e\beta, & e\alpha, & -\tfrac{1}{2}(\alpha^2+\beta^2) \\ -\beta, & -e, & 0, & \beta \\ -\alpha, & 0, & -e\alpha, & \alpha \\ \tfrac{1}{2}(\alpha^2+\beta^2), & e\beta, & e\alpha, & -\tfrac{1}{2}(\alpha^2+\beta^2+2) \end{bmatrix} \quad (e = \pm 1)$$

is a Lorentz group retaining the invariant form (19.19).

§20. Classification of General Gravitational Fields

The theorem on the existence of three types of Einstein space with signature $(- - - +)$, and the corresponding three canonical forms of the curvature tensor relative to a uniquely defined non-holonomic orthonormal tetrad, were both obtained in 1949. These results have found physical application mainly in investigating *empty* spaces where the energy-momentum tensor $T_{\alpha\beta} = 0$ ([258], [263], [299], [265]). The general case when $T_{\alpha\beta} \neq \sigma g_{\alpha\beta}$, hitherto left open, will now be considered [287]. It is natural to assume, firstly, that the classification must lead to the results of §§ 18, 19 when $T_{\alpha\beta} \to \sigma g_{\alpha\beta}$, and secondly, must take into account the algebraic structure of both the curvature and energy-momentum tensors. Spaces corresponding to general gravitational fields are henceforth denoted by the symbol $\boldsymbol{T}$.

We now consider the geometry of a $\boldsymbol{T}$ space defined by a certain distribution and motion of matter, with a metric tensor $g_{\alpha\beta}(x)$ satisfying the Einstein field equations

$$R_{\alpha\beta} - \tfrac{1}{2}Rg_{\alpha\beta} = \lambda T_{\alpha\beta}, \tag{20.1}$$

where λ is a constant and $T_{\alpha\beta}$ is the energy-momentum tensor of the matter. Contracting eqn. (20.1) gives the connection between R and T:

$$\lambda T = -R. \tag{20.2}$$

We now construct the fourth order tensor

$$S_{\alpha\beta\lambda\delta} \equiv \frac{\lambda}{2}(g_{\alpha\gamma}T_{\beta\delta} - g_{\alpha\delta}T_{\beta\gamma} + g_{\beta\delta}T_{\alpha\gamma} - g_{\beta\gamma}T_{\alpha\delta}), \tag{20.3}$$

which, from its definition, has the properties

$$S_{\alpha\beta\gamma\delta} = -S_{\beta\alpha\gamma\delta} = -S_{\alpha\beta\delta\gamma} = S_{\gamma\delta\alpha\beta} \tag{20.4}$$

and

$$S_{\alpha[\beta\gamma\delta]} = 0, \tag{20.5}$$

identical with well-known properties of the curvature tensor of a V_n space. Contracting (20.3) over β, δ, we obtain

$$S_{\alpha\gamma} = \lambda T_{\alpha\gamma} + \frac{\lambda}{2} T g_{\alpha\gamma} = \lambda T_{\alpha\gamma} - \frac{R}{2} g_{\alpha\gamma}. \tag{20.6}$$

We now define a new fourth order tensor

$$P_{\alpha\beta\gamma\delta} \equiv R_{\alpha\beta\gamma\delta} - S_{\alpha\beta\gamma\delta} + \sigma(g_{\alpha\gamma}g_{\beta\delta} - g_{\alpha\delta}g_{\beta\gamma}), \tag{20.7}$$

which we call the *space-matter tensor*. It has the following properties:

1. Relations (20.4) and (20.5), which are also satisfied by the curvature tensor, show that

$$P_{\alpha\beta\gamma\delta} = -P_{\beta\alpha\gamma\delta} = -P_{\alpha\beta\delta\delta} = P_{\gamma\delta\alpha\beta} \tag{20.8}$$

and

$$P_{\alpha[\beta\gamma\delta]} = 0. \tag{20.9}$$

2. Contracting (20.7) over β, δ and using (20.6) leads to the relation

$$P_{\alpha\gamma} = R_{\alpha\gamma} - \lambda T_{\alpha\gamma} + \frac{R}{2} g_{\alpha\gamma} + 3\sigma g_{\alpha\gamma}.$$

Substituting from the field equations (20.1), we obtain

$$P_{\alpha\gamma} = (R+3\sigma)g_{\alpha\gamma}. \tag{20.10}$$

3. If the distribution and motion of the matter, i.e. the energy-momentum tensor $T_{\alpha\beta}$ and the space-matter tensor $P_{\alpha\beta\gamma\delta}$ are given, the curvature of the space is determined to within the scalar σ.

4. Given the metric tensor, the scalar σ and the tensor $P_{\alpha\beta\gamma\delta}$, the energy-momentum tensor can be uniquely determined.

5. If the energy-momentum tensor $T_{\alpha\beta} = 0$ and $\sigma = 0$, $P_{\alpha\beta\gamma\delta}$ is the curvature tensor of the empty space-time.

6. As the tensor $P_{\alpha\gamma}$ satisfies equations similar to the gravitational field equations $R_{\alpha\beta} = \varkappa g_{\alpha\beta}$ for Einstein spaces, it is in this sense the analogue of the Ricci tensor.

The first term of the space-matter tensor (20.7) represents the curvature of the space, and the second the distribution and motion of the matter. We base the classification of gravitational fields on the classification of the algebraic structure of the space-matter tensor.

The fact that $P_{\alpha\beta\gamma\delta}$ has properties (20.8), (20.9) and (20.10) enables the classification problem to be solved by the methods used in classifying Einstein spaces in §§, 18, 19. As indeed we then considered an *algebraic* classification, we need not consider the differential properties of the tensor $P_{\alpha\beta\gamma\delta}$. The tensor $P_{\alpha\beta\gamma\delta}$ has all the algebraic properties of the curvature tensor, and $P_{\alpha\gamma}$ those of the Ricci tensor. The results obtained in §§ 18, 19 can be automatically extended to any *bitensor* with properties (20.8), (20.9) and (20.10), and in particular to the tensor $P_{\alpha\beta\gamma\delta}$; we need therefore only replace $\varkappa$ in eqns. (19.6), (19.7), (19.12), (19.13) and (19.17) by $\omega = R+3\sigma$.

We thus obtain the following theorems covering the classification of general gravitational fields based on the algebraic structure of the space-matter tensor [287].

THEOREM 1. *The matrix* (P_{ab}) *in the bivector space corresponding to any orthonormal tetrad at a point in a* $\boldsymbol{T}$ *space defined by a gravitational field is doubly symmetric:*

$$\left.\begin{aligned}(P_{ab}) &= \begin{pmatrix} M & N \\ N & -M \end{pmatrix},\\ M = (m_{ij}),\quad N &= (n_{ij}) \qquad (i, j = 1, 2, 3),\end{aligned}\right\} \tag{20.11}$$

where the 3×3 *matrices M and N are symmetric,*

$$m_{[ij]} = n_{[ij]} = 0, \tag{20.12}$$

and satisfy the conditions

$$\sum_{i=1}^{3} m_{ii} = -\omega, \quad \sum_{i=1}^{3} n_{ii} = 0. \tag{20.13}$$

Applying three Lorentz and three spatial rotations shows that, as in § 19, the following two theorems are also valid.

THEOREM 2. *There are only three types of general gravitational field as defined by the algebraic structure of the space-matter tensor. These types correspond respectively to the three types of Segre characteristics of the matrix* $(P_{ab}-\lambda g_{ab})$ *for the bivector space:*

(1) $\boldsymbol{T}_1[1\ 1\ 1, \overline{1\ 1\ 1}]$, (2) $\boldsymbol{T}_2[2\ 1, \overline{2\ 1}]$, (3) $\boldsymbol{T}_3[(3, 3)]$.

A characteristic of $\boldsymbol{T}_1$ gravitational fields is the existence of *linear* elementary divisors and six corresponding *mutually orthogonal non-isotropic bivectors* $\underset{s}{W}^{\alpha\beta} \leftrightarrow \underset{s}{W}^a$ (two complex conjugate triads), which are the eigenvectors of P_{ab} in R_6;

$$P_{ab}\underset{s}{W}^b = \underset{s}{\lambda}\underset{s}{W}_a, \quad \underset{s}{\lambda} = \alpha_s \pm i\beta_s \quad (s = 1, \ldots, 6).$$

$\boldsymbol{T}_2$ fields have two *non-linear* elementary divisors of the λ-matrix $(P_{ab}-\lambda g_{ab})$ with complex conjugate divisors $(\lambda-\lambda_s)^2$, $(\lambda-\bar{\lambda}_s)^2$, corresponding to complex conjugate, mutually orthogonal, *isotropic* eigenvectors $\underset{1}{W}^a$ and $\underset{4}{W}^a$, i.e. $\underset{s}{W}_a\underset{s}{W}^a = 0$, and two *non-isotropic* complex conjugate eigenvectors $\underset{2}{W}^a$ and $\underset{5}{W}^a$ orthogonal to each other and to the vectors $\underset{1}{W}^a$ and $\underset{4}{W}^a$, such that

$$P_{ab}\underset{s}{W}^b = \underset{s}{\lambda}\underset{s}{W}_a \qquad (s = 1, 2, 4, 5)$$

and

$$\underset{s}{\lambda} = \alpha_s \pm i\beta_s, \quad \underset{i}{\lambda} = \underset{i+3}{\bar{\lambda}} \qquad (i = 1, 2).$$

Finally, T_3 fields have only two non-linear elementary divisors $(\lambda-\lambda_1)^3$, $(\lambda-\lambda_1)^3$, with the same *real* base $\lambda_1 = \omega/3$, corresponding to two mutually orthogonal *isotropic* eigenvectors $\underset{1}{W^a}$ and $\underset{4}{W^a}$.

THEOREM 3. *The matrix* (P_{ab}) *for* T_1 *gravitational fields has the following form for some non-holonomic, uniquely defined orthonormal tetrad:*

$$\left.\begin{aligned} &(P_{ab}) = \begin{pmatrix} M & N \\ N & -M \end{pmatrix}, \\ &M = \begin{pmatrix} \alpha_1 & 0 & 0 \\ 0 & \alpha_2 & 0 \\ 0 & 0 & \alpha_3 \end{pmatrix}, \quad N = \begin{pmatrix} \beta_1 & 0 & 0 \\ 0 & \beta_2 & 0 \\ 0 & 0 & \beta_3 \end{pmatrix}, \end{aligned}\right\} \tag{20.14}$$

where

$$\sum_{i=1}^{3} \alpha_i = -\omega, \quad \sum_{i=1}^{3} \beta_i = 0; \tag{20.15}$$

T_2 *fields have*

$$M = \begin{pmatrix} \alpha_1 & 0 & 0 \\ 0 & \alpha_2+1 & 0 \\ 0 & 0 & \alpha_2-1 \end{pmatrix}, \quad N = \begin{pmatrix} \beta_1 & 0 & 0 \\ 0 & \beta_2 & 1 \\ 0 & 1 & \beta_2 \end{pmatrix}, \tag{20.16}$$

where

$$\alpha_1+2\alpha_2 = -\omega, \quad \beta_1+2\beta_2 = 0, \tag{20.17}$$

whilst T_3 *fields have*

$$M = \begin{pmatrix} -\frac{\omega}{3} & 1 & 0 \\ 1 & -\frac{\omega}{3} & 0 \\ 0 & 0 & -\frac{\omega}{3} \end{pmatrix}, \quad N = \begin{pmatrix} 0 & 0 & 0 \\ 0 & 0 & -1 \\ 0 & -1 & 0 \end{pmatrix}. \tag{20.18}$$

The canonical forms for the components of the curvature tensor of an Einstein space, (19.6), (19.7), (19.12), (19.13), (19.17), are obtained from (20.14), (20.15), (20.16), (20.17) and (20.18) respectively when $\lambda T_{\alpha\beta} = \sigma g_{\alpha\beta}$, and in this sense these components can be considered as a *limiting case*. In particular when $T_{\alpha\beta} = 0$, $\sigma = 0$ we obtain the appropriate classification for gravitational fields in *empty* space; for this purpose it is sufficient to put $\omega = 0$ in (20.15), (20.17) and (20.18).

These canonical forms are the embodiment of theorem 2, and might be expected to play a part in the *invariant* characterization of gravitational fields in the real world.

The three given types T_i of gravitational field clearly admit of a more detailed classification; thus the cases of *repeated* or *real* λ_s could be distinguished, and we could supplement the above classification with one of $(T_{\alpha\beta}-\lambda g_{\alpha\beta})$ and consider the logical product of the two classifications; such a classification can be carried out using the results of § 9.

The following comments must be made about the classification given above.

In solving only the *algebraic* problem, the value assigned to the *scalar* σ is quite immaterial. Moreover, instead of the tensor $P_{\alpha\beta\gamma\delta}$, defined by (20.7), one can investigate the tensor

$$\boldsymbol{P}_{\alpha\beta\gamma\delta} \equiv \varrho R_{\alpha\beta\gamma\delta}+\boldsymbol{\lambda} S_{\alpha\beta\gamma\delta}+\sigma g_{\alpha[\gamma}g_{\delta]\beta},$$

where ϱ, $\boldsymbol{\lambda}$, σ are *arbitrary scalars*, the choice of which must be based on physical or geometric considerations. However, the formulae are considerably simplified by putting $\varrho = 1$, $\boldsymbol{\lambda} = -1$, $\sigma = -\frac{2}{3}R$ when, as can easily be seen, $\boldsymbol{P}_{\alpha\beta\gamma\delta} \equiv C_{\alpha\beta\gamma\delta}$ – Weyl's *conformal curvature* tensor (see § 34); to avoid errors of sign, note that $R_{\alpha\beta} \equiv R^{\sigma}_{\alpha\sigma\beta}$. We must then put $\omega = 0$ in (20.15), (20.17) and (20.18). Weyl's tensor is convenient because we can use the appropriate geometric identities and because the resulting equations are simpler, as are the canonical forms for the curvature tensor in empty space. Accordingly, we have

$$C_{\alpha\beta\gamma\delta} = R_{\alpha\beta\gamma\delta}+L_{\alpha\beta\gamma\delta}, \tag{20.19}$$

where

$$L_{\alpha\beta\gamma\delta} = -\frac{\lambda}{2}(g_{\alpha\gamma}T_{\beta\delta}-g_{\alpha\delta}T_{\beta\gamma}+g_{\beta\delta}T_{\alpha\gamma}-g_{\beta\gamma}T_{\alpha\delta})$$
$$+\frac{\lambda}{3}T(g_{\alpha\gamma}g_{\beta\delta}-g_{\alpha\delta}g_{\beta\gamma}), \tag{20.20}$$

and the tensor $R_{\alpha\beta}$ and scalar curvature R depend respectively on the energy-momentum tensor $T_{\alpha\beta}$ and the contraction $T = T^{\sigma}{}_{\sigma}$ according to the field equation. Differentiating (20.19) covariantly, we obtain

$$C_{\alpha\beta\gamma\delta,\,\lambda} = R_{\alpha\beta\gamma\delta,\,\lambda}+L_{\alpha\beta\gamma\delta,\,\lambda} \tag{20.21}$$

and, applying Bianchi's identity,

$$C_{\alpha\beta[\gamma\delta,\,\lambda]} = L_{\alpha\beta[\gamma\delta,\,\lambda]}. \tag{20.22}$$

On the other hand, it follows directly from (20.21) that

$$C^{\sigma}_{\beta\gamma\delta,\,\sigma} = \frac{\lambda}{2}\,(T_{\beta\delta,\,\gamma}-T_{\beta\gamma,\,\delta})+\frac{\lambda}{6}\,(T_{,\,\delta}g_{\beta\gamma}-T_{,\,\gamma}g_{\beta\delta}). \tag{20.23}$$

The relation (20.22) and one analogous to (20.23) hold for any n (if $T_{\alpha\beta}$ and T are replaced by $R_{\alpha\beta}$ and R using the field equations) and for any signature. If, however, $n = 4$ and the metric signature $(---+)$, it can be shown that these two relations are equivalent and we need, for example, only consider eqns. (20.23) relating the divergence of Weyl's tensor with the covariant derivatives of the energy-momentum tensor (see example 3 of this section).

Moreover, by using Weyl's tensor (20.19) with (20.14), (20.16), (20.18), in classifying gravitational fields, we derive the following expressions for the curvature tensor of the three possible types:

$$(R_{ab}) = \lambda\begin{pmatrix} M & N \\ N^* & P\end{pmatrix} \qquad (a,\, b = 1,\, \ldots,\, 6)$$

Type I

$$M = \begin{pmatrix} \alpha_1+\frac{1}{2}(T_{11}-T_{44})+\frac{1}{3}T, & \frac{1}{2}T_{12}, & \frac{1}{2}T_{13} \\ \frac{1}{2}T_{12}, & \alpha_2+\frac{1}{2}(T_{22}-T_{44})+\frac{1}{3}T, & \frac{1}{2}T_{23} \\ \frac{1}{2}T_{13}, & \frac{1}{2}T_{23}, & \alpha_3+\frac{1}{2}(T_{33}-T_{44})+\frac{1}{3}T \end{pmatrix}$$

$$P = \begin{pmatrix} -\alpha_1-\frac{1}{2}(T_{22}+T_{33})-\frac{1}{3}T, & \frac{1}{2}T_{12}, & \frac{1}{2}T_{13} \\ \frac{1}{2}T_{12}, & -\alpha_2-\frac{1}{2}(T_{11}+T_{22})-\frac{1}{3}T, & \frac{1}{2}T_{23} \\ \frac{1}{2}T_{13}, & \frac{1}{2}T_{23}, & -\alpha_3-\frac{1}{2}(T_{11}+T_{22})-\frac{1}{3}T \end{pmatrix} \tag{20.24}$$

$$N = \begin{pmatrix} \beta_1, & \frac{1}{2}T_{34}, & -\frac{1}{2}T_{24} \\ -\frac{1}{2}T_{34}, & \beta_2, & \frac{1}{2}T_{14} \\ \frac{1}{2}T_{24}, & -\frac{1}{2}T_{14}, & \beta_3 \end{pmatrix}, \quad \sum_1^3 \alpha_i = \sum_1^3 \beta_i = 0.$$

N^* is the transposed matrix obtained from N by reflection about the principal diagonal.

Type II

$$M = \begin{pmatrix} \alpha_1+\frac{1}{2}(T_{11}-T_{44})+\frac{1}{3}T, & \frac{1}{2}T_{12}, & \frac{1}{2}T_{13} \\ \frac{1}{2}T_{12}, & \alpha_2+\frac{1}{2}(T_{22}-T_{14})+\frac{1}{3}T+1, & \frac{1}{2}T_{23} \\ \frac{1}{2}T_{13}, & \frac{1}{2}T_{23}, & \alpha_2+\frac{1}{2}(T_{33}-T_{44})+\frac{1}{3}T-1 \end{pmatrix}$$

$$P = \begin{pmatrix} -\alpha_1-\frac{1}{2}(T_{22}+T_{33})-\frac{1}{3}T, & \frac{1}{2}T_{12}, & \frac{1}{2}T_{13} \\ \frac{1}{2}T_{12}, & -\alpha_2-\frac{1}{2}(T_{11}+T_{33})-\frac{1}{3}T-1, & \frac{1}{2}T_{23} \\ \frac{1}{2}T_{13}, & \frac{1}{2}T_{23}, & -\alpha_2-\frac{1}{2}(T_{11}+T_{22})-\frac{1}{3}T+1 \end{pmatrix} \tag{20.25}$$

$$N = \begin{pmatrix} \beta_1 & \frac{1}{2}T_{34}, & -\frac{1}{2}T_{24} \\ -\frac{1}{2}T_{34}, & \beta_2, & \frac{1}{2}T_{14}+e \\ \frac{1}{2}T_{24}, & -\frac{1}{2}T_{14}+e, & \beta_2 \end{pmatrix}, \quad \begin{matrix} \alpha_1+2\alpha_2 = \beta_1+2\beta_2 = 0 \\ e = \pm 1 \end{matrix}$$

Type III

$$M = \begin{pmatrix} \frac{1}{2}(T_{11}-T_{44})+\frac{1}{3}T, & \frac{1}{2}T_{12}+e_1, & \frac{1}{2}T_{13} \\ \frac{1}{2}T_{12}+e_1, & \frac{1}{2}(T_{22}-T_{44})+\frac{1}{3}T, & \frac{1}{2}T_{23} \\ \frac{1}{2}T_{13}, & \frac{1}{2}T_{23}, & \frac{1}{2}(T_{33}-T_{44})+\frac{1}{3}T \end{pmatrix}$$

$$P = \begin{pmatrix} -\frac{1}{2}(T_{22}+T_{33})-\frac{1}{3}T, & \frac{1}{2}T_{12}-e_1, & \frac{1}{2}T_{13} \\ \frac{1}{2}T_{12}-e_1, & -\frac{1}{2}(T_{11}+T_{33})-\frac{1}{3}T, & \frac{1}{2}T_{23} \\ \frac{1}{2}T_{13}, & \frac{1}{2}T_{23}, & -\frac{1}{2}(T_{11}+T_{22})-\frac{1}{3}T \end{pmatrix}$$

$$N = \begin{pmatrix} 0 & \frac{1}{2}T_{34}, & -\frac{1}{2}T_{24} \\ -\frac{1}{2}T_{34}, & 0, & \frac{1}{2}T_{14}+e_2 \\ \frac{1}{2}T_{24}, & -\frac{1}{2}T_{14}+e_2, & 0 \end{pmatrix} \quad \begin{matrix} e_1 = \pm 1 \\ e_2 = \pm 1 \end{matrix}$$

e_1 and e_2 are independent of each other.

(20.26)

N^* is the transpose of N.

These equations show explicitly how the space-time curvature depends on functions defining the physical state, i.e. energy-momentum tensor $T_{\alpha\beta}$ and T, and invariants α_i, β_i (for types I and II) defining the principle space-time curvatures; the number of independent invariants does not exceed four.

Examples

1. Determine the type of space found inside an incompressible sphere of radius r_0 and mass M with metric

$$ds^2 = -dx^{1^2}+dx^{2^2}-dx^{3^2}-\frac{(x^1\,dx^1+x^2\,dx^2+x^3\,dx^3)^2}{a^2-r^2}+\left(\frac{3h-h_0}{2hh_0}\right)^2 dx^{4^2},$$

where $a = r_0\sqrt{(r_0/2m)}$, $m = kM/c^2$, $1/h^2 = 1-(2mr^2/r_0^3)$, $1/h_0^2 = 1-(2m/r_0)$.

The energy-momentum tensor is given by

$$T_{\alpha\beta} = \left(\mu_0+\frac{p}{c^2}\right)u_\alpha u_\beta - pg_{\alpha\beta},$$

where μ_0 is the rest mass density and p the pressure.

2. Consider the external problem for the conditions of example 1.

3. Show that with $n = 4$ and signature $(- - - +)$, (20.22) is equivalent to (20.23).

4. Show that the metrics

$$(g_{\alpha\beta}) = \begin{pmatrix} 0 & 0 & 0 & \alpha \\ 0 & -B^2 & 0 & \beta \\ 0 & 0 & -C^2 & \gamma \\ \alpha & \beta & \gamma & 0 \end{pmatrix}, \tag{20.27}$$

where

$$\alpha = \mu(px^2+nx^2+s)$$

$$\beta = \mu(x^3+\int p\,dx^1+\nu)$$

$$\gamma = \mu(x^2+\int n\,dx^1+\omega)$$

and $\mu(x^4)$, $p(x^1)$, $n(x^1)$, $s(x^1)$, $\nu(x^4)$, $\omega(x^4)$ are arbitrary functions of their arguments, and $B(x^4)$ and $C(x^4)$ are determined by ordinary second-order equations, define: (a) a field of type I, (b) an energy-momentum tensor of form $T_{\alpha\beta} = \pm l_\alpha l_\beta$, where $l^\alpha l_\alpha = 0$, $l_\alpha = f(x^4)\delta^4_\alpha$, (c) a field l^α which is *isotropic*, *geodesic* and a *normal congruence*. We have thus obtained the case of electromagnetic radiation (singular case, see § 47). Find $f(x^4)$ and the equations determining $B(x^4)$ and $C(x^4)$. Show also that the electromagnetic field tensor is

$$F_{\alpha\beta} = 2(\varrho l_{[\alpha}h_{\beta]} + \sigma l_{[\alpha}k_{\beta]});\ \varrho^2+\sigma^2 = \text{constant}, \tag{20.28}$$

where h_α and k_α are conformal gradiant vector fields, i.e. gradients apart from a scalar multiplier.

5. Show that if in (20.27) we put

$$\alpha = \mu(x^4)(r'x^2+t'x^3+s)$$
$$\beta = \mu(x^4)(r+C_1x^3e^{x^4}+\nu)$$
$$= \mu(x^4)(t+C_1x^2e^{x^4}+\lambda), \tag{20.29}$$

where $\mu(x^4)$, $r(x^1)$, $t(x^1)$, $s(x^1)$, $\nu(x^4)$, $\lambda(x^4)$ are arbitrary functions of their arguments, $r' \equiv \partial_4 r$, $t' \equiv \partial_4(t)$, we obtain a type II field on the basis of the algebraic structure of the conformal curvature tensor. As in example 4, the energy-momentum tensor corresponds to electro-magnetic radiation; $F_{\alpha\beta}$ has the same structure and l^α has all the properties indicated above.

6. Show that if the energy-momentum tensor $T_{\alpha\beta} = \pm l_\alpha l_\beta$, where $l_\alpha l^\alpha = 0$, which corresponds to the singular case of electromagnetic radiation (see § 47), then

(a) $T_{\alpha\beta}$ has three eigenvectors $\underset{1}{u}^\alpha$, $\underset{2}{u}^\alpha$, $\underset{3}{u}^\alpha$; the first two are non-isotropic (norm < 0), the third isotropic;

(b) as the eigenvalues are zero, the eigenvectors can be chosen to within the transformation:

$$\overset{*}{\underset{1}{u}}{}^{\alpha} = \varrho \cos \varphi \underset{1}{u}{}^{\alpha} + \varrho \sin \varphi \underset{2}{u}{}^{\alpha} + c \underset{3}{u}{}^{\alpha}$$
$$\overset{*}{\underset{2}{u}}{}^{\alpha} = -\sigma \sin \varphi \underset{1}{u}{}^{\alpha} + \sigma \sin \varphi \underset{2}{u}{}^{\alpha} + p \underset{3}{u}{}^{\alpha}$$
$$\overset{*}{\underset{3}{u}}{}^{\alpha} = \gamma \underset{3}{u}{}^{\alpha}, \tag{20.30}$$

where ϱ, σ, c, p, γ and φ are arbitrary parameters;

(c) the electromagnetic field tensor $F_{\alpha\beta}$ can be expressed in the form (20.28). If, in addition, the structure of the conformal curvature tensor gives a type III field, then

(d) an orthonormal tetrad exists at each point relative to which $\underset{3}{u}{}^{\alpha} = l^{\alpha}$, $\underset{1}{u}{}^{\alpha} = h^{\alpha}$, $\underset{2}{u}{}^{\alpha} = k^{\alpha}$ and $\underset{4}{u}{}^{\alpha} = r^{\alpha}$, where $\bar{l}$, $\bar{h}$, $\bar{k}$, $\bar{r}$ are given by equations analogous to example 4 in § 19, replacing the curvature tensor by Weyl's tensor.

(e) l^{α} is an *isotropic, geodesic* normal congruence.

§ 21. Complex Representation of Minkowski Space Tensors

The canonical forms of the curvature and space matter tensors were obtained in §§ 19, 20 by mapping the bivector space on to a three-dimensional metric, complex, flat manifold. Such a mapping has its own interest, as it permits the investigation of a wider range of problems. Thus we can introduce the problem of the algebraic classification of arbitrary bitensors in a bivector space.

Such problems are considered in references [208] and [316] which investigate so-called *biplanar* spaces. As shown below, these are essentially identical with bivector manifolds, and their investigation thus closely parallels that in §§ 19, 20, but with a somewhat different terminology. The results obtained are now outlined.

A vector space B_{2n} with an even number of dimensions is called *biplanar* if it contains a tensor ε_a^b, called the *absolute involution* tensor, which satisfies the conditions

$$\varepsilon_a^c \varepsilon_c^b = -\delta_a^b \tag{21.1}$$

and has the characteristic equation

$$(\varepsilon^2 + 1)^n = 0,$$

with n-fold roots i and $-i$. Hence $\varepsilon_a^a = 0$.

This tensor enables us to provide any given tensor with a *conjugate*, indicated by a bar:

$$\bar{x}^b = \varepsilon_a^b x^a, \quad \bar{\xi}_b = \varepsilon_b^a \xi_a, \quad \bar{a}_{ab} = \varepsilon_a^c \varepsilon_b^d a_{cd}.$$

Vectors a_a can always be so chosen that together with their conjugates they form a basis of independent vectors, henceforth called a *canonical basis*. We also consider the following complex combinations

$$\Delta_a^b = \tfrac{1}{2}(\delta_a^b + i\varepsilon_a^b), \quad \bar{\Delta}_a^b = \tfrac{1}{2}(\delta_a^b - i\varepsilon_a^b),$$

which, as can easily be seen, satisfy the relation

$$\Delta_a^b \varepsilon_b^c = -i\Delta_a^c, \quad \bar{\Delta}_a^b \varepsilon_b^c = i\bar{\Delta}_a^c,$$
$$\Delta_a^b \Delta_b^c = \Delta_a^c, \quad \bar{\Delta}_a^b \bar{\Delta}_b^c = \bar{\Delta}_a^c.$$

These complex constructions can be used as connecting quantities in mapping bitensors of the space B_{2n} on to a complex space of n dimensions. This was done in [261] (p. 247), where the tensor $p^{ab} = g^{ab} \pm i\varepsilon^{ab}$ introduced in the bivector space R_6 is identical with $\Delta^{ab} = g^{bc}\Delta_c^a$. The mapping follows the scheme

$$A^a = \Delta_b^a a^b, \quad \bar{A}^a = -iA^a.$$

To convince ourselves that the complex space on to which B_{2n} is mapped is n-dimensional, we need only note that the linear dependence between vectors in B_{2n},

$$\sum_{k=1}^{m} \lambda^k \underset{k}{a}^c + \sum_{k=1}^{m} \mu^k \underset{k}{\bar{a}}^c = 0,$$

changes to

$$\sum_{k=1}^{m} (\lambda^k - i\mu^k) \underset{k}{A}^c = 0$$

after the mapping. If the $\underset{k}{a}, \underset{k}{\bar{a}}$ are canonical basis vectors they are independent, as are the corresponding vectors $\underset{k}{A}$. Hence any $n+1$ complex vectors $\underset{k}{A}$ are interdependent, and the mapping takes place on a n-dimensional complex space, which we denote by A_n.

Δ_b^a allows us to associate a tensor $T_{ab\ldots}^{c\ldots}$ in A_n with any tensor $t_{ab\ldots}^{c\ldots}$ in the space B_{2n}, where

$$T_{ab\ldots}^{c\ldots} = \Delta_a^h \Delta_b^f \ldots \Delta_g^c \ldots t_{hf\ldots}^{g\ldots}.$$

If, however, one or more of the connecting quantities Δ_b^a in this formula are replaced by $\bar{\Delta}_b^a$ then the tensor $t_{bc\ldots}^{a\ldots}$ is associated with a quantity which can be considered as belonging to both mutually conjugate spaces A_n and

$\bar{A}_n$. These quantities are henceforth referred to as *Hermitians*, because of the similarity in transformation law to Hermitian forms.

Hence we can associate a tensor and a set of Hermitians in A_n with each tensor in B_{2n}. In particular we obtain the following *tensor mode* in A_n for a second order tensor

$$A_{ab} = \tfrac{1}{2}\Delta^c_a(a_{cb} - \bar{a}_{cb})$$

and a *Hermitian mode*

$$A_{a\bar{b}} = \tfrac{1}{2}\Delta^c_a(a_{cb} + \bar{a}_{cb}),$$

where

$$\bar{a}_{cb} = \varepsilon^l_c \varepsilon^f_b a_{lf}.$$

We call a tensor a *protensor* if its Hermitian mode is zero after the mapping; the necessary and sufficient conditions are

$$a_{ab} = -\bar{a}_{ab}.$$

If the tensor mode is zero, we refer to a *prohermitian*—the necessary and sufficient condition being

$$a_{ab} = \bar{a}_{ab}.$$

In the case of Minkowski spaces, the signatures $(+++-)$ or $(---+)$ can be fully described by the condition

$$e_{ijkl}e_{pqrs} = -\begin{vmatrix} g_{ip} & g_{jp} & g_{kp} & g_{lp} \\ g_{iq} & g_{jq} & g_{kq} & g_{lq} \\ g_{ir} & g_{jr} & g_{kr} & g_{lr} \\ g_{is} & g_{js} & g_{ks} & g_{ls} \end{vmatrix} \qquad (i, j = 1, 2, 3, 4),$$

linking the components of the antisymmetric permutation tensor e_{ijkl} with a single independent component $e_{1234} = \sqrt{|g|}$.

As was stated in § 15, the mapping of any bivector in a Minkowski space on to the vectors of the local bivector space E_6 can be expressed by means of connecting quantities

$$u^a = \eta^a_{ij}u^{ij} \quad (a = 1, \ldots, i;\ i, j = 1, \ldots, n), \quad \eta^a_{ij} = -\eta^a_{ji}.$$

As this mapping is one–one

$$u^{ij} = \eta^{ij}_a u^a,$$

where

$$\eta^{ij}_a \eta^b_{ij} = \delta^b_a, \quad \eta^{ij}_a \eta^a_{kl} = \delta^{ij}_{kl} = \delta^i_{[k}\delta^j_{l]}.$$

The metrized bivector space R_6 is obtained by introducing the tensor

$$g_{ab} = \eta^{ij}_a \eta^{kl}_b g_{ijkl}, \quad g_{ijkl} = g_{[i[k}g_{l]j]}$$

(cf. § 15) and the corresponding bitensor for e_{ijkl}

$$2\varepsilon_{ab} = 2\varepsilon_{ba} = \eta_a^{ij}\eta_b^{kl}e_{ijkl}.$$

As

$$\varepsilon_a^b\varepsilon_b^c = -\delta_a^c,$$

ε_b^a can be regarded as the absolute involution tensor in R_6. Clearly such a B_{2n} is identical with the *bivector* space introduced in § 15, and

$$\varepsilon_a^c\varepsilon_b^d g_{cd} = -g_{ab}, \quad \bar{g}_{ac} = -g_{ac},$$

i.e. the metric tensor is a protensor. The tensor mode G_{ab}, defined by

$$G_{ab} = \Delta_a^c\Delta_b^d g_{cd},$$

is regarded as the metric tensor of an A_3 space, which thus becomes a complex R_3.

We can construct a conjugate for any bitensor in a Minkowski space, and prove the following identity,

$$a_{pqij}+\bar{a}_{ijpq} = 4\overset{0}{a}_{[p[i}g_{j]q]}, \tag{21.2}$$

where

$$\overset{0}{a}_{pi} = a_{pi}-\tfrac{1}{4}g_{pi}a^q{}_q, \quad a_{pi} = a^q{}_{pqi}, \quad \overset{0}{a}{}^r{}_r = 0. \tag{21.3}$$

If $\overset{0}{a}_{jp} = 0$, (21.2) shows that

$$\bar{a}_{ijpq} = -a_{pqij}.$$

Bitensors satisfying these conditions are called *isotropic*. A bitensor satisfying the conditions

$$a_{pqij} = \bar{a}_{ijpq}$$

is said to be *simple*; the conditions

$$a_{ijkl} = a_{klij}, \quad a_{ijkl} = -a_{klij}$$

define *symmetric* and *antisymmetric* bitensors respectively.

After mapping on to the bivector space, the conditions for symmetry, antisymmetry, isotropy and simplicity and the conditions characterizing a protensor and prohermitian transform to

$$a_{ab} = a_{ba}, \quad a_{ab} = -a_{ba}, \quad a_{ab} = -\bar{a}_{ba},$$
$$a_{ab} = \bar{a}_{ba}, \quad a_{ab} = -\bar{a}_{ab}, \quad a_{ab} = \bar{a}_{ab}.$$

The relations between these properties can be formulated by means of the following elegant table: *if a bitensor satisfies any two conditions in a column it also satisfies the third condition in the same column.*

symmetric		antisymmetric	
simple	isotropic	simple	isotropic
prohermitian	protensor	protensor	prohermitian

(21.4)

The curvature tensor of a Riemannian space with a local Minkowski metric, is clearly a symmetric bitensor. If the V_4 is also an Einstein space, its Ricci tensor $R_{ij} = \frac{1}{4}R^k_k g_{ij}$, and (21.3) shows that the bitensor R_{ijkl} is isotropic. Then the rule based on table (21.4) shows that *the necessary and sufficient condition that a V_4 with a local Minkowski metric be an Einstein space is that its curvature tensor should be a protensor*, i.e. should map into a second order symmetric tensor in a complex R_3.

The problem of reducing the curvature tensor of an Einstein space to canonical form can thus be given the following alternative interpretation to that in § 19. We reduce to their canonical form in a complex R_3, the pair of symmetric tensors R_{ab} and G_{ab}, which are the tensor modes of the curvature tensor $R_{\alpha\beta\gamma\delta}$ and the tensor $g_{\alpha\gamma}g_{\beta\delta} - g_{\alpha\delta}g_{\beta\gamma}$ respectively for this mapping. We can use the well-known classification in [90], p. 309, for R_3, according to which the two forms

$$\psi = G_{ab}x^a x^b, \quad \varphi = R_{ab}x^a x^b,$$

where the first is non-singular, are reduced to one of the three pairs

$$\begin{aligned}
&\text{I. } \psi = x^{1^2}+x^{2^2}+x^{3^2}, && \varphi = ax^{1^2}+bx^{2^2}+cx^{3^2},\\
&\text{II. } \psi = 2x^1x^2+x^{3^2}, && \varphi = 2ax^1x^2 + bx^{2^2}+cx^{3^2},\\
&\text{III. } \psi = 2x^1x^2+x^{3^2}+x^{3^2}, && \varphi = 2ax^1x^2+ax^{3^2}+2bx^1x^3.
\end{aligned}$$

After inverse mapping from R_3 to R_6, we reduce to the canonical forms given in § 19 to within some Lorentz transformations and three-dimensional rotations, which correspond to certain linear transformations in R_6.

Examples

1. Show that the group of transformations in A_n, corresponding to the group of transformations in B_{2n}, is the general affine group of n-dimensional complex space.

2. Show that the unit tensor δ_a^b is a protensor because of (21.1) and that the corresponding quantity in an A_n space to this tensor is Δ_a^b which is also a unit tensor in A_n [316].

3. Prove identity (21.2) by using the definition of a conjugate tensor.

§22. Basis of the Complete System of Second Order Invariants of a V_4 Space

In investigating general gravitational fields it is of interest to consider the problem of the invariant characterization of the field by a set of rational invariant functions of the metric tensor components $g_{\alpha\beta}$ and their derivatives $\partial_{\gamma_1}g_{\alpha\beta},\ldots,\partial_{\gamma_1}\ldots_{\gamma_k}g_{\alpha\beta}$, which we call kth *order invariants*. The problem has not yet been solved in this form. The first stage in the solution is to determine the basis of the complete system of second order invariants; naturally the latter implies that the components of the metric tensor and the various permissible transformations are C^2 class functions. This first stage of the problem has been solved by Géhéniau and Debever [250], and also independently by P. I. Petrov [270]. In our exposition of the problem for the metric signature $(- - - +)$, we incorporate improvements in method introduced in reference [318].

We consider points in a V_4 space and the corresponding bivector space R_6, and associate the tensors g_{ab}, R_{ab}, and e_{ab} in the R_6 space with the tensors $g_{\alpha\gamma}g_{\beta\delta}-g_{\alpha\delta}g_{\beta\gamma}$, $R_{\alpha\beta\gamma\delta}$ and $e_{\alpha\beta\gamma\delta}$, where $e_{\alpha\beta\gamma\delta}$ is the permutation tensor of the V_4 space. We can in turn associate quantities in the three-dimensional complex space R_3 with these quantities in the R_6 space. The mapping can be carried out in two ways, and most simply if we use the following tensor in the V_4 space:

$$\Delta^{\alpha\beta\gamma\delta} = g^{\alpha[\gamma}g^{\beta]\delta} \pm ie^{\alpha\beta\gamma\delta}$$

([261], pp. 247–50); the corresponding tensors in R_6, taken with mixed components, are

$$\Delta_{\mathrm{a}}^{\mathrm{c}} \equiv \tfrac{1}{2}(\delta_{\mathrm{a}}^{\mathrm{c}}+ie_{\mathrm{a}}^{\mathrm{c}}), \quad \bar{\Delta}_{\mathrm{a}}^{\mathrm{c}} \equiv \tfrac{1}{2}(\delta_{\mathrm{a}}^{\mathrm{c}}-ie_{\mathrm{a}}^{\mathrm{c}}). \tag{22.1}$$

The tensor $\Delta_{\mathrm{a}}^{\mathrm{c}}$ allows us to associate with each tensor a_{ab} $(a, b, = 1, \ldots, 6)$ a *tensor mode*

$$A_{ab} \equiv \Delta_{\mathrm{a}}^{\mathrm{c}}\Delta_{\mathrm{b}}^{d}a_{cd} = \tfrac{1}{2}\Delta_{\mathrm{a}}^{\mathrm{c}}(a_{cb}-\bar{a}_{cb}) \tag{22.2}$$

and a *Hermitian mode*

$$A_{a\bar{b}} = \Delta_{\mathrm{a}}^{\mathrm{c}}\bar{\Delta}_{\mathrm{b}}^{d}a_{cd} = \tfrac{1}{2}\Delta_{\mathrm{a}}^{\mathrm{c}}(a_{cb}+\bar{a}_{cb}), \tag{22.3}$$

where

$$\bar{a}_{cb} = e^d_c e^l_b a_{dl}. \tag{22.4}$$

The tensor Δ^c_a is a unit tensor in the complex space R_3. We recall that a bitensor is *isotropic* (or *protensor*) if it satisfies the condition

$$a_{\alpha\beta\gamma\delta} = a_{\gamma\delta\alpha\beta} \tag{22.5}$$

and has Hermitian mode $A_{a\bar{b}} = 0$ (cf. § 21). A bitensor is *simple* (or prohermitian) if the conditions (22.5) and $A_{ab} = 0$ are both satisfied. An arbitrary bitensor can be resolved into invariant parts by representing it as a sum

$$a_{\alpha\beta,\,\gamma\delta} = \overset{*}{a}_{\alpha\beta,\,\gamma\delta} + \overset{0}{a}_{\alpha\beta,\,\gamma\delta} \tag{22.6}$$

of isotropic and simple bitensors respectively; these terms are completely characterized by the conditions

$$\overset{*}{a}{}^{\alpha}{}_{\beta\alpha\gamma} = \frac{A}{4}\, g_{\beta\gamma} \tag{22.7}$$

or

$$\overset{*}{A}_{ab} = \Delta^c_a a_{cb} \tag{22.8}$$

and

$$\overset{0}{a}_{\alpha\beta,\,\gamma\delta} = 2\overset{0}{a}_{[\alpha[\beta} g_{\gamma]\delta]}, \tag{22.9}$$

where $\overset{0}{a}{}^c_c = 0$; as a result of this $\overset{0}{a}{}^{\alpha}{}_{\beta\alpha\gamma} = \overset{0}{a}_{\beta\gamma}$, and

$$\overset{0}{A}_{a\bar{b}} = \Delta^c_a a_{cb}. \tag{22.10}$$

The required basis must consist of invariants, each obtainable by complete contraction of some tensor by means of the metric and permutation tensors of the V_4 space. We first consider quantities a_{ab} and b_{ab} in the bivector space which are symmetric, isotropic or simple tensors. Corresponding to these there are tensors A_{ab} and B_{ab} in R_3 or Hermitians modes $A_{a\bar{b}}$ and $B_{a\bar{b}}$ respectively. Equation (22.8) shows that

$$A_{ab} B^f_c = \Delta^e_a \Delta^h_c a_{eb} b^f_h,$$

and hence the bitensor

$$C_{\alpha\beta,\,\gamma\delta} = a_{\alpha\beta,\,\sigma\tau} b_{\gamma\delta}{}^{\sigma\tau},$$

if a and b are isotropic bitensors, is mapped in R_3 by the tensor

$$C_{ab} = A_{ac} B^c_b. \tag{22.11}$$

Similarly, when a and b are simple tensors and therefore mapped by Hermitians in R_3, the tensor mode $C_{\alpha\beta,\,\gamma\delta}$ in R_3 is

$$C_{ab} = A_{a\bar{c}}B^{\bar{c}}_{\bar{b}}. \tag{22.12}$$

Equation (22.9) shows that

$$C_{\alpha\beta,\,\gamma\delta} = 2a_{\sigma[\alpha}g_{\beta]\tau}b^{\sigma\tau}{}_{\gamma\delta} = 2a_{\sigma[\alpha}b^{\sigma}{}_{\beta]\gamma\delta}. \tag{22.13}$$

But as

$$b^{\sigma}{}_{\alpha\beta\gamma} = 2\delta^{[\sigma}{}_{[\beta}b^{\tau]}{}_{\gamma]}g_{\tau\alpha},$$

$$C_{\alpha\beta,\,\gamma\delta} = 4a^{\sigma}_{\alpha}b_{[\gamma[\sigma}g_{\beta]\delta]} \qquad [\alpha\beta], \tag{22.14}$$

where the symbol $[\alpha\beta]$ indicates that the final result must be alternated over α, β.

If it is further assumed that $a_{\alpha\beta} = b_{\alpha\beta}$,

$$C_{\alpha\beta,\,\gamma\delta} = 2(a_{[\alpha\,|\,\sigma\,|}\,a^{\sigma}{}_{[\gamma}g_{\delta]\beta]} + a_{\alpha[\gamma}a_{\delta]\beta}). \tag{22.15}$$

We now consider (cf. example 1) the Hermitian

$$\overset{2}{A}_{c\bar{c}} \equiv \varepsilon_{abc}\varepsilon_{\bar{a}\bar{b}\bar{c}}A^{a\bar{a}}A^{b\bar{b}}, \tag{22.16}$$

obtained by contracting the trivector $\varepsilon_{abc} \equiv \Delta^{f}_{a}l_{fbc}$ in the R_3 space, corresponding to an antisymmetric tensor l_{fbc} in R_6, and the complex conjugate trivector $\varepsilon_{\bar{a}\bar{b}\bar{c}}$, by means of the Hermitian. Example 1 and eqn. (22.10) show that the Hermitian mode $A_{c\bar{c}}$ in V_4 has the form

$$\overset{2}{a}_{\alpha\beta,\,\bar{\alpha}\bar{\beta}} = g_{\alpha\beta,\,\gamma\delta,\,\sigma\tau}g_{\bar{\alpha}\bar{\beta},\,\bar{\gamma}\bar{\delta},\,\bar{\sigma}\bar{\tau}}a^{\gamma\delta,\,\bar{\gamma}\bar{\delta}}a^{\sigma\tau,\,\bar{\sigma}\bar{\tau}}$$

or

$$\overset{2}{a}_{\alpha\beta,\,\bar{\alpha}\bar{\beta}} = a_{\bar{\alpha}}{}^{\bar{\sigma}}{}_{\nu[\alpha}a_{\beta]}{}^{\bar{\nu}}{}_{\bar{\sigma}\bar{\beta}}. \tag{22.17}$$

As this bitensor is prohermitian it is simple and can be calculated by (22.9) provided we can evaluate the tensor

$$\overset{2}{a}_{\beta\bar{\beta}} = \overset{2}{a}{}^{\alpha}{}_{\beta,\,\alpha\bar{\beta}} = \tfrac{1}{2}(a^{\alpha\bar{\tau}}{}_{\sigma\alpha}a^{\sigma}{}_{\bar{\beta}\bar{\tau}\bar{\beta}} - a^{\alpha\bar{\tau}}{}_{\sigma\beta}a^{\sigma}{}_{\alpha,\,\bar{\tau}\bar{\beta}}) = -\tfrac{1}{4}(a_{\sigma\bar{\tau}}a^{\sigma\bar{\tau}}{}_{\beta\bar{\beta}} + a_{\alpha\bar{\beta}}a^{\alpha}{}_{\beta}).$$

Noting that

$$a_{\sigma\alpha}b^{\sigma}{}_{\beta,\,\gamma\delta} = \tfrac{1}{2}a_{\sigma\alpha}(b^{\sigma}{}_{\gamma}g_{\delta\beta} - b^{\sigma}{}_{\delta}g_{\gamma\beta} - b_{\beta\gamma}\delta^{\sigma}_{\delta} + b_{\delta\beta}\delta^{\sigma}_{\gamma}), \tag{22.18}$$

it is easy to show that

$$a_{\sigma\bar{\alpha}}a^{\sigma}{}_{\beta}{}^{\bar{\alpha}}{}_{\bar{\beta}} = a_{\alpha\bar{\beta}}a^{\alpha}{}_{\beta} - \tfrac{1}{2}g_{\beta\bar{\beta}}a_{\sigma\tau}a^{\sigma\tau},$$

and therefore

$$a^{2}_{\beta\bar{\beta}} = -\tfrac{1}{2}(a_{\beta\sigma}a^{\sigma}{}_{\bar{\beta}} - \tfrac{1}{4}a_{\sigma\tau}a^{\sigma\tau}b_{\beta\bar{\beta}}). \tag{22.19}$$

We can now construct the set of invariants defining the required basis by using (22.19).

We divide the curvature tensor into its isotropic and simple parts

$$R_{\alpha\beta,\,\gamma\delta} = \overset{*}{R}_{\alpha\beta,\,\gamma\delta} + \overset{0}{R}_{\alpha\beta,\,\gamma\delta}. \tag{22.20}$$

Corresponding to the curvature tensor there is the symmetric tensor R_{ab} in the bivector space; we then obtain the tensor mode B_{ab} and the Hermitian mode $S_{a\bar{b}}$ in the complex R_3 space from eqns. (22.2) and (22.3). The metric tensor g_{ab} in R_6 is isotropic and defines only the tensor mode G_{ab} in R_3. We now consider the quantities B_{ab}, $S_{a\bar{b}}$, G^{ab} and ε_{abc} and by their complete contraction form the set of invariants

$$\left.\begin{aligned} &B_1 = B^a_a, \quad B_2 = \overset{2}{B}{}^a_a, \quad B_3 = \overset{2}{B}{}^a_b B^b_a,\\ &S_2 = \overset{2}{S}{}^a_a, \quad S_3 = \varepsilon_{abc}\varepsilon_{\bar{a}\bar{b}\bar{c}} S^{a\bar{a}} S^{b\bar{b}} S^{c\bar{c}}, \quad S_4 = \overset{2}{S}{}^a_b \overset{2}{S}{}^b_a,\\ &T_1 = B^a_b \overset{2}{S}{}^b_a, \quad T_2 = \overset{2}{B}{}^a_b \overset{2}{S}{}^b_a, \quad T_3 = \overset{2}{B}{}^a_b \overset{2}{S}{}^b_c S^c_a, \end{aligned}\right\} \tag{22.21}$$

where

$$\overset{2}{B}{}^a_b = B^a_c B^c_b, \quad \overset{2}{S}{}^a_b = S^a_{\bar{c}} S^{\bar{c}}_b$$

and the indices are raised by means of the tensor G^{ab} or its complex conjugate $G^{\bar{a}\bar{b}}$. The invariants S_2, S_3 and S_4 in set (22.21) are real, since taking the complex conjugates of their defining equations transforms them into themselves; as will be shown later, the invariant B_1 is also real. The remaining invariants are complex, so that taking the complex conjugates into into account, we obtain a set of fourteen invariants.

To prove that the set (22.21) is complete, we must show that these invariants are independent. Here we must make the following reservation: the number of independent invariants of a V_4 space can always be linked with the bases of the elementary divisors of the λ-matrix of the bivector space $(R_{ab} - \lambda g_{ab})$, which is not in general doubly symmetric. Clearly if we base the classification of V_4 on the Segre characteristics of the λ-matrix, then in addition to the case $\{111, 111\}$ *corresponding to the maximum number of independent invariants*, we obtain Segre characteristics corresponding to *non-linear* elementary divisors, e.g. $\{21111\}$, etc., which *cannot be considered as special cases of V_4* corresponding to the simple characteristic, but as essentially distinct types of space. This argument becomes particularly obvious for Einstein spaces with the three types of T_i space $(i = 1, 2, 3)$. It is clear, however, that a V_4 with a *non-simple type* of characteristic has less than fourteen second order invariants. We therefore consider completeness only for V_4 admitting the *maximum* number of independent invariants.

The quantities (22.21) are algebraic functions of the twenty independent components of the tensors $B_{ab}, \bar{B}_{ab}$ and $S_{a\bar{b}}$. To prove that the invariants are independent in the case when their number is a maximum, we must consider the Jacobian of these functions in a space with point coordinates defined by the independent components of the tensors $B_{ab}, \bar{B}_{ab}$ and $S_{a\bar{b}}$. If a point is found at which a fourteenth order determinant differs from zero, then we can always find another point in the neighbourhood of this point with the same property. Hence the independence is proved if we can show that a single fourteenth order determinant is not zero at a single point. It is not in fact difficult to construct such a determinant (cf. example 2). Thus (22.21) forms the required basis.

The invariants of a V_4 space can be set down by using the following equations, which are easily shown to be valid (cf. example 3):

$$\left.\begin{aligned} B^a_b &\sim \overset{*}{R}{}^{\alpha\beta}{}_{\gamma\delta}, \\ \overset{2}{B}{}^a_b &\sim \overset{*}{R}{}^{\alpha\beta}{}_{\sigma\tau}\overset{*}{R}{}^{\sigma\tau}{}_{\gamma\delta}, \\ \overset{2}{S}{}^a_b &\sim \overset{0}{R}{}^{\sigma\varrho}_{\alpha\beta}\overset{0}{R}{}_{\sigma\varrho}{}^{\gamma\delta} = 4\overset{0}{R}{}^{\sigma}_{\alpha}\overset{0}{R}{}^{[\gamma}_{[\sigma}\delta^{\delta]}_{\beta]}, \end{aligned}\right\} \tag{22.22}$$

where the symbol $\sim$ denotes association between quantities in R_3 and quantities at a point in the V_4 space. Furthermore, corresponding to an invariant C^a_a in R_3 there are two invariants at a point in the V_4 space:

$$C^a_a \sim C^{\alpha\beta}_{\alpha\beta} + i\varepsilon^{\alpha\beta}{}_{kl}C^{kl}_{\alpha\beta}. \tag{22.23}$$

The real and imaginary parts of this expression differ in that the real part is contracted by means of the metric bitensor, the imaginary part by a quadrivector. Applying (22.23) to the quantities (22.22) we obtain the basis of the set of invariants in the following form:

$$\left.\begin{aligned} &B_1 = \overset{*}{R}{}^{\sigma\tau}{}_{\sigma\tau}, \quad B_2 = \overset{*}{R}{}^{\alpha\beta}{}_{\gamma\delta}\overset{*}{R}{}^{\gamma\delta}{}_{\alpha\beta}, \quad B_3 = \overset{*}{R}{}^{\alpha\beta}{}_{\gamma\delta}\overset{*}{R}{}^{\gamma\delta}{}_{\sigma\tau}\overset{*}{R}{}^{\sigma\tau}{}_{\alpha\beta}, \\ &B^1_2 = e^{\alpha\beta}{}_{\gamma\delta}\overset{*}{R}{}^{\gamma\delta}{}_{\sigma\tau}\overset{*}{R}{}^{\sigma\tau}{}_{\alpha\beta}, \quad B^1_3 = e^{\alpha\beta}{}_{\gamma\delta}\overset{*}{R}{}^{\gamma\delta}{}_{\sigma\tau}\overset{*}{R}{}^{\sigma\tau}{}_{\lambda\nu}\overset{*}{R}{}^{\lambda\nu}{}_{\alpha\beta}, \end{aligned}\right\} \quad (\alpha)$$

$$S_2 = \overset{0}{R}{}^{\alpha}{}_{\beta}\overset{0}{R}{}^{\beta}{}_{\alpha}, \quad S_3 = \overset{0}{R}{}^{\alpha}{}_{\beta}\overset{0}{R}{}^{\beta}{}_{\gamma}\overset{0}{R}{}^{\gamma}{}_{\alpha}, \quad S_4 = \overset{0}{R}{}^{\alpha}{}_{\beta}\overset{0}{R}{}^{\beta}{}_{\gamma}\overset{0}{R}{}^{\gamma}{}_{\delta}\overset{0}{R}{}^{\delta}{}_{\alpha}, \quad (\beta)$$

$$\left.\begin{aligned} &T_1 = \overset{*}{R}{}^{\alpha\beta}{}_{\gamma\delta}\overset{0}{R}{}^{\gamma}{}_{\alpha}\overset{0}{R}{}^{\delta}{}_{\beta}, \quad T_2 = \overset{*}{R}{}^{\alpha[\beta}{}_{\gamma\delta}\overset{0}{R}{}^{\sigma]}{}_{\alpha}\overset{*}{R}{}^{\gamma\delta}{}_{\sigma\tau}\overset{0}{R}{}^{\tau}{}_{\beta}, \quad T_3 = \overset{*}{R}{}^{\alpha\beta}A^{\gamma\delta}{}_{\alpha\beta}, \\ &T^1_1 = e^{\alpha\beta}{}_{\gamma\delta}\overset{*}{R}{}^{\gamma\delta}{}_{\sigma\tau}\overset{0}{R}{}^{\sigma}{}_{\alpha}\overset{0}{R}{}^{\tau}{}_{\beta}, \quad T^1_2 = e^{\alpha[\beta}{}_{\gamma\delta}\overset{0}{R}{}^{\nu]}{}_{\alpha}\overset{*}{R}{}^{\gamma\varrho}{}_{\sigma\tau}R^{\sigma\tau}{}_{\nu\lambda}\overset{0}{R}{}^{\lambda}{}_{\beta}, \\ &T^1_3 = e^{\alpha\beta}{}_{\sigma\tau}\overset{*}{R}{}^{\sigma\tau}{}_{\gamma\delta}A^{\gamma\delta}{}_{\alpha\beta}, \end{aligned}\right\} \quad (\gamma)$$

where

$$A^{\gamma\delta}{}_{\alpha\beta} \equiv 4\overset{0}{R}{}^{\gamma}{}_{\sigma}\overset{0}{R}{}^{\sigma}{}_{\tau}\overset{0}{R}{}^{\tau}{}_{\alpha}\overset{0}{R}{}^{\delta}{}_{\beta}+3\overset{0}{R}{}^{\gamma}{}_{\sigma}\overset{0}{R}{}^{\sigma}{}_{\alpha}\overset{0}{R}{}^{\delta}{}_{\tau}\overset{0}{R}{}^{\tau}{}_{\beta}.$$

The invariant $B_{\bar{1}}^{1} = e^{\sigma\tau}{}_{\lambda\nu}\overset{*}{R}{}^{\lambda\nu}$ is missing from this set as it is zero because of the Ricci indentity. The invariants $T_{\bar{1}}^{1}$ and $T_{\bar{3}}^{1}$ have further terms of the form $e^{\alpha\beta}{}_{\sigma\tau}\overset{*}{R}{}^{\sigma\tau}{}_{\alpha\lambda}\overset{0}{R}{}^{\delta}{}_{\beta}\overset{0}{R}{}^{\lambda}{}_{\delta}$ and $e^{\alpha\beta}{}_{\gamma\delta}\overset{*}{R}{}^{\gamma\delta}{}_{\alpha\sigma}\overset{0}{R}{}^{\tau}{}_{\beta}\overset{0}{R}{}^{\lambda}{}_{\tau}\overset{0}{R}{}^{\nu}{}_{\lambda}\overset{0}{R}{}^{\sigma}{}_{\nu}$ which also reduce to zero because of this identity. Invariants T_1 and T_3 are also simplified since terms containing the tensor $\overset{*}{R}{}^{\alpha}{}_{\beta}$ have been expressed in terms of $R\delta^{\alpha}_{\beta}$ by means of (22.7) where R is the scalar curvature, so that these terms are expressed by invariants of lower degree. Thus the required basis is given by the fourteen invariants (α), (β) and (γ).

EXAMPLES

1. Show that the metric *tritensor*

$$g_{\alpha\beta,\,\gamma\delta,\,\lambda\nu} \equiv g_{[\alpha[\gamma}g_{\delta]\lambda}g_{\nu]\beta}$$

is mapped in R_6 by a trivector l_{abc}, and in R_3 by the trivector

$$\varepsilon_{abc} = \Delta^{f}_{a} l_{fbc}.$$

2. Prove that the system (22.21) is independent by investigating the fourteenth order Jacobian determinant, the rows containing the dependent variables $S_2, S_3, S_4, B_1, B_2, B_3, \bar{B}_2, \bar{B}_3, T_1, T_2, T_3, \bar{T}_1, \bar{T}_2, \bar{T}_3$, and the columns the independent variables $\bar{B}_{12}, \bar{B}_{13}, \bar{B}_{23}, R_{12}, R_{13}, R_{23}, \bar{B}_{11}, \bar{B}_{22}, B_{11}, B_{22}, B_{33}, S_{1\bar{1}}, S_{2\bar{2}}, S_{3\bar{3}}$, under the condition $B_{12}, = B_{13} = B_{23} = 0$ [318].

3. Prove formulae (22.22) and (22.23) [318].

4. Show that the simplest basis for Einstein spaces is formed by the set of invariants $B_2, B_{\bar{2}}^{1}, B_3, B_{\bar{3}}^{1}$. Compare this with the results of § 19.

5. Show that the basis for conformally flat spaces is determined by the invariants B_1, S_2, S_3, S_4. If we have a space of constant curvature, the basis reduces to one invariant, e.g. R.

CHAPTER 4

MOTIONS IN EMPTY SPACE

IN THIS chapter we consider gravitational fields in empty space-time ($R_{\alpha\beta} = 0$) admitting various groups of motions. The theory of Lie groups permits an invariant treatment of this important special case. The problem is to find which groups of motion are admitted by gravitational fields of all three possible types. It should be noted that the analogous problem for general gravitational fields has not yet been fully solved, although there does exist a complete classification of gravitational fields admitting groups of motion of all possible orders (cf. Chapter 5); whilst it is known which metrics [$n = 4$, signature $(- - - +)$] admit a given group of motions, the inverse problem as to which groups of motion correspond to a given type of gravitational field remains unsolved.

For the correct formulation of the problem of groups of motion in empty space, the more general Einstein spaces ($R_{\alpha\beta} = \varkappa g_{\alpha\beta}$, $\varkappa \neq 0$) must be considered. As the type of space-time may alter (III $\to$ II $\to$ I) as $\varkappa \to 0$, the classification considered in this chapter is not complete. However, it is sufficiently comprehensive and includes the most interesting, i.e. physically significant cases.

§23. Classification of T_i by Groups of Motions

Let us consider a gravitational field in empty space, i.e. a region where the energy-momentum tensor is zero. The geometry of the field is determined by the components of the metric tensor $g_{\alpha\beta}(x)$ relative to some system of coordinates. These components describe not only the physical, "purely gravitational" properties of the space-time, but also reflect the peculiarities of the particular reference system. Attempts to introduce "inertial" systems of coordinates which separate out the effects of the particular coordinate system ([225], p. 346; [259]) have not yet won general acceptance.

It is therefore natural to base the study of gravitational fields on the invariant properties of the metric defined by the field.

Suppose that the metric of a certain space, defined in some system of coordinates, admits a Lie group G_r of continuous transformations. Then if the metric is given in any other reference system, it still admits the same group. The transformations in this group map the space into itself; such a mapping is called an *automorphism*. The special transformations preserving the metric are *motions* in the V_n. In this sense there is an automorphism corresponding to every gravitational field. It is obvious that $0 \leq r \leq n(n+1)/2$ (≤ 10 for $n = 4$). G_0 is interpreted as the group consisting of the unit element (the identity transformation); when r is a maximum, the space has constant curvature [it is a Minkowski space, when $n = 4$ with signature $(- \ - \ - \ +)$ and zero scalar curvature].

This idea of associating the geometry of a given automorphism with a field, has been developed by several authors; the problem is reduced to determining the possible groups of motion and the corresponding space-time metrics. These considerations apply not only to empty spaces but also to general gravitational fields (cf. Chapter 5). In this chapter we consider V_4 with signature $(- \ - \ - \ +)$ and field equations $R_{\alpha\beta} = 0$ ([261], Ch. II; [180]).

In addition to studying space-times from the point of view of the groups they admit, we can consider the conformal infinitesimal transformations admitted by the manifold [233], [236], or holonomy groups [325], [328].

In order that a T_i space ($i = 1, 2, 3$,) should admit a motion with the operator $X = \xi^\alpha \partial_\alpha$ it is necessary and sufficient (§ 10) that ξ^α should satisfy the Killing equations

$$\underset{L}{\delta} g_{\alpha\beta} = \xi_{\alpha,\beta} + \xi_{\beta,\alpha} = 0. \tag{23.1}$$

In order that a V_n should admit a group G_r, it is necessary and sufficient that a positive integer k should exist, such that the rank of the matrix of the combined sets of linear algebraic equations (10.14) up to the kth in the unknowns ξ_α and $\xi_{\alpha,\beta}$ satisfying condition (23.1), should be $\frac{1}{2}n(n+1)-r$, and that the addition of further sets of equations from the $(k+1)$th onwards should not alter this rank (§ 10).

In classifying T_i ($i = 1, 2, 3$) by groups of motion, the following two questions are of primary importance. (1) Suppose that T_i admits a G_r group of motions; what values can r take for a given i? (2) What is the group structure for a given r and i?

The following theorem determines the upper limit of r: *there are no T_i spaces ($i = 1, 2, 3$) other than spaces of constant curvature which admit groups of motions G_r with $r \geqslant 7$.*†

A V_n cannot admit a group of motions G_r with $r > n(n+1)/2$ (for $n = 4$, $r \leqslant 10$). If $r = 10$, we have a space of constant curvature, in this case a flat Minkowski space; this will always be a T_1 and can be obtained from (19.18) by putting $\alpha_i = \beta_i = 0$. We omit this trivial case. The possibility of or $r = 9$ is excluded by the general theorem of Fubini [18], p 54), which states that a V_n with $n > 2$ does not admit a complete group of motions order $\frac{1}{2}n(n+1)-1$. The order r of a complete group cannot be equal to eight either, as was shown by Yegorov ([230], p 202) for V_4 with the signature $(- - - +)$. We will now show that for $T_i (i = 1, 2, 3)$, the order of maximum mobility can be still further reduced to six.

In order that r should equal 7, the matrix of the set

$$\underset{L}{\delta} R_{\alpha\beta\gamma\delta} = 0 \tag{23.2}$$

or of the expanded set of homogenous algebraic equations

$$\underset{L}{\delta} R_{\alpha\beta\gamma\delta} \equiv B^{\sigma\tau}{}_{\gamma\delta\alpha\beta}\xi_{\sigma,\tau} + A^{\sigma}{}_{\gamma\alpha\beta\delta}\xi_{\sigma} = 0, \tag{23.3}$$

in the unknowns $\xi_{\alpha,\beta}$ and ξ_{α} satisfying condition (23.1), must be of rank 3. Here

$$A^{\sigma}{}_{\lambda\alpha\beta\delta} \equiv R_{\beta\nu\alpha\delta,}{}^{\sigma}, \; B^{\sigma\tau}{}_{\gamma\delta\alpha\beta} \equiv 2\delta^{\sigma}_{[\delta}R^{\tau}{}_{\gamma]\alpha\beta} + 2\delta^{\sigma}_{[\beta}R^{\tau}{}_{\alpha]\gamma\delta}. \tag{23.4}$$

Let us consider the matrix formed from the coefficients A and B for an orthonormal tetrad with components of the curvature tensor defined by eqns. (19.18) and (19.19) (19.20). Neglecting those lines which are identical in virtue of the double symmetry of the matrices (R_{ab}) and $(R_{ab,\lambda})$, we obtain from (19.18)

$$(T_1)\begin{pmatrix} 0 & 0 & 0 & 0 & 0 & 0 & R_{1414,\sigma} \\ 0 & 0 & \beta_1-\beta_2 & 0 & 0 & \alpha_1-\alpha_2 & R_{1424,\sigma} \\ 0 & \beta_3-\beta_1 & 0 & 0 & \alpha_3-\alpha_1 & 0 & R_{1434,\sigma} \\ 0 & 0 & 0 & 0 & 0 & 0 & R_{1423,\sigma} \\ 0 & 0 & \alpha_2-\alpha_1 & 0 & 0 & \beta_1-\beta_2 & R_{1431,\sigma} \\ 0 & \alpha_1-\alpha_3 & 0 & 0 & \beta_3-\beta_1 & 0 & R_{1412,\sigma} \\ 0 & 0 & 0 & 0 & 0 & 0 & R_{2424,\sigma} \\ \beta_2-\beta_3 & 0 & 0 & \alpha_2-\alpha_3 & 0 & 0 & R_{2434,\sigma} \\ 0 & 0 & 0 & 0 & 0 & 0 & R_{2431,\sigma} \\ \alpha_3-\alpha_2 & 0 & 0 & \beta_2-\beta_3 & 0 & 0 & R_{2412,\sigma} \end{pmatrix}; \tag{23.5}$$

† This theorem and its proof also hold, *mutatis mutandis*, for the case $R_{\alpha\beta} = \varkappa g_{\alpha\beta}$ ($\varkappa \neq 0$),

The first six columns correspond to $\sigma\tau$'s 14, 24, . . ., in (23.4) and the last column represents the four corresponding to $\sigma = 1, 2, 3, 4$. Any fourth order minor determinant of this matrix must be zero. Considering for example, the determinant formed from rows and columns 2, 3, 5 and 6, we obtain

$$[(\alpha_1-\alpha_2)^2+(\beta_1-\beta_2)^2][(\alpha_1-\alpha_3)^2+(\beta_1-\beta_3)^2] = 0.$$

Putting $\alpha_s+i\beta_s = K_s$, $\alpha_s-i\beta_s = \bar{K}_s$ ($s = 1, 2, 3,$) we obtain

$$(K_1-K_2)(\bar{K}_1-\bar{K}_2)(K_1-K_3)(\bar{K}_1-\bar{K}_3) = 0.$$

Hence

$$(K_1-K_2)(K_1-K_3) = 0.$$

As K_1, K_2 and K_3 are essentially equivalent we can obtain the other determinants by a cyclic interchange of the indices 1, 2, and 3 in the last equation. We thus obtain

$$(K_3-K_1)(K_1-K_2) = (K_2-K_3)(K_3-K_1) = (K_1-K_2)(K_2-K_3) = 0.$$

The conditions $\Sigma\alpha_i = \Sigma\beta_i = \Sigma K_i = 0$, show that $K_s = 0$ ($s = 1, 2, 3$), corresponding to a Minkowski space.

For T_2, we obtain the following matrix from (19.19)

$$(T_2)\quad \left(\begin{array}{ccccccc}
0, & 0, & 0, & 0, & 0, & 0, & R_{1414,\sigma} \\
0, & -1, & \beta_2-\beta_1, & 0, & 0, & \alpha_2-\alpha_1+1, & R_{1424,\sigma} \\
0, & -\beta_2+\beta_1, & 1, & 0, & \alpha_1-\alpha_2+1, & 0, & R_{1434,\sigma} \\
0, & 0, & 0, & 0, & 0, & 0, & R_{1423,\sigma} \\
0, & 0, & \alpha_1-\alpha_2-1, & 0, & -1, & \beta_2-\beta_1, & R_{1431,\sigma} \\
0, & \alpha_2-\alpha_1-1, & 0, & 0, & \beta_1-\beta_2, & 1, & R_{1412,\sigma} \\
2, & 0, & 0, & 0, & 0, & 0, & R_{2424,\sigma} \\
0, & 0, & 0, & -2, & 0, & 0, & R_{2434,\sigma} \\
0, & 0, & 0, & 2, & 0, & 0, & R_{2431,\sigma} \\
2, & 0, & 0, & 0, & 0, & 0, & R_{2412,\sigma} \\
-2, & 0, & 0, & 0, & 0, & 0, & R_{3434,\sigma} \\
0, & 0, & 0, & -2, & 0, & 0, & R_{3412,\sigma}
\end{array}\right). \quad (23.6)$$

It is not difficult to see that equating any fourth order determinant to zero leads to a contradiction.

For T_3 we obtain from (19.20)

$$(T_3)\begin{pmatrix} 0 & 0 & 0 & 0 & 0 & 2 & R_{1414,\,\sigma} \\ 0 & -1 & 0 & 0 & 0 & 0 & R_{1424,\,\sigma} \\ 0 & 0 & 1 & -1 & 0 & 0 & R_{1434,\,\sigma} \\ 0 & 0 & -2 & 0 & 0 & 0 & R_{1423,\,\sigma} \\ 0 & 0 & 0 & 0 & -1 & 0 & R_{1431,\,\sigma} \\ 1 & 0 & 0 & 0 & 0 & 1 & R_{1412,\,\sigma} \\ 2 & 0 & 0 & 0 & 0 & -2 & R_{2424,\,\sigma} \\ 0 & 0 & 0 & 0 & 1 & 0 & R_{2434,\,\sigma} \\ 0 & 0 & 2 & 2 & 0 & 0 & R_{2431,\,\sigma} \\ 0 & -1 & 0 & 0 & 0 & 0 & R_{2412,\,\sigma} \\ -2 & 0 & 0 & 0 & 0 & 0 & R_{3434,\,\sigma} \\ 0 & 0 & 0 & -2 & 0 & 0 & R_{3412,\,\sigma} \end{pmatrix}. \tag{23.7}$$

A direct examination of this matrix shows the existence of a sixth order non-zero determinant; hence $r < 7$, which proves the theorem.

The matrices (23.5), (23.6) and (23.7) enable us to specify the order of the complete group, (i.e. group of maximum order), for all three types of T_i. We use the theorem: *if T_1 admits a group of motions G_r with $r > 4$ it also admits a group G_{10} and is a space corresponding to the special theory of relativity (a Minkowski space); T_2 has maximum mobility $r = 6$, and T_3 $r \leqslant 4$.*

The compact bivector-space indices replace $R_{\alpha\beta\gamma\delta,\,\sigma}$ by $R_{ab,\,\sigma}$ $(a, b = 1, \ldots, 6; \sigma = 1, 2, 3, 4)$. Using the notation

$$R_{ab,\,\sigma}+iR_{ab+3,\,\sigma} = \Pi_{ab,\,\sigma} \qquad (a,\ b \leqslant 3) \tag{23.8}$$

and allowing for the double symmetry of $R_{ab,\,\sigma}$ which follows from (5.13), we can restate the Bianchi identities (5.16) and relations (23.8) in the form of the twelve identities

$$\left.\begin{aligned} \Pi_{1a,\,1}+\Pi_{2a,\,2}+\Pi_{3a,\,3} &= 0, \\ \Pi_{3a,\,1}-\Pi_{1a,\,3}+i\Pi_{2a,\,4} &= 0, \\ \Pi_{2a,\,1}-\Pi_{1a,\,2}-i\Pi_{3a,\,4} &= 0, \\ \Pi_{3a,\,2}-\Pi_{2a,\,3}-i\Pi_{1a,\,4} &= 0, \end{aligned} \qquad (a = 1, 2, 3). \right\} \tag{23.9}$$

Moreover, (5.13) shows that

$$\Pi_{11,\,\sigma}+\Pi_{22,\,\sigma}+\Pi_{33,\,\sigma} = 0. \tag{23.10}$$

Assuming that T_1 admits a group of motions G_5, then all the sixth order minor determinants of the matrix (23.5) must be zero. It is certainly true that among the fourth order determinants formed from the first six columns, there is at least one which is not zero, since otherwise the T_1 would be flat. As the three sets of α_i and β_i are essentially equivalent, it can be assumed without loss of generality that, for example, the determinant formed from rows and columns 2, 3, 5 and 6, equal to

$$[(\alpha_1 - \alpha_3)^2 + (\beta_1 - \beta_3)^2][(\alpha_1 - \alpha_2)^2 + (\beta_1 - \beta_2)^2]$$

is not zero. As all the other elements of these rows and columns (up to column six) are zero, all the second order minors formed from $R_{1414,\sigma}$, $R_{1423,\sigma}$, $R_{2424,\sigma}$, $R_{2431,\sigma}$, $R_{3434,\sigma}$, $R_{3412,\sigma}$ $(\sigma = 1, 2, 3, 4)$, and from the matrices

$$\begin{pmatrix} \beta_2 - \beta_3 & \alpha_2 - \alpha_3 & R_{2434,\sigma} \\ \alpha_3 - \alpha_2 & \beta_2 - \beta_3 & R_{2412,\sigma} \end{pmatrix} \qquad (\sigma = 1, 2, 3, 4) \tag{23.11}$$

must be zero. Combining the rows of these matrices in pairs as in (23.8), we see that any minor of the matrix

$$\begin{pmatrix} \Pi_{11,\sigma} & 0 \\ \Pi_{22,\sigma} & 0 \\ \Pi_{33,\sigma} & 0 \\ \Pi_{23,\sigma} & K_2 - K_3 \end{pmatrix} \qquad (\sigma = 1, 2, 3, 4) \tag{23.12}$$

is zero. It then follows from (23.10) that $\Pi_{aa,\sigma} = 0$ $(a = 1, 2, 3)$ whether $K_2 - K_3$ is zero or not. Let us consider the quantities $\Pi_{23,\sigma}$. If $K_2 \neq K_3$, it follows from (23.11) that $\Pi_{23,\sigma} = 0$ and the Bianchi identities (23.9) show that all $\Pi_{ab,\sigma} = 0$, i.e. all the covariant derivatives of the components of the curvature tensor are zero; the space is *symmetric*.

If, however, $K_2 = K_3$, we write the equations $\underset{L}{\delta} R_{\alpha\beta\gamma\delta} = 0$ for the orthonormal tetrad; introducing the notation

$$\left.\begin{aligned} \underset{s}{W}_1 = -\, i(\underset{s}{\xi}_{1,4} + i\underset{s}{\xi}_{2,3}), \quad \underset{s}{W}_2 = -\, i(\underset{s}{\xi}_{2,4} + i\underset{s}{\xi}_{3,1}), \\ \underset{s}{W}_3 = -i(\underset{s}{\xi}_{3,4} + i\underset{s}{\xi}_{1,2}) \end{aligned}\right\} \tag{23.13}$$

and combining the twelve equations thus obtained as in (23.8), we arrive at the set of equations

$$\left.\begin{array}{l} \underset{s}{\xi}^{\sigma}\Pi_{11,\sigma} = 0, \quad \underset{s}{\xi}^{\sigma}\Pi_{12,\sigma} = (K_1-K_2)\underset{s}{W}_5, \\ \underset{s}{\xi}^{\sigma}\Pi_{13,\sigma} = (K_3-K_1)\underset{s}{W}_2, \quad \underset{s}{\xi}^{\sigma}\Pi_{22,\sigma} = 0, \\ \underset{s}{\xi}^{\sigma}\Pi_{23,\sigma} = (K_2-K_3)\underset{s}{W}_1, \quad \underset{s}{\xi}^{\sigma}\Pi_{33,\sigma} = 0 \\ (\sigma = 1, 2, 3, 4). \end{array}\right\} \tag{23.14}$$

Hence

$$\underset{s}{\xi}^{\sigma}\Pi_{23,\sigma} = 0.$$

We first assume that the G_5 group of motions admitted by the T_1 space is intransitive, so that the space is filled with hypersurfaces of transitivity. If these surfaces are *isotropic*, the metric of the V_4 can be reduced ([260], p. 219) to the form

$$ds^2 = 2dx^1dx^4 + g_{22}(x^4)dx^{2^2} + 2g_{23}(x^4)dx^2dx^3 + g_{23}(x^4)dx^{3^2}.$$

However, it is easy to show that the λ-matrix $(R_{ab} - \lambda g_{ab})$ for this metric has non-simple divisors and cannot therefore define a T_1. If on the other hand the hypersurfaces are non-isotropic, then since they admit a G_5 they must also admit a G_6 group, because the group G_5 cannot be the complete group for a V_3. The V_4 admits a system of geodesically parallel hypersurfaces of transitivity, of constant curvature. If these are taken as coordinate surfaces, the appropriate field equations clearly correspond to a flat Minkowski space.

If, however, the G_5 group is transitive, four of the vectors $\underset{s}{\xi}^{\alpha}$ will be linearly independent. Hence we can put $\underset{s}{\xi}^{\alpha} = \delta^{\alpha}_{s}$ for the orthonormal tetrad, so it again follows from (23.14) that $\Pi_{23,\sigma} = 0$; we again have *symmetrical* T_1 spaces.

Let us investigate the conditions of integrability for the set of equations

$$R_{\alpha\beta\gamma\delta,\lambda} = 0, \tag{23.15}$$

defining symmetrical V_n. Differentiating these equations covariantly and alternating, we obtain

$$R_{\alpha\beta\gamma\delta,[\lambda\nu]} = R_{\sigma\lambda\nu[\alpha}R^{\sigma}{}_{\beta]\gamma\delta} + R_{\sigma\lambda\nu[\gamma}R^{\sigma}{}_{\delta]\alpha\beta} = 0.$$

Writing this equation for the orthonormal tetrad, we have

$$\alpha_1\beta_2+\alpha_2\beta_1-\alpha_1\beta_3-\alpha_3\beta_1 = 0, \quad \alpha_1\alpha_2-\alpha_1\alpha_3-\beta_1\beta_2+\beta_1\beta_3 = 0$$

and another four equations obtained by cyclic interchange of the indices 1, 2 and 3. It follows that

$$K_1(K_2-K_3) = K_2(K_3-K_1) = K_3(K_1-K_2) = 0.$$

Since $\Sigma K_s = 0$, $K_s = 0$ ($s = 1, 2, 3$) and the T_1 is flat, i.e. it admits the group of motions G_{10}. We will show later that T_1 exist with a G_4 group, so there is a *gap* from $r = 4$ to $r = 10$ in the distribution of the order of *complete* G_r groups for T_1.

Let us turn to T_2 spaces. We subsequently show that there exists a T_2 with G_6. Since a T_2 does not admit a G_r with $r > 6$, the T_2 space of maximum mobility admits a G_6. *T_2 and T_3 spaces cannot be spaces of constant curvature.*†

Finally, let us consider T_3. If $r > 4$, any minor determinant of the matrix (23.7) of order $\frac{1}{2}n(n+1) - r + 1 = 11 - r \leqslant 6$, must be zero. But there is a sixth order non-zero determinant in this matrix. Hence for T_3, $r \leqslant 4$. This completes the proof of the theorem.

This theorem shows that we need only consider G_r groups with $r \leqslant 4$, for T_1 and T_3 and groups with $r \leqslant 6$ for T_2.

The first step in the practical determination of groups of motions is to decide which *dissimilar* structures are admitted by the T_i space. We will see later that only certain of the structures given in (10.16), (10.17), and (10.19) are admitted by T_i spaces. Subgroups of stability become important and the following theorem should be noted here; *if a T_1 gravitational field admits a group of motions with a subgroup of stability H_p, then* (1) $p = 0$ *if K_1, K_2 and K_3 are all different;* (2) $p \leqslant 2$ *if* $K_1 = K_2 \neq K_3$*;* (3) $p = 6$ *if* $K_1 = K_2 = K_3$*; for T_2 we have:* (1) $p = 0$ *if K_1 and* $K_2 \neq 0$*;* (2) $p = 2$ *if* $K_1 = K_2 = 0$*; for T_3 p is always zero* [228].

Here $K_s = \alpha_s + i\beta_s$ is defined by formulae (19.18) and (19.19). We introduce a *normal* system of coordinates (§ 6) into the T_i space, specifying some point as the origin. The metric then takes the form

$$ds^2 = \{\overset{0}{g}_{\alpha\beta} - \tfrac{1}{2}\overset{0}{R}_{\lambda\alpha\mu\beta}x^\lambda x^\mu + \dots\}\, dx^\alpha dx^\beta. \qquad (23.16)$$

It follows from the definition that motions leave these coefficients unchanged. The invariance of the first coefficient shows that the subgroup of stability is also a subgroup of the group of Lorentz transformations which leave the form $-dx^{1^2} - dx^{2^2} - dx^{3^2} + dx^{4^2}$ invariant.

The invariance of the second coefficient shows that every H_p leaves

† This theorem also holds for $R_{\alpha\beta} = \varkappa g_{\alpha\beta}$ $(\varkappa \neq 0)$.

the coefficients $\overset{0}{R}_{\alpha(\beta\gamma)\delta}$ unchanged. Thus (5.8) shows that the components of this tensor are invariant under the transformations H_p. The group of Lorentz transformations is isomorphic with the group of rotations in a complex three-dimensional Euclidean space ([205], p. 423). In order that the $\overset{0}{R}_{\alpha\beta\gamma\delta}$ remain unchanged, the Π_{ab} must remain invariant under some rotation in complex R_3; we allow for the double symmetry of the matrix (R_{ab}). We must find the orthogonal complex transformations which convert the following two forms in R_3 into themselves:

$$\varphi = \sum_1^3 z^{i^2} \quad \text{and} \quad \psi = \sum \Pi_{ij} z^i z^j.$$

Matrix (19.18) shows that for T_1, $\psi = \sum_1^3 K_i z^{i^2}$. If the K_i ($i = 1, 2, 3$) are all different, it is clear that there is no orthogonal non-degenerate transformation, apart from an identity, which preserves these two forms. If $K_1 = K_2 \neq K_3$, then such a transformation does exist, and it depends on one parameter; since $\sum K_s = 0$, we can put $K_1 = K_2 = \lambda$, $K_3 = -2\lambda$, $\lambda \neq 0$, and we then see that a permissible transformation is a rotation in the plane $\{z^1, z^2\}$. One complex parameter gives two real ones and so $p \leqslant 2$. If $K_1 = K_2 = K_3$, we obtain the trivial case of a flat T_1.

For T_2, (19.19) shows that $\psi = K_1 z^{1^2} + (K_2+1) z^{2^2} + (K_2-1) z^{3^2} + 2iz^1z^2$ where $K_1 + 2K_2 = 0$. If K_1 and $K_2 \neq 0$, H_r reduces to the identity transformation. If $K_1 = K_2 = 0$, we obtain a single-parameter transformation, which for the real case gives $p \leqslant 2$.

For T_3 it follows from (19.20) that $\psi = 2z^2(z^1 - iz^3)$, and it is easy to see that H_p is always identical with the unit element of the group.

§ 24. Non-isomorphic Structures of Groups of Motions Admitted by Empty Spaces

Two groups are said to be *isomorphic* if it is possible to so choose their bases that their structures (structure constants) become identical.

The determination of space-time metrics admitting some group of motions reduces essentially to the problem of finding *non-isomorphic* structures of groups of motion, a problem usually involving lengthy algebraic calculations. Moreover, as pointed out in the introduction to this chapter, the correct formulation of this problem requires a general approach without assuming that the space-time is empty. In view of this, we will *seek only those more restricted G_r groups of motion which are subgroups of the complete group of motions of the corresponding spaces T_i' of maximum mobility* ($i = 1, 2, 3$).

In this way we obtain almost all the possible cases, including all the currently available particular solutions (centro-symmetric, axially symmetric, etc.) which have interesting physical applications. In addition, completely new families of solutions in *closed* form containing many arbitrary functions, are obtained by this method. Quite apart from the *fundamental* problem that arises of these arbitrary functions, it must be emphasized that the method of Lie groups gives a more exact idea of the reality behind the Einstein field equations, and of the truly colossal range of possibilities which lies hidden in the expression "space-time metric". From the point of view of the historical development of the theory of relativity, it is clear that Einstein, Hilbert, and their contemporaries had a rather primitive idea of what in fact is meant by "space-time metric" and of its scope. They possessed only a few of the simplest examples (for example, Schwarzschild's centro-symmetric solution, the solution of Weyl and Levi-Civita with axial symmetry, cosmological metrics) on the one hand, and an intuitive and somewhat indefinite conception of the "Riemann metric in general" on the other. They did not realize what a powerful instrument they were forging. However, a theory always has practical applications, and in our view the essential task in the present stage of development of the theory, is to give relativity a more solid basis. By "basis" we mean further experiments and a more refined mathematical analysis of the principal aspects of the general theory.

The above formulation of the problem shows the necessity of finding the space of maximum mobility for each of the three types of gravitational field in empty space (T_i). Note that if instead of the field equations $R_{\alpha\beta} = 0$, we consider the equations $R_{\alpha\beta} = \varkappa g_{\alpha\beta}$ ($\varkappa \neq 0$), we obtain fundamentally different solutions. Thus, for example, the order of maximum mobility is $r = 2$ for T_3 and $r = 4$ for T_3^*. But as $\varkappa \to 0$, T_3^* tends to T_3 with the appropriate group of motions. Without going into the details of the reason for this, we will simply repeat that the complete solution requires a consideration of both (a) $\varkappa = 0$ and (b) $\varkappa \neq 0$.

In the following section we show that the T_1' space of maximum mobility is a flat space with a ten-parameter group of general Lorentz transformations which includes three Lorentz rotations, three normal rotations and four displacements. T_2' is a space which, in a special system of coordinates, has metric

$$ds^2 = -\sinh^2 x^4\, dx^{2^2} - \sin^2 x^4\, dx^{3^2} + 2dx^1\, dx^4, \tag{24.1}$$

with a *transitive group* of motions G_6, operators

$$\left.\begin{aligned} X_1 = p_1,\quad X_2 = p_2,\quad X_3 = p_3,\quad X_4 = x^2p_1 - \coth x^4 p_2,\\ X_5 = x^3 p_1 - \cot x^4 p_3,\quad X_6 = \tfrac{1}{2}(x^{2^2} + x^{3^2})p_1\\ -x^2 \coth x^4 p_2 - x^3 \cot x^4 p_3 + p_4 \end{aligned}\right\} \tag{24.2}$$

where $p_i \equiv \dfrac{\partial}{\partial x_i}$, and structure

$$\left.\begin{aligned} [X_1X_k] = 0 \qquad (k = 2, \ldots, 6),\\ [X_2X_3] = [X_2X_5] = [X_4X_5] = [X_3X_4] = 0,\\ [X_2X_4] = X_1,\quad [X_2X_6] = X_4,\quad [X_3X_5] = X_1,\\ [X_3X_6] = X_5,\quad [X_4X_6] = X_2,\quad [X_5X_6] = -X_3. \end{aligned}\right\} \tag{24.3}$$

For T_3' spaces we must first determine the order of the maximum group. Suppose that there is an automorphic group of motions G_r in T_3'. According to § 19, there is a unique canonical orthonormal tetrad corresponding to each point in T_3', with matrix $(R_{ab} - \lambda g_{ab})$ $(a, b = 1, \ldots, 6)$ determined by the canonical formulae (19.20). The coordinate vectors of each tetrad are functions of the coordinates of the points of T_3': $\underset{s}{i}^\alpha = \underset{s}{i}^\alpha(x)$. The orthonormal tetrad is not altered by multiplying the λ-matrix $(R_{ab} - \lambda g_{ab})$ by any constant complex quantity ν.

The group of motions of T_3' does not contain a subgroup of stability, and so every motion in T_3' maps any point of this manifold on some other point with a different orthonormal tetrad. Mapping all these tetrads on to the one at the point P, we see each motion corresponds to a certain angular displacement of the tetrads at the point P. Mapping all the orthonormal tetrads and the corresponding bitensors on to a three-dimensional complex metric flat space, we obtain the following theorem; *the group of motions of T_3' is isomorphic with the group of linear homogeneous transformations of a three-dimensional complex plane under which the forms*

$$\varphi = \sum_1^3 z^{i^2}, \quad \psi = 2z^2(z^1 - iz^3)$$

transform into themselves to within an arbitrary complex factor ν.

It is easy to verify that such transformations contain not more than one arbitrary complex coefficient (namely ν), i.e. two real ones.

Hence T_3' *cannot admit a greater than two-parameter group of motions.* If such a group were Abelian, the transformations corresponding to the two real parameters would commute, but $v \neq i\bar{v}$ and hence the group is *non-Abelian*.

If T_1 spaces admit a G_r with $r > 4$, they are flat: we have therefore to determine the structures of the following G_r groups—for T_1, $r = 1, 2, 3, 4$; for T_2, $r = 1, \ldots, 6$; and for T_3, $r = 1, 2$.

The structures can be found by means of the following algorithm. To find a G_r subgroup of the complete group of the $T_i{}'$ space of maximum mobility, with base operators $X_1, \ldots, X_{r_i}$, where $r_1 = 10$, $r_2 = 6$ and $r_3 = 2$, we assume that the operators of the unknown subgroup have the form

$$Y_j = \sum_{k=1}^{r_i} C_j^k X_k, \quad C_j^k = \text{const.}$$

We take into account the structure of the complete group $[X_i, X_j] = C_{ij}^k X_k$, write down all the possible structure equations (10.18), (10.17) and (10.19), and determine directly in each case when the structure equations are consistent, thus finding the structure constants. Once the structure has been obtained, the base of the group can be changed by means of non-degenerate linear transformations. For G_5 groups with canonical real structures not given in explicit form, the Jacobi relations (10.18) must be used.

There is a second method of solving the problem. In the next chapter we classify V_4 spaces with signatures $(- - - +)$ of Minkowski type at any given point, obtaining in a special system of coordinates, the metrics corresponding to all the possible groups of motion.

To determine a $T_i (i = 1, 2, 3)$ with a group of given structure, we take the corresponding V_4 space, impose the condition that the metric should satisfy the field equations ($R_{\alpha\beta} = 0$), and then obtain the λ-matrix $(R_{ab} - \lambda g_{ab})$ $(a, b = 1, \ldots, 6)$ in the bivector space with characteristic corresponding to the specified value of i.

This method is simple in principle but suffers from the following disadvantages; (1) it leads to awkward calculations even in the first stage of writing down the field equations, and (2) the conditions defining the *type* of space are, like the field equations, second order differential equations, but *non-linear* even in the lowest derivatives. These difficulties are particularly pronounced for spaces of low mobility ($r = 1, 2$). It is impossible to avoid imposing these conditions, since in general (and specially when $r = 1, 2$) different types of space can admit a group of the same structure (cf. § 57). In view of this the following third method, in which the conditions defining the type of space directly determine the group structure, is preferable.

Suppose that a T_i admits a group G_r ($r \geqslant 2$). According to (10.6) the Killing vectors must satisfy the structure equations

$$\underset{p}{\xi}{}^{\sigma}\partial_{\sigma}\underset{q}{\xi}{}^{\alpha} - \underset{q}{\xi}{}^{\sigma}\partial_{\sigma}\underset{p}{\xi}{}^{\alpha} = C^{s}_{pq}\underset{s}{\xi}{}^{\alpha}. \tag{24.4}$$

As the Riemann connectivity (cf. § 3) has zero torsion, $\Gamma^{\alpha}_{[\beta\gamma]} = 0$ so that

$$\underset{p}{\xi}{}^{\sigma}\underset{q}{\xi}{}^{\alpha}{}_{,\sigma} - \underset{q}{\xi}{}^{\sigma}\underset{p}{\xi}{}^{\alpha}{}_{,\sigma} = C^{s}_{pq}\underset{s}{\xi}{}^{\alpha}, \tag{24.5}$$

the invariant form of the *structure equations*. As the Riemann metric is preserved in motions so must the corresponding Riemann connectivity, i.e. the Lie derivative of a connectivity coefficient must be zero for any Killing vector admitted by the G_r group:

$$\underset{L}{\delta}\Gamma^{\alpha}_{\beta\gamma} = \xi^{\alpha}{}_{,\beta\gamma} + \xi^{\sigma}R^{\alpha}{}_{\beta\sigma\gamma} = 0. \tag{24.6}$$

Differentiating (24.5) covariantly, substituting for $\xi_{\alpha,\beta\gamma}$, and taking the Ricci identities (5.8) into account, we obtain

$$\underset{p}{\xi}{}_{\alpha,}{}^{\sigma}\underset{q}{\xi}{}_{\beta,\sigma} - \underset{p}{\xi}{}_{\beta,}{}^{\sigma}\underset{q}{\xi}{}_{\alpha,\sigma} + \underset{p}{\xi}{}^{\sigma}\underset{q}{\xi}{}^{\tau}R_{\sigma\tau\alpha\beta} = C^{s}_{pq}\underset{s}{\xi}{}_{\alpha,\beta}, \tag{24.7}$$

where $\xi_{\alpha,}{}^{\sigma} \equiv g^{\beta\sigma}\xi_{\alpha,\beta}$.

Differentiating (24.7) covariantly and making the same substitution again, we obtain one further relation:

$$\begin{aligned} R_{\sigma\alpha\tau\gamma}(\underset{q}{\xi}{}_{\beta,}{}^{\sigma}\underset{p}{\xi}{}^{\tau} - \underset{p}{\xi}{}_{\beta,}{}^{\sigma}\underset{q}{\xi}{}^{\tau}) + R_{\sigma\beta\tau\gamma}(\underset{p}{\xi}{}_{\alpha,}{}^{\sigma}\underset{q}{\xi}{}^{\tau} - \underset{q}{\xi}{}_{\alpha,}{}^{\sigma}\underset{p}{\xi}{}^{\tau}) \\ + (R_{\sigma\alpha\tau\beta} - R_{\sigma\beta\tau\alpha})(\underset{p}{\xi}{}^{\sigma}{}_{,\gamma}\underset{q}{\xi}{}^{\tau} + \underset{q}{\xi}{}^{\tau}{}_{,\gamma}\underset{p}{\xi}{}^{\sigma}) \\ + \underset{p}{\xi}{}^{\sigma}\underset{q}{\xi}{}^{\tau}(R_{\sigma\alpha\tau\beta,\gamma} - R_{\sigma\beta\tau\alpha,\gamma}) = -C^{s}_{pq}\underset{s}{\xi}{}^{\sigma}R_{\sigma\gamma\alpha\beta}. \end{aligned} \tag{24.8}$$

Formulae (24.5), (24.7) and (24.8) are valid in any holonomic system of coordinates. If in particular, the T_i space is provided with a normal system of coordinates, with the components $R_{\alpha\beta\gamma\delta}$ in canonical form at the origin, we obtain a set of equations into which we can substitute the curvature tensor components from (19.18), (19.19) and (19.20) These equations determine the values of $\overset{0}{\underset{s}{\xi}}{}^{\alpha}$, $\overset{0}{\underset{s}{\xi}}{}_{\alpha,\beta}$ and C^{s}_{pq}; the components $\overset{0}{R}_{\alpha\beta\gamma\delta,\lambda}$ in eqns. (24.8) must either be determined or eliminated. Note that substituting for the $\overset{0}{R}_{\alpha\beta\gamma\delta,\lambda}$ in (24.8) from eqns. (23.3) for all the Killing vectors produces an identity, i.e. eqns. (23.3) are equivalent to (24.8).

We must also take into account the following considerations:

(1) any linear combination of the vectors $\overset{0}{\underset{s}{\xi}}{}^{\alpha}$ with constant coefficients can be used; hence r components $\overset{0}{\underset{s}{\xi}}{}^{\alpha}$ can first be reduced to zero or unity;

(2) it is impossible that $\overset{0}{\underset{s}{\xi}}{}^{\alpha} = \overset{0}{\underset{s}{\xi}}{}_{\alpha,\beta} = 0$ *simultaneously* for any Killing vector since this would imply that $\underset{s}{\xi}{}^{\alpha} = 0$ in any system of coordinates ([147], §§ 18, 53);

(3) if the group admits a stationary subgroup H_p, then p of the vectors $\overset{0}{\underset{s}{\xi}}{}^{\alpha} = 0$ $(s = 1, \ldots, p)$.

We thus obtain the following algorithm: (1) we reduce certain of the components $\overset{0}{\underset{s}{\xi}}{}^{\alpha}$ or $\overset{0}{\underset{s}{\xi}}{}_{\alpha,\beta}$ to 0 or 1; (2) we solve the set of equations (24.5), (24.7); (3) we add the set (24.8), if necessary, having first eliminated the $\overset{0}{R}_{\alpha\beta\gamma\delta,\lambda}$, and verify the Jacobi conditions (10.18) for the constants obtained.

It should be noted that not only is the structure of the group important, but also the rank of the matrix $(\overset{0}{\underset{s}{\xi}}{}^{\alpha})$ and other geometrical properties of the operators; it is of particular importance to deduce from the $\overset{0}{\underset{s}{\xi}}{}^{\alpha}$ or $\overset{0}{\underset{s}{\xi}}{}_{\alpha,\beta}$ at the origin whether the surfaces of transitivity are *isotropic* or no t. It is there fore just as essential to determine the initial values of $\overset{0}{\underset{s}{\xi}}{}^{\alpha}$ and $\overset{0}{\underset{s}{\xi}}{}_{\alpha,\beta}$ as to find the structure constants.

An examination of the possible real non-isomorphic structures of G_3, G_4 and G_5 groups shows that every G_5 admits a G_4 subgroup and every G_4 a G_3 subgroup, but there is a G_3 group of rotations [the insoluble G_3 of type IX in the classification of (10.16)] which does not contain a G_2 subgroup although this is the only G_3 possessing this property. In the subsequent discussion we first apply the algorithm to G_3 groups. We need then only investigate those G_4 containing the non-isomorphic real G_3 structures already found, and so on.

The application of this algorithm is based on the existence of the canonical forms (19.18), (19.20) for the three type of T_i gravitational field.

We thus obtain the following possible structures:

T_1 SPACES

$$r = 2$$

Both Abelian and non-Abelian groups are possible:

$$1. \quad [X_1X_2] = 0, \qquad 2. \ [X_1X_2] = X_1. \tag{24.9}$$

$$r = 3$$

The possible structures depend on the degeneracy of the principal curvatures K_i ($i = 1, 2, 3$). If $K_1 = K_3 = K_3$, the space is flat (cf. § 19) and admits a ten-term group of motions. If $K_1 = K_2 \neq K_3$, the structures

$$1. \quad [X_1X_2] = X_3, \quad [X_2X_3] = eX_1, \quad [X_3X_1] = X_2, \quad e = \pm 1 \tag{24.10}$$

are possible corresponding to *insoluble* groups of types VIII and IX in the classification (10.16). In addition, we can choose coordinates so that $\overset{0}{\underset{1}{\xi}}{}^{\alpha} = 0$, $\overset{0}{\underset{2}{\xi}}{}^{\alpha} = \delta_1^{\alpha}$, and $\overset{0}{\underset{3}{\xi}}{}^{\alpha} = \delta_2^{\alpha}$ or δ_4^{α}.

We can determine the metric tensor of the two-dimensional plane which is the pencil of the set of null vectors $\{\overset{0}{\underset{2}{\xi}}{}^{\alpha}, \overset{0}{\underset{s}{\xi}}{}^{\alpha}\}$, and therefore the signature of the surface of transitivity. Hence *a single parameter subgroup of stability exists and two-dimensional non-isotropic surfaces of transitivity with signature* $(- -)$ or $(- +)$.

$$2. \quad [X_1X_2] = 0, \quad [X_1X_3] = X_2, \quad [X_2X_3] = -X_1, \tag{24.11}$$

a soluble group of type VII under the same conditions as before.

3. The Abelian group $[X_iX_j] = 0$ ($i, j = 1, 2, 3$) with the initial conditions

$$\left.\begin{aligned}
&\overset{0}{\underset{1}{\xi}}{}^{\alpha} = \delta_2^{\alpha}, \quad \overset{0}{\underset{2}{\xi}}{}^{\alpha} = \delta_3^{\alpha}, \quad \overset{0}{\underset{3}{\xi}}{}^{\alpha} = \delta_4^{\alpha},\\
&\overset{0}{\underset{1}{\xi}}{}_{\alpha,\beta} = \begin{cases} k & \alpha\beta = 12,\\ 0, \end{cases} \quad \overset{0}{\underset{2}{\xi}}{}_{\alpha,\beta} = \begin{cases} -a & \alpha\beta = 13,\\ -b & \alpha\beta = 14,\\ 0, \end{cases}\\
&\overset{0}{\underset{3}{\xi}}{}_{\alpha,\beta} \begin{cases} a & \alpha\beta = 14,\\ b & \alpha\beta = 13,\\ 0. \end{cases}
\end{aligned}\right\} \tag{24.12}$$

Hence the subgroup of stability is identical with the unit element of the group. The group coincides with its own centre. The rank of the matrix

$(\underset{s}{\xi}^{\alpha})$ is equal to the order of the group and so *the surfaces of transitivity are three-dimensional and non-isotropic;*

$$K_1, \quad K_2, \quad K_3 \neq 0. \tag{24.13}$$

4. The Abelian group with the initial conditions

$$\overset{0}{\underset{1}{\xi}}{}^{\alpha} = \delta_1^{\alpha}, \quad \overset{0}{\underset{2}{\xi}}{}^{\alpha} = \delta_2^{\alpha} - \delta_3^{\alpha}, \quad \overset{0}{\underset{3}{\xi}}{}^{\alpha} = \delta_4^{\alpha}, \tag{24.14}$$

corresponding to *non-isotropic surfaces of transitivity with indefinite metric.*

5. A group with a structure of type VII

$$\left.\begin{aligned} &[X_1X_2] = 0, \quad [X_1X_3] = -\frac{K}{2}X_1 - \frac{K\sqrt{3}}{2}X_2, \\ &[X_2X_3] = \frac{K\sqrt{3}}{2}X_1 - \frac{K}{2}X_2 \qquad (K \neq 0), \end{aligned}\right\} \tag{24.15}$$

acting on a hypersurface of transitivity with signature $(--+)$.

$$r = 4$$

1. $[X_1X_2] = X_3, \quad [X_2X_3] = X_1, \quad [X_3X_1] = X_2, \quad [X_iX_4] = 0$
$$(i = 1,\ 2,\ 3), \tag{24.16}$$

2. $[X_1X_2] = X_3, \quad [X_2X_3] = -X_1, \quad [X_3X_1] = X_2, \quad [X_iX_4] = 0$
$$(i = 1,\ 2,\ 3), \tag{24.17}$$

3.
$$\left.\begin{aligned} &[X_1X_2] = X_3, \quad [X_1X_3] = X_2, \quad [X_2X_3] = KX_1, \\ &[X_1X_4] = 0, \quad [X_2X_4] = \varepsilon X_2, \quad [X_3X_4] = \varepsilon X_3, \end{aligned}\right\} \tag{24.18}$$

where $\varepsilon = 0$ or 1, and

$$\overset{0}{\underset{4}{\xi}}{}^{\alpha} = p\delta_1^{\alpha} + q\delta_2^{\alpha}, \quad \overset{0}{\underset{4}{\xi}}{}_{\alpha,\beta} = \begin{cases} 2 & \alpha\beta = 12, \\ 0. \end{cases}$$

4. $[X_iX_j] = 0 \quad (i,\ j = 1,\ 2,\ 3), \quad [X_1X_4] = -\frac{K}{2}X_1 - \frac{K\sqrt{3}}{2}X_3,$

$$[X_2X_4] = KX_2, \quad [X_3X_4] = \frac{K\sqrt{3}}{2}X_1 - \frac{K}{2}X_3, \quad K \neq 0,$$

and the matrix $(\underset{s}{\xi}^{\alpha})$ is of rank 4, i.e. the group is *simple-transitive.*

T_2 SPACES

The group structures for T_2 spaces can be found by the first method discussed above, i.e. investigating all the possible subgroups of the complete six-parameter group. The third method can also be used and leads quite rapidly to the same results.

$$r = 2$$

Both Abelian and non-Abelian groups of motions are possible.

1. $[X_1X_2] = 0,$ 2. $[X_1X_2] = X_1.$

$$r = 3$$

1. The Abelian group acting on an *isotropic* hypersurface of transitivity.

$$[X_iX_j] = 0 \qquad (i, j = 1, 2, 3). \tag{24.19}$$

As the rank of the matrix for this intransitive group is equal to the order of the group, any operators compatible with the structure can be taken, and so the groups are defined to within a *similarity*. The operators $X_i = p_i$ can be chosen. The subgroup of stability of this G_3 is identical with the unit element of the group.

2. A G_3 group of type II with the structure

$$[X_1X_2] = 0, \quad [X_1X_3] = 0, \quad [X_2X_3] = X_1, \tag{24.20}$$

acting on *isotropic* hypersurfaces. The group operators

$$X_1 = p_1, \quad X_2 = p_2+p_3, \quad X_3 = x^2p_1-\coth x^4p_2 \tag{24.21}$$

may be chosen.

3. A G_3 group of type II, but acting on *isotropic two-dimensional* surfaces of transitivity. The operators of this group are of the form

$$X_1 = p_1, \quad X_2 = p_2, \quad X_3 = x^2p_1-\coth x^4p_2. \tag{24.22}$$

4. A G_3 group of type III with the structure

$$[X_1X_2] = 0, \quad [X_1X_3] = X_1, \quad [X_2X_3] = 0, \tag{24.23}$$

acting on a *non-isotropic* V_3 hypersurface of transitivity; the operators may be chosen as follows:

$$\left.\begin{aligned} X_1 &= x^2p_1+(1-\coth x^4)p_2, \quad X_2 = p_1, \\ X_3 &= \tfrac{1}{2}(x^{2^2}+x^{3^2})p_1-x^2\coth x^4p_2-x^3\cot x^4p_3+p_4. \end{aligned}\right\} \tag{24.24}$$

This exhausts all the possible G_3 structures admitted by a T_2 space with $r \leqslant 3$. G_4 and G_5 structures are omitted as it is more convenient to determine them by a different method.

As was shown above, the only possible structure for T_3 spaces is the non-Abelian group G_2 II; $[X_1X_2] = X_1$.

This section shows the need for investigating the spaces of maximum mobility for each for the three possible types of T_i ($i = 1, 2, 3$).

The solution of the problem for T_1 is obvious and more complicated for T_2; as far as T_3 are concerned, the problem is complex in both theory and application due to the low mobility in such empty spaces. For this reason not a single example of such fields has been given in the literature, and the solution given in the following section defines empty spaces of a new type.

§ 25. Spaces of Maximum Mobility T_1, T_2 and T_3

We now prove the following theorem: *the spaces of maximum mobility are:* (1) *for T_1, a Minkowski space;* (2) *for T_2, spaces admitting a six-parameter soluble transitive group with a metric of the form*

$$ds^2 = 2\,dx^1\,dx^4 + g_{22}\,dx^{2^2} + 2g_{23}\,dx^2\,dx^3 + g_{33}\,dx^{3^2},$$

where g_{22}, g_{23}, g_{33} are determined by conditions (25.12), (25.18)*;* (3) *for T_3, spaces admitting a non-Abelian G_2 group of motions with the following metric in a special system of coordinates*

$$ds^2 = e(\alpha\,dx^{1^2} + dx^{2^2}) + 2\,dx^3\,dx^4 + \lambda\,dx^{4^2},$$

where $e = \pm 1$, and α and λ are functions defined by (25.69).

In proving this theorem the three possible types of space are considered separately.

T_1 Space

As pointed out above, this case is trivial since T_1 manifolds include a flat Minkowski space, and conversely, every T_1 admitting a G_{10} group of motions corresponds to the space-time of special relativity with signature $(- - - +)$.

T_2 Space

Suppose that the T_2 admits a G_6 as the group of maximum order. If a G_6 group in a V_4 is intransitive, the V_4 is a Kagan subprojective space (Kruchkovich [303]), i.e. a conformal-Euclidean manifold. As conformal-Euclidean Einstein spaces (cf. example 2, § 13) are always flat they also admit a G_{10} group corresponding to a Minkowski space.

The G_6 group is therefore transitive. Referring the Killing vectors $\underset{s}{\xi}^\alpha$ of the group to the orthonormal tetrad in which (19.19) apply, shows t ha the rank of the matrix $(\underset{s}{\xi}^\alpha)_0$ is 4; by using linear combinations, the $\underset{s}{\xi}^\alpha$ at a given point can be reduced to the form

$$\underset{1}{\xi}^{\alpha} \doteq \delta_1^{\alpha}+\delta_4^{\alpha}, \quad \underset{2}{\xi}^{\alpha} \doteq \delta_2^{\alpha}, \quad \underset{3}{\xi}^{\alpha} \doteq \delta_3^{\alpha}, \quad \underset{4}{\xi}^{\alpha} \doteq \delta_4^{\alpha}, \quad \underset{5}{\xi}^{\alpha} \doteq \underset{6}{\xi}^{\alpha} \doteq 0,$$

the last two vectors forming a subgroup of stability G_2; $G_2 \subset G_6$ ($\doteq$ denotes equality in the canonical orthonormal tetrad).

Equating all fifth order minor determinants of matrix (23.6) to zero shows that all the $\alpha_i = \beta_i = 0$, and in addition

$$\begin{gathered} R_{11,\sigma} = R_{14,\sigma} = 0, \quad R_{12,\sigma} = R_{16,\sigma}, \quad R_{13,\sigma} = -R_{15,\sigma} \\ R_{22,\sigma} = R_{26,\sigma} = -R_{33,\sigma}, \quad R_{23,\sigma} = -R_{25,\sigma} = R_{36,\sigma} \end{gathered} \tag{25.1}$$

where $\sigma = 1, 2, 3, 4$, whilst the other indices are bivector-space indices, e.g. $R_{11,\sigma} \rightarrow R_{1414,\sigma}$.

Let

$$R_{12,\sigma} = A_\sigma, \quad R_{13,\sigma} = B_\sigma, \quad R_{22,\sigma} = C_\sigma, \quad R_{23,\sigma} = D_\sigma:$$

the Bianchi indentities show that

$$\begin{gathered} A_1+A_4 = B_1+B_4 = A_2+B_3 = A_3-B_2 = 0, \quad A_1+C_2+D_3 = 0, \\ C_3-B_1-D_2 = 0, \quad C_1+C_4-A_2 = 0, \quad B_2-D_1-D_4 = 0. \end{gathered} \tag{25.2}$$

Applying eqns. (23.3) to $\underset{5}{\xi}^{\alpha}$, $\underset{6}{\xi}^{\alpha}$ shows that

$$\underset{s}{\xi}_{1,4} \doteq \underset{s}{\xi}_{2,3} \doteq 0, \quad \underset{s}{\xi}_{2,4} \doteq \underset{s}{\xi}_{1,2}, \quad \underset{s}{\xi}_{3,4} \doteq \underset{s}{\xi}_{1,3} \qquad (s = 5, 6). \tag{25.3}$$

As any linear combination of $\underset{s}{\xi}^{\alpha}$ ($s = 5, 6$) with constant coefficients is permissible, we can always take

$$\underset{5}{\xi}_{\alpha,\beta} \doteq \begin{pmatrix} 0 & 1 & 0 & 0 \\ -1 & 0 & 0 & 1 \\ 0 & 0 & 0 & 0 \\ 0 & -1 & 0 & 0 \end{pmatrix}, \quad \underset{6}{\xi}_{\alpha,\beta} \doteq \begin{pmatrix} 0 & 0 & 1 & 0 \\ 0 & 0 & 0 & 0 \\ -1 & 0 & 0 & 1 \\ 0 & 0 & -1 & 0 \end{pmatrix},$$

$$\underset{k}{\xi}_{\alpha,\beta} \doteq \begin{pmatrix} 0, & 0, & 0, & \underset{k}{\xi}_{1,4} \\ 0, & 0, & \underset{k}{\xi}_{2,3}, & \underset{k}{\xi}_{2,4} \\ 0, & -\underset{k}{\xi}_{2,3}, & 0, & \underset{k}{\xi}_{3,4} \\ -\underset{k}{\xi}_{1,4} & -\underset{k}{\xi}_{2,4}, & -\underset{k}{\xi}_{3,4}, & 0 \end{pmatrix} \qquad (k \neq 5, 6).$$

If the above expressions for $\underset{s}{\xi}^{\alpha}$, $\underset{s}{\xi}_{\alpha,\beta}$ (s = 1,..., 6) are inserted into the structure equations (24.6) and (24.8), the matrix (23.6) determines the structure constants C^{p}_{qr}; these are expressed in terms of $\xi_{\alpha,\beta}$, and we thus obtain the following relationships:

$$\underset{1}{\xi}_{\alpha,\beta} \doteq 0, \quad \underset{2}{\xi}_{\alpha,\beta} = \begin{pmatrix} 0 & 0 & 0 & \underset{2}{\xi}_{1,4} \\ 0 & 0 & \underset{2}{\xi}_{2,3} & 0 \\ 0 & -\underset{2}{\xi}_{2,3} & 0 & 0 \\ -\underset{2}{\xi}_{1,4} & 0 & 0 & 0 \end{pmatrix},$$

$$\underset{3}{\xi}_{\alpha,\beta} \doteq \begin{pmatrix} 0 & 0 & 0 & \underset{3}{\xi}_{1,4} \\ 0 & 0 & \underset{3}{\xi}_{2,3} & 0 \\ 0 & -\underset{3}{\xi}_{2,3} & 0 & 0 \\ -\underset{3}{\xi}_{1,4} & 0 & 0 & 0 \end{pmatrix}, \quad \underset{4}{\xi}_{\alpha,\beta} \doteq \begin{pmatrix} 0 & 0 & 0 & \underset{4}{\xi}_{1,4} \\ 0 & 0 & \underset{4}{\xi}_{2,3} & \underset{4}{\xi}_{2,4} \\ 0 & -\underset{4}{\xi}_{2,3} & 0 & \underset{4}{\xi}_{3,4} \\ -\underset{4}{\xi}_{1,4} & -\underset{4}{\xi}_{2,4} & -\underset{4}{\xi}_{3,4} & 0 \end{pmatrix} \tag{25.4}$$

where

$$\underset{4}{\xi}_{2,4} \doteq \underset{2}{\xi}_{1,4} \doteq -\underset{3}{\xi}_{2,3}, \quad \underset{4}{\xi}_{3,4} \doteq \underset{2}{\xi}_{2,3} \doteq \underset{3}{\xi}_{1,4}. \tag{25.5}$$

If these $\underset{s}{\xi}_{\alpha,\beta}$ are inserted into the integrability conditions (23.3) for the orthonormal tetrad, the following additional relations are obtained for the $R_{\alpha\beta\gamma\delta,\sigma}$:

$$A_1+A_4 = B_1+B_4 = 0, \quad A_i = B_i \quad (i = 2, 3),$$
$$C_2+2A_4 = C_3-2B_4 = D_2+2B_4 = D_3+2A_4 = 0.$$

Combining these equations with (25.2) shows that

$$A_\sigma = B_\sigma = 0, \quad C_i = D_i, \quad C_1+C_4 = D_1+D_4 = 0$$
$$(\sigma = 1, \ldots, 4), \qquad (i = 2, 3).$$

Hence, by using (25.1), all the components of the tensor $R_{\alpha\beta\gamma\delta,\sigma}$ in the orthonormal tetrad can be expressed in terms of the parameters a and b, where

$$a \doteq C_4 \equiv -2\underset{4}{\xi}_{1,4}, \quad b \doteq D_4 \equiv 2\underset{4}{\xi}_{3,4}. \tag{25.6}$$

If the operators of the G_6 group are renumbered according to the scheme

$$X_1 \to -X_2, \quad X_2 \to X_1, \quad X_3 \to X_5, \quad X_4 \to X_6, \quad X_5 \to X_4, \quad X_6 \to -X_3,$$

the structure equations take the form

$$\left.\begin{aligned}
&[X_1X_i] = 0\ (i = 2, 3, 5),\ [X_1X_4] = X_2,\quad [X_1X_6] = X_4+bX_5,\\
&[X_2X_j] = 0\ (j = 3, 4, 5),\ [X_2X_6] = -aX_2,\\
&[X_3X_4] = 0,\quad [X_3X_5] = X_2,\quad [X_3X_6] = -aX_3+bX_4-X_5,\\
&[X_4X_5] = 0,\ [X_4X_6] = X_1-bX_3-aX_4,\ [X_5X_6] = -bX_1+X_3.
\end{aligned}\right\} \tag{25.7}$$

Knowing the group structure, we now construct the group operators in a holonomic system of coordinates, and determine the metric from the Killing and field equations. The values of $\underset{s}{\xi^\alpha}$ in the orthonormal tetrad show that the group (25.7) admits a G_5 subgroup acting on an isotropic surface $\overset{*}{V}_3$. But under these conditions, the metric of any V_4 (whatever the signature [260]) has the following form in a special system of coordinates:

$$ds^2 = 2dx^1\,dx^4+g_{22}\,dx^{2^2}+2g_{23}\,dx^2\,dx^3+g_{33}\,dx^{3^2};\quad g_{\alpha\beta}(x^4) \tag{25.8}$$

and the operators of the G_5 subgroup the form

$$X_1 = p_2,\quad X_2 = p_1,\quad X_3 = p_3,\quad X_4 = x^2p_1+\varphi p_2+\psi p_3,$$
$$X_5 = x^3p_1+\psi p_2+\theta p_3,\quad \varphi(x^4),\quad \psi(x^4),\quad \theta(x^4). \tag{25.9}$$

The Killing equations for these operators and the metric (25.8) provides the following conditions for the functions φ, ψ, θ:

$$\varphi' = \frac{g_{33}}{g},\quad \psi' = -\frac{g_{23}}{g},\quad \theta' = \frac{g_{22}}{g},\quad g = |g_{\alpha\beta}| = g_{23}^2-g_{22}g_{33} < 0. \tag{25.10}$$

Using the structure equations (25.7) for commutators of the type $[X_\alpha X_6]$, we determine the sixth operator in the form

$$X_6 = [\tfrac{1}{2}(x^{2^2}-x^{3^2})+bx^2x^3-ax^1+\lambda]p_1+[(b\psi+\varphi)x^2+(b\varphi-\psi)x^3]p_2$$
$$+[(\psi+b\theta)x^2+(b\psi-\theta-a)x^3]p_3+\mu p_4, \tag{25.11}$$

where λ and μ are functions of x^4, and φ, ψ, θ, satisfy the additional conditions

$$\begin{aligned}
-\mu\varphi' &= \psi^2-\varphi^2-2b\varphi\psi-a\varphi+1,\\
-\mu\psi' &= -b(1+\psi^2)-\varphi\psi+\theta\psi-b\varphi\theta,\\
-\mu\theta' &= \theta^2-\psi^2-2b\psi\theta+a\theta+1.
\end{aligned} \tag{25.12}$$

Writing the Killing equations for the sixth operator and the metric (25.8), we obtain

$$\mu = ax^4+\alpha,\quad \lambda' = 0,\quad \alpha = \text{constant}, \tag{25.13}$$

and the set of differential equations

$$\mu g'_{22}+2(\varphi+b\psi)g_{22}+2(\psi+b\theta)g_{23} = 0,$$
$$\mu g'_{23}+(b\varphi-\psi)g_{22}+(2b\psi-\theta+\varphi-a)g_{23}+(\psi+b\theta)g_{33} = 0,$$
$$\mu g'_{33}+2(b\varphi-\psi)g_{23}+2(b\psi-\theta-a)g_{33} = 0,$$

which are satisfied identically in virtue of (25.10) and (25.12). Equation (25.10) shows that

$$g' = -\frac{2g}{\mu}(2b\psi-\theta+\varphi-a). \tag{25.14}$$

All the field equations $R_{\alpha\beta} = 0$ are satisfied identically for the metric (25.8) except $R_{44} = 0$, which can be written in the expanded form

$$g_{33}g''_{22}-2g_{23}g''_{23}+g_{22}g''_{33}+\frac{1}{2g}[g^2_{33}g'^2_{22}+g^2_{23}g'^2_{33}+2g_{22}g_{33}g'^2_{23}$$
$$+2g^2_{23}g'_{33}g'_{22}+2g^2_{23}g'^2_{23}-4g_{33}g_{23}g'_{22}g'_{23}-4g_{22}g_{23}g'_{23}g'_{33}] = 0. \tag{25.15}$$

Expressing the derivatives g' and g'' in terms of the derivatives of the metric tensor components $g'_{\alpha\beta}$, $g''_{\alpha\beta}$ allows the field equation (25.15) to be written in the form

$$gg''-\tfrac{1}{2}g'^2+g(g'_{22}g'_{33}-g'^2_{23}) = 0. \tag{25.16}$$

Differentiating eqns. (25.12) we obtain

$$\left.\begin{aligned} \varphi'' &= \frac{2}{\mu}[\psi'(b\varphi-\psi)+\varphi'(b\psi+\varphi)], \\ \psi'' &= \frac{1}{\mu}[\psi'(2b\psi+\varphi-\theta-a)+\varphi'(\psi+b\theta)+\theta'(b\varphi-\psi)], \\ \theta'' &= \frac{2}{\mu}[\psi'(\psi+b\theta)+\theta'(b\psi-\theta-a)]. \end{aligned}\right\} \tag{25.17}$$

As $g = (\varphi'\theta'-\psi'^2)^{-1}$ and g' is determined by eqns. (25.14), the field equation (25.16) can be used to show that

$$4\mu^2b^2 = g[ab\psi^2+2(\psi+b\theta)-ab\varphi\theta]^2. \tag{25.18}$$

This can be checked by differentiating and substituting for φ', θ' and ψ' from eqns. (25.12); the equation is satisfied identically. The field equation (25.18) is a first integral of the set of Killing equations.

Physically, the essential fact is that the space-time metric for the case T_2, involves two insoluble co-ordinate transformations with parameters a and b. As the general sets of differential equations (25.12) or (25.17) cannot be integrated by means of elementary functions, the following two special cases are of interest:

$1^0, \quad a \neq 0, \quad b = 0.$

In this case $\psi = 0$, and it can be shown that only the following metrics are possible:

$$ds^2 = 2dx^1\,dx^4 + \frac{\mu}{a^2}\left[\frac{1}{p_1^2}\cosh^2(\ln \bar{c}\mu^{-p_1})\,dx^{2^2} + \frac{1}{q_1^2}\cos^2(\ln \bar{\bar{c}}\mu^{-q_1})\,dx^{3^2}\right], \tag{25.19}$$

$$ds^2 = 2dx^1\,dx^4 + \frac{\mu}{a^2}\left[\frac{1}{p_1^2}\cosh^2(\ln \bar{c}\mu^{-p_1})\,dx^{2^2} + \frac{1}{q^2}\sinh^2(\ln \bar{\bar{c}}\mu^{q_2})\,dx^{3^2}\right], \tag{25.20}$$

$$ds^2 = 2dx^1\,dx^4 - \frac{\mu}{a^2}\left[\frac{1}{p_1^2}\sinh^2(\ln \bar{c}\mu^{-p_1})\,dx^{2^2} + \frac{1}{q_1^2}\cosh^2(\ln \bar{\bar{c}}\mu^{q_2})\,dx^{3^2}\right], \tag{25.21}$$

$$ds^2 = 2dx^1dx^4 + x^4\left\{\sinh^2\left[\ln \bar{c}(2x^4)^{\pm 1/\sqrt{2}}\right]dx^{2^2} + \tfrac{1}{2}\ln \bar{\bar{c}}x^4\,dx^{3^2}\right\}, \tag{25.22}$$

where a, $\bar{c}$, $\bar{\bar{c}}$ are constants,

$$\mu = ax^4 + \alpha, \quad \alpha \text{ constant},$$

$$p_1 = \frac{1}{2a}\sqrt{4+a^2}; \quad q_1 = \frac{1}{2a}\sqrt{4-a^2}; \quad q_2 = \frac{1}{2a}\sqrt{a^2-4},$$

$2^0, \quad a = b = 0.$

This case corresponds to symmetric space-time, as all the components of $R_{\alpha\beta\gamma\delta,\lambda}$ are zero. There are two possible metrics:

$$\left.\begin{aligned} ds^2 &= 2dx^1\,dx^4 + \alpha\cosh^2\left(\frac{x^4}{\alpha}+\bar{c}\right)dx^{2^2} + \alpha\cos^2\left(\frac{x^4}{\alpha}+\bar{\bar{c}}\right)dx^{3^2}, \\ ds^2 &= 2dx^1\,dx^4 - \alpha\sinh^2\left(\frac{x^4}{\alpha}+\bar{c}\right)dx^{2^2} - \alpha\sin^2\left(\frac{x^4}{\alpha}+\bar{\bar{c}}\right)dx^{3^2}, \end{aligned}\right\} \tag{25.23}$$

where a, $\bar{c}$, $\bar{\bar{c}}$ are constants.

A particular noteworthy metric belonging to (25.23) is

$$ds^2 = 2dx^1\,dx^4 - \sinh^2(c_1x^4)\,dx^{2^2} - \sin^2(c_2\,x^4)\,dx^{3^2}.$$

An examination of the G_6 structure (25.7) shows that the group is *soluble* and (as shown above) *transitive*.

The theorem is thus true for T_2.

T_3 SPACE

This problem is considerably more complicated.

The following theorem, due to Bel [299], is useful in studying T_3; *in every T_3 an isotropic geodesic vector field $l^\alpha(x)$ exists which satisfies the equations*

$$R_{\alpha\beta\gamma\delta}l^\beta l^\delta = e_{\alpha\beta\sigma\tau}R^{\sigma\tau}{}_{\gamma\delta}l^\beta l^\delta = 0, \tag{25.24}$$

$$g_{\alpha\beta}l^\alpha l^\beta = 0, \quad l^\sigma l^\alpha{}_{,\sigma} = al^\alpha, \tag{25.25}$$

where $e_{\alpha\beta\gamma\delta}$ is the permutation tensor of the space (see example 7, § 3; [91], p. 366).

It is sufficient to prove the existence of this field and properties (25.25), (25.24) for the non-holonomic orthonormal tetrad appropriate to the canonical matrices (19.20). In this reference system $l^\alpha = \nu(\delta^\alpha_2 + \delta^\alpha_4)$, and the indicated properties follow directly from this equation; the last relation in (25.25) is a consequence of the Bianchi identities (5.16) (see examples 2 and 4, § 19).

Again, *if a T_3 admits a group of motions G_r, the Lie derivative of the vector l^α in the direction of any Killing vector $\underset{s}{\xi}{}^\alpha$ $(s = 1, \ldots, r)$ satisfies the equation*

$$\underset{L}{\delta}{}_s l^\alpha = \underset{s}{\omega} l^\alpha. \tag{25.26}$$

Bearing in mind that the Lie derivatives of the tensors $R_{\alpha\beta\gamma\delta}$ and $e_{\alpha\beta\gamma\delta}$ are zero, we differentiate (25.24), refer the result to the orthonormal tetrad (19.20), and thus obtain

$$\underset{L}{\delta}{}_s l^1 \stackrel{*}{=} \underset{L}{\delta}{}_s l^3 \stackrel{*}{=} \underset{L}{\delta}{}_s l^2 - \underset{L}{\delta}{}_s l^4 \stackrel{*}{=} 0,$$

Hence

$$\underset{L}{\delta}{}_s l^\alpha \stackrel{*}{=} \underset{s}{\omega} l^\alpha.$$

But this tensor equation must be true in any system of coordinates and therefore (25.26) is true.

As T_3 can admit as its maximum group only the non-Abelian G_2, the structure of this group can be reduced to the form $[X_1X_2] = X_1$.

This intransitive group must act in two-dimensional surfaces of transitivity, and whether the surfaces are isotropic or not, a system of coordinates can be chosen with the following Killing vectors defining the group operators:

$$\underset{1}{\xi}{}^\alpha = \delta^\alpha_1, \quad \underset{2}{\xi}{}^\alpha = x^1\delta^\alpha_1 + \delta^\alpha_4. \tag{25.27}$$

The system of coordinates is then defined to within the transformation

$$\left.\begin{aligned} x^{1'} &= x^1 + e^{x^4} f(x^2, x^3), \\ x^{2'} &= \varphi(x^2, x^3), \\ x^{3'} &= \psi(x^2, x^3), \\ x^{4'} &= x^4 + \theta(x^2, x^3), \end{aligned}\right\} \tag{25.28}$$

where f, φ, ψ and θ are analytic functions in some region, such that the transformation is non-degenerate $\varphi_2\psi_3 - \varphi_3\psi_2 \neq 0$ $\left(\varphi_i \equiv \dfrac{\partial\varphi}{\partial x_i}, \ldots\right)$, but otherwise arbitrary. Transformations which do not alter the form of the given operators X_s are said to be *admissible for the* X_s; it is possible to simplify some of the metric tensor components by a convenient choice of f, φ, ψ and θ.

As the metric signature on the surfaces of transitivity is not yet significant, x^4 cannot in general be identified with the *time* coordinate.

Integrating the Killing equations for the vectors (25.27) we obtain the metric of the required space in the form

$$\left.\begin{aligned} g_{1\alpha} &= a_{1\alpha}(x^2, x^3)e^{-kx4}, \quad k = \begin{cases} 2 & \text{for} \quad \alpha = 1, \\ 1 & \text{for} \quad \alpha = 2, 3, 4, \end{cases} \\ g_{ij} &= a_{ij}(x^2, x^3) \quad (i, j \neq 1). \end{aligned}\right\} \tag{25.29}$$

Let us determine the isotropic geodesic congruence l^α. Integrating eqns. (25.26) for the vectors (25.27), we obtain

$$l^\alpha = \nu(e^{x^4}\lambda_1\delta_1^\alpha + \lambda_2\delta_2^\alpha + \lambda_3\delta_3^\alpha + \lambda_4\delta_4^\alpha),$$

where $\nu = \nu(x^1, x^2, x^3, x^4) \neq 0$, and $\lambda_i = \lambda_i(x^2, x^3)$ $(i = 1, 2, 3, 4)$.

Applying the admissible transformation (25.28) we obtain the following expressions for the components l^α in the new system of coordinates:

$$\left.\begin{aligned} l^{1'} &= \nu e^{-\theta} e^{x^{4'}}(\lambda_1 + f_{2'}\lambda_2 + f_{3'}\lambda_3 + f\lambda_4), \\ l^{2'} &= \nu(\varphi_{2'}\lambda_2 + \varphi_{3'}\lambda_3), \\ l^{3'} &= \nu(\psi_{2'}\lambda_2 + \psi_{3'}\lambda_3), \\ l^{4'} &= \nu(\theta_{2'}\lambda_2 + \theta_{3'}\lambda_3 + \lambda_4). \end{aligned}\right\} \tag{25.30}$$

Hence the possibility of using the arbitrary functions f, φ, ψ and θ depends essentially on whether λ_2, λ_3 and λ_4 are identically equal to zero under *any* admissible transformation. We can show that λ_4 is not identically equal to zero. If it were, $\lambda_2 = \lambda_3 \equiv 0$ and $l^\alpha = \sigma\delta_1^\alpha$, i.e. the congruence l^α

is identical (apart from a scalar factor) with the Killing vector $\underset{1}{\xi}^{\alpha}$. Differentiating the first equation of (25.25) covariantly, we obtain

$$\underset{1}{\xi}^{\sigma}\underset{1}{\xi}_{\sigma} = 0, \quad \underset{1}{\xi}^{\sigma}{}_{,\alpha}\underset{1}{\xi}_{\sigma} = 0, \quad \underset{1}{\xi}^{\sigma}{}_{,\beta}\underset{1}{\xi}_{\sigma,\alpha}+\underset{1}{\xi}^{\sigma}\underset{1}{\xi}_{\sigma,\alpha\beta} = 0. \tag{25.31}$$

As the Lie derivative of a Riemann connectivity coefficient is zero

$$\underset{L}{\delta_s} \Gamma^{\alpha}_{\beta\gamma} = \underset{s}{\xi}^{\alpha}{}_{,\beta\gamma}+\underset{s}{\xi}^{\sigma}R_{\sigma\gamma}{}^{\alpha}{}_{\beta} = 0. \tag{25.32}$$

The last equation in (25.31) can therefore be put into the form

$$\underset{1}{\xi}^{\sigma}{}_{,\beta}\underset{1}{\xi}_{\sigma,\alpha} = \underset{1}{\xi}^{\sigma}\underset{1}{\xi}^{\tau}R_{\sigma\alpha\tau\beta} \equiv 0 \tag{25.33}$$

on using (25.24) with $\underset{1}{\xi}^{\sigma}$ expressed in terms of l^{σ}.

For the non-holonomic orthonormal tetrad corresponding to formulae (19.20), $l^{\alpha} \doteq \nu(\delta^{\alpha}_{2}+\delta^{\alpha}_{4})$; then writing (25.31) and (25.33) for this tetrad, we obtain $\underset{1}{\xi}_{\alpha,\beta} \doteq 0$. But this is a tensor equation and must be true in any other system of coordinates, i.e. $\underset{1}{\xi}_{\alpha,\beta} = \underset{1}{\xi}_{\alpha,\beta\gamma} = 0$; it follows from (25.32) that $\underset{1}{\xi}^{\alpha}R_{\sigma\gamma\alpha\beta} = 0$. However, for the canonical orthonormal tetrad, this equation leads to a contradiction and so the original assumption cannot hold; λ_4 cannot be identically equal to zero. Hence γ_4 can be taken as non-zero without loss of generality.

We can also show that is is impossible to assume that $\lambda_2 = \lambda_3 \equiv 0$ *under any* admissible transformation. If this assumption is valid, (25.27) and (25.30) show that $l^{2'} = l^{3'} = 0$, and the l^{α} can be expressed as linear combinations (with variable coefficients) of the $\underset{1}{\xi}^{\alpha}$ and $\underset{2}{\xi}^{\alpha}$; since $l^{4'} \neq 0$, the coefficient of $\underset{2}{\xi}^{\alpha}$ is not zero and

$$l^{\alpha} = \sigma(a\underset{1}{\xi}^{\alpha}+\underset{2}{\xi}^{\alpha}), \tag{25.34}$$

where σ and a are functions of x^{α} ($\alpha = 1, 2, 3, 4$). As the congruence $l^{\alpha}(x)$ is defined to within a scalar factor, we choose $\sigma = 1$. As $\lambda_4 \neq 0$ in (25.30) and the choice of f is unrestricted, we can take $\lambda_1 = 0$ and

$$\left.\begin{aligned} l^{\alpha} = a\delta^{\alpha}_{4}, \quad a = -x^{1}, \quad l^{\sigma}l_{\sigma,\alpha} = -\delta^{1}_{4} = 0, \\ \underset{1}{\xi}^{\sigma}\xi_{\alpha,\sigma} = -1, \quad \underset{2}{\xi}^{\sigma}\xi_{\alpha,\sigma} = -a. \end{aligned}\right\} \tag{25.35}$$

Moreover the first equation (25.25) shows that

$$l^{\sigma}l_{\sigma} = 0, \quad l^{\sigma}l_{\sigma,\alpha} = 0. \tag{25.36}$$

The first of these equations shows that $g_{44} = 0$, and the second that

$$\partial_4 g_{\alpha 4} = a g_{4\alpha}, \tag{25.37}$$

or by (25.29)

$$\partial_4 g_{14} = a g_{14}, \quad a g_{24} = a g_{34} = 0.$$

As $a \neq 0$, $g_{24} = g_{34} = 0$, and $a_{14} \neq 0$, since otherwise the metric would be degenerate; (25.29) shows that $a = -1$, and

$$l^\sigma l_{\alpha,\sigma} = -l_\alpha. \tag{25.38}$$

Setting down the results obtained for the non-holonomic canonical orthonormal tetrad with $l^\alpha \doteq \nu(\delta^\alpha_2 + \delta^\alpha_4)$, we note that the structure equation of the non-Abelian group $[X_1 X_2] = X_1$ is not altered by the substitution $Y_1 = X_1$, $Y_2 = cX_1 + X_2$, where c is an arbitrary constant. Therefore if at a given point of the V_4 space we choose $c = a(\overset{0}{x}{}^\alpha)$, we can put $l^\alpha \doteq \sigma \underset{2}{\xi}{}^\alpha$ and $a \doteq 0$. We thus have

$$\underset{1}{\xi}{}^\alpha(0, b, c, d), \quad \underset{2}{\xi}{}^\alpha(0, \sigma, 0, \sigma), \qquad \sigma \neq 0, \tag{25.39}$$

and it follows from (25.35), (25.36) and (25.38) that

$$\left.\begin{aligned} a_2 + a_4 \doteq 0, \quad l_{2,\alpha} + l_{4,\alpha} \doteq 0, \\ l_{\alpha,2} + l_{\alpha,4} \doteq \sigma(-\delta^\alpha_2 + \delta^\alpha_4). \end{aligned}\right\} \tag{25.40}$$

As

$$l_{\alpha,\beta} \doteq a_{,\beta} \underset{1}{\xi}_\alpha + \underset{2}{\xi}_{\alpha,\beta}, \tag{25.41}$$

eqn. (25.40) shows that

$$\left.\begin{aligned} &a_{,1} \doteq a_{,3} \doteq a_{,2} + a_{,4} \doteq 0, \quad b - d \neq 0, \quad a_{,2} \doteq \frac{1}{b-d}, \\ &\underset{2}{\xi}_{2,4} \doteq -1, \quad \underset{2}{\xi}_{2,3} \doteq \underset{2}{\xi}_{3,4}, \quad \underset{2}{\xi}_{1,2} \doteq -\underset{2}{\xi}_{1,4}. \end{aligned}\right\} \tag{25.42}$$

Putting $\sigma = -1$ in (25.34) shows that

$$l_{\sigma,\alpha\beta} = a_{,\alpha\beta} \underset{1}{\xi}_\sigma + a_{,\alpha} \underset{1}{\xi}_{\sigma,\beta} + a_{,\beta} \underset{1}{\xi}_{\sigma,\alpha} + a \underset{1}{\xi}_{\sigma,\alpha\beta} + \underset{2}{\xi}_{\sigma,\alpha\beta}.$$

Substituting for the second derivatives of the Killing vectors from (24.32) we obtain

$$l_{\sigma,\alpha\beta} = a_{,\alpha\beta} \underset{1}{\xi}_\sigma + a_{,\alpha} \underset{1}{\xi}_{\sigma,\beta} + a_{,\beta} \underset{1}{\xi}_{\sigma,\alpha} - a \underset{1}{\xi}{}^\tau R_{\tau\beta\sigma\alpha} - \underset{2}{\xi}{}^\tau R_{\tau\beta\alpha\sigma}. \tag{25.43}$$

Differentiating (25.38) covariantly, we obtain

$$l^\sigma l_{\alpha,\sigma\beta} + l^\sigma{}_{,\beta} l_{\alpha,\sigma} + l_{\alpha,\beta} = 0. \tag{25.44}$$

Equation (25.34) shows that $l^{\sigma}l_{\alpha,\sigma\beta} = 0$ in a holonomic system of coordinates; for the non-holonomic orthonormal tetrad, (25.44) can be written

$$l^{\sigma}\underset{1}{\xi}_{\alpha,\sigma\beta}+l^{\sigma}{}_{,\beta}l_{\alpha,\sigma}+l_{\alpha,\beta} \doteq 0.$$

Letting α and β take all possible values produces a set of equations implying that $\underset{2}{\xi}_{\alpha,\beta} \doteq 0$, $a_2(b-d) \doteq 0$, which contradicts (25.42). Hence λ_2 and λ_3 are not identically zero under all admissible transformations.

Thus all four indeterminate functions f, φ, ψ and θ may be used in (25.30). As we have proved that l^{α} cannot lie in the tangent plane to the surface of transitivity, we can impose the condition that all the components l^{α}, except l^2 and l^3, must not be zero. Only in the case of a degenerate transformation can l^2 and l^3 be reduced to zero simultaneously.

Imposing the condition $l^{1'} = l^{2'} = l^{4'} = 0$ provides a consistent set of differential equations of the Cauchy type for analytic functions f, φ and θ; in the new system of coordinates

$$l^{\alpha} = \nu\delta^{\alpha}_{3}, \tag{25.45}$$

where ν is an arbitrary function of x.

The admissible transformations are those leaving (25.45) unchanged

$$\left.\begin{aligned} x^1 &= x^{1'}+e^{x^{4'}}f(x^{2'}), \quad x^2 = \varphi(x^{2'}), \quad x^3 = \psi(^{2'}, x^{3'}),\\ x^4 &= x^{4'}+\theta(x^{2'}), \quad \varphi'\psi_3 \neq 0. \end{aligned}\right\} \tag{25.46}$$

Since the congruence l^{α} is isotropic, $a_{33} = 0$. Integrating the second of eqns. (25.25), we obtain

$$a_{13} = \varrho b_1(x^2)e^{-x^4}, \quad a_{23} = \varrho b_2(x^2), \quad a_{33} = 0, \quad a_{34} = \varrho b_4(x^2)$$

where ϱ an arbitrary function of x^2 and x^3 cannot be zero, as the metric would then be degenerate. After the transformation (25.46), we obtain

$$\left.\begin{aligned} g^{1'3'} &= \varrho b_1 e^{-\theta}\psi_{3'}e^{-x^{4'}},\\ g_{2'3'} &= \varrho\psi_{3'}(e^{-\theta}b_1 f'+b_2\varphi'+b_4\theta'),\\ g_{3'4'} &= \varrho\psi_{3'}(b_1 e^{-\theta}f+b_4). \end{aligned}\right\} \tag{25.47}$$

The $b_i(x^2)$ can be simplified by suitably choosing the functions f, φ, ψ and θ; the following three distinct cases

$$(1)\ b_1 \neq 0, \quad (2)\ b_1 = 0, \quad b_4 \neq 0, \quad (3)\ b_1 = b_4 = 0 \tag{25.48}$$

must be considered separately.

We choose the function ν in (25.45) to be 1, so that $l^{\alpha} = \delta^{\alpha}_{3}$ and hence $l_{\alpha} = g_{3\alpha}$.

Then *in any admissible system of coordinates* (25.46)

$$l_{\alpha,\beta} = \Gamma_{\alpha,3\beta}$$

and

$$l_{\alpha,\beta}l^{\beta} = \partial_2 g_{3\alpha}. \tag{25.49}$$

Consider case (1), $b_1 \neq 0$. Equations (25.47) show that we can, without loss of generality, assume that b_1, b_2 and b_4 are not zero in some admissible system of coordinates. Choosing f, θ and ψ so that $g_{2'3'} = g_{3'4'} = 0$, we obtain $b_1 = 1$, $b_2 = b_4 = 0$, $\varrho = 1$ in the new system of coordinates; (25.49) shows that

$$l_{\alpha,\beta}l^{\beta} = 0. \tag{25.50}$$

This tensor equation holds for any system of coordinates and in particular for any admissible system, where it has the form

$$\partial_2 g_{3\alpha} = 0. \tag{25.51}$$

We now investigate the special admissible transformations, which permit $b_2 = b_4 = 0$ to remain invariant. Equation (25.47) shows that these transformations are obtained from (25.46) by putting $f = 0$, and retaining $\varphi(x^2)$, $\psi(x^{2'}, x^{3'})$ and $\theta(x^{2'})$ as arbitrary functions restricted only by the inequality $\varphi'\psi_{,3'} \neq 0$. Applying such a transformation to eqns. (25.47) corresponding to the old metric tensor components $g_{13} = e^{-x^4}$, $g_{23} = g_{33} = g_{34} = 0$, and condition (25.51) to the new coordinate system, we obtain

$$\partial_{3'} g_{3'\alpha'} = \frac{\partial}{\partial x^{3'}} (e^{-\theta}\psi_{3'})\, \delta^1_{\alpha'} = 0.$$

Hence when $\alpha = 1$,

$$\frac{\partial^2}{\partial x^{3'^2}} \psi = 0,$$

which is inconsistent with the arbitrary nature of ψ, and so case (1) is impossible.

In case (3), $b_1 = b_4 = 0$, in any admissible system of coordinates; g_{33} can be reduced to 1 by an appropriate choice of $\psi_{3'}(x^{2'}, x^{3'})$. Condition (25.50) holds in any coordinate system and (25.51) in any admissible system.

The latter equation reduces to an identity for $\alpha = 1$, 3 or 4, and we obtain

$$\frac{\partial}{\partial x^{3'}} (\varphi'\psi_{,3'}) = 0, \quad \text{i.e.} \quad \psi_{,3'3'} = 0$$

for $\alpha = 2$ in the new system of coordinates, a condition not satisfied by any arbitrary ψ. Case (3) is also impossible.

Thus only case (2) remains: $b_1 \equiv 0, b_4 \not\equiv 0$. Equations (25.47) show that if we put $b_2\varphi' + b_4\theta' = 0$, $g_{1'3'} = g_{2'3'} = g_{3'3'} = 0$, $g_{3'4'} = \varrho b_4 \psi_{3'} e^{-\theta}$; ϱ and ψ are functions of $x^{2'}$, $x^{3'}$, and $b_4 = b_4(x^{2'})$. ψ can be chosen to reduce $g_{3'4'}$ to unity. Condition (25.50) is valid in this system of coordinates and (25.51) in any admissible coordinate system. Taking $b_1 = b_2 = 0$, $\varrho b_4 = 1$, we find those admissible transformations which maintain these conditions. Equations (25.47) show that this occurs when $\psi_{,3'} = 1$ and $\psi_{,3'3'} = 0$, and there is no inconsistency as in the two previous cases. Even if we do not reduce the component $g_{3'4'}$ to unity, obtaining the condition

$$\frac{\partial}{\partial x^{3'}}(\varrho\psi_{,3'}) = 0,$$

which implies that when $\nu = 1$ the factor ϱ is fully determined by the coordinate $x^3 = \psi$, there is no inconsistency.

We now assume that the transformation has been carried out, but that $\nu \neq 1$ and $g_{13} = g_{23} = g_{33} = 0$, $g_{34} = g_{34}(x^{2'}, x^{3'})$.

We select from (25.46) those transformations leaving these conditions invariant. Equations (25.47) show that these have the form

$$\left.\begin{aligned} x^1 &= x^{1'} + f(x^{2'})e^{x^{4'}}, \quad x^2 = \varphi(x^{2'}), \\ x^3 &= \psi(x^{2'}, x^{3'}), \quad x^4 = x^{4'} + C, \end{aligned}\right\} \tag{25.52}$$

where $C = \text{constant}$, $\varphi'\psi_{,3'} \neq 0$. We note that after such a transformation, the new components of the metric tensor have the form (25.29) and the $a_{\alpha'\beta'}$ can be expressed in terms of the $a_{\alpha\beta}$ as follows:

$$\left.\begin{aligned} a_{1'1'} &= a_{11}e^{-2\theta}, \quad a_{1'2'} = e^{-\theta}(a_{11}e^{-\theta}f' + a_{12}\varphi'), \\ a_{1'3'} &= 0, \quad a_{1'4'} = e^{-\theta}(a_{11}e^{-\theta}f + a_{14}), \\ a_{2'2'} &= a_{11}e^{-2\theta}f'^2 + 2a_{12}e^{-\theta}f'\varphi' + a_{22}\varphi'^2, \\ a_{2'3'} &= 0, \quad a_{2'4'} = a_{11}e^{-2\theta}ff' + a_{12}e^{-\theta}f\varphi' \\ &\qquad + a_{14}e^{-\theta}f' + a_{24}\varphi' + a_{34}\psi_{12'}, \\ a_{3'3'} &= 0, \quad a_{3'4'} = a_{34}\psi_{,3'}, \\ a_{4'4'} &= a_{11}e^{-2\theta}f^2 + 2a_{14}e^{-\theta}f + a_{44}. \end{aligned}\right\} \tag{25.53}$$

The component a_{34} cannot be zero for otherwise the metric would be degenerate. Since apart from the condition $\psi_{,3'} \neq 0$, ψ is a completely arbitrary function of $x^{2'}, x^{3'}$, the component $a_{2'4'}$ in (25.33) can always be taken as zero. In addition, according to (25.25) $a_{3'4'} = \varrho b_4(x^{2'})$, where ϱ can be expressed in terms of the arbitrary scalar factor $\nu\,(l^\alpha = \nu g_3^\alpha)$, restricted only by the inequality $\nu \neq 0$: $\varrho = \omega(x^2)\nu$. Hence ν can be chosen so that $a_{3'4'} = 1$, independently of the values of ψ and $\psi_{,3'} \neq 0$. The remaining components $a_{\alpha'\beta'}$ in (25.53) are independent of ν and ψ; the arbitrary functions $\varphi(x^{2'})$ ($\varphi' \neq 0$) and f, will be used when some $a_{\alpha\beta}$, or ratios between them, are independent of x^3.

Using the following compact notation for the components in the new system of coordinates

$$a_{11} = \alpha, \quad a_{12} = \beta, \quad a_{14} = \delta, \quad a_{22}\ \gamma, \quad a_{44} = \lambda.$$

the required metric matrix is

$$(g_{\alpha\beta}) = \begin{pmatrix} \alpha e^{-2x^4} & \beta e^{-x^4} & 0 & \delta e^{-x^4} \\ \beta e^{-x^4} & \gamma & 0 & 0 \\ 0 & 0 & 0 & 1 \\ \delta e^{-x^4} & 0 & 1 & \lambda \end{pmatrix}. \tag{25.54}$$

$$|g_{\alpha\beta}| = g = -\sigma e^{-2x^4}, \quad \sigma = \alpha\gamma - \beta^2 > 0, \quad \alpha,\ \gamma \neq 0.$$

The primes have been dropped and α, β, γ and δ are unknown functions of x^2 and x^3.

The unknown components must satisfy eqns. (25.24) and the field equations $R_{\alpha\beta} = 0$. Equation (25.45) shows that the first set of eqns. (25.24) reduce to

$$R_{\alpha 3\beta 3} = 0 \quad (\alpha, \beta = 1, 2, 3, 4); \tag{25.55}$$

and the second set to

$$e_{3\beta\sigma\tau} R^{\sigma\tau}{}_{3\gamma} = 0,$$

or to the equivalent set

$$R^{24}{}_{3\gamma} = R^{14}{}_{3\gamma} = R^{12}{}_{3\gamma} = 0, \tag{25.56}$$

since the only non-zero components of the permutation tensor $e_{\alpha\beta\gamma\delta}$ are those where all the indices are different. The contravariant metric matrix is

$$(g^{\alpha\beta}) = \begin{pmatrix} \dfrac{\gamma e^{2x^4}}{\sigma} & \dfrac{-\beta e^{x^4}}{\sigma} & \dfrac{-\gamma\delta e^{x^4}}{\sigma} & 0 \\ \dfrac{-\beta e^{x^4}}{\sigma} & \dfrac{\alpha}{\sigma} & \dfrac{\beta\delta}{\sigma} & 0 \\ \dfrac{-\gamma\delta e^{x^4}}{\sigma} & \dfrac{\beta\delta}{\sigma} & \dfrac{\omega}{\sigma} & 1 \\ 0 & 0 & 1 & 0 \end{pmatrix}. \tag{25.57}$$

$$\omega = \beta^2\lambda - \alpha\gamma\lambda + \gamma\delta^2,$$

Equation (25.55) shows that

$$R^{24}{}_{3\gamma} = g^{2\sigma}g^{4\tau}R_{\sigma\tau 3\gamma} \equiv 0.$$

Similarly, the second of eqns. (25.56) is identically zero; the third leads to the set of equations

$$R_{12\,3\gamma} = 0, \qquad (\gamma = 1, 2, 3, 4). \tag{25.58}$$

The field equations $g^{\sigma\tau}R_{\sigma\alpha\tau\beta} = 0$ for the metric (25.57) and eqs. (25.55) and (25.58) show that

$$R_{1212} = R_{1431} = R_{1423} = R_{2431} = R_{2423} = 0, \tag{25.59}$$

$$g^{11}R_{1412}+g^{12}R_{2412} = R_{2434},\; g^{12}R_{1412}+g^{22}R_{2412} = R_{1434}, \tag{25.60}$$

$$g^{11}R_{1414}+2g^{12}R_{1424}+2g^{13}R_{1434}+g^{22}R_{2424}+2g^{23}R_{2434} = 0. \tag{25.61}$$

Equations (25.59) are not obtained directly from the field equations but via (25.58) and the Ricci identity

$$R_{1[234]} = 0.$$

Before integrating the equations obtained, we must determine under what conditions the particular manifold considered defines a T_3 space.

It is not difficult to show by using *elementary transformations*, that *the Segre characteristic of the matrix* $(R_{ab}-\lambda g_{ab})$ $(a, b = 1, \ldots, 6)$ *written in the bivector space form, with metric* (25.54) *is* $(3^\circ,3)$ *only if at least one of the components* R_{1412} *or* R_{2412} *satisfying* (25.55), (25.58) *and* (25.60) *is not zero.* (*Shirokov* [91]). Equation (25.60) shows that this is only possible if R_{1434} and R_{2434} are not simultaneously zero.

We are thus faced with the problem of integrating the overdetermined *system of eighteen second order partial differential equations* (25.55), (25.58), (25.59), (25.60), (25.61) *with the five unknown functions* α, β, γ, δ and λ *of the two arguments* x^2 *and* x^3; the unknown functions can be simplified by appropriate choice of the arbitrary functions, in accordance with (25.53).

We first consider the equation $R_{3423} = 0$. For the metric (25.54) this can be written in the form

$$\delta_3(\gamma\beta_3-\beta\gamma_3) = 0. \tag{25.62}$$

Assuming initially that $\delta_3 \equiv \dfrac{\partial}{\partial x^3}\delta = 0$, (25.53) shows that since f and θ are arbitrary functions of x^2, $\dfrac{\partial}{\partial x^3}a_{11} = \dfrac{\partial}{\partial x^3}a_{14} = 0$, i.e. $\alpha_3 = 0$. Similarly, the equation $R_{3131} = 0$ leads to $\alpha\beta_3 = 0$. Since, according to (25.54), α

cannot be zero, $\beta_3 \equiv \dfrac{\partial}{\partial x^3} a_{12} = 0$ and (25.53) shows that β can be reduced to zero by choosing f correctly. It is also possible to reduce δ to zero by appropriate choice of θ. If, however, the second factor in (25.62) is zero, (25.53) shows that

$$e^{-2\theta} f'^2 \varphi' \left(a_{12} \frac{\partial a_{11}}{\partial x^3} - a_{11} \frac{\partial a_{12}}{\partial x^3} \right) + e^{-\theta} f' \varphi'^2 \left(a_{22} \frac{\partial a_{11}}{\partial x^3} - a_{11} \frac{\partial a_{22}}{\partial x^3} \right)$$
$$+ \varphi'^3 \left(a_{22} \frac{\partial a_{12}}{\partial x^3} - a_{12} \frac{\partial a_{22}}{\partial x^3} \right) = 0.$$

As f, φ and θ are arbitrary functions ($\varphi' \neq 0$) of the variable x^2, each bracket must be zero and in particular $\dfrac{\partial}{\partial x^3} \left(\dfrac{a_{12}}{a_{11}} \right) = 0$. Hence by an appropriate choice of f, β can be reduced to zero. Similarly, the equation $R_{1423} = 0$ leads to

$$\gamma \alpha_2 \delta_3 = 0.$$

Equation (25.54) shows that $\gamma \neq 0$; similarly $\alpha_2 \neq 0$, for otherwise $\dfrac{\partial}{\partial x^2} (a_{11} e^{-2\theta}) = 0$, and since θ is an arbitrary function of x^2, $a_{11} = 0$; but this is impossible as $\sigma > 0$. Hence $\delta_3 = 0$ as initially assumed. It is therefore always possible to reduce β and δ to zero and the metric of the space to

$$(g_{\alpha\beta}) = \begin{pmatrix} \alpha e^{-2x^4} & 0 & 0 & 0 \\ 0 & \gamma & 0 & 0 \\ 0 & 0 & 0 & 1 \\ 0 & 0 & 1 & \lambda \end{pmatrix}, \qquad (25.63)$$

$$g = -\alpha\gamma e^{-2x^4}, \quad \alpha, \ \gamma \neq 0,$$

where α, γ and λ are unknown functions of x^2 and x^3. Moreover, (25.53) shows that the $\varphi(x^{2'})$ is subject only to the condition $\varphi' \neq 0$, and is not yet uniquely defined.

Equations (25.55), (25.58) and (25.59) for the metric (25.63) lead to the set of differential equations

$$\left.\begin{aligned}
&\alpha_3(\lambda_3+1) = 0, \quad \alpha_{33}-\frac{\alpha_3^2}{2\alpha} = 0,\\
&\gamma_{33}-\frac{\gamma_3^2}{2\gamma} = 0, \quad \lambda_{33} = 0, \quad \gamma_3\lambda_3 = 0,\\
&\alpha_{23}-\frac{\alpha_2}{2}\left(\frac{\gamma_3}{\gamma}+\frac{\alpha_3}{\alpha}\right) = 0,\\
&\alpha_{22}-\frac{\alpha_2}{2}\left(\frac{\gamma_2}{\gamma}+\frac{\alpha_2}{\alpha}\right)+\frac{\gamma_3}{2}(\alpha-\lambda\alpha_3) = 0,
\end{aligned}\right\} \tag{25.64}$$

all the other equations being satisfied identically.

If we assume that $\gamma_3 \neq 0$ in the fifth equation of this set, $\lambda_3 = 0$. The first equation then shows that $\alpha_3 = 0$, the sixth that $\alpha_2 = 0$, and the last equation that $\alpha\gamma_3 = 0$, contradicting the condition $\alpha \neq 0$; hence $\gamma_3 = 0$. If we put $f = 0$, $\theta = 0$, $\psi_3 = 1$, $\psi_2 = 0$ in the transformation (25.53) the form of the metric (25.63) is left unaltered but the component $a_{2'2'} = \gamma\varphi'^2$, where $\varphi' \neq 0$. As φ is an arbitrary function of x^2, it can be so chosen that $a_{2'2'} = e = \pm 1$. Thus we can put $\gamma = e = \pm 1$ in (25.63) and (25.64) without loss of generality, and the set (25.64) can be integrated directly to give

$$\alpha = e(c_1+c_2x^2+c_3x^3)^2, \quad \lambda = a(x^2)x^{3^2}+b(x^2) \quad \gamma = e = \pm 1, \tag{25.65}$$

where a and b are unknown functions of x^2 and

$$ac_3 = 0, \quad c_i = \text{const} \qquad (i = 1, 2, 3). \tag{25.66}$$

As pointed out above, both curvature tensor components R_{1434} or R_{2434} cannot be zero. For the metric of (25.63) subject to condition (25.65), R_{1434} is identically zero, and $R_{2434} = -\frac{1}{2}\lambda_{23}$, so that $\lambda_3 = a \neq 0$; according to the first of eqns. (25.66) $c_3 = 0$ and $\alpha = e\,(c_1+c_2x^2)^2$.

The component $R_{2412} \equiv 0$, and $R_{1412} = \alpha_2 e^{-2x^4}/2$; as both cannot be zero $\alpha_2 \neq 0$, i.e. $c_2 \neq 0$.

Equations (25.60) and (25.61) reduce to

$$\lambda_{23}+\frac{\alpha_2}{\alpha} = 0, \quad \lambda_{22}+\frac{\alpha_2\lambda_2}{2\alpha}+e(\lambda_3+2) = 0. \tag{25.67}$$

Replacing λ and α by the expressions given above, and equating the coefficients of x^3 to zero, we obtain

$$\lambda = \ln\frac{c_6}{(c_1+c_2x^2)^2}\left[x^3-\frac{e}{4c_2^2}(c_1+c_2x^2)^2+c_4\right]-\frac{e}{c_2^2}(c_1+c_2x^2)^2+c_5, \tag{25.68}$$

where the c_i are constants. Putting

$$x^1 = \frac{1}{c_2}x^{1'}, \quad x^2 = \pm x^{2'} - \frac{c_1}{c_2}, \quad x^3 = -x^{3'} - c_4, \quad x^4 = -x^{4'}$$

shows that c_1 and c_4 can be zero without changing the structure of the matrix (25.63). Hence the required $g_{\alpha\beta}$ components can be derived by putting

$$\alpha = ex^{2^2}, \ \gamma = e, \ \lambda = \left(x^3 + \frac{e}{4}x^{2^2}\right)\ln(px^{2^2}) - ex^{2^2} + q, \ e = \pm 1, \quad (25.69)$$

where p and q are arbitrary constants which cannot be eliminated by *admissible* transformations.

This solution was first obtained in a different system of coordinates by the author ([261], p. 201); no shorter method of solving the problem is available.

The metric of the space is

$$ds^2 = e(\alpha\, dx^{1^2} + dx^{2^2}) + 2dx^3\, dx^4 + \lambda\, dx^{4^2}, \quad (25.70)$$

where the potentials α and λ are defined by formulae (25.69).

The general integral of the Killing equations (with *arbitrary p and q*) for the metric (25.70) shows that the Killing vector is

$$\xi^\alpha = (c_1 + c_2x^1)\delta_1^\alpha + c_2\delta_2^\alpha$$

corresponding to a non-Abelian group of motions G_2 acting on non-isotropic surfaces of transitivity; the space does not admit any group of higher mobility; the group is a maximum—as proved above from other considerations.

§26. T_1 Spaces Admitting Motions

We turn to the determination of T_i spaces admitting various groups of motions.

The cases where T_1 admit only G_1 or G_2 do not present any difficulties. In the case of G_1 a system of coordinates must exist in which all the $g_{\alpha\beta}(x)$ are independent of some coordinate x^α (cf. § 10, example 1), the parametric curves x^α being trajectories of the motion; the field equations differ from the general equations in only this particular.

In the case of G_2, a system of coordinates can be chosen with Killing vectors in a form (cf. example 1) convenient for integrating the Killing equations, so that the field equations can be immediately written down.

In view of this we need only consider G_r with $r \geqslant 3$. A T_i space admitting a group of motions G_r will henceforward be indicated by the notation $T_{i,r}$.

The general method of finding the metric of T_i spaces admitting a particular group of motions $G_r (r \geqslant 3)$ is as follows; we determine the rank of $(\underset{s}{\xi}^\alpha)$ for each of the non-isomorphic G_3 structures obtained in § 24, and check whether the surface of transitivity is isotropic or not; a holonomic system of coordinates is chosen with the Killing vector in its simplest form. Finally we integrate the Killing equations and the field equations $R_{\alpha\beta} = 0$, and thus obtain the metrics of all the possible T_i with G_3.

Since every G_4 admits a G_3 subgroup, all $T_{i,4}$ must be contained among the set of $T_{i,3}$. We can therefore find all the $T_{i,4}$ by imposing on $T_{i,3}$ the additional requirement that they should admit G_4 of the proper structure. The position is exactly the same with $T_{i,5}$. The only $T_{i,6}$ (apart from the $T_{2,6}$ obtained in § 25) are flat spaces.

We select the admissible coordinate transformations for the given G_r, i.e. those which *convert the components of the Killing vectors* $\underset{s}{\xi}^\alpha(x)$ *into the same functions of the new variables.*

$T_{1,3}$ Spaces

1. Let us consider the structures (24.10) corresponding to the two insoluble G_3 groups when they contain subgroups of stability G_1, and thus two-dimensional *non-isotropic* surfaces of transitivity with indefinite or negative-definite metrics.

We consider an insoluble G_3 acting transitively on two-dimensional non-isotropic surfaces of transitivity. Each surface admits the group of full mobility for a V_2 space (§ 10, example 9), and is therefore a surface of constant curvature. We determine the set of operators inducing the G_3 group with given structure on such a V_2. Section 10, p. 59 showed that if the rank of the matrix $(\underset{s}{\xi}^\alpha)$ is equal to the order of the group, then two groups are *similar* when their structures are identical.

In the present case however, the rank of the matrix $(\underset{s}{\xi}^\alpha)$ is less than the order of the group. Two distinct cases arise when the structures are identical, depending on the sign of K, the curvature of the surface; K cannot be zero as G_3 contains an Abelian G_2 subgroup for a Euclidean or pseudo-Euclidean plane and cannot be insoluble. The operators for $K \neq \pm 1$ can be found by solving the Killing equations for $K = \pm 1$, and dividing all the variables in $\underset{s}{\xi}^{\alpha}(x)$ by $\sqrt{\pm K}$. As it is always possible to eliminate this factor, only the cases $K = \pm 1$ need be considered. However this is true only if we assume that the operators of the surfaces of transitivity do not vary from surface to surface. In general, if a V_4 with an indefinite metric admits an intransitive group of motions G_r, and the generic rank q of the matrix $(\underset{s}{\xi}^{\alpha})$ is less than r, it is possible to so renumber the ξ^{α} that

$$\underset{k}{\xi}^{\alpha} = \varphi_k^t \underset{t}{\xi}^{\alpha} \qquad (t = 1, \ldots, q;\ k = q+1, \ldots, r),$$

where the φ_k are functions of x. Suppose that p of the functions φ are independent. The following two cases then arise ([147], p. 226): (1) if in the coordinate system in which $\underset{s}{\xi}^{\nu} = 0$ for $\nu = q+1, \ldots, n$, the rank of the Jacobian matrix of the functions φ with respect to $x^1, \ldots, x^q$ is equal to p, then coordinate systems exist in which the $\underset{s}{\xi}^{\alpha}$ are at most functions of $x^1, \ldots, x^q$; (2) if the rank of the Jacobian is less than p, such coordinate systems do not, in general, exist.

It is sufficient to show that the Jacobian is non-zero at the origin of a normal system of coordinates, since continuity considerations extend the result over a finite region. In the case under consideration $q = 2$ and $\underset{1}{\xi}^{\alpha} = \overset{2}{\varphi}\underset{2}{\xi}^{\alpha} + \overset{3}{\varphi}\underset{3}{\xi}^{\alpha}$. Lowering the index α, differentiating covariantly and applying the result to the origin where the initial conditions are

$$\overset{0}{\underset{1}{\xi}}{}^{\alpha} = 0, \quad \overset{0}{\underset{2}{\xi}}{}^{\alpha} = \delta_1^{\alpha}, \quad \overset{0}{\underset{3}{\xi}}{}^{\alpha} = \delta_3^{\alpha} \quad (\text{or} \quad \delta_4^{\alpha})$$

$$\overset{0}{\underset{1}{\xi}}{}_{\alpha,\beta} = \begin{cases} \alpha_3 & \alpha\beta = 12 \\ -\beta_3 & \alpha\beta = 34 \\ 0 & \end{cases} \qquad \overset{0}{\underset{2}{\xi}}{}_{\alpha,\beta} = \overset{0}{\underset{3}{\xi}}{}_{\alpha,\beta} = 0$$

corresponding to (24.10).

Putting $\alpha, \beta = 1, 2, 3, 4$, we obtain $\beta_3 = 0$, $\alpha_3 \neq 0$; as $\underset{2}{\xi}_{\alpha,\beta}$ is asymmetric,

$$(\overset{2}{\varphi}_2)_0 = 1, \quad (\overset{3}{\varphi}_1)_0 = 1, \quad (\overset{3}{\varphi}_2)_0 = 0,$$

i.e. the required Jacobian is

$$\begin{vmatrix} 0 & 1 \\ 1 & 0 \end{vmatrix} \neq 0.$$

The analysis for case $\overset{0}{\underset{3}{\xi}}{}^{\alpha} = \delta_4^{\alpha}$ is similar.

Thus there exists a system of coordinates relative to which the surfaces of transitivity have the equations $x^3 =$ constant, $x^4 =$ constant, while $\underset{s}{\xi}^\alpha = \underset{s}{\xi}^\alpha(x^1, x^2)$.

A study of the operators on these surfaces with metric signature $(--)$ or $(-+)$ can be restricted to the metrics

$$\begin{aligned} &1^\circ. \quad -dx^{2^2} - \cos^2 x^2\, dx^{3^2}, \\ &2^\circ. \quad dx^{1^2} - \cos^2 x^1\, dx^{4^2}, \\ &3^\circ. \quad dx^{4^2} - \cosh^2 x^4\, dx^{1^2}, \end{aligned} \tag{26.1}$$

referred to semi-geodesic coordinates, since all other cases can be reduced to these three by means of non-degenerate real transformations. The corresponding operators are defined by the Killing vectors

$$(1) \quad \left.\begin{aligned} &\underset{1}{\xi}^\alpha = \delta_3^\alpha, \quad \underset{2}{\xi}^\alpha = \sin x^3 \delta_2^\alpha + \cos x^3 \cot x^2 \delta_3^\alpha, \\ &\underset{3}{\xi}^\alpha = \cos x^3 \delta_2^\alpha - \sin x^3 \cot x^2 \delta_2^3; \qquad G_3 \text{ type IX.} \end{aligned}\right\} \tag{26.2}$$

$$(2) \quad \left.\begin{aligned} &\underset{1}{\xi}^\alpha = \delta_4^\alpha, \quad \underset{2}{\xi}^\alpha = \sinh x^4 \delta_1^\alpha + \cosh x^4 \tan x^1\, \delta_4^\alpha, \\ &\underset{3}{\xi}^\alpha = \cosh x^4\, \delta_1^\alpha + \sinh x^4 \tan x^1\, \delta_4^\alpha; \quad G_3 \text{ type VIII.} \end{aligned}\right\} \tag{26.3}$$

$$(3) \quad \left.\begin{aligned} &\underset{1}{\xi}^\alpha = \delta_1^\alpha, \quad \underset{2}{\xi}^\alpha = \sinh x^4 \delta_1^\alpha + \cosh x^4 \tan x^1\, \delta_4^\alpha, \\ &\underset{3}{\xi}^\alpha = \cosh x^4 \delta_1^\alpha + \sinh x^4 \tan x^1 \delta_4^\alpha; \quad G_3 \text{ type VIII.} \end{aligned}\right\} \tag{26.4}$$

Let us consider (26.2). The Killing equations $\xi_{(\alpha,\,\gamma)} = 0$ for $\underset{1}{\xi}^\alpha$ show that the $g_{\alpha\beta}$ are independent of x^3, and those for $\underset{2}{\xi}^\alpha$ and $\underset{3}{\xi}^\alpha$ that

$$g_{11} = \alpha, \quad g_{14} = \beta, \quad g_{22} = \gamma, \quad g_{33} = \gamma \sin^2 x^2, \quad g_{44} = \delta, \tag{26.5}$$

where α, β, γ and δ are functions of x^1 and x^4, all the other components $g_{\alpha\beta}$ being zero.

We now determine the general form of the *admissible* transformation,

$$x^{\alpha'} = \varphi^{\alpha'}(x^1, \ldots, x^n),$$

for the Killing vectors (26.2); by definition,

$$\underset{1}{\xi}^{\alpha'} \equiv \partial_3 \varphi^{\alpha'} = \delta_3^{\alpha'},$$

$$\underset{2}{\xi}^{\alpha'} \equiv \sin x^3 \partial_2 \varphi^{\alpha'} + \cos x^3 \cot x^2 \partial_3 \varphi^{\alpha'} = \sin \varphi^3 \delta_2^{\alpha'} + \cos \varphi^3 \cot \varphi^2 \delta_3^{\alpha'},$$

$$\underset{3}{\xi}^{\alpha'} \equiv \cos x^3 \partial_2 \varphi^{\alpha'} - \sin x^3 \cot x^2 \partial_3 \varphi^{2'} = \cos \varphi^3 \delta_2^{\alpha'} - \sin \varphi^3 \cot \varphi^2 \delta_3^{\alpha'}.$$

Putting $\alpha = 1, 2, 3, 4$ provides a set of first order partial differential equations with four unknown functions. The general integral of this set, which determines the admissible transformation, has the form

$$x^{1'} = \overset{*}{f}(x^1, x^4), \qquad x^{2'} = x^2 + \overset{*}{k}\pi,$$
$$x^{3'} = x^3 + \overset{*}{s}\pi, \qquad x^{4'} = \overset{*}{\theta}(x^1, x^4),$$

where $\overset{*}{k}$, $\overset{*}{s}$ are integers, and $\overset{*}{f}$ and $\overset{*}{\theta}$ are arbitrary functions of their arguments. This transformation is clearly non-degenerate with an inverse given by the same equations with x^α and $x^{\alpha'}$ interchanged and without the asterisk. This transformation does not alter the $\underset{s}{\xi}^\alpha$ or the equations of the surfaces of transitivity, $x^2 =$ constant, $x^3 =$ constant, and hence it leaves the structure of the integrals (26.5) unchanged. For the $g'_{\alpha\beta}$ we obtain

$$g'_{11} = \alpha(\partial_1 f)^2 + 2\beta\, \partial_1 f \partial_1 \theta + \delta(\partial_1\theta)^2,$$
$$g'_{14} = \alpha\, \partial_1 f\, \partial_4 f + \beta(\partial_1 f\, \partial_4\theta + \partial_4 f\, \partial_1\theta) + \delta\, \partial_1\theta\, \partial_4\theta,$$
$$g'_{44} = \alpha(\partial_4 f)^2 + 2\beta\, \partial_4 f\, \partial_4\theta + \delta(\partial_4\theta)^2.$$

As the functions f and θ are arbitrary, conditions

$$g'_{14} = 0, \quad g'_{11}g'_{44} = -1,$$

can be imposed, leading to the set of differential equations

$$\partial_1 f = -\frac{(\beta\partial_4 f + \delta\partial_4\theta)}{\alpha\partial_4 f + \beta\partial_4\theta}\,\partial_1\theta, \quad (\partial_1\theta)^2 = -\frac{(\alpha\partial_4 f + \beta\partial_4\theta)^2}{(\alpha\delta - \beta^2)g'^2_{44}},$$

where g'_{44} depends only on the derivatives $\partial_4 f$ and $\partial_4\theta$. We now show that this is a consistent set of equations leading to a *real* system of coordinates. If the coordinate system is real, the right-hand side of the second equation must be positive, i.e. $\alpha\delta - \beta^2 < 0$.

But for the metric (26.5)

$$g \equiv |g_{\alpha\beta}| = (\alpha\delta - \beta^2)\gamma^2 \sin^2 x^2 < 0,$$

and hence $\alpha\delta - \beta^2 < 0$. Substituting for $\partial_1\theta$ from the second equation into the first, and assuming that the functions are *holomorphic* in some region, leads to a set of equations of the Cauchy–Kowalewski type (Finikov [160], § 1):

$$\partial_1 f = F(x; \partial_4 f, \partial_4\theta), \quad \partial_1\theta = \Psi(x; \partial_4 f, \partial_4\theta).$$

Hence the set of equations is consistent; by taking f and θ as the integrals of this set, the conditions imposed above are satisfied and the metric is reduced to the form

$$(1)\qquad (g_{\alpha\beta}) = \begin{pmatrix} \alpha & 0 & 0 & 0 \\ 0 & \gamma & 0 & 0 \\ 0 & 0 & \gamma \sinh x^2 & 0 \\ 0 & 0 & 0 & -\dfrac{1}{\alpha} \end{pmatrix}, \qquad (26.6)$$

where α and γ are functions of x^1 and x^4. The system of coordinates is 4-orthogonal.

Six of the ten field equations $R_{\alpha\beta} = 0$ for this metric are satisfied identically, and the other four are

$$\left.\begin{aligned} &\frac{2}{\gamma}\left(\gamma_{11}-\frac{\gamma_1^2}{2\gamma}\right)-\frac{1}{\gamma}\left(\frac{\alpha_1\gamma_1}{\alpha}+\alpha\alpha_4\gamma_4\right)-\alpha\left(\alpha_{44}+\frac{\alpha_{11}}{\alpha^2}-\frac{2\alpha_1^2}{\alpha^3}\right) = 0, \\ &\frac{2}{\gamma}\left(\gamma_{44}-\frac{\gamma_4^2}{2\gamma}+\frac{\alpha_1\gamma_1}{2\alpha^3}+\frac{\alpha_4\gamma_4}{2\alpha}\right)+\frac{1}{\alpha}\left(\alpha_{44}+\frac{\alpha_{11}}{\alpha^2}-\frac{2\alpha_1^2}{\alpha^3}\right) = 0, \\ &\gamma_{14}-\frac{\gamma_1\gamma_4}{2\gamma}+\frac{1}{2\alpha}(\alpha_1\gamma_4-\alpha_4\gamma_1) = \frac{\gamma_{11}}{\alpha}-\alpha\gamma_{44}-\frac{\alpha_1\gamma_1}{\alpha^2}-\alpha_4\gamma_4-2 = 0, \end{aligned}\right\} \qquad (26.7)$$

using the notation $\partial_i f = f_i (i = 1, 4)$. This set is consistent, and actual solutions are given later for the case of $T_{1,4}$; the problem of the arbitrariness in the solutions of this set of equations is not dealt with, but the problem can be solved completely by well-known methods (Finikov [160], pp. 19–23).

As the same investigation can be carried out for the Killing vectors (26.3) and (26.4), the actual analysis will not be repeated; (cf. example 2) the other two solutions of the field equations of $T_{1,3}$ with the *structure* VIII are

$$(2)\qquad (g_{\alpha\beta}) = \begin{pmatrix} \alpha & 0 & 0 & 0 \\ 0 & \beta & 0 & 0 \\ 0 & 0 & \dfrac{1}{\beta} & 0 \\ 0 & 0 & 0 & -\alpha \cos^2 x^1 \end{pmatrix}, \qquad (26.8)$$

where α and β are functions of x^2 and x^3 satisfying the set of equations

$$\left.\begin{aligned}
&\frac{1}{\alpha}\left(\alpha_{22}-\frac{\alpha_2^2}{2\alpha}-\frac{\alpha_2\beta_2}{2\beta}+\frac{\beta\alpha_3\beta_3}{2}\right)+\beta\left(\frac{\beta_{33}}{2}-\frac{\beta_{22}}{2\beta^2}+\frac{\beta_2^2}{\beta^3}\right)=0,\\
&\frac{1}{\alpha}\left(\alpha_{33}-\frac{\alpha_3^2}{2\alpha}-\frac{\alpha_2\beta_2}{2\beta^3}+\frac{\alpha_3\beta_3}{2\beta}\right)+\frac{1}{\beta}\left(\frac{\beta_{33}}{2}-\frac{\beta_{22}}{2\beta^2}+\frac{\beta_2^2}{\beta^3}\right)=0,\\
&\alpha_{23}-\frac{\alpha_2\beta_3}{2\alpha}-\frac{\beta_2\beta_3}{2\beta}+\frac{\alpha_3\beta_2}{2\beta}=\alpha_{22}+\beta^2\alpha_{33}-\frac{\alpha_2\beta_2}{\beta}+\beta\alpha_3\beta_3-2\beta=0,
\end{aligned}\right\} \tag{26.9}$$

and

(3)
$$(g_{\alpha\beta})=\begin{pmatrix}\alpha & 0 & 0 & 0\\ 0 & \beta & 0 & 0\\ 0 & 0 & \beta\cosh^2 x^2 & 0\\ 0 & 0 & 0 & -\dfrac{1}{\alpha}\end{pmatrix}, \tag{26.10}$$

where α and β are functions of x^1 and x^4 satisfying the field equations

$$\left.\begin{aligned}
&\alpha\left(\frac{\alpha_{11}}{\alpha^2}+\alpha_{44}-\frac{2\alpha_1^2}{\alpha^3}\right)+\frac{2}{\beta}\left(-\beta_{11}+\frac{\beta_1^2}{2\beta}+\frac{\alpha_1\beta_1}{2\alpha}-\alpha\frac{\alpha_4\beta_4}{2}\right)=0,\\
&\frac{1}{\alpha}\left(\frac{\alpha_{11}}{\alpha^3}+\alpha_{44}-\frac{2\alpha_4^2}{\alpha^3}\right)+\frac{2}{\beta}\left(\beta_{44}-\frac{\beta_4^2}{2\beta}+\frac{\alpha_1\beta_4}{2\alpha^3}-\frac{\alpha_4\beta_4}{2\alpha}\right)=0,\\
&\beta_{14}-\frac{\beta_1\beta_4}{2\beta}+\frac{1}{2\alpha}(\alpha_1\beta_4-\alpha_4\beta_1)=0,\\
&\beta_{11}-\alpha^2\beta_{44}-\frac{1}{\alpha}\alpha_1\beta_1+\alpha\alpha_4\beta_4+2\alpha=0.
\end{aligned}\right\} \tag{26.11}$$

These three solutions exhaust all the possible $T_{1,3}$ with insoluble G_3 groups of motions in the case where these G_3 *act on two-dimensional surfaces of transitivity of constant curvature.*

2. Let us consider the structure (24.11); The corresponding operators act on pseudo-Euclidean two-dimensional surfaces; the G_3 has a one-parameter subgroup of stability. Assuming Killing vectors of the form

$$\underset{1}{\xi}^\alpha=\delta_1^\alpha,\quad \underset{2}{\xi}^\alpha=\delta_4^\alpha,\quad \underset{3}{\xi}^\alpha=x^4\delta_1^\alpha+x^1\delta_4^\alpha \tag{26.12}$$

the set of Killing equations are

$$\partial_1 g_{\alpha\beta}=\partial_4 g_{\alpha\beta}=0,$$
$$g_{\alpha\sigma}\,\partial_\beta(x^4\delta_1^\sigma+x^1\delta_4^\sigma)+g_{\beta\sigma}\partial_\alpha(x^4\delta_1^\sigma+x^1\delta_4^\sigma)=0.$$

Hence the $g_{\alpha\beta}$ are independent of x^1 and x^4 and, moreover,

$$g_{4i} = 0, \qquad (i = 1, 2, 3), \quad g_{12} = g_{13} = g_{11}+g_{44} = 0.$$

Thus the metric tensor components have the form

$$g_{11} = \alpha, \quad g_{22} = \beta, \quad g_{23} = \gamma, \quad g_{33} = \delta, \quad g_{44} = -\alpha,$$

where α, β, γ and δ are functions of x^2 and x^3, and all the other components are zero. The general form of the *admissible transformation* for these operators is

$$x^1 = x^{1'}, \quad x^2 = \varphi(x^{2'}, x^{3'}), \quad x^3 = \psi(x^{2'}, x^{3'}), \quad x^4 = x^{4'},$$

and by choosing φ and ψ as the integrals of the consistent set of equations

$$\partial_2\varphi = -\frac{\partial_2\psi(\partial_3\varphi+\gamma\,\partial_3\psi)}{\beta\,\partial_3\varphi+\gamma\,\partial_3\psi}, \quad (\partial_2\psi)^2 = \frac{(\beta\,\partial_3\varphi+\gamma\,\partial_3\psi)^2}{(\beta\delta-\gamma^2)g_3'^2},$$

we satisfy the conditions that for real coordinate systems

$$g'_{23} = 0, \quad g'_{22}g'_{33} = 1.$$

The metric reduces to the form

$$(g_{\alpha\beta}) = \begin{pmatrix} \alpha & 0 & 0 & 0 \\ 0 & \beta & 0 & 0 \\ 0 & 0 & \dfrac{1}{\beta} & 0 \\ 0 & 0 & 0 & -\alpha \end{pmatrix}, \tag{26.13}$$

where α and β are functions of x^2 and x^3 satisfying the field equations

$$\left.\begin{aligned} \alpha_{22}-\beta^2\alpha_{33}-\frac{\alpha_2^2}{2\alpha}+\frac{\beta^2\alpha_3^2}{2\alpha} &= 0, \\ \frac{1}{\alpha}\left(\alpha_{33}-\frac{\alpha_3^2}{2\alpha}-\frac{\alpha_2\beta_2}{2\beta^3}+\frac{\alpha_3\beta_3}{2\beta}\right) & \\ +\frac{1}{2\beta}\left(\beta_{33}+\frac{\beta_{22}}{\beta^2}+\frac{4\beta_2^2}{\beta^3}\right) &= 0, \\ \alpha_{33}-\frac{\alpha_3^2}{4\alpha}-\frac{\alpha_2\beta_2}{2\beta^3}+\frac{\alpha_3\beta_3}{2\beta}+\frac{\alpha_2^2}{4\alpha\beta^2} &= 0, \\ \alpha_{23}-\frac{\alpha_2\alpha_3}{2\alpha}-\frac{\alpha_2\beta_3}{2\beta}+\frac{\alpha_3\beta_2}{2\beta} &= 0. \end{aligned}\right\} \tag{26.14}$$

3. Suppose we have an Abelian G_3 subgroup corresponding to (24.12); the subgroup of stability is simply the unit element of the group, and the group acts on non-isotropic hypersurfaces of transitivity. As the hypersurfaces are non-isotropic, they are geodesically parallel, and a system of coordinates can be chosen in which these surfaces are defined by $x^4 =$ constant, hence

$$g_{i4} = 0 \quad (i = 1, 2, 3), \quad g_{44} = e_4 = \pm 1.$$

As the rank of the matrix $(\underset{s}{\xi}^\alpha)$ is equal to the order of the group, any Killing vectors may be chosen providing they define an Abelian structure, are of rank 3 and are independent of x^4; the group is thus defined to within a similarity. If we take Killing vectors $\underset{s}{\xi}^\alpha = \delta^\alpha_s$ $(s = 1, 2, 3;\ \alpha = 1, 2, 3, 4)$, the Killing equations show that the $g_{\alpha\beta}$ are functions of x^4 only.

The initial conditions (24.12) show that

$$\underset{s}{\overset{0}{\xi}}_{[\alpha}\underset{s}{\overset{0}{\xi}}_{\beta,\gamma]} = 0, \tag{26.15}$$

for any orthonormal tetrad, i.e. the vectors $\underset{s}{\overset{0}{\xi}}_\alpha$ are field gradients to within a scalar factor. This situation will obviously hold in any holonomic system of coordinates. The vectors $\underset{s}{\overset{0}{\xi}}_\alpha$, in the orthonormal tetrad and the vectors $\underset{s}{\xi}^\alpha$ in a holonomic system of coordinates are in general not necessarily identical; they are connected by linear equations with constant coefficients: but they can be made identical for any Abelian structure without changing the structure. Hence the vectors $\underset{s}{\xi}^\alpha$ may be taken as coordinates. Equation (26.15) reduces to the set of equations

$$g_{ij}\, g'_{jk} + g_{jk} g'_{ij} = 0 \qquad (i, j, k = 1, 2, 3),$$

where at least two of i, j and k are different. Hence

$$g'_{ij} = 0 \qquad (i \neq j),$$

so that the angles between the $\underset{s}{\xi}^\alpha$ are independent of the choice of the surface $x^4 =$ constant, and these vectors can be chosen such that

$$g_{ij} = 0 \qquad (i \neq j).$$

The space thus admits a 4-orthogonal system

$$ds^2 = \sum_1^3 g_{ii}\, dx^{i^2} + e_4\, dx^{4^2}, \quad e_4 = \pm 1, \tag{26.16}$$

where the $g_{ii} = \alpha_i(x^4)$.

Let us write down the field equations for this metric. The curvature tensor components are

$$R_{ijij} = -\frac{e_4\alpha_i'\alpha_j'}{4}, \quad R_{i4i4} = -\tfrac{1}{2}\alpha_i''+\frac{\alpha_i'^2}{4\alpha_i} \qquad (i, j = 1, 2, 3;\ i \neq j),$$

all the others being zero. The Ricci tensor components are

$$R_{ii} = \frac{1}{\alpha_j} R_{ijij}+\frac{1}{\alpha_k} R_{ikik}+e_4R_{i4i4},$$

$$R_{44} = \sum_1^3 R_{i4i4}, \quad R_{ij} = 0 \qquad (i, j, k \neq),$$

and the field equations reduce to

$$\alpha_i''+\tfrac{1}{2}\alpha_i'\left(\frac{\alpha_j'}{\alpha_j}+\frac{\alpha_k'}{\alpha_k}-\frac{\alpha_i'}{\alpha_i}\right) = 0, \quad \sum_1^3\left(\alpha_i''-\frac{1}{2}\frac{\alpha_i'^2}{\alpha_i}\right) = 0.$$

Putting $\alpha_i = e_ie^{2\tau_i}$, $e_i = \pm 1$, we obtain

$$\tau_i''+\tau_i'\sum_1^3 \tau_j' = 0 \qquad (i = 1, 2, 3), \qquad \sum_1^3 \tau_i''+\sum_1^3 \tau_i'^2 = 0. \tag{26.17}$$

Adding the first three equations we obtain

$$\sum_{j=1}^3 \tau_j''+(\sum_{j=1}^3 \tau_j')^2 = 0.$$

Denoting $\sum \tau_j$ by ω we obtain

$$\omega''+\omega'^2 = 0.$$

Hence either

$$\omega = \ln(c_1x^4+c_2), \quad c_1,\ c_2 = \text{const}, \tag{26.18}$$

or $\omega =$ constant, a singular solution.

In the first case, (26.17) shows that

$$\tau_i = \ln(c_1x^4+c_2)^{\beta_i}+\gamma_i, \quad \sum\beta_i = 1, \quad \sum\beta_i^2 = 1, \quad \sum\gamma_i = 0. \tag{26.19}$$

As the singular solution leads to a flat space, it is not considered. Putting $2\beta_i = \lambda_i$, we obtain

$$\alpha_i = e_ie^{2\gamma_i}(c_1x^4+c_2)^{\lambda_i}, \quad e_i = \pm 1, \quad \prod_1^3 e_i = -1,$$

where the constants λ_i and γ_i satisfy the conditions

$$\sum_1^3 \gamma_i = 0, \quad \sum_1^3 \lambda_i = 2, \quad \sum \lambda_i\lambda_j = 0 \qquad (i \neq j). \tag{26.20}$$

4. We now consider the case when the principal curvatures are not all equal, the group structure is Abelian, and eqs. (24.14) are applicable. The initial conditions show that the subgroup of stability is identical with the unit element of the group, the rank of the matrix $(\underset{s}{\xi}^{\alpha})$ is three, and the group acts on non-isotropic hypersurfaces of transitivity with an indefinite metric of signature $(- - +)$.

Taking these geodesically parallel hypersurfaces as coordinate surfaces with equations $x^1 = \text{constant}$, so that $g_{11} = -1$, $g_{ii} = 0$ $(i \neq 0)$, we can choose the following Killing vectors:

$$\underset{1}{\xi}^{\alpha} = \delta^{\alpha}_{2}, \quad \underset{2}{\xi}^{\alpha} = \delta^{\alpha}_{3}, \quad \underset{3}{\xi}^{\alpha} = \delta^{\alpha}_{4}.$$

Hence $g_{\alpha\beta} = g_{\alpha\beta}(x^1)$ and the metric is

$$(g_{\alpha\beta}) = \begin{pmatrix} -1 & 0 \\ 0 & A \end{pmatrix}, \quad A = (g_{ij}) \qquad (i, j = 2, 3, 4), \tag{26.21}$$

where the g_{ij} are functions of x^1.

Three of the field equations for (26.21) are satisfied identically and the rest reduce to a set of ordinary equations:

$$\left.\begin{aligned} a^{ij}(a''_{ij} - \tfrac{1}{2}a^{pq}a'_{pi}a'_{qj}) &= 0, \\ a''_{ij} + \tfrac{1}{2}a^{pq}(a'_{pq}a'_{ij} - 2a'_{pi}a'_{qj}) &= 0, \\ a_{ip}a^{pq} = \delta^{q}_{i} \qquad (i, j, p, q &= 2, 3, 4). \end{aligned}\right\} \tag{26.22}$$

5. We now consider the structure (24.15). The subgroup of stability is identical with the unit element of the group. The group acts as a simple transitive group on non-isotropic hypersurfaces of transitivity with metric signature $(- - +)$. It is easily seen that the structure (24.15) can be converted by means of a linear non-degenerate transformation to a G_3 structure of type VII (with $q = 1$):

$$[X_1X_2] = 0, \quad [X_1X_3] = X_2, \quad [X_2X_3] = -X_1 + X_2.$$

As the surfaces of transitivity are three-dimensional, they can be taken as coordinate surfaces as in the previous case, with equations $x^1 = \text{constant}$; the group operators can be chosen as follows:

$$X_1 = p_3, \quad X_2 = p_4, \quad X_3 = p_2 - x^4p_3 + (x^3 + x^4)p_4.$$

In this semi-geodesic system of coordinates $g_{11} = -1$, $g_{1i} = 0$ $(i = 1, 2, 3)$;

integrating the Killing equations shows that the metric is

$$\left.\begin{aligned} g_{11} &= -1, \quad g_{1i} = 0 \qquad (i = 2, 3, 4), \\ g_{22} &= a_{22}, \quad g_{23} = e^{-\frac{x^2}{2}} A \cos\omega, \\ g_{24} &= \tfrac{1}{2} e^{-\frac{x^2}{2}} A (\cos\omega + \sqrt{3}\sin\omega), \\ g_{33} &= e^{-x^2}(C + B\cos\nu), \\ g_{34} &= \tfrac{1}{2} e^{-x^2}[C + B(\cos\nu + \sqrt{3}\sin\nu)], \\ g_{44} &= \tfrac{1}{2} e^{-x^2}[2C - B(\cos\nu - \sqrt{3}\sin\nu)], \\ \text{where} \qquad \omega &= \frac{\sqrt{3}x^2}{2} + \varphi, \quad \nu = \sqrt{3}x^2 + \psi, \end{aligned}\right\} \tag{26.23}$$

and a_{22}, A, B, C, φ and ψ are functions of x^1. Equating the Ricci tensor components to zero, we obtain a set of field equations for the unknown functions (cf. example 1 below).

T_1 Spaces Admitting G_4

We consider all the above T_1 admitting a G_3 group of motions, and select those also admitting a G_4.

1. Consider the metric (26.8) with an insoluble group G_4 of type VIII. There is only one G_4 structure for T_1 spaces which includes this G_3 as a subgroup—namely (24.11). Thus for the Killing vectors (26.3), we define X_4 so that the commutators $[X_i X_4]$ $(i = 1, 2, 3)$ vanish; this leads to the equations

$$\left.\begin{aligned} \partial_1 \underset{4}{\xi}^\alpha &= 0, \\ A\cosh x^4 + B\sinh x^4 &= 0, \\ A\sinh x^4 - B\cosh x^4 &= 0, \end{aligned}\right\} \tag{26.24}$$

where

$$A = \tan x^1 \partial_4 \underset{4}{\xi}^\alpha - \frac{1}{\cos^2 x^1} \underset{4}{\xi}^1 \delta_4^\alpha - \underset{4}{\xi}^4 \delta_1^\alpha$$

and

$$B = \underset{4}{\xi}^4 \tan x^1 \, \delta_4^\alpha.$$

Hence $A = B = 0$, i.e.

$$\underset{4}{\xi}^4 = 0, \quad \tan x^1 \partial_4 \underset{4}{\xi}^\alpha - \frac{1}{\cos^2 x^1} \underset{4}{\xi}^1 \delta_4^\alpha = 0.$$

As $\underset{4}{\xi}^{\alpha}$ is independent of x^1, $\underset{4}{\xi}^1 = 0$, $\partial_4 \underset{4}{\xi}^{\alpha} = 0$, or

$$\underset{4}{\xi}^{\alpha} = \underset{4}{\xi}^2(x^2, x^3)\, \delta_2^{\alpha} + \underset{4}{\xi}^3(x^2, x^3)\, \delta_3^{\alpha}.$$

Using the admissible transformation

$$x^{1'} = x^1 + k\pi, \quad x^{2'} = \varphi(x^2, x^3), \quad x^{3'} = \psi(x^2, x^3), \quad x^{4'} = x^4,$$

we obtain

$$\underset{4}{\xi}^{\alpha} = \delta_3^{\alpha},$$

i.e. $g_{\alpha\beta}$ is independent of x^3. Applying the admissible transformation

$$x^{1'} = x^1 + k\pi, \quad x^{2'} = \varphi(x^2), \quad x^{3'} = \psi(x^3), \quad x^{4'} = x^4,$$

for all four operators, again leads to the metric (26.8), with the additional conditions that $\partial_3 \alpha = \partial_3 \beta = 0$; the set of field equations (26.9) now reduces to

$$\alpha_{22} - \frac{1}{2\alpha}\, \alpha_2^2 = 0, \quad \beta_{22} - \frac{2}{\beta}\, \beta_2^2 + \frac{1}{\alpha}\, \alpha_2 \beta_2 = 0, \quad \alpha_{22} - \frac{1}{\beta}\, \alpha_2 \beta_2 = 2\beta,$$

which are integrable by quadrature. The resulting metric is

$$ds^2 = -(c_1 x^2 + c_2)^2\, dx^{1^2} - \frac{c_1(c_1 x^2 + c_2)}{x^2 + c_3}\, dx^{2^2} - \frac{x^2 + c^3}{c_1(c_1 x^2 + c_3)}\, dx^{3^2} + (c_1 x^2 + c_2)^2 \cos^2 x^1 dx^{4^2}, \qquad (26.25)$$

where the c_i are constants. We have obtained a *static* solution with an insoluble G_4.

Any further increase of mobility for (26.25) leads to a flat space.

In exactly the same way, we obtain the following spaces for the metrics (26.8) and (26.10):

$$ds^2 = -\frac{c_1(c_1 x^1 + c_2)}{x^1 + c_3}\, dx^{1^2} - (c_1 x^1 + c_2)^2\, dx^{2^2} - (c_1 x^1 + c_2)^2 \sin^2 x^2\, dx^{3^2} + \frac{x^1 + c_3}{c_1(c_1 x^1 + c_2)}\, dx^{4^2}, \qquad (26.26)$$

$$ds^2 = -\frac{c_1(c_1 x^1 + c_2)}{c_3 - x^1}\, dx^{1^2} - (c_1 x^1 + c_2)^2\, dx^{2^2} - (c_1 x^1 + c_2)^2 \cosh^2 x^2\, dx^{3^2} + \frac{c_3 - x^1}{c_1(c_1 x^1 + c_2)}\, dx^{4^2}, \qquad (26.27)$$

Putting $c_1 = 1$, $c_2 = 0$, $c_3 = -2a$ in the first equation leads to the *Schwarzschild solution*, $c_1 = 1$, $c_2 = -2a$, $c_3 = 0$ to the "reciprocal" *solution of Buchdahl* (cf. § 14). In these three solutions the number of essential parameters can be reduced to one by linear displacement and deformation transformations.

2. Applying the same method to the metric (26.10) with the structure VIII, the only one admitting the given G_3 as a subgroup, shows that

$$ds^2 = -c_1 \frac{(c_1x^1+c_2)}{c_3-x^1} dx^{1^2} - (c_1x^1+c_2)^2\, dx^{2^2} - (c_1x^1+c_2)^2 \cosh^2 x^2\, dx^{3^2} + \frac{c^3-x^1}{c_1(c_1x^1+c_2)}\, dx^{4^2}. \tag{26.28}$$

3. Consider the metric (26.13) with a soluble structure. The only G_4 structure with such a G_3 subgroup has the three additional commutators

$$[X_1X_4] = 0,\quad [X_2X_4] = sX_2,\quad [X_3X_4] = sX_3.$$

Using a linear substitution reduces the problem to determining $\underset{4}{\xi}^\alpha$, with the following structure:

$$[X_1X_2] = 0,\quad [X_1X_3] = X_2,\quad [X_2X_3] = X_1,\quad [X_1X_4] = \varepsilon X_1,$$
$$[X_2X_4] = \varepsilon X_2,\quad [X_3X_4] = 0,\quad \varepsilon = \begin{cases} 0 \\ 1 \end{cases}$$

and

$$\underset{1}{\xi}^\alpha = \delta_1^\alpha,\quad \underset{2}{\xi}^\alpha = \delta_4^\alpha,\quad \underset{3}{\xi}^\alpha = x^4\delta_1^\alpha + x^1\delta_4^\alpha.$$

The last three commutators reduce to

$$\partial_1 \underset{4}{\xi}^\alpha = \varepsilon\delta_1^\alpha,\quad \partial_4 \underset{4}{\xi}^\alpha = \varepsilon\delta_4^\alpha,\quad (x^4\varepsilon - \underset{4}{\xi}^4)\delta_1^\alpha + (x^1\varepsilon - \underset{4}{\xi}^1)\,\delta_4^\alpha = 0,$$

whence

$$\underset{4}{\xi}^\alpha = \varepsilon(x^1\delta_1^\alpha + x^4\delta_4^\alpha) + \underset{4}{\xi}^2(x^2, x^3)\,\delta_2^\alpha + \underset{4}{\xi}^3(x^2, x^3)\,\delta_3^\alpha.$$

Applying the admissible transformation

$$x^{1'} = x^1,\quad x^{2'} = \varphi(x^2, x^3),\quad x^{3'} = \psi(x^2, x^3),\quad x^{4'} = x^4$$

for the first three operators reduces $\underset{4}{\xi}^\alpha$ to the form

$$\underset{4}{\xi}^{1'} = \varepsilon x^{1'},\quad \underset{4}{\xi}^{4'} = \varepsilon x^{4'},\quad \underset{4}{\xi}^{2'} = \partial_2\varphi\underset{4}{\xi}^2 + \partial_3\varphi\underset{4}{\xi}^3,$$
$$\underset{4}{\xi}^{3'} = \partial_2\psi\underset{4}{\xi}^2 + \partial_3\psi\partial_4\underset{4}{\xi}^3.$$

Hence provided the functions involved are holomorphic, the fourth Killing vector can take the form

$$\underset{4}{\xi}^{\alpha} = \varepsilon x^1\delta_1^\alpha + \varepsilon x^4\delta_4^\alpha + \delta_2^\alpha\varepsilon_1, \quad \varepsilon_1 = 0 \quad \text{or } 1.$$

Once again, the Killing equations for the first three vectors show that $b_{11} = \alpha$, $g_{22} = \beta$, $g_{23} = \gamma$, $g_{33} = \delta$, $g_{44} = -\alpha$, all the other components geing zero. The fourth Killing equation shows that

$$\varepsilon_1\alpha_2 + 2\varepsilon\alpha = 0, \quad \varepsilon_1\beta_2 = 0, \quad \varepsilon_1\gamma_2 = 0, \quad \varepsilon_1\delta_2 = 0.$$

If $\varepsilon_1 = 0$, $\varepsilon\alpha = 0$; but $\alpha \neq 0$, for otherwise the metric becomes degenerate. Hence $\varepsilon = 0$, making $\underset{4}{\xi}^{\alpha} = 0$ which is absurd. Thus $\varepsilon_1 = 1$, $\partial_2\alpha + 2\varepsilon\alpha = 0$, and β, γ, δ are functions of the variable x^2 only. Clearly the space is degenerate if $\varepsilon = 1$; thus $\varepsilon = 0$, and α must be a function of x^3 only.

Applying the following admissible transformation for all four operators

$$x^{1'} = x^1, \quad x^{2'} = x^2 + \varphi(x^3), \quad x^{3'} = \psi(x^3), \quad x^{4'} = x^4$$

reduces the component g'_{23} to zero, and we can then satisfy the equation $g'_{22}g'_{33} = 1$. Putting $\partial_2\alpha = \partial_2\beta = 0$ in eqns. (26.14), we obtain

$$\alpha_{33} - \frac{\alpha_3^2}{2\alpha} = 0, \quad \beta_{33} + \frac{\alpha_3\beta_3}{\alpha} = 0, \quad \alpha_{33} - \frac{\alpha_3^2}{4\alpha} + \frac{\alpha_3\beta_3}{2\beta} = 0.$$

Neglecting flat spaces, we obtain the metric

$$(g_{\alpha\beta}) = \begin{pmatrix} -(kx^3+1)^2 & 0 & 0 & 0 \\ 0 & -\dfrac{1}{kx^3+1} & 0 & 0 \\ 0 & 0 & -(kx^3+1) & 0 \\ 0 & 0 & 0 & (kx^3+1)^2 \end{pmatrix}, \tag{26.29}$$

i.e. a *static* solution.

4. Consider the metric (26.16) subject to (26.10), with an Abelian G_3 group. As the metric is defined to within certain constants, the Killing equations can be directly integrated and the vector $\underset{4}{\xi}^{\alpha}$ of the required G_4 group determined. Applying the transformation

$$x^{1'} = \sqrt{|c_i^{\lambda_i}|}\, e^{\gamma_i} x^i, \quad x^{4'} = x^4 + \frac{c_2}{c_1}$$

(clearly $c_1 \neq 0$) reduces the metric (26.16) to

$$g_{44} = e_i(x^4)^{\lambda_i}, \quad g_{44} = e_4, \quad g_{\alpha\beta} = 0 \quad (\alpha \neq \beta), \tag{26.30}$$

where

$$\left.\begin{aligned} &\sum \gamma_i = 0, \quad \sum \lambda_i = 2, \quad \sum \lambda_i \lambda_j = 0 \qquad (i \neq j), \\ &\prod_1^4 e_i = -1, \quad e_\alpha = \pm 1. \end{aligned}\right\} \tag{26.31}$$

The Killing equations for (26.30) show that

$$\lambda_i \underset{4}{\xi}^4 + 2x^4\, \partial_i \underset{4}{\xi}^i = 0,$$

$$e_i x^{4\lambda_i}\, \partial_j \underset{4}{\xi}^i + e_j x^{4\lambda_j}\, \partial_i \underset{4}{\xi}^j = 0,$$

$$e_i x^{4\lambda_i}\, \partial_4 \underset{4}{\xi}^i + e_4\, \partial_i \underset{4}{\xi}^4 = 0,$$

$$\partial_4 \underset{1}{\xi}^4 = 0.$$

According to the fourth equation

$$\underset{1}{\xi}^4 = \varphi(x^1, x^2, x^3).$$

The third set leads to

$$\underset{1}{\xi}^i = -e_4 e_i\, \partial_i \varphi \int (x^4)^{-\lambda_i}\, dx^4 + c^i(x^1, x^2, x^4).$$

The first set reduces to

$$\lambda_i \varphi + 2x^4 \left\{ -e_i e_4\, \partial_i \varphi \int (x^4)^{-\lambda_i}\, dx^4 + \partial_i c^i \right\} = 0$$

(no summation over i).

If $\lambda_i = 2$, $\lambda_j = \lambda_k = 0$ ($j, k \neq i$; $j \neq k$), but the space is then flat. If, however, $\lambda_i \neq 2$, the fact that φ and c^i are independent of x^4 implies that

$$\lambda_i \varphi = 0, \quad -e_i e_4\, \partial_i \varphi \int (x^4)^{-\lambda_i}\, dx^4 + \partial_i c^i = 0.$$

If one $\lambda_i = 0$, (26.31) shows that one other must be zero, and the third equal to 2, which again leads to a flat space. Hence $\lambda_i \neq 0$ ($i = 1, 2, 3$), and therefore

$$\varphi = 0, \quad \partial_i c^i = 0, \quad c^i = c^i(x^j, x^k) \qquad (j, k \neq i;\ i, j, k = 1, 2, 3).$$

As $\xi^4 = 0$, the required group is *intransitive* and $\xi^{i'} = c^i(x^j, x^k)$. The second set of equations reduces to

$$e_i (x^4)^{\lambda_i}\, \partial_j c^i + e_j (x^4)^{\lambda_j}\, \partial_i c^j = 0 \qquad (i \neq j)$$

(no summation over i, j).

If no two λ_i are equal, the solution is c^i = constant, defining the trivial operators $X_i = p_i$, whose existence was assumed anyway, but there is no G_4 group. The assumption $\lambda_1 = \lambda_2 = \lambda_3$ contradicts (26.31). Hence $\lambda_1 = \lambda_2$; eqn. (26.31) shows that

$$\lambda_1 = \lambda_2 = \tfrac{4}{3}, \quad \lambda_3 = -\tfrac{2}{3},$$

and $c^1 = c^1(x^2)$, $c^2 = c^2(x^1)$, c^3 = constant. c^3 can be reduced to zero by combination with the operator X_3, the structure remaining unaltered; c^1 and c^2 must be linear functions of their arguments, since their derivatives being identical with the structural constants must themselves be constants. Clearly, they can be reduced to the form

$$c^1 = kx^2, \quad c^2 = sx^1, \quad e_1k+e_2s = 0.$$

Thus the fourth Killing vector

$$\underset{4}{\xi^\alpha} = -e_2x^2\delta_1^\alpha+e_1x^1\delta_2^\alpha$$

corresponds to either *a Euclidean or a Lorentz rotation, depending on the signature of the form* $g_{11}\, dx^{12} + g_{22}\, dx^{22}$.

Consider the space defined by the metric

$$ds^2 = e_1(x^4)^{4/3}\, dx^{1^2}+e_2(x^4)^{4/3}dx^{2^2}+e_3(x^4)^{-2/3}\, dx^{3^2}+e_4\, dx^{4^2}. \quad (26.32)$$

The solutions of Kasner and Narlikar and Karmarkar (cf. § 14) belong to this type. This metric gives a *static* solution if the index 4 is ascribed to a spatial direction, and a *non-static* one if x^4 corresponds to time. The metric always corresponds to the structure

$$[X_iX_j] = 0 \qquad (i, j = 1, 2, 3), \qquad [X_1X_4] = e_1X_2,$$
$$[X_2X_4] = -e_2X_1, \quad [X_3X_4] = 0.$$

If $e_1 = e_2$, we have case VI_4 in the classification of (10.19)—with an Abelian G_3 subgroup; *if* $e_1 = -e_2$ we have case VI_1 type (5).

5. Let us consider the case when the principal curvatures are not equal, and the metric is given by (26.21) subject to (26.22). This metric is only possible for an Abelian structure, when the initial conditions admit the only G_4 group which includes an Abelian G_3 subgroup. The structure

of this G_4 is

$$\left.\begin{aligned} &[X_iX_k] = 0 \qquad (i, k = 1, 2, 3),\\ &[X_1X_4] = -\frac{k}{2}X_1 - \frac{k\sqrt{3}}{2}X_3,\\ &[X_2X_4] = kX_2, \quad [X_3X_4] = \frac{k\sqrt{3}}{2}X_1 - \frac{k}{2}X_3 \quad (k \neq 0). \end{aligned}\right\} \tag{26.33}$$

The group is simply transitive; as

$$\underset{1}{\xi}^{\alpha} = \delta_1^{\alpha}, \quad \underset{2}{\xi}^{\alpha} = \delta_3^{\alpha}, \quad \underset{3}{\xi}^{\alpha} = \delta_4^{\alpha},$$

then $\underset{4}{\xi}^4 \neq 0$. If on the other hand, we take the spaces corresponding to the structure (24.10), the only G_4 structure for T_1 containing this G_3 is again (26.33) (apart from a different system of numbering). These two cases lead to the same class of space if the mobility is increased. As (24.10) was earlier reduced to the metric (26.23), only those spaces (26.23) admitting one simple-transitive group with the structure

$$\left.\begin{aligned} &[X_1X_2] = 0, \quad [X_1X_3] = X_2, \quad [X_2X_3] = -X_1 - X_2,\\ &[X_1X_4] = X_1, \quad [X_2X_4] = X_2, \quad [X_3X_4] = 0, \end{aligned}\right\} \tag{26.34}$$

need be considered.

The same first three operators can be taken as for (26.23). The last three commutators of (26.34) then define $\underset{4}{\xi}^{\alpha}$ in the form

$$\begin{aligned} \underset{4}{\xi}^{\alpha} = {} & \underset{4}{\xi}^1(x^1)\delta_1^{\alpha} + \underset{4}{\xi}^2(x^1)\delta_2^{\alpha} + \left[x^3 + e^{x^2/2}\left(p\cos\frac{\sqrt{3}x^2}{2} + q\sin\frac{\sqrt{3}x^2}{2}\right)\right]\delta_3^{\alpha}\\ & + \left\{x^4 - \frac{1}{2}e^{x^2/2}\left[(p + q\sqrt{3})\cos\frac{\sqrt{3}x^2}{2}\right.\right.\\ & \left.\left. + (q - p\sqrt{3})\sin\frac{\sqrt{3}x^2}{2}\right]\right\}\delta_4^{\alpha}. \end{aligned} \tag{26.35}$$

We do not now assume the hypersurfaces of transitivity on which the operators X_1, X_2 and X_3 act to be coordinate surfaces, and we can therefore use the following admissible transformations which do not alter the form of these three operators:

$$\begin{aligned} x^{1'} &= f(x^1),\\ x^{3'} &= x^3 + e^{x^2/2}[M\cos\nu + N\sin\nu],\\ x^{2'} &= x^2 + \varphi(x^1),\\ x^{4'} &= x^4 - \tfrac{1}{2}e^{x^2/2}[(M + N\sqrt{3})\cos\nu + (N - M\sqrt{3})\sin\nu], \end{aligned}$$

where f, φ, M and N are arbitrary functions of their arguments, and

$$\nu = \frac{\sqrt{3}x^2}{2}.$$

These transformations allow $\underset{4}{\xi}^\alpha$ in the new system of coordinates to take the form

$$\underset{4}{\xi}^\alpha = \delta_1^\alpha + x^3\delta_3^\alpha + x^4\delta_4^\alpha \tag{26.36}$$

and the following admissible transformations:

$$\left.\begin{aligned} x^{1'} &= x^1 + c_1, \\ x^{3'} &= x^3 + e^{x^1+(1/2)x^2}(c_3 \cos \nu + c_4 \sin \nu), \\ x^{2'} &= x^2 + c_2, \\ x^{4'} &= x^4 - \tfrac{1}{2} e^{x^1+(1/2)x^2}[(c_3 + c_4\sqrt{3}) \cos \nu + (c_4 - c_3\sqrt{3}) \sin \nu], \\ c_i &= \text{const} \qquad (i = 1, 2, 3, 4), \end{aligned}\right\} \tag{26.37}$$

leave all four operators unaltered.

Integrating the Killing equations for all four operators, we obtain the metric tensor components

$$\left.\begin{aligned} &g_{11} = c_{11}, \quad g_{12} = c_{12}, \quad g_{13} = c_{13}e^{-x^1} \cos \varrho, \\ &g_{14} = \tfrac{1}{2} e^{-x^1-(1/2)x^2} c_{13}(\cos \varrho + \sqrt{3} \sin \varrho), \\ &g_{22} = c_{22}, \quad g_{23} = c_{23}e^{-x^1-(1/2)x^2} \cos \omega, \\ &g_{24} = \tfrac{1}{2} e^{-x^1-(1/2)x^2} c_{23}(\cos \omega + \sqrt{3} \sin \omega), \\ &g_{33} = e^{-2x^1-x^2}(c_{33} + c_{34} \cos \nu), \\ &g_{34} = \tfrac{1}{4} e^{-2x^1-x^2}[c_{33} + c_{34}(\cos \nu + \sqrt{3} \sin \nu)], \\ &g_{44} = \tfrac{1}{2} e^{-2x^1-x^2}[2c_{33} - c_{34}(\cos \nu - \sqrt{3} \sin \nu)], \\ &\varrho = \frac{\sqrt{3}x^2}{2} + c_{14}, \quad \omega = \frac{\sqrt{3}x^2}{2} = c_{24}, \quad \nu = \sqrt{3}x^2 + c_{44}, \end{aligned}\right\} \tag{26.38}$$

where c_{ij} = constant. The four arbitrary constants in the admissible transformation (26.37) can be used to simplify this metric. We can choose c_3 and c_4 so that $c_{1'3'} = c_{2'3'} = 0$ in the new system of coordinates; putting $c_2 = c_{44}/\sqrt{3}$ reduces $c_{4'4'}$ to zero. Finally, c_1 can be used to simplify the coefficient $c_{3'4'}$; $c_{44} \neq 0$, as this implies a metric with a non-Lorentz signature; a suitable choice of c_1 reduces c_{34} to $\varepsilon = \pm 1$. The metric now has the form

$$(g_{\alpha\beta}) = \begin{pmatrix} \alpha & \beta & 0 & 0 \\ \beta & \gamma & 0 & 0 \\ 0 & 0 & e^{-2x^1-x^2}(\delta+\varepsilon\cos\nu) & \tfrac{1}{2}e^{-2x^1-x^2}[\delta+\varepsilon(\cos\nu+\sqrt{3}\sin\nu)] \\ 0 & 0 & \tfrac{1}{2}e^{-2x^1-x^2}[\delta + \varepsilon(\cos\nu+\sqrt{3}\sin\nu)] & \tfrac{1}{4}e^{-2x^1-x^2}[2\delta - \varepsilon(\cos\nu-\sqrt{3}\sin\nu)] \end{pmatrix}. \quad (26.39)$$

On integrating the field equations for this metric, we can further specify the constants in the metric. We omit the detailed analysis (cf. example 2, § 26), but note that after a transformation of coordinates (not an admissible one), the required metric can take the form

$$ds^2 = -\alpha\,(dx^{1^2}+dx^{4^2})-2\beta\,dx^1\,dx^4-dx^{2^2}-e^{-2\nu}\,dx^{3^2}, \quad (26.40)$$

where

$$\alpha = e^\nu \cos(\sqrt{3}\nu),\quad \beta = e^\nu \sin(\sqrt{3}\nu),\quad \nu = kx^2,\quad k = \text{const.}$$

This space is a T_1 manifold *admitting a simple-transitive* G_4 *group of motions*; it is also characterized by *constant principal curvatures* (the constant bases of the elementary divisors of the λ-matrix $(R_{ab} - \lambda g^{ab})$ $(a, b = 1, \ldots, 6)$.

It can also be shown (cf. example 3, § 26) that *the principal curvatures are proportional to* ε_s, *the three roots of the cyclotomic equation* $\varepsilon_s^3 = 1$.

Since a T_1 cannot admit a group of motions G_r with $r > 4$ if it is not a space of constant curvature, a further increase of the mobility only leads to the trivial case of a flat space.

Examples

1. Write down the field equations $R_{\alpha\beta} = 0$ for the metric (26.23).

2. Do the same for (26.39). Determine the transformation of coordinates which converts the metric thus obtained into (26.40).

3. Determine the principal curvatures for the metric (26.40) and show that they are (1) constants, (2) proportional to the roots ε_s $(s = 1, 2, 3)$ of the cyclotomic equation $\varepsilon_s^3 = 1$.

§ 27. T_2 and T_3 Spaces Admitting Motions

For the reasons given in the case of T_1 spaces, we begin classifying T_2 spaces by means of G_3 groups (cf. example 1, § 27).

$T_{2,3}$ Spaces

As shown in § 24, there are three types of G_3 structure corresponding to T_2 manifolds, and one of these leads to two alternatives.

Let us consider these structures.

1. The *Abelian* group with the structure (24.19) acts on isotropic hypersurfaces of transitivity. There are, as will be shown in § 31, three possible cases, corresponding to the metrics (31.27), (31.28) and (31.29).

Nine of the field equations for (31.27), i.e.

$$(g_{\alpha\beta}) = \begin{pmatrix} 0 & 0 & 0 & 1 \\ 0 & \alpha & \beta & 0 \\ 0 & \beta & \gamma & 0 \\ 1 & 0 & 0 & 0 \end{pmatrix}, \tag{27.1}$$

$$g_{\alpha\beta} = g_{\alpha\beta}(x^4), \quad X_i = p_i (i = 1, 2, 3),$$

are satisfied identically. The remaining equation $R_{44} = 0$ can be written in the expanded form

$$\gamma\alpha'' - 2\beta\beta'' + \alpha\gamma'' + \frac{1}{2g}(\gamma^2\alpha'^2 + \alpha^2\gamma'^2 + 2\alpha\gamma\beta'^2$$
$$+ 2\beta^2\alpha'\gamma' + 2\beta^2\beta'^2 - 4\beta\gamma\alpha'\beta' - 4\alpha\beta\beta'\gamma') = 0, \tag{27.2}$$

where $g = |g_{\alpha\beta}| = \beta^2 - \alpha\gamma < 0$.

Thus *the field potentials depend on two arbitrary functions of one argument.*

X_1 is an exceptional operator acting on each hypersurface of transitivity $x^4 =$ constant; the complete group on the hypersurface is $G_{3+\infty}$ indicating the presence of an exceptional operator which can be multiplied by an *arbitrary function.* It is not difficult however to verify (cf. example 3, § 27) that the metric (27.1) also admits a G_5 group with operators

$$X_i = p_i \quad (i = 1, 2, 3), \qquad X_4 = x^2 p_1 + \omega^s p_s,$$
$$X_5 = x^3 p_1 + \sigma^s p_s \qquad (s = 2, 3), \tag{27.3}$$

where

$$\omega^2 = \int \frac{\gamma\, dx^4}{g}, \quad \omega^3 = -\int \frac{\beta\, dx^4}{g}, \quad \sigma^2 = -\int \frac{\beta\, dx^4}{g}, \quad \sigma^3 = \int \frac{\alpha\, dx^4}{g}, \tag{27.4}$$

but not a G_6 group, as it contains arbitrary functions and cannot be a space of maximum mobility for T_2. Thus in our group classification, the metric (27.1) must belong to T_2 spaces admitting G_5. Equations (27.3) and (27.4) show that the G_5 group acts on *isotropic* hypersurfaces of transitivity and that the group structure is

$$\left.\begin{aligned} &[X_1X_i] = 0 \quad (i = 1, 2, 3), \quad [X_2X_3] = [X_2X_5] = [X_3X_4] = 0, \\ &[X_2X_4] = [X_3X_5] = X_1, \quad [X_4X_5] = 0. \end{aligned}\right\} \tag{27.5}$$

The metrics (31.28) and (31.29) cannot define T_2 spaces (example 4, § 27). If we calculate the curvature tensor components for metric (27.1), and the elementary divisors of the λ-matrix $(R_{ab}-\lambda g_{ab})$ $(a, b = 1, \ldots, 6)$ using bivector space indices, we obtain the Segre characteristic [21°, 21]; hence this metric does define a T_2 space.

2. The G_3 group with the structure (24.20) of type II acts on *isotropic* hypersurfaces of transitivity, and has one exceptional operator [cf. (24.21)]. We will show in § 31 that the metric can be represented in a special system of coordinates by the matrix (31.30), i.e.

$$(g_{\alpha\beta}) = \begin{pmatrix} 0 & 0 & 0 & 1 \\ 0 & \alpha & \beta - x^3 & \\ 0 & \beta & \gamma & 0 \\ 1 - x^3 & 0 & 0 & \end{pmatrix}, \quad g = \beta^2 - \alpha\gamma < 0, \ \alpha, \ \gamma \neq 0, \tag{27.6}$$

where α, β and γ are functions of x^4, and the operators of the G_3 group have the form

$$X_1 = p_1, \quad X_2 = p_2, \quad X_3 = x^2 p_1 + p_3. \tag{27.7}$$

Calculating the matrix $(g^{\alpha\beta})$ and the curvature tensor components shows that only three of the latter, R_{2424}, R_{2434} and R_{3434} are non-zero; the set of field equations reduces to the single equation

$$\gamma R_{2424} - 2\beta R_{2434} + \alpha R_{3434} = 0,$$

all the others being satisfied identically. Moreover, the λ-matrix $(R_{ab}-\lambda g_{ab})$ $(a, b = 1, \ldots, 6)$ for the metric (27.6) has the Segre characteristic [21°, 21], corresponding to a T_2 space. However, a direct integration of the Killing equations for this metric gives the following general expression for the unknown Killing vector:

$$\xi^\alpha = (c_3 x^2 + x^3 u + c_1)\, \delta_1^\alpha + (u + c_2)\, \delta_2^\alpha + (c_3 + v)\, \delta_3^\alpha,$$

where the c_i $(i = 1, 2, 3)$ are constants, and u and v are functions of the variable x^4 only, satisfying the ordinary linear homogeneous equations

$$u' = -\frac{1}{g}(\beta u+\gamma v), \quad v' = \frac{1}{g}(\alpha u-\beta v). \tag{27.8}$$

As their coefficients are continuous in some region, the equations are soluble in the region; the general solution is

$$u = c_4\underset{1}{u}+c_5\underset{2}{u}, \quad v = c_4\underset{1}{v}+c_5\underset{2}{v},$$

where the fundamental set $\underset{1}{u}, \underset{2}{u}, \underset{1}{v}, \underset{2}{v}$ satisfies the condition

$$\begin{vmatrix} \underset{1}{u} & \underset{1}{v} \\ \underset{2}{u} & \underset{2}{v} \end{vmatrix} \neq 0 \tag{27.9}$$

in the above region. Thus the space admits a G_5 group of motions with operators

$$X_i = p_i \quad (i = 1, 2), \quad X_3 = x^2p_1+p_3, \quad X_4 = \underset{1}{u}x^3p_1+\underset{1}{u}p_2+\underset{1}{v}p_3,$$
$$X_5 = \underset{2}{u}x^3p_1+\underset{2}{u}p_2+\underset{2}{v}p_3,$$

where $\underset{i}{u}, \underset{i}{v}$ $(i = 1, 2)$ form a set of two independent solutions of the original linear set (27.8). Calculating all the possible commutators for these operators, we obtain the structure

$$\left.\begin{aligned} [X_1X_i] = 0 \quad (i = 2, 3, 4, 5), \quad [X_2X_3] = X_1, \\ [X_2X_j] = 0 \quad (j = 4, 5), \\ [X_3X_k] = 0 \quad (k = 4, 5), \quad [X_4X_5] = kX_1, \end{aligned}\right\} \tag{27.10}$$

where

$$k = \underset{1}{v}\underset{2}{u}-\underset{2}{v}\underset{1}{u}.$$

Here k is a constant, for differentiating this equation and substituting for $\underset{i}{u}'$ and $\underset{i}{v}'$ $(i = 1, 2)$ from (27.8) gives $k' = 0$. According to (27.9) $k \neq 0$, so that by multiplying the operator X_5, for example, by $1/k$, we can reduce k to unity. This makes it clear that the structure (27.10) is isomorphic with (27.5) already considered under $T_{2,3}$ with an Abelian G_3 group acting on isotropic hypersurfaces of transitivity; renumbering the operators by means of the permutation $\begin{pmatrix} 12345 \\ 12435 \end{pmatrix}$ transforms either structure into the other. The Abelian G_3 subgroup of the G_5 group (27.10) has basic operators X_1, X_2, X_4 and acts on isotropic hypersurfaces of transitivity according to (27.6), i.e. the metric (27.6) can be reduced to (27.1) by a non-degenerate transformation, so that this case does not merit special attention.

3. Consider a G_3 group of the same type II, but acting on *isotropic two-dimensional* surfaces of transitivity; the group operators may be given, for example, in the form (24.22), i.e. there is one *exceptional* operator X_1, since it is also an exceptional operator for the space of maximum mobility. A system of coordinates can always be chosen (cf. § 30, G_3 II in $\dot{V}_2$) relative to which the operators are (apart from the actual numbering)

$$X_1 = p_1, \quad X_2 = p_2, \quad X_3 = x^2p_1+x^4p_2, \tag{27.11}$$

and the metric matrix is

$$(g_{\alpha\beta}) = \begin{pmatrix} 0 & 0 & 0 & \beta \\ 0 & -\beta & 0 & 0 \\ 0 & 0 & \alpha & 0 \\ \beta & 0 & 0 & 0 \end{pmatrix}, \tag{27.12}$$

$$g_{\alpha\beta} = g_{\alpha\beta}(x^3, x^4), \quad g = \alpha\beta^3 < 0, \quad \alpha, \beta < 0.$$

The field equations for this metric show that β is independent of x^3, so that the functions α and β are subject only to the two conditions

$$\beta_3 = 0, \quad \frac{1}{\beta}\left(\beta_{44}-\frac{3\beta_4^2}{2\beta}\right)+\frac{1}{\alpha}\left(\alpha_{44}-\frac{\alpha_4^2}{2\alpha}-\frac{\alpha_4\beta_4}{\beta}\right) = 0, \tag{27.13}$$

where $f_i \equiv \partial f/\partial x^i$. Hence *the field potentials* (27.1) *depend on three arbitrary functions, two of which depend on x^3 and the third on x^4.*

It is not difficult to verify (cf. example 2, § 27) that the Segre characteristic of the λ-matrix is [21°, 21]; hence if the condition $\beta_{44} - (3\beta_4^2/2\beta) \neq 0$ is satisfied for arbitrary values of the functions α and β, it is impossible to obtain T_1 or T_3 spaces, as the corresponding Segre characteristics are of a different type.

On integrating the Killing equations for the metric (27.12), with α and β subject to condition (27.13), we simply obtain the operators (27.11), i.e. in general this metric admits only a three-parameter group of motions.

4. Let us consider the case of a G_3 group of type III with the structure (24.23) acting on *non-isotropic hypersurfaces* of transitivity.

As will be shown in § 32, the system of coordinates in this case can always be chosen so that the metric matrix of any V_4 space with this group of motions is (31.7), i.e.

$$(g_{\alpha\beta}) = \begin{pmatrix} a_{11} & a_{12}e^{x^1} & a_{13} & 0 \\ a_{12}e^{x^1} & a_{22}e^{2x^1} & a_{23}e^{x^1} & 0 \\ a_{13} & a_{23}e^{x^1} & a_{33} & 0 \\ 0 & 0 & 0 & e_4 \end{pmatrix}, \quad e_4 = \pm 1, \tag{27.14}$$

and the group operators are

$$X_1 = p_2, \quad X_2 = p_3, \quad X_3 = -p_1 + x^2 p_2. \tag{27.15}$$

We note that the structure (24.23) is unchanged, apart from a linear non-degenerate substitution

$$Y_1 = AX_1, \quad Y_2 = BX_2, \quad Y_3 = CX_1 + DX_2 + X_3 \quad (A, B \neq 0),$$

i.e. the operators X_1 and X_2 for this structure are defined to within constant factors. As previously pointed out in § 24, a system of coordinates can be found for this structure, with operators defined by (24.24). The particular operator $X_2 = p_1$ is also admitted by the space of maximum mobility in the same system of coordinates [see metric (25.3) and operators (25.4) with index 1 changed to 2], where it corresponds to an isotropic Killing vector field; this must also be true for the space under consideration. Hence the operator X_2 is isotropic, and $a_{33} = 0$ by (27.14) and (27.15). Moreover, (24.24) and (25.3) show that the operators $\{X_1, X_2\}$ form an Abelian subgroup acting on isotropic two-dimensional surfaces of transitivity. This property also holds for the metric (27.14) with the operators (27.15), for X_1 and X_2 are uniquely defined for a given structure (apart from non-essential constant factors). In other words the determinant

$$\begin{vmatrix} a_{22} & a_{23} \\ a_{23} & 0 \end{vmatrix} = 0,$$

i.e. $a_{23} = 0$. Putting $a_{11} = \alpha$, $a_{12} = \beta$, $a_{13} = \gamma$, $a_{22} = \delta$, the metric reduces to

$$(g_{\alpha\beta}) = \begin{pmatrix} \alpha & \beta e^{x^1} & \gamma & 0 \\ \beta e^{x^1} & \delta e^{2x^1} & 0 & 0 \\ \gamma & 0 & 0 & 0 \\ 0 & 0 & 0 & e_4 \end{pmatrix}, \tag{27.16}$$

$$g = -e_4 \delta \gamma^2 e^{2x^1}, \quad e_4 \delta > 0, \quad e_4 = \pm 1,$$

where α, β, γ and δ are functions of x^4. If we require that the admissible transformations for the operators (27.15) should also leave the structure

(27.16) invariant, these transformations must have the form

$$\left.\begin{aligned} x^1 = x^{1'}+c_1, \quad x^2 = x^{2'}+c_2e^{-x^{1'}}, \\ x^3 = x^{3'}+c_3, \quad x^4 = \pm x^{4'}+c_4, \quad c_i = \text{const} \end{aligned}\right\} \tag{27.17}$$

and therefore not contain arbitrary functions. Integrating the field equations $R_{\alpha\beta} = 0$ for the metric (27.16) we obtain

$$\gamma' = \delta' = \beta'' = 0, \quad \alpha'' - \frac{\beta'^2}{\delta} + 2e_4 = 0,$$

i.e. γ and δ are constants, and

$$\beta = px^4+q, \quad \alpha = \frac{1}{2}\left(\frac{p^2}{\delta} - 2e_4\right)x^{4^2}+rx^4+s,$$

where p, q, r and s are constants. By an appropriate choice of the constants c_1, c_2 and c_4 in (27.17), we can convert δ into e_4, and reduce q and r to zero.

The metric becomes

$$(g_{\alpha\beta}) = \begin{pmatrix} \frac{e_4}{2}(p^2-2)x^{4^2}+s & px^4e^{x^1} & \gamma & 0 \\ px^4e^{x^1} & e_4e^{2x} & 0 & 0 \\ \gamma & 0 & 0 & 0 \\ 0 & 0 & 0 & e_4 \end{pmatrix}, \tag{27.18}$$

where $e_4 = \pm 1$, and p, s and γ are arbitrary constants. If we calculate the elementary divisors of the λ-matrix $(R_{ab} - \lambda g_{ab})$ $(a, b = 1, \ldots, 6)$ for this metric, we find that its Segre characteristic is [21°, 21] providing only that $p \neq 0$ or $\pm\sqrt{2}$. Hence under this condition the metric defines a T_2 space. If, however, we integrate the Killing equations of the metric (27.18), we obtain a Killing vector ξ^α depending on six independent variables, corresponding to the T_2 space of maximum mobility, though in a different system of coordinates.

T_2 Spaces Admitting G_4 and G_5 Groups of Motion

These spaces can be determined by selecting those T_2 admitting G_3, which also admit a G_4, since every G_4 group contains a G_3 subgroup. As the metrics (27.1) and (27.6) correspond to the groups of motions G_5, we can determine T_2 spaces admitting G_4 by finding for what values of the functions α and β the metric (27.1) admits a four-parameter group of motions.

If the space also admits an X_4 operator in addition to the G_3 group operators (27.11), then in accordance with the second basic theorem of Lie [cf. § 10, (10.18)], the commutators of X_4 are

$$[X_i X_4] = A_i X_1 + B_i X_2 + C_i X_3 + D_i X_4 \qquad (i = 1, 2, 3), \qquad (27.19)$$

where A_i, B_i, C_i and D_i are the structure constants. Writing down the structure equations and eliminating the partial derivatives of the components of the Killing vector $\underset{4}{\xi}^\alpha$, corresponding to the operator ($X_4 \equiv \underset{4}{\xi}^\alpha \partial_\alpha$), we obtain

$$\underset{4}{\xi}^3 Q = \underset{4}{\xi}^4 Q = 0, \quad Q = D_1 x^3 + D_2 x^4 - D_3.$$

If $Q \neq 0$, the operator X_4 acts on those isotropic two-dimensional surfaces of transitivity along which the points of the T_2 space move under the action of the operators X_i ($i = 1, 2, 3$); each such surface must be an *isotropic* $\overset{*}{V}_2$ manifold admitting a G_4 group of motions transitive on the $\overset{*}{V}_2$, but it is easy to see that this is impossible. Hence $Q=0$, and the constants $D_i = 0$. Integrating the remaining eqs. (27.19), we obtain $A_1 = B_2 + C_3$, $B_1 = C_1 = 0$, and the Killing vector

$$\underset{4}{\xi}^\alpha = [A_2 x^2 + \tfrac{1}{2} C_2 x^{2^2} + M(x^3, x^4)]\, \delta_1^\alpha + (B_2 x^2 + A_2 x^4 + C_2 x^2 x^4 - A_3)\delta_2^\alpha$$
$$+ N(x^3, x^4)\, \delta_3^\alpha + [C_2 x^{4^2} + (B_2 - C_3) x^4 - B_3]\, \delta_4^\alpha, \qquad (27.20)$$

where M and N are arbitrary functions of x^3, x^4. Direct substitution shows that the Jacobi conditions (10.18) are satisfied identically for the structure constants A_i, B_i and C_i. The Killing equations for $\underset{4}{\xi}^\alpha$ (27.20) and the metric (27.12) show that $A_2 = B_2 + C_3 = 0$, $M = M(x^3)$ and

$$\partial_4(\underset{4}{\xi}^4 \beta) = 0, \quad \alpha N_4 + \beta M' = 0, \quad \alpha_3 N + \frac{k}{\beta}\alpha_4 + 2\alpha N_3 = 0. \qquad (27.21)$$

Moreover, the single field equation (27.13) reduces to

$$\alpha_{44} - \frac{\alpha_4^2}{2\alpha} - \frac{2C_2 x^4 \alpha_4}{(C_2 x^{4^2} - B_3)} - \frac{2C_2(2C_2 x^{4^2} + B_3)\alpha}{(C_2 x^{4^2} - B_3)^2} = 0. \qquad (27.22)$$

The structure constants may be further specified as follows.

The classification of the possible real G_4 group structures, (10.17) and (10.19), shows that only structures of type I, II, III or VI can contain a G_3 subgroup of type II. As the structure constanst C^i_{12}, C^i_{13} and C^i_{23} have canonical form, it can be assumed without loss of generality that the C^i_{j4} ($j = 1, 2, 3$) for the structure dealt with above must also have canonical form.

If, however, the structure corresponds to either I or II in the classification (10.17), the constant C_2 is zero, and (27.12) shows that $\beta_4 = 0$. But then, as pointed out above, the Segre characteristic of the λ-matrix $(R_{ab}-\lambda g_{ab})$ $(a, b = 1, \ldots, 6)$ in the bivector space cannot be [21°, 21], which is essential for a T_2 space. If the G_4 group had one of the possible VI type structures, the manifold would admit an Abelian G_3. But, as shown above, the T_2 would then admit a G_5 group as the complete group of motions, and would therefore belong to $T_{2,5}$.

The only alternative remaining is that the G_4 structure is of type III. Thus $q = 0$, $A_1 = A_2 = A_3 = B_2 = C_3 = 0$, $C_2 = 1$, $B_3 = -1$, and (27.20) and (27.21) show that

$$\beta = \frac{k}{x^{4^2}+1}, \quad k = \text{const} \neq 0,$$

$$N_4\alpha + M'B = 0, \quad N\alpha_3 + \frac{k}{\beta}\alpha_4 + 2\alpha N_3 = 0, \tag{27.23}$$

and (27.22) becomes

$$a_{44} - \frac{\alpha_4^2}{2\alpha} + \frac{2x^4\alpha_4}{(x^{4^2}+1)} - \frac{2(2x^{4^2}-1)\alpha}{(x^{4^2}+1)^2} = 0. \tag{27.24}$$

The metric (27.12) and the operators (27.11) remain unchanged in form under any transformation of the type

$$x^{1'} = x^1 + f(x^3), \quad x^{2'} = x^2, \quad x^{3'} = \psi(x^3), \quad x^{4'} = x^4.$$

The functions f and ψ are arbitrary; an appropriate choice of f reduce $M(x^3)$ to zero; as (27.23) shows that $N = N(x^3)$ when $M = 0$, we can reduce N to unity by an appropriate choice of $\psi(\psi_3 \neq 0$, for otherwise the transformation becomes degenerate.) We then obtain the following equation for α from (27.23):

$$\alpha_3 + (x^{4^2}+1)\alpha_4 = 0.$$

The general integral is $\alpha = \alpha(x^3 - \tan^{-1}x^4)$. It is not difficult to see that this equation together with (27.24) implies that $\alpha = 0$, corresponding to a degenerate metric.

T_3 Spaces Admitting Motions

The T_3 of maximum mobility is defined by a G_2 group of motions as was shown in § 24.

The trivial problem of determining the T_3 admitting a G_1 group reduces to writing down the field equations so that the components $g_{\alpha\beta}(x)$ are independent of one variable, and the algebraic structure of the curvature tensor satisfies certain conditions (cf. example 6 below).

EXAMPLES

1. Write down the field equations for T_2 admitting G_1 or G_2 groups of motion of both possible structures (cf. § 29).

2. Show that the Segre characteristic of the λ-matrix $(R_{ab}-\lambda g_{ab})$ $(a, b = 1, \ldots 6)$ for the metrics (27.2), (27.3) is $[21^{\circ}, 21]$.

3. Verify that the metric (27.4) admits a G_5 group of motions with the operators (27.6).

4. Show that the metrics (31.28) and (31.29) cannot define T_2 spaces.

5. Show that G_4 groups of types IV, V, VII and VIII cannot contain a G_3 subgroup of type II.

6. Write down the field equations and the structure conditions for T_3 spaces admitting a G_1 group of motions.

7. Show that the metric

$$ds^2 = -e^{x^2x^4}(dx^{1^2}+dx^{2^2})-2dx^3\,dx^4$$
$$+\left[x^2x^3+2e^{x^2x^4}\left(\frac{x^{2^2}}{x^{4^2}}-\frac{4x^2}{x^{4^3}}+\frac{6}{x^{4^4}}\right)\right]dx^{4^2}$$

defines a T_3 space admitting the one parameter group of motions with the operator $X = p_1$.

§ 28. Summary of Results. Survey of Well-known Solutions of the Field Equations

The solutions of the field equations ($R_{\alpha\beta} = 0$) which admit groups of motions are naturally the most interesting from the physical point of view. All the well-known solutions to be found in the literature belong to this class.

Of particular interest are spaces of high mobility corresponding to the simplest, and physically most important, models of the distribution and motion of matter. These spaces may be said to define the geometry of an empty space-time continuum under the most characteristic conditions.

As all the *point* transformations included in a given group of motions define automorphisms (mapping of a space-time on itself) which leave unchanged the equations describing the geometry of the space-time and the distribution of matter, a gravitational field can only be determined to within automorphisms of motion. In this sense *the geometry of any gravitational field is the geometry of the automorphisms of motion of the corresponding space-time*. To every gravitational field there corresponds a group of motions G_r $(0 \leqslant r \leqslant 10)$. If $r = 10$, we have a space of constant curvature (a Minkowski space when $R_{\alpha\beta} = 0$ and the system of coordinates is suitably chosen); if $r = 0$ the space admits only the identity transformation $\overset{*}{x}^\alpha = x^\alpha$ and the group of automorphisms coincides with the unit element of the group.

All the well-known solutions found in the literature can be considered from this point of view. Some of the spaces, especially T_2 and T_3, obtained in this chapter have not been derived before. Even some of the T_1 spaces were hitherto unknown and have a number of interesting properties, the T_1 spaces admitting a simple-transitive G_4 group with unequal complex principal curvatures, for example.

The classification of gravitational fields in empty space-time can conveniently be represented by Table 1.

TABLE 1

Type / Group	T_1	T_2	T_3
G_{10}	Minkowski Space $T_{1,10}$		
G_7, G_8, G_9			
G_6	Gap	Space of Max. Mobility $T_{2,6}$	
G_5		$T_{2,5}$	
G_4	$T_{1,4}$	$T_{2,4}$	
G_3	$T_{1,3}$	$T_{2,3}$	
G_2	$T_{1,2}$	$T_{2,2}$	Space of Max. Mobility $T_{3,2}$
G_1	$T_{1,1}$	$T_{2,1}$	$T_{3,1}$
G_0	$T_{1,0}$	$T_{2,0}$	$T_{3,0}$

The light shading (gap) indicates that the T_i with the given group also admits the next highest group (with T_1 this means a G_{10} group correspo nd ing to a flat space-time). The heavy shading indicates that there is no space with a group of that particular order. Each entry gives the type of space and the order of the possible complete group. It follows directly from this table that *T_2 and T_3 spaces cannot be flat* (the curvature tensor is always non-zero).

The class admitting the largest variety of groups of motion is the first type T_1. These spaces with the most symmetrical curvature tensor structure are the most frequently used in problems of relativistic cosmology, the principal field in which the theory of gravitation has been applied. This may explain why most well-known solutions belong to the class $T_{1,r}$ $(r = 2, 3, 4)$.

We now classify the principal solutions of the field equations

$$R_{\alpha\beta} = \varkappa g_{\alpha\beta},$$

giving the type of space and the order of the group of motions admitted (cf. § 14).

1. The Schwarzschild solution (14.1) for a spherically symmetric field. As shown in § 26, this solution is defined by the metric (26.26) with

$$c_1 = 1, \quad c_2 = 0, \quad c_3 = -2a,$$

corresponding to $T_{1,4}$ spaces with an insoluble G_4 of type VIII. An insoluble group could be expected as the space has two-dimensional non-isotropic surfaces of transitivity with constant curvature and the group of motions is always insoluble for these (providing the curvature is non-zero).

2. Kottler's solution (14.2) is a generalization of Schwarzschild's for the case when the field equations are

$$R_{\alpha\beta} = \varkappa g_{\alpha\beta}.$$

The group of motions is the same as before.

3. Weyl and Levi-Civita derived a static solution with axial symmetry (14.3). As this solution depends on only two variables x^3 and $\varrho = (x^{1^2} + x^{2^2})^{1/2}$, there are two operators which take the form

$$X_1 = p_1, \quad X_2 = p_2$$

in some system of coordinates, and which therefore define an *Abelian* G_2 acting on non-isotropic surfaces of transitivity. Moreover, as this solution is subject to the condition that μ and ν vanish at infinity, the space can degenerate into a flat space. As only T_1 spaces possess this property, we have a $T_{1,2}$ *with an Abelian group acting transitively on non-isotropic two-dimensional surfaces of transitivity.*

4. Brinkman's solution for $n = 4$ corresponds to a $T_{1,10}$

5. The Kasner solution defines $T_{1,4}$ spaces with an insoluble G_4 of type VI_4 having an Abelian G_3 subgroup. The group is *intransitive*.

6. Mitter's solution is a special case of Weyl and Levi-Civita's without increase of mobility. We thus have a $T_{1,2}$ with an Abelian group. Levi-Civita's solution can be reduced to this case.

7. Delsarte's solution is a particular integral of Weyl and Levi-Civita's field equations. The mobility is not increased. The Segre characteristic of the λ-matrix is [(11)1, (11)1].

8. The solution (14.6) defines a $T_{1,4}$ with an insoluble G_4 of type VIII.

9. Every reducible space (14.7), or (14.8) has a six-parameter group of motions when $R_{\alpha\beta} = \varkappa g_{\alpha\beta}$ and $\varkappa \neq 0$; if, however, $\varkappa = 0$, we have a flat $T_{1,10}$.

10. The Narlikar–Karmarkar solution (14.12) reduces to Kasner's.

11. The Einstein–Rosen solution admits two explicit operators $X_1 = p_2$ and $X_2 = p_3$, defining an Abelian G_2 group. The Killing equations for this metric admit a solution other than

$$\underset{1}{\xi^\alpha} = c_1\, \delta_2^\alpha + c_2\, \delta_3^\alpha, \quad c_1, c_2 = \text{const},$$

only when the metric tensor satisfies additional conditions not derivable from the field equations. These are: (1) the space admits a 4-orthogonal system of coordinates; (2) it admits an Abelian G_2 group of motions acting on non-isotropic surfaces; (3) $g_{11} = -g_{44}$; (4) the metric signature is $(- - - +)$. A more detailed discussion of these spaces can be found in § 57.

Clearly any solution can be treated in this way; to sum up, we may say that the general classification of gravitational fields in empty space according to the structure of the curvature tensor, combined with the classification of groups of motion, leads to *an invariant characterization of gravitational fields.*

The case of gravitational fields with $T_{\alpha\beta} \neq 0$ is more complicated, but a solution exists and will be considered in the following chapter.

CHAPTER 5

CLASSIFICATION OF GENERAL GRAVITATIONAL FIELDS BY GROUPS OF MOTIONS

WHEN $n = 4$ and the gravitational field equations have the form

$$R_{\alpha\beta}-\tfrac{1}{2}Rg_{\alpha\beta} = \lambda T_{\alpha\beta},$$

the classification of spaces with arbitrary energy-momentum tensors $T_{\alpha\beta}$ according to the groups of motion they admit, reduces to the problem of classifying arbitrary V_4 spaces subject to the single condition

$$|g_{\alpha\beta}| < 0.$$

This is equivalent to the condition that the metric of the V_4 space at any point has a signature $(---+)$ (multiplying the metric by -1 is disregarded).

Developing the idea put forward in the last chapter from the physical point of view, we consider each general gravitational field as a phenomenon invariant with respect to some r-parameter group of motions $G_r (0 \leqslant r \leqslant n(n+1)/2 = 10)$, and the geometry of the space-time as the geometry of the automorphisms corresponding to the given G_r group.

The problem of classifying V_4 spaces by groups of motions has been solved by various authors (cf. § 29) starting from different additional assumptions. A general solution is given in the joint papers [337], [357], [358], by Petrov, Kaigorodov and Abdullin, which contain a systematic summary of previous results, fill gaps in the theory, and provide a solution for simple and multiple transitive groups.

This classification permits the study of gravitational fields from a group, and hence an invariant, point of view [321].

Fubini considered the systematic classification of V_4 spaces with *positive-definite* metrics by groups of motion, in two papers, seeking V_4 spaces with groups $G_r (r = 1, 2, 3, 4)$ in [18], and $r \geqslant 5$ in [20]; there is an error in this last paper. In establishing the allowed group structures, Fubini proved the following theorem but with a slightly incorrect formulation: every G_r group ($r \leqslant 8$) in a V_4 admits a G_{r-1} subgroup. However, as Yegorov has shown, this statement needs modifying [230]. Fubini first proved the theorem for $r \leqslant 7$, and also for $r = 8$ when the G_8 is not isomorphic with the complete projective group. If the G_8 is not thus restricted, the theorem no longer holds and V_4 spaces with a G_8 group can be found which do not admit a G_7 [230]. Fubini also incorrectly used Lie's classification of G_r group structures over the field of complex numbers [7]; the number of structures is greater *over the field of real numbers*, and so V_4 spaces with transitive groups are excluded; moreover the group of rotations is a real G_3 group which does not contain a G_2 subgroup.

A different method of finding V_4 spaces with a positive-definite metric which admit multiply transitive groups G_5, G_6 and G_7 was first used by Vranceanu [200].

If the metric is indefinite, V_4's can also exist admitting G_r acting on isotropic surfaces of transitivity with a singular metric, as was first noted by Yegorov. Kruchkovich suggested a method for finding such V_4's and gave the appropriate classification for intransitive G_r ($r = 3, 4$) [260]. Due to an oversight detected by Petrov, several classes of space were omitted in this analysis; in addition, V_4's admitting G_2 and transitive G_4 were not considered. These cases are very important in the study of gravitational fields.

The classification is based on a consideration of the canonical real structures (§ 10).

If we are to restrict the analysis to V_4 spaces corresponding to real gravitational fields, the condition $|g_{\alpha\beta}| < 0$ must be imposed on all classes of space obtained. The investigation is confined to the class of analytic functions.

§ 29. Gravitational Fields Admitting a G_r Group ($r \leqslant 2$)

The Killing vectors defining the operators $\underset{s}{X} = \underset{s}{\xi}^{\alpha}(x)\,\partial_{\alpha}$ will be denoted by $\underset{s}{\xi}^{\alpha}$, and the derivative $\partial\varphi/\partial x^{\alpha} \equiv \partial_{\alpha}\varphi$ for brevity by φ_{α}.

Group G_0

This group is identical with the unit element of the group. We have a V_4 space with metric tensor components $g_{\alpha\beta}(x)$ subject only to the condition $|g_{\alpha\beta}| < 0$.

GROUP G_1

The Killing vector ξ^α may be either *isotropic* or *non-isotropic.*

Non-isotropic ξ^α ($\xi_\alpha \xi^\alpha \neq 0$). The V_4 admits a system of coordinates with $\underset{1}{\xi}^\alpha = \delta_1^\alpha$ and $g_{\alpha\beta} = g_{\alpha\beta}(x^2, x^3, x^4)$; the coordinate curves with parameter x^1 are trajectories of the motion.

Isotropic $\underset{1}{\xi}^\alpha$ ($\underset{1}{\xi}_\alpha \underset{1}{\xi}^\alpha = 0$); conditions as in the previous case, but in addition $g_{11} = 0$.

GROUP G_2

Manifolds with a singular metric will henceforward be denoted by an asterisk, thus $\overset{*}{V}_m$.

G_2 I *(Abelian group) acting on a* V_2. As the rank of $(\underset{s}{\xi}^\alpha)$ ($s = 1, 2$; $\alpha = 1, 2, 3, 4$) must be 2, a system of coordinates can be chosen in which the Killing vectors have components $\underset{s}{\xi}^3 = 0$, $\underset{s}{\xi}^4 = 0$, and the equations of the surfaces of transitivity are $x^3 =$ constant, $x^4 =$ constant. Using the structure equations and the most general coordinate transformation which leaves these properties invariant,

$$x^{1'} = f(x^1, x^2, x^3, x^4), \quad x^{2'} = \varphi(x^1, x^2, x^3, x^4),$$
$$x^{3'} = \psi(x^3, x^4), \quad x^{4'} = \theta(x^3, x^4),$$

we choose f and φ so as to reduce the Killing vectors to the form

$$\underset{1}{\xi}^\alpha = \delta_1^\alpha, \quad \underset{2}{\xi}^\alpha = \delta_2^\alpha.$$

The required general coordinate transformation must be of the form

$$x^1 = x^{1'} + f(x^{3'}, x^{4'}), \quad x^2 = x^{2'} + \varphi(x^{3'}, x^{4'}),$$
$$x^3 = \psi(x^{3'}, x^{4'}), \quad x^4 = \theta(x^{3'}, x^{4'}),$$

where f, φ, ψ and θ are new arbitrary functions satisfying the condition $\psi_3\theta_4 - \psi_4\theta_3 \neq 0$ for a non-singular transformation. The Killing equations

$$\xi^\sigma \partial_\sigma g_{\alpha\beta} + g_{\alpha\sigma} \partial_\beta \xi^\sigma + g_{\beta\sigma} \partial_\alpha \xi^\sigma = 0$$

for $\underset{1}{\xi}^\alpha$ and $\underset{2}{\xi}^\alpha$ show that $g_{\alpha\beta} = g_{\alpha\beta}(x^3, x^4)$. The arbitrary functions f, φ, ψ and θ are used to reduce the metric to one of the canonical forms. We choose functions which satisfy the set of equations

$$f_4 = \sigma(g^{13}\theta_3 - g^{14}\psi_3), \quad \varphi_4 = \sigma(g^{23}\theta_3 - g^{24}\psi_3),$$
$$\psi_4 = \sigma(g^{33}\theta_3 - g^{34}\psi_3), \quad \theta_4 = \sigma(g^{34}\theta_3 - g^{44}\psi_3),$$

where $\sigma = e_4/\sqrt{e_4\omega}$, $e_4 = \pm 1$, $\omega = g^{33}\theta_3^2 - 2g^{34}\psi_3\theta_3 + g^{44}\psi_3^2$, and e_4 is chosen so that σ is real in some domain in the V_4. The variables x^α in $g_{\alpha\beta}$ and $g^{\alpha\beta}$ must be replaced by $x^{\alpha'}$, since f, φ, ψ and θ are functions of the $x^{\alpha'}$. When these conditions are satisfied and the $g_{\alpha\beta}$ are analytic in some domain, the set of differential equations is of the Cauchy–Kowalewski type, soluble in the domain by means of analytic functions (Finikov [160]). But this set of equations is equivalent in the new system of coordinates to the set

$$g_{i'4'} = 0 \quad (i = 1, 2, 3), \quad g_{4'4'} = e_4.$$

Hence in a special holonomic system of coordinates, the metric is

$$(g_{\alpha\beta}) = \left(\begin{array}{c|c} A & 0 \\ \hline 0 & e_4 \end{array}\right), \tag{29.1}$$

$$e_4 = \pm 1, \quad A = (g_{ij}) \qquad (i, j = 1, 2, 3),$$
$$g_{ij} = g_{ij}(x^3, x^4), \quad X_1 = p_1, \quad X_2 = p_2.$$

G_2 I *(Abelian group) acting on a* $\overset{*}{V}_2$. As $\overset{*}{V}_2$ is an isotropic manifold its metric is singular, i.e. if the $\overset{*}{V}_2$ is referred to coordinates x^1, x^2, $g_{11}g_{22} - g_{12}^2 = 0$. If, however, the rank of the matrix (g_{ij}) $(i, j = 1, 2)$ were zero, $g_{11} = g_{12} = g_{22} = 0$ in some system of coordinates, and $|g_{\alpha\beta}| > 0$ $(\alpha, \beta = 1, 2, 3, 4)$; but this is impossible for real gravitational fields. Hence the rank of the metric on $\overset{*}{V}_2$ is one, and there is a single family of *special* curves on $\overset{*}{V}_2$ which are isotropic and orthogonal to the $\overset{*}{V}_2$: $g_{ij}\xi^i = 0$ $(i, j = 1, 2)$. If G_2 admits an operator $X = \xi^i \partial_i$, it also admits the operator $Y = \sigma(x)\xi^i \partial_i$ where σ is an arbitrary scalar, i.e. we have two operators with a common trajectory; conversely, if two operators acting on $\overset{*}{V}_2$ have a common trajectory, this trajectory is a special curve of $\overset{*}{V}_2$.

The special curves clearly represent a *system of imprimitivity* of G_2. There are two possibilities; (a) one of the G_2 operators is special; (b) there is no special operator.

(a) $\underset{1}{\xi}^\alpha$ *is a special Killing vector*. As the rank of $(\underset{s}{\xi}^\alpha)$ must be 2, we can introduce a system of coordinates for which $\underset{s}{\xi}^3 = \underset{s}{\xi}^4 = 0$ $(s = 1, 2)$ ([147], p. 223), and the equations of $\overset{*}{V}_2$ are $x^3 =$ constant, $x^4 =$ constant. These properties are invariant with respect to the transformations $x^{1'} = f$, $x^{2'} = \varphi$,

$x^{3'} = \psi$, $x^{4'} = \theta$, where f and φ depend on x^1, x^2, x^3, x^4, and ψ and θ on x^3, x^4, subject only to the condition $(f_1\varphi_2 - f_2\varphi_1) \times (\psi_3\theta_4 - \psi_4\theta_3) \neq 0$ for a non-singular transformation; f and φ can always be chosen so that $\underset{1}{\xi}^{\alpha'} = \underset{1}{\delta}^{\alpha'}$, and then the most general form of the allowed transformation is

$$x^{1'} = x^1 + f(x^2, x^3, x^4), \quad x^{2'} = \varphi(x^2, x^3, x^4),$$
$$x^{3'} = \psi(x^3, x^4), \quad x^{4'} = \theta(x^3, x^4)$$

provided that $\varphi_2(\psi_3\theta_4 - \psi_4\theta_3) \neq 0$. As $\underset{1}{\xi}^{\alpha}$ is a special operator, $g_{11} = g_{12} = 0$ and $\partial_1 \underset{2}{\xi}^2 = 0$ in this system of coordinates. The structure equation $[X_1, X_2] = 0$ shows that $\underset{2}{\xi}^{\alpha} = \underset{2}{\xi}^{\alpha}(x^2, x^3, x^4)$; applying the allowed transformation, we obtain

$$\underset{2}{\xi}^{1'} = \underset{2}{\xi}^1 + f_2 \underset{2}{\xi}^2, \quad \underset{2}{\xi}^{2'} = \varphi_2 \underset{2}{\xi}^2.$$

As the rank of the $\mathring{V}_2$ metric is one, $\underset{2}{\xi}^2 \neq 0$, and a non-degenerate transformation with a suitable φ_2 and f_2 can result in $\underset{2}{\xi}^{\alpha'} = \delta_2^{\alpha'}$. The allowed transformation is then $x^{1'} = x^1 + f$, $x^{2'} = x^2 + \varphi$, $x^{3'} = \psi$, $x^{4'} = \theta$, where f, φ, ψ and θ are arbitrary functions of x^3, x^4, provided that $\psi_3\theta_4 - \psi_4\theta_3 \neq 0$. Under this transformation $g_{1'4'} = g_{13}\psi_4 + g_{14}\theta_4$, and since g_{13} and g_{14} cannot both vanish simultaneously we can take $g_{14} \neq 0$ without loss of generality. The component g_{22} is invariant under the transformation, and cannot vanish for otherwise $|g_{\alpha\beta}| > 0$. We can therefore put

$$\psi_4 = \frac{\sqrt{-e^4 g_{14}^2 g_{22}}}{g}, \quad e_4 = \pm 1,$$

$$\theta_4 = -\frac{g_{13}}{g_{14}}\psi_4, \quad \varphi_4 = -\frac{1}{g_{22}}(g_{13}\psi_{14} + g_{24}\theta_4).$$

$$f_4 = \frac{1}{g_{13}\psi_3 + g_{14}\theta_3}[g_{23}\varphi_4\psi_3 + g_{24}\varphi_4\theta_3 + g_{33}\psi_3\psi_4 + g_{34}(\psi_3\theta_4 + \psi_4\theta_3) + g_{44}\theta_3\theta_4].$$

leading to a compatible set of equations of the Cauchy–Kowalewski type. Taking f, φ, ψ and θ as the integrals of this set, we obtain

$$(g_{\alpha\beta}) = \begin{pmatrix} 0 & 0 & g_{13} & 0 \\ 0 & g_{22} & g_{23} & 0 \\ g_{13} & g_{23} & g_{33} & 0 \\ 0 & 0 & 0 & e_4 \end{pmatrix}, \tag{29.2}$$

$$g_{\alpha\beta} = g_{\alpha\beta}(x^3, x^4), \quad X_1 = p_1, \quad X_2 = p_2.$$

(b) *G_2 has no special operators.* The whole discussion can be repeated almost word for word except that a Killing vector $\xi^\alpha = \delta_1^\alpha$ cannot exist if the x^1 are special curves. There are then two possibilities:

$$(g_{\alpha\beta}) = \begin{pmatrix} 0 & 0 & 0 & g_{14} \\ 0 & g_{22} & -x^1g_{22}+g_{23} & 0 \\ 0 & -x^1g_{22}+g_{23} & x^{1^2}g_{22}-2x^1g_{23}+g_{33} & g_{34} \\ g_{14} & 0 & g_{34} & 0 \end{pmatrix}, \tag{29.3}$$

$$g_{\alpha\beta} = g_{\alpha\beta}(x^3, x^4), \quad X_1 = p_2, \quad X_2 = p_1+x^3p_2;$$

$$(g_{\alpha\beta}) = \begin{pmatrix} 0 & 0 & g_{13} & 0 \\ 0 & g_{22} & -x^1g_{22}+g_{23} & 0 \\ g_{13} & -x^1g_{22}+g_{23}, & x^{1^2}g_{22}-2x^1g_{23}+g_{33} & 0 \\ 0 & 0 & 0 & e_4 \end{pmatrix}, \tag{29.4}$$

$$e_4 = \pm 1, \quad g_{\alpha\beta} = g_{\alpha\beta}(x^3, x^4), \quad X_1 = p_2, \quad X_2 = p_1+x^3p_2.$$

G_2 II *(non-Abelian group) acting on V_2*. Repeating the analysis used for G_2 I acting on V_2, but noting that the structure equations are now of type II, we obtain

$$(g_{\alpha\beta}) = \begin{pmatrix} e^{-2x^2}a_{11} & e^{-x^2}a_{12} & e^{-x^2}a_{13} & 0 \\ e^{-x^2}a_{12} & a_{22} & a_{23} & 0 \\ e^{-x^2}a_{13} & a_{23} & a_{33} & 0 \\ 0 & 0 & 0 & e_{44} \end{pmatrix}, \tag{29.5}$$

$$e_4 = \pm 1, \quad a_{\alpha\beta} = a_{\alpha\beta}(x^3, x^4), \quad X_1 = p_1, \quad X_2 = x^1p_1+p_2.$$

G_2 II *acting on $\mathring{V}_2$*. There are again two possibilities:

(a) *G_2 contains a special operator.* The analysis for case (a) of G_2 I acting on $\mathring{V}_2$ holds with appropriate modifications for difference in structure equations. The resulting canonical form is

$$(g_{\alpha\beta}) = \begin{pmatrix} 0 & 0 & e^{-x^2}a_{13} & 0 \\ 0 & a_{22} & a_{23} & 0 \\ e^{-x^2}a_{13} & a_{23} & a_{33} & 0 \\ 0 & 0 & 0 & e_4 \end{pmatrix}, \tag{29.6}$$

$$a_{\alpha\beta} = a_{\alpha\beta}(x^3, x^4), \quad X_1 = p_1, \quad X_2 = x^1p_1+p_2.$$

(b) *G has no special operators.* It is always possible to choose a system of coordinates in which $\mathring{V}_2$ has the equations x^3 = constant, x^4 = constant, $\underset{s}{\xi}^3 = \underset{s}{\xi}^4 = 0$ ([147], p. 223). We choose the special curves on $\mathring{V}_2$ as the coordinate curves x^1. Then $g_{11} = g_{12} = 0$ and $\partial_1\underset{s}{\xi}^2 = 0$ follow from the

Killing equations; the most general transformation of coordinates preserving these relations is

$$x^{1'} = f(x^1, x^2, x^3, x^4), \quad x^{2'} = \varphi(x^2, x^3, x^4),$$
$$x^{3'} = \psi(x^3, x^4), \quad x^{4'} = \theta(x^3, x^4)$$

provided that $f_1\varphi_2(\psi_3\theta_4-\psi_4\theta_3) \neq 0$. The component $\underset{1}{\xi}^2 \neq 0$, since there is no special operator. Therefore putting $f_2 = (-\underset{1}{\xi}^1/\underset{1}{\xi}^2)f_1$, $\varphi_2 = 1/\underset{1}{\xi}^2$, we obtain a compatible set of equations of the Cauchy–Kowalewski type, and reduce the Killing vector to $\underset{1}{\xi}^{\alpha'} = \delta_2^{\alpha'}$. The allowed transformations are now

$$x^{1'} = f(x^1, x^3, x^4), \quad x^{2'} = x^2+\varphi(x^3, x^4),$$
$$x^{3'} = \psi(x^3, x^4), \quad x^{4'} = \theta(x^3, x^4),$$

where $f_1(\psi_3\theta_4-\psi_4\theta_3) \neq 0$. The structure equations and $\partial_1\underset{2}{\xi}^2 = 0$ show that $\underset{2}{\xi}^\alpha = A(x^1, x^3, x^4)\, \delta_1^\alpha+[x^2+B(x^3, x^4)]\, \delta_2^\alpha$ where $A \neq 0$, since otherwise there would be a special operator. We can thus reduce $\underset{2}{\xi}^{\alpha'}$ to the form $\delta_1^{\alpha'}+x^2\delta_2^{\alpha'}$ by means of non-singular allowed transformations. The direct transformation

$$x^1 = x^{1'}+f(x^{3'}, x^{4'}), \quad x^2 = x^{2'},$$
$$x^3 = \psi(x^{3'} x^{4'}), \quad x^4 = \theta(x^{3'}, x^{4'})$$

is still allowed provided that $\psi_3\theta_4-\psi_4\theta_3 \neq 0$.

The Killing equations for these operators show that

$$g_{11} = g_{12} = 0, \quad g_{13} = a_{13}, \quad g_{14} = a_{14},$$
$$g_{22} = e^{-2x^1}a_{22}, \quad g_{23} = e^{-x^1}a_{23}, \quad g_{24} = e^{-x^1}a_{24},$$
$$g_{33} = a_{33}, \quad g_{34} = a_{34}, \quad g_{44} = a_{44},$$

where $a_{\alpha\beta} = a_{\alpha\beta}(x^3, x^4)$.

Applying the allowed transformation shows that $g_{1'4'} = a_{13}\psi_4+a_{14}\theta_4$, and since a_{13}, a_{14} cannot be zero simultaneously ($| g_{\alpha\beta} | \neq 0$), we can take $a_{14} \neq 0$ without loss of generality. Then

$$g_{2'4'} = e^{-x^{1'}}e^{-f}(a_{23}\psi_4+a_{24}\theta_4),$$
$$g_{3'4'} = f_3 g_{1'4'}+f_4 g_{1'3'}+\psi_3(a_{33}\psi_4+a_{34}\theta_4)+\theta_3(a_{33}\psi_4+a_{44}\theta_4).$$

Imposing the conditions $g_{1'4'} = g_{2'4'} = g_{3'4'} = 0$ leads to a set of equations of the Cauchy type: $f_4 = F, \psi_4 = \Phi, \theta_4 = \Psi$. If $g_{1'4'} = 0$, then $g_{1'3'} \neq 0$, and the three equations can be solved explicitly for f_4, φ_4 and θ_4. We note that if $a_{14}\,a_{23}-a_{13}\,a_{24}=0$, the system reduces to $f_4=F$, $\varphi_4=\Phi$, with θ_4 arbitrary, leading to the same results. The canonical form is therefore

$$(g_{\alpha\beta}) = \begin{pmatrix} 0 & 0 & a_{13} & 0 \\ 0 & e^{-2x^1}a_{22} & e^{-x^1}a_{23} & 0 \\ a_{13} & e^{-x^1}a_{23} & a_{33} & 0 \\ 0 & 0 & 0 & a_{44} \end{pmatrix}, \tag{29.7}$$

$$a_{\alpha\beta} = a_{\alpha\beta}(x^3, x^4), \quad X_1 = p_2, \quad X_2 = p_1 + x^2 p_2.$$

This exhausts the gravitational fields admitting a Lie group of motions G_2.

§ 30. Gravitational Fields Admitting a G_3 Group of Motions Acting on a V_2 or $\dot{V}_2$

If a V_4 admits a G_3 group of motions with a matrix $(\underset{s}{\xi^\alpha})$ ($s = 1, 2, 3$; $\alpha = 1, 2, 3, 4$) of rank 2, this intransitive group acts on two-dimensional surfaces of transitivity which may be either *non-isotropic* (V_2) or *isotropic* ($\dot{V}_2$). We consider these two cases separately.

G_3 acting on a V_2. As G_3 is the maximum group, such a V_2 must have a constant curvature K. Three cases are possible; (1) $K=0$, (2) $K>0$ and (3) $K<0$.

In the first case ($K=0$), the V_2 is a flat two-dimensional space with either $(++)$ or $(+-)$ as signature, neglecting possible simultaneous change of the signs. In both cases the G_3 admits an Abelian G_2 subgroup, and $\underset{1}{\xi^\alpha} \neq \sigma \underset{2}{\xi^\alpha}$ for otherwise there would be a special operator acting on V_2, which is excluded. Hence the required V_4 are contained in type (29.1). It is, however, more convenient to use a method not based on the metric (29.1).

In general, if a V_n with an indefinite metric admits a continuous group of motions G_r, and the rank q of the matrix $(\underset{s}{\xi^\alpha})$ is less than r, the $\underset{s}{\xi^\alpha}$ can be renumbered so that

$$\underset{h}{\xi^\alpha} = \varphi^t_h \underset{t}{\xi^\alpha} \qquad (t = 1, \ldots, q;\ h = q+1, \ldots, r), \tag{30.0}$$

where the φ^t_h are functions of x. Suppose that p of the functions φ^t_h are independent; two alternatives arise ([147], p. 226).

(1) If in the coordinate system for which $\underset{s}{\xi}^{\nu} = 0$ $(\nu = q+1, \ldots, n)$ the rank of the Jacobian of the functions φ with respect to $x^1, \ldots, x^q$ is p, coordinate systems exist in which $\underset{s}{\xi}^{\alpha}$ are at most functions of $x^1, \ldots, x^q$.

(2) If the rank of the Jacobian $< p$, this is no longer true.

Let us consider the case $K = 0$. It is not difficult to show that the two possible signatures correspond to two different real G_3 structures, (a) $(+\,+)$ to G_3 VII with $q = 0$, and (b) $(+\,-)$ to G_3 VI with $q = -1$. We first assume that the rank of the Jacobian is p and consider a G_3 VII; a system of coordinates then exists in the V_4 in which the V_2 has the equations

$$x^3 = \text{const}, \quad x^4 = \text{const}, \quad \underset{s}{\xi}^3 = \underset{s}{\xi}^4 = 0 \qquad (s = 1, 2, 3) \quad \text{and} \quad \underset{s}{\xi}^{\alpha}(x^1, x^2).$$

The required operators can therefore be found by integrating the Killing equations in the two-dimensional space with metric $ds^2 = e_1\, dx^{1^2} + e_2\, dx^{2^2}$. We thus obtain

$$\underset{1}{\xi}^{\alpha} = \delta_1^{\alpha}, \quad \underset{2}{\xi}^{\alpha} = \delta_2^{\alpha}, \quad \underset{3}{\xi}^{\alpha} = -x^2\delta_1^{\alpha} + x^1\delta_2^{\alpha}.$$

These operators admit the transformation $x^1 = x^{1'}$, $x^2 = x^{2'}$, $x^3 = \psi(x^{2'}, x^{4'})$, $x^4 = \theta(x^{3'}, x^{4'})$.

Integrating the Killing equations for these operators and using ψ and θ, we obtain the following canonical form for the metric:

$$G_3 \text{ VII} \quad (g_{\alpha\beta}) = \begin{pmatrix} a_{11} & 0 & 0 & 0 \\ 0 & a_{22} & 0 & 0 \\ 0 & 0 & a_{33} & 0 \\ 0 & 0 & 0 & e_4 \end{pmatrix} \tag{30.1}$$

$$a_{\alpha\beta} = a_{\alpha\beta}(x^3, \; x^4), \quad a_{11} = a_{22}, \; e_4 = \pm 1,$$
$$X_1 = p_1, \; X_2 = p_2, \; X_3 = -x^2p_1 + x^1p_2.$$

Similarly in the case of G_3 VI, we obtain

$$(g_{\alpha\beta}) = \begin{pmatrix} 0 & g_{12} & 0 & 0 \\ g_{12} & 0 & 0 & 0 \\ 0 & 0 & g_{33} & 0 \\ 0 & 0 & 0 & e_4 \end{pmatrix}, \tag{30.2}$$

$$g_{\alpha\beta} = g_{\alpha\beta}(x^3, x^4), \quad e_4 = \pm 1,$$
$$X_1 = p_1, \quad X_2 = p_2, \quad X_3 = x^1p_1 - x^2p_2.$$

If, on the other hand, we assume the second alternative for the Jacobian, i.e. rank < 2, we obtain two operators with a common trajectory, which must be excluded.

A V_2 with a non-zero curvature ($K \neq 0$) cannot admit a G_3 containing an Abelian G_2 subgroup, and hence the G_3 must be insoluble, i.e. G_3 VIII or G_3 IX.

Suppose that the Jacobian has rank 2. As in the previous case, the operators can be found by considering the metric of the V_2 space, referred to a semi-geodesic system of coordinates:

$$ds^2 = e_1 dx^{1^2} + e_2\sigma^2 dx^{2^2},$$
$$e_1 = \pm 1, \quad e_2 = \pm 1, \quad \sigma = \sigma(x^1, x^2).$$

As the curvature of this V_2 is constant but non-zero, integrating the Killing equations shows that after simplification the operators are

$$\left.\begin{array}{ll}
\text{I. } G_3 \text{ IX} & e_1 = e_2, \quad e_1 k > 0, \\
& \underset{1}{\xi^\alpha} = \cos x^2 \delta^\alpha_1 + \sin x^2 \tan x^1 \, \delta^\alpha_2, \\
& \underset{2}{\xi^\alpha} = \delta^\alpha_2, \quad \underset{3}{\xi^\alpha} = \sin x^2 \, \delta^\alpha_1 - \cos x^2 \tan x^1 \, \delta^\alpha_2, \\
\text{II. } G_3 \text{ VIII} & e_1 = -e_2, \quad e_1 k > 0, \\
& \underset{1}{\xi^\alpha} = \cosh x^2 \delta^\alpha_1 + \sinh x^2 \tan x^1 \delta^\alpha_2, \quad \underset{2}{\xi^\alpha} = \delta^\alpha_2, \\
& \underset{3}{\xi^\alpha} = \sinh x^2 \, \delta^\alpha_1 + \cosh x^2 \tan x^1 \, \delta^\alpha_2, \\
\text{III. } G_3 \text{ VIII} & e_1 = -e_2, \quad e_1 k < 0, \\
& \underset{1}{\xi^\alpha} = \cos x^2 \, \delta^\alpha_1 - \sin x^2 \tanh x^1 \, \delta^\alpha_2, \quad \underset{2}{\xi^\alpha} = \delta^\alpha_2, \\
& \underset{3}{\xi^\alpha} = \sin x^2 \, \delta^\alpha_1 + \cos x^2 \tanh x^1 \, \delta^\alpha_2 \\
\text{IV. } G_3 \text{ VIII} & e_1 = e_2, \quad e_1 k < 0, \\
& \underset{1}{\xi^\alpha} = \cosh x^2 \, \delta^\alpha_1 - \sinh x^2 \tanh x^1 \, \delta^\alpha_2, \\
& \underset{2}{\xi^\alpha} = \delta^\alpha_2, \quad \underset{3}{\xi^\alpha} = \sinh x^2 \, \delta^\alpha_1 - \cosh x^2 \tanh x^1 \, \delta^\alpha_2.
\end{array}\right\} \tag{30.3}$$

Two groups G_r and H_r are said to be *similar* if there is a set of r independent functions $\varphi^s(a)$ ($s = 1, \ldots, r$) such that a non-singular transformation of coordinates exists under which the finite equations of G_r with parameters a transform to the finite equations of H_r with parameters $a'^s = \varphi^s(a)$. We are only interested in defining groups in V_4 to within a similarity, and two groups are similar if their structures can be made identical by suitable choice of bases. However, if the ranks q of the matrices $(\underset{s}{\xi}^{\alpha})$ are less then r, as in the present case, the last condition is not sufficient. Applying (30.3) to G_r and H_r we find ([147], p. 80) that G_r and H_r are similar only when the matrices $(\underset{s}{\xi}^{\alpha})$ for the two groups have the same rank, each pair of corresponding minors of order q have the same rank, and the corresponding set of equations $\varphi'^{h}_{p}(x') = \varphi^{h}_{p}(x)$ $(h = 1, \ldots, q; p = q+1, \ldots, r)$ is consistent and does not reduce to relations between the variables of either set. The problem of which group (30.3) of types II, III and IV are similar must be solved for *real parameters.* If $\underset{1}{\xi}^{\alpha}$ is expressed in terms of $\underset{2}{\xi}^{\alpha}$ and $\underset{3}{\xi}^{\alpha}$, eqs. (30.3) for the three types show that

$$\text{II.}\quad \varphi^1 = -\frac{\tan x^1}{\sinh x^2}, \qquad \varphi^2 = \coth x^2,$$

$$\text{III.}\quad \varphi^1 = -\frac{\tanh x^1}{\sin x^2}, \qquad \varphi^2 = \cot x^2,$$

$$\text{IV.}\quad \varphi^1 = \frac{\tanh x^1}{\sinh x^2}, \qquad \varphi^2 = \coth x^2.$$

Applying equations $\varphi'^{h}(x') = \varphi^{h}(x)$ to groups II and III shows that, $\tan x^1 = \tanh x^{1'}$, $\sin x^2 = \sinh x^{2'}$, $\tan x^2 = \tanh x^{2'}$; eliminating x^2 we have $x^{2'} = 0$, which is a relation between the $x^{\alpha'}$. We obtain the same result for groups III–IV, but groups II–IV are similar for real parameters. Hence only types I, II and IV of operators (30.3) need be considered. Using the Killing equations and the allowed transformations as before, we obtain the following canonical metric forms:

$$G^3 \quad \text{VIII} \quad (g_{\alpha\beta}) = \begin{pmatrix} a_{11} & 0 & 0 & 0 \\ 0 & -a_{11}\cos^2 x^1 & 0 & 0 \\ 0 & 0 & a_{33} & 0 \\ 0 & 0 & 0 & e_4 \end{pmatrix}, \tag{30.4}$$

$$a_{\alpha\beta} = a_{\alpha\beta}(x^3, x^4), \quad X_1 = \cosh x^2 p_1 + \sinh x^2 \tan x^1 p_2, \quad X_2 = p_2,$$

$$X_3 = \sinh x^2 p_1 + \cosh x^2 \tan x^1 p_2, \quad e_4 = \pm 1$$

$$G_3 \quad \text{VIII} \quad (g_{\alpha\beta}) = \begin{pmatrix} a_{11} & 0 & 0 & 0 \\ 0 & -a_{11}\cosh^2 x^1 & 0 & 0 \\ 0 & 0 & a_{33} & 0 \\ 0 & 0 & 0 & e_4 \end{pmatrix}, \tag{30.5}$$

$$a_{\alpha\beta} = a_{\alpha\beta}(x^3, x^4), X_1 = \cos x^2 p_1 - \sin x^2 \tanh x^1 p_2, \quad X_2 = p_2,$$
$$X_3 = \sin x^2 p_1 + \cos x^2 \tanh x^1 p_2, \quad e_4 = \pm 1;$$

$$G_3 \quad \text{IX} \quad (g_{\alpha\beta}) = \begin{pmatrix} a_{11} & 0 & 0 & 0 \\ 0 & a_{11}\cos^2 x^1 & 0 & 0 \\ 0 & 0 & a_{33} & 0 \\ 0 & 0 & 0 & e_4 \end{pmatrix}, \tag{30.6}$$

$$a_{\alpha\beta} = a_{\alpha\beta}(x^3, x^4), \quad X_1 = \cos x^2 p_1 + \sin x^2 \tan x^1 p_2,$$
$$X_2 = p_2, \quad X_3 = \sin x^2 p_1 - \cos x^2 \tan x^1 p_1, \quad e_4 = \pm 1.$$

These metrics have been obtained by assuming that the rank p of Jacobian $\partial(\varphi^1, \varphi^2)/\partial(x^1, x^2)$ is two. If, however, $p < 2$, then in case (30.6) and $p = 1$ for example, we have $\underset{3}{\xi}^\alpha = a\underset{1}{\xi}^\alpha + f(a, x^3, x^4)\underset{2}{\xi}^\alpha$, where a is some function. The structure equation $[X_2\ X_3] = X_1$ shows that a is zero and $f(x^3, x^4)$, so that the trajectories of the two operators on V_2 coincide. The case $p = 0$ leads to the same result, which also applies to the two types (30.4) and (30.6). The above cases exhaust all the V_4 admitting a G_3 acting on a V_2.

G_3 acting on a $\overset{*}{V}_2$. In this case, G_3 must be either a soluble group of types II, III or an *insoluble* group of type VIII, as shown by Kruchkovich ([203], p. 15). We first consider G_3 II. If alternative (1) holds for the Jacobian, the equations of $\overset{*}{V}_2$ must be $x^3 = \text{constant}$, $x^4 = \text{constant}$, in some system of coordinates and $\underset{s}{\xi}^\alpha = \underset{s}{\xi}^\alpha(x^1, x^2)$. It is then possible to choose a coordinate system on $\overset{*}{V}_2$ such that $X_1 = p_1, X_2 = p_2, X_3 = x^2 p_1$. However, the Killing equations for these operators show that the V_4 has a singular metric. Hence the rank of the Jacobian of the functions φ defined by eqns. (30.0) is less than 2. We can therefore choose a system of coordinates such that

$$\underset{1}{\xi}^\alpha = \delta^\alpha_1, \ \underset{2}{\xi}^\alpha = \delta^\alpha_2, \quad \underset{3}{\xi}^\alpha = a\delta^\alpha_1 + \lambda(a, x^3, x^4)\, \delta^\alpha_2,$$

where

$$a = a(x^1, x^2, x^3, x^4).$$

The structure equations show that

$$a = x^2 + \sigma(x^3, x^4), \quad \lambda = \lambda(x^3, \ x^4).$$

The allowed transformations for the operators X_1 and X_2 are of the form

$$x^{1\prime} = x^1 + f(x^3, x^4), \quad x^{2\prime} = x^2 + \varphi(x^3, x^4),$$
$$x^{3\prime} = \psi(x^3, x^4), \quad x^{4\prime} = \theta(x^3, x^4).$$

As λ cannot be a constant because the metric would then be singular, $\underset{3}{\xi}^\alpha$ can be reduced by the above transformation to the form $\underset{3}{\xi}^\alpha = x^2\,\delta_1^\alpha + x^4\,\delta_2^\alpha$. The following transformation is then still allowed:

$$x^1 = x^{1'} + f(x^{3'}, x^{4'}), \quad x^2 = x^{2'}, \quad x^3 = \psi(x^{3'}, x^{4'}), \quad x^4 = x^{4'}. \tag{30.7}$$

The Killing equations show that

$$g_{11} = g_{12} = g_{13} = g_{24} = g_{23} = 0, \quad g_{14} + g_{22} = 0, \quad g_{\alpha\beta} = g_{\alpha\beta}(x^3, x^4).$$

As $|g_{\alpha\beta}| \neq 0$, $g_{33} \neq 0$, $g_{14} \neq 0$. Equations (30.7) show that

$$g_{1'4'} = g_{14}, \quad g_{3'3'} = g_{33}\psi_{3'}^2, \quad g_{3'4'} = g_{14}f_{3'} + g_{33}\psi_{3'}\psi_{4'} + g_{34}\psi_{3'},$$
$$g_{4'4'} = 2g_{14}f_{4'} + g_{33}\psi_{4'}^2 + 2g_{34}\psi_{4'} + g_{44}.$$

Imposing the conditions $g_{3'4'} = g_{4'4'} = 0$ leads to a Cauchy–Kowalewski set for f_4 and ψ_4. Thus,

$$G_3 \ \text{II on} \ \overset{*}{V}_2 \quad (g_{\alpha\beta}) = \begin{pmatrix} 0 & 0 & 0 & g_{14} \\ 0 & g_{22} & 0 & 0 \\ 0 & 0 & g_{33} & 0 \\ g_{14} & 0 & 0 & 0 \end{pmatrix}, \tag{30.8}$$

$$g_{14} + g_{22} = 0, \quad g_{\alpha\beta} = g_{\alpha\beta}(x^3, x^4);$$
$$X_1 = p_1, \quad X_2 = p_2, \quad X_3 = x^2 p_1 + x^4 p_2.$$

We now consider G_3 III on $\overset{*}{V}_2$ with alternative (1) for the Jacobian. The operators can be reduced to the form $X_1 = p_1$, $X_2 = p_2$, $X_3 = x^1 p_1$ ([203], p. 15), and the Killing equations show that $g_{1\alpha} = 0\,(\alpha = 1, 2, 3, 4)$, i.e. the metric of the V_4 space is singular. If, however, we assume alternative (2) for the Jacobian, then $X_1 = p_1$, $X_2 = p_2$, $X_3 = x^1 p_2 + x^4 p_2$ as before. But the Killing equations show that

$$g_{1\alpha} = 0 \quad (\alpha = 1, 2, 3, 4),$$

which is impossible. Hence there are no V_4 with a G_3 III acting on $\overset{*}{V}_2$.

Let us now consider G_3 VIII on $\overset{*}{V}_2$, with alternative (1) for the Jacobian. The operators can be reduced to the form

$$X_1 = e^{-x^2}(p_1 + p_2), \quad X_2 = p_2, \quad X_3 = e^{x^2}(-p_1 + p_2),$$

but here again $|g_{\alpha\beta}| = 0$. Suppose now that the second alternative holds; then $\underset{3}{\xi}^\alpha = a\underset{1}{\xi}^\alpha + \lambda(a, x^3, x^4)\underset{2}{\xi}^\alpha$, where a is some function. G_3 VIII contains a G_2 II subgroup acting transitively on $\overset{*}{V}_2$. But as the order is equal to the rank of the matrix $(\underset{s}{\xi}^\alpha)(s = 1, 2)$, every $\overset{*}{V}_2$ which admits a non-

Abelian G_2 group admits operators

$$X_1 = e^{-x^2}(p_1+\varepsilon p_2), \quad X_2 = p_2, \quad \varepsilon = 0,\ 1,$$

in a system of coordinates in which the equation of $\overset{*}{V}_2$ are x^3 = constant, x^4 = constant. If $\varepsilon = 0$, however, the resulting G_2 cannot be a subgroup of G_3 VIII. Hence

$$X_1 = e^{-x^2}(p_1+p_2), \quad X_2 = p_2, \quad X_3 = ae^{-x^2}(p_1+p_2)+\lambda p_2.$$

The structure equations for these operators show that $a = 0$, $\lambda = 2e^{x^2}$, and the Killing equations that $g_{2\alpha} = 0$ ($\alpha = 1, 2, 3, 4$). Hence there are no gravitational fields corresponding to V_4 with a G_3 VIII on $\overset{*}{V}_2$.

The above classes of space exhaust all the V_4 admitting a G_3 acting on V_2 or $\overset{*}{V}_2$. It should be noted that these classes correspond to physically significant gravitational fields.

§ 31. Gravitational Fields Admitting a G_3 Group of Motions Acting on a V_3 or $\overset{*}{V}_3$

We now consider cases when G_3 acts as a simple-transitive group on V_3 or $\overset{*}{V}_3$ hypersurfaces of transitivity.

The rank of the matrix $(\underset{s}{\xi}^\alpha)$ is three.

G_3 acting on V_3. As the G_3 acts on hypersurfaces of transitivity, the hypersurfaces are geodesically parallel ([147], p. 210), and it is convenient to introduce a semi-geodesic system of coordinates in which the surfaces are given by x^4 = constant.

Fubini's investigation was confined to V_4's with a definite metric; as the metric on a non-isotropic V_3 can be indefinite if the V_4 metric is indefinite, Fubini's analysis must be modified. As the rank of the matrix $(\underset{s}{\xi}^\alpha)$ for the operators of the intransitive G_3 under investigation equals the order of the group, there are coordinate systems in V_4 in which the components $\underset{s}{\xi}^\alpha$ of the Killing vectors are at most functions of x^1, x^2, x^3, the hypersurfaces of transitivity being defined by x^4 = constant ([147], p. 223). Hence the only operators to be considered correspond to V_3 with an indefinite metric admitting a simple-transitive G_3 group. This problem has been solved by Kruchkovich [203].

Using the methods developed above simplifies the analysis. We consider soluble G_3 of types I, II, III, IV, V, VI, and VII and introduce a semi-geodesic system of coordinates in which the equations of the hypersurfaces V_3 are x^4 = constant, so that x^4 is the arc length of the non-isotropic geodesics orthogonal to the hypersurfaces. In this system of coordinates, the metric tensor components $g_{14} = g_{24} = g_{34} = 0$, and $g_{44} = e_4 = \pm 1$, depending on the norm of the tangent vector to the geodesic x^4. The operators for all seven structures are of the form

$$X_1 = p_2, \quad X_2 = p_3,$$
$$X_3 = -p_1 + (ax^2 + cx^3)p_2 + (bx^2 + dx^3)p_3. \tag{31.1}$$

The constants a, b, c, d, for these structures are given in the following table.

G_3	a	b	c	d
I	0	0	0	0
II	0	0	1	0
III	1	0	0	0
IV	1	0	1	1
V	1	0	0	1
VI	1	0	0	$q \neq 0$ or 1
VII	0	1	-1	$q(q^2 < 4)$

(31.2)

The Killing equations for the first two operators (31.1) show that $g_{\alpha\beta} = g_{\alpha\beta}(x^1, x^4)$; the equation for the third operator gives the set of equations

$$\left.\begin{aligned} \partial_1 g_{11} &= 0, \\ \partial_1 g_{12} &= ag_{12}+bg_{13}, \\ \partial_1 g_{13} &= cg_{12}+dg_{13}, \\ \partial_1 g_{22} &= 2ag_{22}+2bg_{23}, \\ \partial_1 g_{23} &= cg_{22}+(a+d)g_{23}+bg_{33}, \\ \partial_1 g_{33} &= 2cg_{23}+2dg_{33}, \\ g_{14} &= g_{24} = g_{34} = 0,\ g_{44} = e_4. \end{aligned}\right\} \tag{31.3}$$

The canonical form of the metric for each type of structure is found from this set, the appropriate values having been inserted from (31.2). If the most general transformation which leaves the operators (31.1) unchanged is determined for any possible structure, it is found to depend on four arbitrary functions of x^4. If in addition we insist that the equations $g_{4i} = 0$ $(i = 1, 2, 3)$, $g_{44} = e_4$ be preserved, the allowed transformation must depend on four arbitrary constants. But the integrals of the set (31.3) involve arbitrary functions of x^4, and hence cannot be simplified by using allowed transformations. Thus for all soluble G_3 on V_3 we obtain the classification

$$(g_{\alpha\beta}) = \left(\begin{array}{c|c} (g_{ij}) & 0 \\ \hline 0 & e_4 \end{array}\right) (i, j = 1, 2, 3),\ e_4 = \pm 1, \tag{31.4}$$

where the (g_{ij}) for the seven non-isomorphic structures are given by

G_3 I on V_3

$$(g_{ij}) = (a_{ij}),\quad a_{ij} = a_{ij}(x^4);\quad X_i = p_i (i = 1, 2, 3). \tag{31.5}$$

G_3 II on V_3

$$(g_{ij}) = \begin{pmatrix} a_{11} & a_{12} & a_{12}x^1+a_{13} \\ a_{12} & a_{22} & a_{22}x^1+a_{23} \\ a_{12}x^1+a_{13} & a_{22}x^1+a_{23} & a_{22}x^{1^2}+2a_{23}x^1+a_{33} \end{pmatrix}, \tag{31.6}$$

$$a_{ij} = a_{ij}(x^4);$$

$$X_1 = p_2,\quad X_2 = p_3,\quad X_3 = -p_1+x^3p_2.$$

G_3 III on V_3

$$(g_{ij}) = \begin{pmatrix} a_{11} & a_{12}e^{x^1} & a_{13} \\ a_{12}e^{x^1} & a_{22}e^{2x^1} & a_{23}e^{x^1} \\ a_{13} & a_{23}e^{x^1} & a_{33} \end{pmatrix}, \tag{31.7}$$

$$a_{ij} = a_{ij}(x^4);$$

$$X_1 = p_2,\quad X_2 = p_3,\quad X_3 = -p_1+x^2p_2.$$

G_3 IV on V_3

$$(g_{ij})=\begin{pmatrix} a_{11} & a_{12}e^{x^1} & (a_{12}x^1+a_{13})e^{x^1} \\ a_{12}e^{x^1} & a_{22}e^{2x^1} & (a_{22}x^1+a_{23})e^{2x^1} \\ (a_{12}x^1+a_{13})e^{x^1} & (a_{22}x^1+a_{23})\, e^{2x^2} & (a_{22}x^{11}+2a_{23}x^1+a_{33})e^{2x^1} \end{pmatrix}; \quad (31.8)$$

$$X_1 = p_2, \quad X_2 = p_3, \quad X_3 = -p_1+(x^2+x^3)p_2+x^3p_3.$$

G_3 V on V_3

$$(g_{ij}) = \begin{pmatrix} a_{11} & a_{12}e^{x^1} & a_{13}e^{x^1} \\ a_{12}e^{x^1} & a_{22}e^{2x^1} & a_{23}e^{2x^1} \\ a_{13}e^{x^1} & a_{23}e^{2x^1} & a_{33}e^{2x^1} \end{pmatrix}, \quad (31.9)$$

$$a_{ij} = a_{ij}(x^4); \quad X_1 = p_2, \quad X_2 = p_3, \quad X_3 = -p_1+x^2p_2+x^3p_3.$$

G_3 VI on V_3

$$(g_{ij}) = \begin{pmatrix} a_{11} & a_{12}e^{x^1} & a_{13}e^{qx^1} \\ a_{12}e^{x^1} & a_{22}e^{2x^1} & a_{23}e^{(1+q)x^1} \\ a_{13}e^{qx^1} & a_{23}e^{(1+q)x^1} & a_{33}e^{2qx^1} \end{pmatrix}, \quad (31.10)$$

$$a_{ij} = a_{ij}(x^4);$$

$$X_1 = p_2, \quad X_2 = p_3, \quad X_3 = -p_1+x^2p_2+qx^3p_3 \ (q \neq 0 \text{ and } 1).$$

G_3 VII on V_3

$$\left.\begin{aligned} g_{11} &= a_{11}, \quad g_{12} = \frac{1}{2}e^{\frac{q}{2}x^1}\left[(qa_{12}-pa_{13})\cos\frac{px^1}{2}\right. \\ &\qquad \left. +(pa_{12}+qa_{13})\sin\frac{px^1}{2}\right], \\ g_{13} &= e^{(q/2)x^1}\left(a_{12}\cos\frac{px^1}{2}+a_{13}\sin\frac{px^1}{2}\right), \\ g_{22} &= e^{qx^1}[a_{22}+a_{23}\cos px^1+a_{33}\sin px^1], \\ g_{23} &= \tfrac{1}{2}e^{qx^1}[qa_{22}+(qa_{23}+pa_{33})\cos px^1 \\ &\qquad +(qa_{33}-pa_{23})\sin px^1], \\ g_{33} &= e^{qx^1}\left[a_{22}+\left(\frac{q^2-2}{2}a_{23}+\frac{pq}{2}a_{33}\right)\cos px^1\right. \\ &\qquad \left. +\left(\frac{q^2-2}{2}a_{33}-\frac{pq}{2}a_{23}\right)\sin px^1\right]; \end{aligned}\right\} \quad (31.11)$$

$$a_{ij} = a_{ij}(x^4); \quad p = \sqrt{4-q^2}, \quad q^2 < 4;$$

$$X_1 = p_2, \quad X_2 = p_3, \quad X_3 = -p_1-x^3p_2+(qx^3+x^2)p_3.$$

In order to find all the non-isomorphic real structures of insoluble G_3 VIII and G_3 IX acting on V_3, we must first determine the operators in some semi-geodesic system of coordinates. Two intransitive groups G_r and H_r in the same number of variables and with the same structures are similar if their matrices $(\underset{s}{\xi}^{\alpha})$ are of rank r([147], p. 78). Hence only G_3 operators with a matrix of rank 3 need be considered, corresponding to structures VIII and IX; in this way the groups are defined to within a similarity. The signature of the V_4 does not affect the result. We choose the appropriate operators as follows:

$$G_3\ \text{VIII} \qquad \left.\begin{aligned} &X_1 = p_2, \quad X_2 = x^2 p_2 + p_3, \\ &X_3 = e^{x^3} p_1 + x^{2^2} p_2 + 2x^2 p_3. \end{aligned}\right\} \tag{31.12}$$

$$G_3\ \text{IX} \qquad \left.\begin{aligned} X_1 &= p_2, \\ X_2 &= \cos x^2 p_1 - \cot x^1 \sin x^2 p_2 + \frac{\sin x^2}{\sin x^1} p_3, \\ X_3 &= -\sin x^2 p_1 - \cot x^1 \cos x^2 p_2 + \frac{\cos x^2}{\sin x^1} p_3. \end{aligned}\right\} \tag{31.13}$$

Integrating the Killing equations for (31.12), we obtain

G_3 VIII on V_3

$$\left.\begin{aligned} g_{11} &= a_{11}, \\ g_{12} &= e^{-x^3}[a_{11} x^{1^2} - 2a_{13} x^1 + a_{12}), \\ g_{13} &= -a_{11} x^1 + a_{13}, \\ g_{22} &= e^{-2x^3}[a_{11} x^{1^4} - 4a_{13} x^{1^3} + 2(a_{12} + 2a_{33}) x^{1^2} \\ &\qquad - 4a_{23} x^1 + a_{22}], \\ g_{23} &= e^{-x^3}[-a_{11} x^{1^3} + 3a_{13} x^{1^2} \\ &\qquad - (a_{12} + 2a_{33}) x^1 + a_{23}], \\ g_{33} &= a_{11} x^{1^2} - 2a_{13} x^1 + a_{33}, \\ g_{i4} &= 0 \qquad (i = 1,\ 2,\ 3), \\ g_{44} &= e_4 = \pm 1. \end{aligned}\right\} \tag{31.14}$$

Similarly for (31.13) we obtain

G_3 IX on V_3

$$\left.\begin{aligned} g_{11} &= -a_{12} \sin 2x^3 - a_{22} \cos 2x^3 + \tfrac{1}{2} a_{11}, \\ g_{12} &= (a_{13} \cos x^3 - a_{23} \sin x^3) \cos x^1 \\ &\qquad + (a_{12} \cos 2x^3 - a_{22} \sin 2x^3) \sin x^1, \\ g_{13} &= a_{13} \cos x^3 - a_{23} \sin x^3, \\ g_{22} &= a_{33} \cos^2 x^1 + 2(a_{23} \cos x^3 + a_{13} \sin x^3) \sin x^1 \cos x^1 \\ &\qquad + (a_{12} \sin 2x^3 + a_{22} \cos 2x^3 + \tfrac{1}{2} a_{11}) \sin^2 x^1, \\ g_{23} &= a_{33} \cos x^1 + (a_{23} \cos x^3 + a_{13} \sin x^3) \sin x^1, \\ g_{33} &= a_{33}, \\ g_{4i} &= 0 \qquad (i = 1,\ 2,\ 3), \\ g_{44} &= e_4 = \pm 1. \end{aligned}\right\} \tag{31.15}$$

The a_{ij}, as before, are arbitrary functions of x^4. If the most general transformation which leaves the operators unchanges is determined, it must depend only on arbitrary constants; further simplification of (31.14) and (31.15) by such a transformation is therefore impossible.

G_3 acting on $\overset{*}{V}_3$. If a G_3 acts transitively on isotropic surfaces of transitivity $\overset{*}{V}_3$, a semi-geodesic system of coordinates can always be introduced in V_4 in which the $\overset{*}{V}_3$ can be defined by $x^4 =$ constant. By an appropriate choice of coordinates on $\overset{*}{V}_3$, the parametric curves x^1 and the family of special curves of $\overset{*}{V}_3$ can be made to coincide. As the tangent vector to these special curves must satisfy the conditions

$$g_{ij}u^i = 0 \qquad (i,\ j = 1,\ 2,\ 3),$$
$$g_{11} = g_{12} = g_{13} = 0, \quad g_{14} \neq 0,$$

i.e. in this system of coordinates

$$(g_{\alpha\beta}) = \begin{pmatrix} 0 & 0 & 0 & g_{14} \\ 0 & g_{22} & g_{23} & g_{24} \\ 0 & g_{23} & g_{33} & g_{34} \\ g_{14} & g_{24} & g_{34} & g_{44} \end{pmatrix}, \tag{31.16}$$
$$g_{22}g_{33} - g_{23}^2 \neq 0.$$

As pointed out by Kruchkovich ([260], p. 200) (31.16) is invariant under the transformation

$$\left.\begin{aligned} x^{1'} &= \varphi^1(x^1,\ x^2,\ x^3,\ x^4), \\ x^{2'} &= \varphi^2(x^2,\ x^3,\ x^4), \\ x^{3'} &= \varphi^3(x^2,\ x^3,\ x^4), \\ x^{4'} &= \varphi^4(x^4), \end{aligned}\right\} \tag{31.17}$$

where the φ^α are arbitrary functions. That this is the most general form of transformation possessing this property can be shown by integrating the set of differential equations defining the property.

As the trajectories defined by any operator of the G_3 group are assumed to lie on $\overset{*}{V}_3$, the Killing equations for any ξ^α give

$$\partial_1\xi^2 = \partial_1\xi^3 = 0, \quad \xi^4 = 0. \tag{31.18}$$

These conditions are clearly invariant under any transformation (31.17). The procedure for finding G_r acting on isotropic hypersurfaces of transitivity is as follows. As every G_3 group, except G_3 IX, admits a G_2 subgroup, we first look for canonical forms of the G_2 operators among the G_3 acting on $\overset{*}{V}_3$, in the isotropic-geodesic system of coordinates (31.16) and then find the operator X_3 from the structure equations and (31.18). The metric of the V_4 is then determined from the Killing equations. This method cannot be applied to G_3 IX.

As pointed out in § 29, the family of special curves on $\overset{*}{V}_3$ forms a system of imprimitivity for any intransitive group of motions acting on the $\overset{*}{V}_3$; hence if the group admits a special operator corresponding to a displacement along a special curve, it must be an operator of the one-dimensional *invariant subgroup* of the group ([120], pp. 26, 141). As a result its commutator with any other operator of the group differs from the special operator only by a constant factor which may be zero; the G_2 containing a special operator must therefore be isolated. It turns out that all G_3 can be determined by considering an Abelian G_2 with and without a special operator and a non-Abelian G_2 without one.

If an Abelian G_2 *admits a special operator*, it acts on an isotropic $\overset{*}{V}_2$ ([120], p. 201) and we can choose a system of coordinates from among (31.16) in which the G_2 operators have the form

$$X_1 = p_1, \quad X_2 = p_2, \tag{31.19}$$

and the system of coordinates is defined to within the transformation

$$\left.\begin{aligned} x^{1\prime} = x^1+\varphi^1(x^3, x^4), \quad x^{2\prime} = x^2+\varphi^2(x^3, x^4), \\ x^{3\prime} = \varphi^3(x^3, x^4), \quad x^{4\prime} = \varphi^4(x^4). \end{aligned}\right\} \tag{31.20}$$

This is the most general form of the allowed transformation.

Kruchkovich's treatment of Abelian G_2's without special operators is incorrect, and leads to the exclusion of several classes of V_4 [260]. We now consider this case.

Suppose therefore that the Abelian G_2 does not contain a special operator and that $X_1 = \underset{1}{\xi}^i p_i$ $(i = 1, 2, 3)$. Applying (31.17), we obtain

$$\begin{aligned} \underset{1}{\xi}^{1\prime} &= \varphi^1_1 \underset{1}{\xi}^1 + \varphi^1_2 \underset{1}{\xi}^2 + \varphi^1_3 \underset{1}{\xi}^3, \\ \underset{1}{\xi}^{2\prime} &= \varphi^2_2 \underset{1}{\xi}^2 + \varphi^2_3 \underset{1}{\xi}^3, \\ \underset{1}{\xi}^{3\prime} &= \varphi^3_2 \underset{1}{\xi}^2 + \varphi^3_3 \underset{1}{\xi}^3, \\ \xi^{4\prime} &= 0 \, (\varphi^i_j \equiv \partial_j \varphi^i) \end{aligned}$$

in the new system of coordinates. $\underset{1}{\xi}^2$ and $\underset{4}{\xi}^3$ cannot both be zero, for otherwise $\underset{1}{\xi}^\alpha$ would be a special Killing vector, which is excluded. The functions φ^i are restricted by the conditions

$$\varphi^2_1 = \varphi^3_1 = 0 \quad \text{and} \quad \varphi^1_1\varphi^4_4(\varphi^2_2\varphi^3_3 - \varphi^2_3\varphi^3_2) \neq 0.$$

We seek φ^2 and φ^3 satisfying equations

$$\varphi^2_2\underset{1}{\xi}^2 + \varphi^2_3\underset{1}{\xi}^3 = 1: \quad \varphi^3_2\underset{1}{\xi}^2 + \varphi^3_3\underset{1}{\xi}^3 = 0.$$

Equation (31.18) shows that the coefficients $\underset{1}{\xi}^2$ and $\underset{1}{\xi}^3$ like φ^2 and φ^3 depend only on x^2, x^3 and x^4, and both $\underset{1}{\xi}^2$ and $\underset{1}{\xi}^3$ cannot vanish. The set thus reduces to the Cauchy–Kowalewski form and is consistent. Moreover, its integrals clearly satisfy the inequality $\varphi^2_2\varphi^3_3 - \varphi^2_3\varphi^3_2 \neq 0$.

We define the function $\varphi^1(x)$ as a solution of $\varphi^1_i\underset{1}{\xi}^i = 0$ ($i = 1, 2, 3$), also satisfying the condition $\varphi^1_1 \neq 0$. Thus $X_1 = p_2$ in the new system of coordinates.

It is easy to show that the following transformation preserves the form of X_1,

$$\left.\begin{aligned} x^{1'} &= \varphi^1(x^1, x^3, x^4), \quad x^{2'} = x^2 + \varphi^2(x^3, x^4),\\ x^{3'} &= \varphi^3(x^3, x^4), \quad x^{4'} = \varphi^1(x^4), \end{aligned}\right\} \tag{31.21}$$

where $\varphi^1_1\varphi^3_3\varphi^4_4 \neq 0$. The structure equation $[X_1 X_2] = 0$ and (31.18) show that the Killing vector $\underset{2}{\xi}^\alpha$ has the form

$$\underset{2}{\xi}^\alpha = \underset{2}{\xi}^1(x^1, x^3, x^4)\,\delta^\alpha_1 + \underset{2}{\xi}^2(x^3, x^4)\,\delta^\alpha_2 + \underset{2}{\xi}^3(x^3, x^4)\,\delta^\alpha_3.$$

In the transformed coordinates (31.21) we obtain

$$\underset{2}{\xi}^{1'} = \varphi^1_1\underset{2}{\xi}^1 + \varphi^1_3\underset{2}{\xi}^3, \quad \underset{2}{\xi}^{2'} = \underset{2}{\xi}^2 + \varphi^2_3\underset{2}{\xi}^3, \quad \underset{2}{\xi}^{3'} = \varphi^3_3\underset{2}{\xi}^3, \quad \underset{2}{\xi}^{4'} = 0.$$

If $\underset{2}{\xi}^3 \neq 0$, we define the functions φ^1, φ^2, φ^3 as integrals of the consistent set

$$\varphi^1_1\underset{2}{\xi}^1 + \varphi^1_3\underset{2}{\xi}^3 = 0; \quad \underset{2}{\xi}^2 + \varphi^2_3\underset{2}{\xi}^3; = 0, \quad \varphi^3_3\underset{2}{\xi}^3 = 1$$

and hence

$$X_1 = p_2, \quad X_2 = p_3 \tag{31.22}$$

—the case obtained in [260]. If, however, we take $\underset{2}{\xi}^3 = 0$, we obtain the cases excluded by Kruchkovich. The appropriate conditions are

$$\underset{2}{\xi}^{1'} = \varphi^1_1\underset{2}{\xi}^1, \quad \underset{2}{\xi}^{2'} = \underset{2}{\xi}^2, \quad \underset{2}{\xi}^{3'} = 0 \quad \underset{2}{\xi}^{4'} = 0,$$

where $\underset{2}{\xi}^2 \neq 0$, $\underset{2}{\xi}^1 \neq 0$ since a special operator is excluded. Suppose that

$\underset{2}{\xi}^2 = \underset{2}{\xi}^2(x^3, x^4)$; then φ^1 and φ^2 are defined by the conditions

$$\varphi_1^1 \underset{2}{\xi}^1 = 1; \qquad \varphi^3 = \underset{2}{\xi}^2(x^3, x^4)$$

and so

$$X_1 = p_2, \quad X_2 = p_1 + x^3 p_2. \tag{31.23}$$

If, however, $\underset{2}{\xi}^2 = \underset{2}{\xi}^2(x^4)$, we impose the conditions $\varphi_1^1 \underset{2}{\xi}^1 = 1$, $\varphi^4 = \underset{2}{\xi}^2$ so that

$$X_1 = p_2, \quad X_2 = p_1 + x^4 p_2. \tag{31.24}$$

We cannot have constant $\underset{2}{\xi}^2$ for otherwise $\varphi_1^1 \underset{2}{\xi}^1 = 1$ would imply operators $X_1 = p_2$, $X_2 = p_1 + cp_2$, equivalent to $X_1 = p_2$, $X_2 = p_1$, giving a special operator.

Finally, if we have a *non-Abelian G_2 without a special operator* ([260], p. 202), the metric (31.16) can be chosen so that

$$X_1 = p_2, \quad X_2 = p_1 + x^2 p_2 \tag{31.25}$$

or

$$X_1 = p_2, \quad X_2 = x^2 p_1 + p_3. \tag{31.26}$$

We now consider V_4 with G_3 acting on $\overset{*}{V}_3$. Suppose that we have an Abelian $G_3 : [X_i X_j] = 0$ $(i, j = 1, 2, 3)$ containing a G_2 with a special operator. Any operator of the group is a one-dimensional invariant subgroup of the group, and the transformation (31.17) can be chosen so that the operators $\{X_1, X_2\}$ for G_2 are (31.19). The structure equations for G_3 I show that

$$\underset{3}{\xi}^\alpha = \underset{3}{\xi}^1(x^3, x^4)\, \delta_1^\alpha + \underset{3}{\xi}^2(x^3, x^4)\, \delta_2^\alpha + \underset{3}{\xi}^3(x^3, x^4)\, \delta_3^\alpha.$$

Applying the transformation (31.20) we obtain

$$\underset{3}{\xi}^{1\prime} = \underset{3}{\xi}^1 + \varphi_3^1 \underset{3}{\xi}^3, \quad \underset{3}{\xi}^{2\prime} = \underset{3}{\xi}^2 + \varphi_3^2 \underset{3}{\xi}^3, \quad \underset{3}{\xi}^{3\prime} = \varphi_3^3 \underset{3}{\xi}^3, \quad \underset{3}{\xi}^{4\prime} = 0.$$

As $\underset{3}{\xi}^3 \neq 0$ (otherwise G_3 would act on a $\overset{*}{V}_2$), a suitable choice of φ^1, φ^2 and φ^3 reduces $\underset{3}{\xi}^{\alpha\prime}$ to $\delta_3^{\alpha\prime}$ by a non-singular transformation, so that $X_i = p_i$ $(i = 1, 2, 3)$ and $g_{\alpha\beta} = g_{\alpha\beta}(x^4)$. The general form of the allowed transformation is now

$$x^{1\prime} = x^1 + \varphi^1(x^4), \quad x^{2\prime} = x^2 + \varphi^2(x^4),$$
$$x^{3\prime} = x^3 + \varphi^3(x^4), \quad x^{4\prime} = \varphi^4(x^4), \quad \varphi_4^4 \neq 0.$$

It is easy to show that this transformation permits a choice of φ^α so that $g_{1'4'} = 1$, $g_{4'i'} = 0$ $(i = 2, 3, 4)$. Hence

(1) G_3 I on V_3

$$(g_{\alpha\beta}) = \begin{pmatrix} 0 & 0 & 0 & 1 \\ 0 & g_{22} & g_{23} & 0 \\ 0 & g_{23} & g_{33} & 0 \\ 1 & 0 & 0 & 0 \end{pmatrix}, \tag{31.27}$$

$$g_{\alpha\beta} = g_{\alpha\beta}(x^4), \quad X_i = p_i, \qquad i = 1, 2, 3.$$

We now assume that G_2 does not contain a special operator, and that the operators can be reduced to the form (31.22). The G_3 structure equations and (31.18) show that

$$\underset{3}{\xi}^\alpha = \underset{3}{\xi}^1(x^1, x^4)\,\delta_1^\alpha + \underset{3}{\xi}^2(x^4)\delta_2^\alpha + \underset{3}{\xi}^3(x^4)\,\delta_3^\alpha.$$

The allowed transformation for $\underset{1}{\xi}^\alpha$ and $\underset{2}{\xi}^\alpha$ is of form

$$x^{1'} = \varphi^1(x^1, x^4), \qquad x^{2'} = x^2 + \varphi^2(x^4),$$
$$x^{3'} = x^3 + \varphi^3(x^4), \quad x^{4'} = \varphi^4(x^4).$$

In the transformed coordinates $\underset{3}{\xi}^\alpha$ can be reduced to the form

$$\underset{3}{\xi}^{\alpha'} = \delta_1^{\alpha'} + \lambda(x^{4'})\,\delta_2^{\alpha'} + x^{4'}\,\delta_3^{\alpha'}.$$

We note that $\underset{3}{\xi}^\alpha$ = constant (for otherwise G_3 would admit a G_2 acting on an isotropic $\overset{*}{V}_2$, i.e. would admit a special operator). Using the Killing equations and the allowed transformation for all three group operators, we obtain the following metric:

(2) G_3 I on $\overset{*}{V}_3$

$$(g_{\alpha\beta}) = \begin{pmatrix} 0 & 0 & 0 & a_{14} \\ 0 & a_{22} & a_{23} & -(a_{22}\lambda + a_{23})x^1 \\ 0 & a_{23} & a_{23} & -(a_{23}\lambda + a_{33})x^1 \\ a_{14} & -(a_{22}\lambda + a_{23})x^1 & -(a_{23}\lambda + a_{33})x^1 & (a_{22}\lambda^2 + 2a_{23}\lambda + a_{33})x^{1^2} \end{pmatrix} \tag{31.28}$$

$$X_1 = p_2, \; X_2 = p_3, \; X_3 = p_1 + \sigma p_2 + x^4 p_3, \text{ where } \sigma = \int \lambda\, dx^4.$$

If, however, we assume operators (31.23), the structure equations and (31.18) show that $\underset{3}{\xi}^3 = 0$, i.e. a G_3 on $\overset{*}{V}_2$. For operators (31.24), we obtain

(3) G_3 I on $\overset{*}{V}_3$

$$(g_{\alpha\beta}) = \begin{pmatrix} 0 & 0 & 0 & a_{14} \\ 0 & a_{22} & a_{23} & -a_{22}x^1 \\ 0 & a_{23} & a_{33} & -a_{23}x^1 \\ a_{14} & -a_{22}x^1 & -a_{23}x^1 & a_{22}x^{1^2} \end{pmatrix}; \qquad (31.29)$$

$$X_1 = p_2, \; X_2 = p_1 + x^4 p_2, \; X_3 = p_3.$$

Kruchkovich did not consider class (3) of V_4 with G_3 I acting on $\overset{*}{V}_3$. The above scheme can be repeated for all the possible G_3 structures. The classification of G_3 acting on $\overset{*}{V}_3$ was obtained in [260] by this method, but certain cases corresponding to G_2 with operators (31.23) and (31.24) were omitted.

We thus obtain the following classification:

(1) G_3 II on $\overset{*}{V}_3$

$$(g_{\alpha\beta}) = \begin{pmatrix} 0 & 0 & 0 & 1 \\ 0 & g_{22} & g_{23} & -x^3 \\ 0 & g_{23} & g_{33} & 0 \\ 1 & -x^3 & 0 & 0 \end{pmatrix}, \qquad (31.30)$$

$$g_{\alpha\beta} = g_{\alpha\beta}(x^4); \quad X_1 = p_1, \quad X_2 = p_2, \quad X_3 = x^2 p_1 + p_3.$$

(2) G_3 II on $\overset{*}{V}_3$

$$\left.\begin{aligned} g_{11} &= g_{12} = g_{13} = 0, \quad g_{14} = a_{14}, \quad g_{22} = a_{22}, \\ g_{44} &= \varepsilon(\tfrac{1}{4}a_{22}x^{1^4} + a_{23}x^{1^3} + a_{33}x^{1^2} + 2a_{34}x^1), \\ g_{23} &= a_{22}x^1 + a_{23}, \quad g_{33} = a_{22}x^{1^2} + 2a_{23}x^1 + a_{33}, \\ g_{24} &= \varepsilon(\tfrac{1}{2}a_{22}x^{1^2} + a_{23}x^1), \\ g_{34} &= \varepsilon(\tfrac{1}{2}a_{22}x^{1^3} + \tfrac{3}{2}a_{23}x^{1^2} + a_{33}x^1) + a_{34}; \end{aligned}\right\} \qquad (31.31)$$

$$a_{ij} = a_{ij}(x^4);$$

$$X_1 = p_2, \quad X_2 = p_3, \quad X_3 = -p_1 + x^3 p_2 + \varepsilon x^4 p_3 \qquad (\varepsilon = 0 \text{ or } 1).$$

(1) G_3 III on $\overset{*}{V}_3$

$$(g_{\alpha\beta}) = \begin{pmatrix} 0 & 0 & 0 & 1 \\ 0 & a_{22}e^{-2x^3} & a_{23}e^{-x^3} & 0 \\ 0 & a_{23}e^{-x^3} & a_{33} & 0 \\ 1 & 0 & 0 & 0 \end{pmatrix}, \qquad (31.32)$$

$$a_{\alpha\beta} = a_{\alpha\beta}(x^4);$$

$$X_1 = p_1, \quad X_2 = p_2, \quad X_3 = x^2 p_2 + p_3.$$

(2) G_3 III on $\overset{*}{V}_3$

$$(g_{\alpha\beta}) = \begin{pmatrix} 0 & 0 & 0 & e^{-x^3} \\ 0 & a_{22} & a_{23} & 0 \\ 0 & a_{23} & a_{33} & 0 \\ e^{-x^3} & 0 & 0 & 0 \end{pmatrix}, \tag{31.33}$$

$$a_{\alpha\beta} = a_{\alpha\beta}(x^4);$$

$$X_1 = p_2, \quad X_2 = p_1, \quad X_3 = x^1 p_1 + p_3.$$

(3) G_3 III on $\overset{*}{V}_3$

$$(g_{\alpha\beta}) = \begin{pmatrix} 0 & 0 & 0 & a_{14} \\ 0 & a_{22} & a_{23}e^{-x^1} & -\varepsilon a_{22}x^1 \\ 0 & a_{23}e^{-x^1} & a_{33}e^{-2x^1} & -\varepsilon a_{23}x^1 e^{-x^1} \\ a_{14} & -\varepsilon a_{22}x^1 & -\varepsilon a_{23}x^1 e^{-x^1} & \varepsilon a_{22}x^{1^2} + a_{44} \end{pmatrix}; \tag{31.34}$$

$$X_1 = p_2, \quad X_2 = p_3, \quad X_3 = p_1 + \varepsilon x^4 p_2 + x^3 p_3.$$

(4) G_3 III on $\overset{*}{V}_3$

$$(g_{\alpha\beta}) = \begin{pmatrix} 0 & 0 & 0 & 1 \\ 0 & a_{22}x^{3^{-2}} & (a_{23} - a_{22}x^1)x^{3^{-2}} & a_{24}x^{3^{-1}} \\ 0, & (a_{23} - a_{22}x^1 x^{3^{-3}}, & (a_{22}x^{1^2} - 2a_{23}x^1 + a_{33})x^{3^{-2}}, & (a_{34} - a_{24}x^1)x^{3^{-1}} \\ 1 & a_{24}x^{3^{-2}} & (a_{34} - a_{24}x^1)x^{3^{-1}} & 0 \end{pmatrix}; \tag{31.35}$$

$$a_{\alpha\beta} = a_{\alpha\beta}(x^4); \quad X_1 = p_2, \quad X_2 = p_1 + x^3 p_2, \quad X_3 = x^2 p_2 + x^3 p_3.$$

This class of V_4 was omitted in [260].

The groups of types IV, V and VI can be combined into one with the single structure

$$[X_1 X_2] = 0, \quad [X_1 X_3] = X_1, \quad [X_2 X_3] = \varepsilon X_1 + q X_2,$$

where

$$\left.\begin{array}{ll} G_3 \text{ IV:} & \varepsilon = 1, \quad q = 1; \\ G_3 \text{ V:} & \varepsilon = 0, \quad q = 1; \\ G_3 \text{ VI:} & \varepsilon = 0, \quad q \neq 1. \end{array}\right\} \tag{31.36}$$

If there is a special operator, it must be X_1 in G_3 IV, can be taken as X_1 in G_3 V, and can be made to coincide with X_1 in G_3 VI by a suitable change in the basis of the group.

We obtain in this way the following canonical form of the metric for these structures:

(1) G_3 IV, V, VI on $\overset{\bullet}{V}_3$

$$(g_{\alpha\beta}) = \begin{pmatrix} 0 & 0 & 0 & e^{-x^3} \\ 0 & a_{22}e^{-2qx^3} & a_{23}e^{-qx^3} & -\varepsilon x^3 e^{-x^3} \\ 0 & a_{23}e^{-qx^3} & a_{33} & 0 \\ e^{-x^3} & -\varepsilon x^3 e^{-x^3} & 0 & 0 \end{pmatrix}; \qquad (31.37)$$

$$a_{\alpha\beta} = a_{\alpha\beta}(x^4);$$
$$X_1 = p_1, \quad X_2 = p_2, \quad X_3 = (x^1+\varepsilon x^2)p_1+qx^2p_2+p_3.$$

ε, q are defined by (31.36).

If, however, there is a special operator in the group, we obtain

(2) G_3 IV, V, VI on $\overset{\bullet}{V}_3$

$$\left.\begin{aligned} g_{11} = g_{12} = g_{13} = g_{44} = 0, \quad g_{14} = 1, \\ g_{22} = a_{22}e^{-2x^1}, \quad g_{24} = a_{24}e^{-x^1}; \end{aligned}\right\} \qquad (31.38)$$

$$\left.\begin{aligned} g_{23} &= \begin{cases} e^{-2x^1}(a_{23}-a_{22}x^1) & \text{IV} \quad (\varepsilon = q = 1), \\ a_{23}e^{-(q+1)x^1} & \text{V, VI} \quad (\varepsilon = 0); \end{cases} \\ g_{33} &= \begin{cases} e^{-2x^1}(a_{22}x^{1^2}-2a_{23}x^1+a_{33}) & \text{IV}, \\ e^{-2qx^1}a_{33} & \text{V, VI}; \end{cases} \\ g_{34} &= \begin{cases} e^{-x^1}(a_{34}-a_{24}x^1) & \text{IV}, \\ a_{34}e^{-qx^1} & \text{V, VI}; \end{cases} \end{aligned}\right\} \qquad (31.39)$$

$$a_{\alpha\beta} = a_{\alpha\beta}(x^4); \quad X_1 = p_2, \quad X_2 = p_3, \quad X_3 = p_1+(x^2+\varepsilon x^3)p_2+qx^3p_3.$$

Besides these V_4, there are classes of spaces corresponding to the operators (31.23) and (31.24); these can be determined by the same method.

G_3 IV on $\overset{\bullet}{V}_3$

$$(g_{\alpha\beta}) = \begin{pmatrix} 0 & 0 & 0 & e^{x^3} \\ 0 & a_{22}e^{2x^3} & -a_{22}x^1e^{2x^3}+a_{23}e^{x^3} & 0 \\ 0 & -a_{22}x^1e^{2x^3}+a_{23}e^{x^3} & a_{22}x^{1^2}e^{2x^3}-2a_{23}x^1e^{x^3}+a_{33} & a_{34} \\ e^{x^3} & 0 & a_{34} & 0 \end{pmatrix} \qquad (31.40)$$

$$a_{\alpha\beta} = a_{\alpha\beta}(x^4); \quad X_1 = p_2, \quad X_2 = p_1+x^3p_2, \quad X_3 = x^1p_1+x^2p_2-p_3.$$

G_3 V on $\overset{\bullet}{V}_3$

$$(g_{\alpha\beta}) = \begin{pmatrix} 0 & 0 & 0 & e^{x^3} \\ 0 & a_{22}e^{2x^3} & a_{33}e^{x^3} & -a_{22}x^1e^{2x^3} \\ 0 & a_{23}e^{x^3} & a_{33} & -a_{23}x^1e^{x^3}+a_{34} \\ e^{x^3} & -a_{22}x^1e^{2x^3} & -a_{23}x^1e^{x^3}+a_{34} & a_{22}x^{1^2}e^{2x^3} \end{pmatrix}; \qquad (31.41)$$

$$a_{\alpha\beta} = a_{\alpha\beta}(x^4); \quad X_1 = p_2, \quad X_2 = p_1+x^4p_2, \quad X_3 = x^1p_1+x^2p_2-p_3.$$

G_3 VI on $\overset{*}{V}_3$

$$\left.\begin{aligned}
&g_{11} = g_{12} = g_{13} = g_{24} = g_{44} = 0,\\
&g_{14} = (x^3)^{\frac{q}{q-1}}, \quad g_{22} = a_{22}(x^3)^{\frac{2}{q-1}},\\
&g_{23} = -a_{22}x^1(x^3)^{\frac{2}{q-1}} + a_{23}(x^3)^{\frac{2-q}{q-1}},\\
&g_{33} = -a_{22}x^{1^2}(x^3)^{\frac{2}{q-1}} - 2a_{23}x^1(x^3)^{\frac{2-q}{q-1}} + a_{33}(x^3)^{-2},\\
&g_{34} = a_{34}(x^3)^{-1};
\end{aligned}\right\} \qquad (31.42)$$

$$a_{\alpha\beta} = a_{\alpha\beta}(x^4), \qquad q \neq 1, 0;$$

$$X_1 = p_2, \quad X_2 = p_1 + x^3 p_2, \quad X_3 = qx^1 + x^2 p_2 + (1-q)x^3 p_3.$$

There can be no special operators in G_3 VII groups, since there are no one-dimensional invariant subgroups; the metric, obtained in [260], is of the form

G_3 VII on $\overset{*}{V}_3$

$$\left.\begin{aligned}
&g_{11} = g_{12} = g_{13} = g_{44} = 0, \quad g_{14} = 1,\\
&g_{22} = e^{-qx^1}[a_1 + (qa_2 + pa_3)\cos px^1 + (qa_3 - pa_2)\sin px^1],\\
&g_{23} = e^{-qx^1}\left[\frac{q}{2}a_1 + 2a_2\cos px^1 + 2a_3\sin px^1\right],\\
&g_{24} = e^{-(q/2)x^1}\left[a_4\cos\frac{px^1}{2} + a_5\sin\frac{px^1}{2}\right],\\
&g_{33} = e^{-qx^1}[a_1 + (qa_2 - pa_3)\cos px^1 + (qa_3 + pa_2)\sin px^1],\\
&g_{34} = e^{-(q/2)x^1}\left[\left(\frac{p}{2}a_5 - \frac{q}{2}a_4\right)\cos\frac{px^1}{2}\right.\\
&\qquad\qquad \left. -\left(\frac{p}{2}a_4 + \frac{q}{2}a_5\right)\sin\frac{px^1}{2}\right];
\end{aligned}\right\} \qquad (31.43)$$

$$a_\alpha = a_\alpha(x^4), \quad p = \sqrt{4-q^2}, \quad q^2 < 4;$$

$$X_1 = p_2, \quad X_2 = p_3, \quad X_3 = p_1 - x^3 p_2 + (x^2 + qx^3)p_3.$$

There is still one more possible class of V_4; the metric, obtained in [357], is of the form

G_3 VII on $\overset{*}{V}_3$

$$\left.\begin{aligned}
&g_{11} = g_{12} = g_{13} = g_{44} = 0, \quad g_{14} = P\sqrt{R},\\
&g_{22} = a_{22}P^2R^{-1},\\
&g_{23} = -a_{22}P^2R^{-1}x^1 + a_{23}PR^{-3/2},\\
&g_{24} = (a_{24}-x^3)PR^{-1/2},\\
&g_{33} = a_{22}P^2R^{-1}x^{1^2} - 2a_{23}PR^{-3/2}x^1 + a_{33}R^{-2},\\
&g_{34} = -(a_{24}-x^3)PR^{-1}x^1,
\end{aligned}\right\} \qquad (31.44)$$

$$P = e^{-pq\tan^{-1}(2x^3-q)p}, \quad R = x^{3^2} - qx^3 + 1,$$

$$q^2 < 4, \quad p = \frac{1}{\sqrt{4-q^2}}, \quad q = \text{const};$$

$$a_{ij} = a_{ij}(x^4);$$

$$X_1 = p_2, \quad X_2 = p_1 + x^3p_2,$$

$$X_3 = [x^2 + x^1(q-x^3)]p_1 + x^2x^3p_2 + (x^{3^2} - qx^3 + 1)p_3.$$

There are clearly no special operators for insoluble G_3 VIII and IX. G_3 VIII has a non-Abelian subgroup $G_2\{X_1, X_2\}$ which can be assigned operators (31.25) or (31.26). But operators (31.26) lead to a singular metric; the structure equations for (31.25) after some simplification show that

$$X_1 = p_2, \quad X_2 = p_3 + x^2p_2,$$

$$X_3 = -e^{x^3}p_1 + (\varepsilon e^{2x^3} + x^{2^2})p_2 + 2x^2p_3, \quad \varepsilon = 0, \quad \pm 1.$$

Depending on the value of ε, we obtain three classes of space:

(1) G_3 VIII on $\overset{*}{V}_3$

$$\left.\begin{aligned}
&g_{11} = g_{12} = g_{13} = 0, \quad g_{14} = 1, \quad g_{44} = 0,\\
&g_{22} = [\tfrac{1}{2}a_1 + a_2e^{4x^1} + a_3e^{-4x^1}]e^{-2x^3},\\
&g_{23} = [a_2e^{4x^1} - a_3e^{-4x^1}]e^{-x^3},\\
&g_{24} = [\tfrac{1}{2} + a_4e^{2x^1} + a_5e^{-2x^1}]e^{-x^3},\\
&g_{33} = -\tfrac{1}{2}a_1 + a_2e^{4x^1} + a_3e^{-4x^1},\\
&g_{34} = a_4e^{2x^1} - a_5e^{-2x^1}; \quad a_\alpha = a_\alpha(x^4);
\end{aligned}\right\} \qquad (31.45)$$

$$X_1 = p_2, \quad X_2 = x^2p_2 + p_3, \quad X_3 = -e^{x^3}p_1 + (e^{2x^3} + x^{2^2})p_2 + 2x^2p_3.$$

(2) G_3 VIII on $\dot{V}_3$

$$\left.\begin{aligned} g_{11} &= g_{12} = g_{13} = 0, \quad g_{14} = 1, \quad g_{44} = 0,\\ g_{22} &= [-\tfrac{1}{2}a_1+a_2\cos 4x^1+a_3\sin 4x^1]e^{-2x^3},\\ g_{23} &= [a_3\cos 4x^1-a_2\sin 4x^1]e^{-x^3},\\ g_{24} &= [-\tfrac{1}{2}+a_4\cos 2x^1+ a_5\sin 2x^1]e^{-x^3},\\ g_{33} &= [-\tfrac{1}{2}a_1-a_2\cos 4x^1-a_3\sin 4x^1],\\ g_{34} &= a_5\cos 2x^1-a_4\sin 2x^1; \quad a_\alpha = a_\alpha(x^4); \end{aligned}\right\} \tag{31.46}$$

$$X_1 = p_2, \quad X_2 = x^2p_2+p_3,$$
$$X_3 = -e^{x^3}p_1+(-e^{2x^3}+x^{2^2})p_2+2x^2p_3.$$

(3) G_3 VIII on $\dot{V}_3$

$$\left.\begin{aligned} g_{11} &= g_{12} = g_{13} = 0, \quad g_{14} = 1, \quad g_{44} = 0,\\ g_{22} &= (4a_1x^{1^2}+4a_2x^1+a_3)e^{-2x^3},\\ g_{23} &= (2a_1x^1+a_2)e^{-x^3},\\ g_{24} &= (-x^{1^2}+2a_4x^1+a_5)e^{-x^3},\\ g_{33} &= a_1, \quad g_{34} = -x^1+a^4; \quad a_\alpha = a_\alpha(x^4); \end{aligned}\right\} \tag{31.47}$$

$$X_1 = p_2, \quad X_2 = x^2p_2+p_3, \quad X_3 = -e^{x^3}p_1+x^{2^2}p_2+2x^2p_3.$$

We finally consider insoluble G_3 IX which also contains no special operators, so that X_1 can be reduced to $X_1 = p_2$. As G_3 IX is the only G_3 not containing a G_2 subgroup, the operators X_2 and X_3 must be found directly from the structure equations using allowed transformations. The metric is then determined from the Killing equations and reduced by means of allowed transformations ([260], p. 209) to the form

G_3 IX on $\overset{*}{V}_3$

$$\left.\begin{aligned} g_{11} &= g_{12} = g_{13} = g_{44} = 0, \quad g_{14} = 1,\\ g_{22} &= (a_1\sin 2x^1-a_2\cos 2x^1+a^3)\cos^2 x^3,\\ g_{23} &= -(a_1\cos 2x^1+a_2\sin 2x^1)\cos x^3,\\ g_{24} &= \sin x^3+(a_4\cos x^1+a_5\sin x^1)\cos x^3,\\ g_{33} &= a_2\cos 2x^1-a_1\sin 2x^1+a_3,\\ g_{34} &= a_4\sin x^1-a_5\cos x^1; \end{aligned}\right\} \tag{31.48}$$

$$a_i = a_i(x^4);$$
$$X_1 = p_2,$$
$$X_2 = \sec x^3\cos x^2\, p_1-\tan x^3\cos x^2p_2+\sin x^2p_3,$$
$$X_3 = -\sec x^3\sin x^2p_1+\tan x^1\sin x^2p_2+\cos x^2p_3.$$

This completes the classification of gravitational fields admitting a G_r group of motions with $r \leqslant 3$.

It should be noted that to obtain real gravitational fields, the conditions $|g_{\alpha\beta}| < 0$, must be imposed for this ensures that the metric can be reduced to the Minkowski form.

§ 32. Gravitational Fields Admitting a Simply-transitive or Intransitive G_4 Group of Motions

If the order of a complete group G_r acting on V_4 is 4, and the rank of the matrix $(\underset{s}{\xi}^\alpha)$ $(s = 1, 2, 3, 4)$ is less than 4, the G_4 must have V_3 or $\mathring{V}_3$ surfaces of transitivity since V_2 and $\mathring{V}_2$ cannot admit a group of order greater than 3.

The type of structure of any group we consider accords with the classification in § 10.

G_4 ACTING ON V_3

If we choose a system of coordinates in which the hypersurfaces of transitivity V_3 satisfy the equation $x^4 =$ constant, the condition $\underset{s}{\xi}^4 = 0$ $(s = 1, 2, 3, 4)$ holds for all the group operators. We first consider G_4 I groups; the structure is

$$[X_1X_2] = [X_1X_3] = 0, \quad [X_2X_3] = X_1, \quad [X_1X_4] = cX_1, \ [X_2X_4] = X_2,$$
$$[X_3X_4] = (c-1)X_3,$$

where c is a group parameter. This group contains a G_3 II subgroup acting on V_3, dealt with in § 31; hence the required spaces must be among those with metric (31.6) and G_3 operators of the form

$$X_1 = p_2, \quad X_2 = p_3, \quad X_3 = -p_1 + x^3 p_2.$$

If the operator $X_4 = \underset{4}{\xi}^\alpha p_\alpha$, the structure equations show that

$$\left.\begin{aligned} &\partial_2 \underset{4}{\xi}^\alpha = c\delta^\alpha_2, \quad \partial_3 \underset{4}{\xi}^\alpha = \delta^\alpha_3, \\ &-\partial_1 \underset{4}{\xi}^\alpha + cx^3\,\delta^\alpha_2 - \underset{4}{\xi}^3\,\delta^\alpha_2 = (c-1)(-\delta^\alpha_1 + x^3\,\delta^\alpha_2), \end{aligned}\right\} \tag{32.1}$$

so that the Killing vector $\underset{4}{\xi}^\alpha$ must be

$$\underset{4}{\xi}^\alpha = [(c-1)x^1 + \alpha]\,\delta^\alpha_1 + (cx^2 - \gamma x^1 + \beta)\,\delta^\alpha_2 + (x^3 + \gamma)\,\delta^\alpha_3, \tag{32.2}$$

where α, β and γ are functions of x^4. The Killing equations for $\underset{4}{\xi}^\alpha$ with metric (31.6) for the component g_{ij} $(i, j = 1, 2, 3)$ reduce to

$$g_{1i}\alpha' + g_{2i}(-x^1\gamma' + \beta') + g_{3i}\gamma' = 0 \qquad (i = 1, 2, 3).$$

As this set has a non-zero determinant, α, β and γ are constants. An allowed transformation (i.e. one which alters neither the operators X_1, X_2, X_3 nor metric (31.6)) is of the form

$$x^{1'} = x^1 + c_1, \quad x^{2'} = x^2 + e^{-x^1}c_2, \quad x^{3'} = x^3 + c_3, \quad x^{4'} = x^4 + c_4.$$

Applying this to the operator (32.2) gives the following expressions for the parameters α, β and γ in the new system of coordinates:

$$\overset{*}{\alpha} = (1-c)c_1 + \alpha, \quad \overset{*}{\beta} = -c_1c_3 - cc_2 + \gamma c_1 + \beta, \quad \overset{*}{\gamma} = c_3 + \gamma.$$

There are two possible cases: (a) $c \neq 1$, when the c_i can be chosen so that $\overset{*}{\alpha} = \overset{*}{\gamma} = 0$, $\overset{*}{\beta} \neq 0$; (b) $c = 1$, when $\overset{*}{\alpha} \neq 0$, and the c_i can be chosen so that $\overset{*}{\beta} = \overset{*}{\gamma} = 0$; the corresponding operators X_4 for the two cases are

$$(a)\ X_4 = (c-1)x^1p_1 + cx^2p_2 + x^3p_3,$$

$$(b)\ X_4 = \alpha p_1 + x^2p_2 + x^3p_3.$$

Integrating the Killing equations for the metric (31.6) and operator X_4 in case (b) leads to a contradiction, and so does case (a) unless $c = 0$. We thus obtain the metric

G_4 I $(c = 0)$

$$(g_{\alpha\beta}) = \begin{pmatrix} 0 & 0 & a_{13} & 0 \\ 0 & a_{22} & a_{22}x^1 & 0 \\ a_{13} & a_{22}x^1 & a_{22}x^{1^2} & 0 \\ 0 & 0 & 0 & e_4 \end{pmatrix}, \tag{32.3}$$

$$a_{ij} = a_{ij}(x^4), \quad e_4 = \pm 1;$$

$$X_1 = p_2, \quad X_2 = p_3, \quad X_3 = -p_1 + x^3p_2, \quad X_4 = -x^1p_1 + x^3p_3.$$

The remaining G_4 group structures can be investigated by the same method, and in this way we obtain the following spaces:

G_4 II—there are no gravitational fields admitting this intransitive G_4 group of motions.

G_4 III $(q = 0)$

$$(g_{\alpha\beta}) = \begin{pmatrix} a_{11} & 0 & 0 & 0 \\ 0 & a_{22} & a_{22}x^1 & 0 \\ 0 & a_{22}x^1 & a_{22}{x^1}^2+a_{11} & 0 \\ 0 & 0 & 0 & e_4 \end{pmatrix}, \tag{32.4}$$

$$a_{ij} = a_{ij}(x^4); \quad X_1 = p_2, \quad X_2 = p_3, = X_3 = -p_1+x^3p_2,$$
$$X_4 = -x^3p_1+\tfrac{1}{2}({x^3}^2-{x^1}^2)p_2+x^1p_3.$$

G_4 IV

$$(g_{\alpha\beta}) = \begin{pmatrix} a_{11} & 0 & 0 & 0 \\ 0 & 0 & a_{23}e^{x^1} & 0 \\ 0 & a_{23}e^{x^1} & 0 & 0 \\ 0 & 0 & 0 & e_4 \end{pmatrix} \tag{32.5}$$

$$e_4 = \pm 1, \quad a_{\alpha\beta} = a_{\alpha\beta}(x^4);$$
$$X_1 = p_2, \quad X_2 = p_3, \quad X_3 = -p_1+x^2p_2, \quad X_4 = -p_1+x^3p_3.$$

G_4 V

$$(g_{\alpha\beta}) = \begin{pmatrix} a_{11} & 0 & 0 & 0 \\ 0 & a_{22}e^{2x^1} & 0 & 0 \\ 0 & 0 & a_{22}e^{2x^1} & 0 \\ 0 & 0 & 0 & e_4 \end{pmatrix}, \tag{32.6}$$

$$e_4 = \pm 1, \; a_{\alpha\beta} = a_{\alpha\beta}(x^4); \quad X_1 = p_2, \quad X_2 = p_3,$$
$$X_3 = -p_1+x^2p_2+x^3p_3, \quad X_4 = -x^3p_2+x^2p_3.$$

G_4 VII

$$(g_{\alpha\beta}) = \begin{pmatrix} a_{11} & 0 & 0 & 0 \\ 0 & a_{22} & 0 & 0 \\ 0 & 0 & e_3a_{22}\sinh^2 x^2 & 0 \\ 0 & 0 & 0 & e_4 \end{pmatrix}, \tag{32.7}$$

$$X_1 = \cos x^3p_2+(\coth x^2 \sin x^3-1)p_3,$$
$$X_2 = \sin x^3p_2+\coth x^2 \cos x^3\, p_3,$$
$$X_3 = \cos x^3p_2-(\coth x^2 \sin x^3+1)\, p_3, \quad X_4 = p_1.$$

N.B. There is also a V_4 obtained by replacing all the trigonometrical functions in (32.7) by the corresponding hyperbolic functions.

G_4 VII

$$(g_{\alpha\beta}) = \begin{pmatrix} 0 & a_{12} & -a_{12}x^2 & 0 \\ a_{12} & a_{22} & -a_{22}x^2 & 0 \\ -a_{12}x^2 & -a_{22}x^2 & a_{22}x_2^2-a_{12}/2 & 0 \\ 0 & 0 & 0 & e_4 \end{pmatrix}, \tag{32.8}$$

$$X_1 = e^{-x^3}(p_1 - x^{2^2}p_2 - 2x^2p_3), \quad X_2 = p_3, \quad X_3 = e^{x^3}p_2, \quad X_4 = p_1.$$

G_4 VIII

$$(g_{\alpha\beta}) = \begin{pmatrix} a_{11} & 0 & 0 & 0 \\ 0 & a_{22} & 0 & 0 \\ 0 & 0 & a_{22}\sin^2 x^2 & 0 \\ 0 & 0 & 0 & e_4 \end{pmatrix}, \tag{32.9}$$

$$X_1 = p_3, \quad X_2 = \sin x^3 p_2 + \cot x^2 \cos x^3 p_3,$$
$$X_3 = \cos x^3 p_2 - \cot x^2 \sin x^3 p_3, \quad X_4 = p_1.$$

G_4 VIII

$$(g_{\alpha\beta}) = \begin{pmatrix} \alpha+\beta(1-\cos 2x^3) & -\beta \sin 2x^3 \sin x^1 & 0 & 0 \\ -\beta \sin 2x^3 \sin x^1 & \alpha + 2\beta \cos^2 x^3 \sin^2 x^1 & \alpha \cos x^1 & 0 \\ 0 & \alpha \cos x^1 & \alpha & 0 \\ 0 & 0 & 0 & e_4 \end{pmatrix}, \tag{32.10}$$

α and β are functions of x^4. The operators are

$$X_1 = p_2,$$
$$X_2 = \cos x^2 p_1 - \cot x^1 \sin x^2 p_2 + \frac{\sin x^2}{\sin x^1} p_3,$$
$$X_3 = -\sin x^2 p_1 - \cot x^1 \cos x^2 p_2 + \frac{\cos x^2}{\sin x^1} p_3,$$
$$X_4 = \sin x^3 p_1 - \frac{\cos x^3}{\sin x^1} p_2 + \cot x^1 \cos x^3 p_3.$$

For the soluble groups with a G_4 VI$_i$ structure ($i = 1, 2, 3, 4$) containing Abelian G_3 subgroups, there are the following possible spaces:

G_4 VI$_1$, VI$_2$
$$ds^2 = a_{11}\, dx^{1^2} + a_{22}(dx^{2^2} + \varepsilon\, dx^{3^2}) + e_4 dx^4, \tag{32.11}$$

where $a_{ij} = a_{ij}(x^4)$, $\varepsilon = \pm 1$, $e_4 = \pm 1$, and the operators are

$$X_1 = p_2,\ X_2 = p_3,\ X_3 = -p_1,\ X_4 = x^2p_3 \pm x^3p_2.$$

G_4 VI$_3$. There are no gravitational fields admitting this group.

G_4 VI$_4$
$$ds^2 = a_{12}(2dx^1\, dx^2 + dx^{3^2}) + a_{22}\, dx^{2^2} + e_4\, dx^{4^2},$$
$$a_{ij} = a_{ij}(x^4),\ X_1 = p_2,\ X_2 = p_3,\ X_3 = -p_1,\ X_4 = x^2p_3 - x^3p_1. \tag{32.12}$$

The case of G_4 acting on V_3 with the G_3 subgroup acting on $\overset{*}{V}_2$ must now be considered.

It was shown in § 30 that G_3 II is the only group acting on $\overset{*}{V}_2$ admitted by real gravitational fields. A system of coordinates can always be chosen so that (30.8) is transformed to

$$ds^2 = a_{11}(x^1, x^4)\, dx^{1^2} + a_{12}(x^1, x^4)\,(2dx^1\, dx^2 + dx^{3^2}) + e_4\, dx^{4^2},\ e_4 = \pm 1. \tag{32.13}$$

In accordance with our algorithm, we must consider those manifolds in (32.13) which admit groups G_4 I, II and III. We obtain the following metrics:

$G_4 I.$ $\quad c \neq 2$

$$ds^2 = a_{11}\left(\frac{dx^1}{x^1}\right)^2 + a_{12}(x^1)^{\frac{2}{c-2}}(2dx^1\, dx^2 + dx^{3^2}) + e_4\, dx^{4^2}, \tag{32.14}$$

where $\quad a_{ij}(x^4),\ e_4 = \pm 1,\ X_1 = p_2,\ X_2 = p_3,\ X_3 = x^3 p_2 - x^1 p_3,$

$$X_4 = (2-c)x^1 p_1 + c x^2 p_2 + x^3 p_3, \quad c = \text{const.}$$

G_4 II. $ds^2 = a_{11}\, dx^{1^2} + a_{12} e^{-2x^1}(2dx^1\, dx^2 + dx^{3^2}) + e_4\, dx^{4^2},\ a_{ij}(x^4)$ (32.15)

$$X_1 = p_2,\ X_2 = p_3,\ X_3 = x^3 p_2 - x^1 p_3,\ X_4 = p_1 + 2x^2 p_2 + x^3 p_3.$$

$$G_4 \text{ III. } ds^2 = a_{11}\left(\frac{dx^1}{x^{1^2} + q x^1 + 1}\right)^2 + a_{12}\frac{e^{2qp \tan^{-1}(2x^1+q)p}}{x^{1^2} + q x^1 + 1}$$

$$\times (2dx^1\, dx^2 + dx^{3^2}) + e_4\, dx^{4^2},\quad p = \frac{1}{\sqrt{4-q^2}},\quad a_{ij}(x^4). \tag{32.16}$$

The operators are

$$X_1 = p_2,\ X_2 = p_3,\ X_3 = x^3 p_2 - x^1 p_3,$$

$$X_4 = -(x^{1^2} + q x^1 + 1)p_1 + (q x^2 + \tfrac{1}{2} x^{3^2})p_2 - x^1 x^3 p_3.$$

We now investigate all intransitive G_4's acting on an isotropic three-dimensional manifold. These spaces were studied by Kruchkovich [260].

G_4 ACTING ON $\overset{*}{V}_3$

We consider the G_4 group of type I containing a G_3 II subgroup. The system of coordinates can be chosen so that the G_3 operators are of the form (31.25) or of other possible forms.

The structure equations show that in the case of (31.25)

$$\underset{4}{\xi}^1 = 2x^1 + A(x^3, x^4),\quad \underset{4}{\xi}^2 = x^2,\quad \underset{4}{\xi}^3 = c(x^3, x^4).$$

Applying the allowed transformation

$$x^{1'} = x^1+f(x^3, x^4),\ x^{2'} = x^2,\ x^{4'} = x^4,\ x^{3'} = \begin{cases} \displaystyle\int \frac{dx^3}{c}, & c \neq 0 \\ x^3, & c = 0 \end{cases}$$

reduces the operator X_4 to the form

$$X_4 = 2x^1p_1+x^2p_2+\varepsilon p_3, \quad \varepsilon = 0 \quad \text{or } 1.$$

But according to the Killing equations, $\varepsilon = 1$. Hence the allowed transformations which conserve all four operators reduce the metric to the form

G_4 I

$$ds^2 = \alpha(x^4)e^{-2x^3}(2\, dx^1\, dx^4+dx^{2^2})+\beta(x^4)dx^{2^2}. \tag{32.17}$$

In the same way, we obtain for (31.26),

G_4 I

$$ds^2 = 2dx^1\, dx^4-2\alpha(x^4)\, dx^2\, dx^3-2x^3\, dx^2\, dx^4; \tag{32.18}$$

$$X_1 = p_1, \quad X_2 = p_2, \quad X_3 = x^2p_1+p_3, \quad X_4 = x^2p_2-x^3p_3. \tag{32.19}$$

The analysis for the other types is similar to that given above, and a summary of the results is therefore sufficient.

G_4 II cannot be an intransitive group of motions with isotropic hypersurfaces of transitivity.

G_4 III

$$\left.\begin{aligned} ds^2 &= 2dx^1\, dx^4+\alpha(x^4)(dx^{2^2}+dx^{3^2})-2x^3\, dx^2\, dx^3, \\ X_1 &= p_1, \quad X_2 = p_2, \quad X_3 = x^2p_1+p_3, \\ X_4 &= \tfrac{1}{2}(x^{2^2}-x^{3^2})p_1-x^3p_2+x^2p_3. \end{aligned}\right\} \tag{32.20}$$

G_4 IV

$$\left.\begin{aligned} ds^2 &= 2dx^1\, dx^4+2\alpha(x^4)e^{-x^1}\, dx^2\, dx^3, \\ X_1 = p_2, \quad X_2 &= p_3, \quad X_3 = p_1+x^3p_3, \quad X_4 = p_1+x^2p_2. \end{aligned}\right\} \tag{32.21}$$

G_4 V

$$\left.\begin{aligned} ds^2 &= 2dx^1\, dx^4+\alpha(x^4)e^{-x^1}\, (dx^{2^2}+dx^{3^2}), \\ X_1 &= p_2, \quad X_2 = p_3, \quad X_3 = -x^3p_2+x^2p_3, \\ X_4 &= 2p_1+x^2p_2+x^3p_3. \end{aligned}\right\} \tag{32.22}$$

G_4 VII

$$\left.\begin{aligned} ds^2 &= 2dx^1\, dx^4+\alpha(x^4)(e^{-2x^3}\, dx^{2^2}-e_2\, dx^{3^2}), \\ e_2 &= \pm 1; \quad X_1 = p_2, \quad X_2 = x^2p_2+p_3, \\ X_3 &= (x^{2^2}+e_2e^{2x^3})p_2+2x^2p_3, \quad X_4 = p_1. \end{aligned}\right\} \tag{32.23}$$

G_4 VII

$$\left.\begin{aligned} ds^2 &= 2dx^1\,dx^4+\alpha(x^4)(e^{-2x^3}\,dx^{2^2}-e_2\,dx^{3^2}) \\ &\qquad\qquad +e_2e^{-x^3}\,dx^2\,dx^4, \\ e_2 &= \pm 1;\quad X_1 = p_2,\quad X_2 = x^2p_2+p_3, \\ X_3 &= -e^{x^2}p_1+(x^{2^2}+e_2e^{2x^3})p_2+2x^3p_3,\quad X_4 = p_1. \end{aligned}\right\} \tag{32.24}$$

G_4 VIII

$$\left.\begin{aligned} ds^2 &= 2dx^1\,dx^4+\alpha(x^4)\,(\cos^2 x^3\,dx^{2^2}+dx^{3^2}) \\ &\qquad +2\varepsilon \sin x^3\,dx^2\,dx^4, \\ \varepsilon &= 0 \quad\text{or}\quad 1;\quad X_1 = p_2, \\ X_2 &= \varepsilon \sec x^3 \cos x^2p_1-\tan x^3 \cos x^2p_2+\sin x^2p_3, \\ X_4 = p_1,\quad X_3 &= -\varepsilon \sec x^3 \sin x^2\,p_1 \\ &\qquad +\tan x^3 \sin x^2\,p_2+\cos x^2\,p_3. \end{aligned}\right\} \tag{32.25}$$

We now consider the case when V_4 admits an intransitive G_4 acting on $\mathring{V}_3$ and contains an Abelian G_3 subgroup. All these groups are combined in structures of the type G_3 IV$_i$ ($i = 1, 2, 3, 4$). Analysing these types by the above method results in the following two possibilities:

G_4 VI$_3$†

$$\left.\begin{aligned} ds^2 &= 2dx^1\,dx^4+\alpha(x^4)dx^{2^2} \\ &\qquad +2\beta(x^4)\,dx^2\,dx^3+\gamma(x^4)\,dx^{3^2}, \\ X_1 = p_1,\quad X_2 &= p_2,\quad X_3 = p_3,\quad X_4 = x^2p_1+\omega p_2+\lambda p_3, \end{aligned}\right\} \tag{32.26}$$

where $\quad \omega(x^4) = -\int \frac{\gamma\,dx^4}{\sigma},\quad \lambda(x^4) = \int \frac{\beta\,dx^4}{\sigma},\quad \sigma = \alpha\gamma-\beta^2.$

G_4 VI$_2$

$$\left.\begin{aligned} ds^2 &= 2dx^1\,dx^4+\alpha(x^4)(dx^{2^2}+e\,dx^{3^2}),\quad e = \pm 1, \\ X_1 &= p_1,\quad X_2 = p_2,\quad X_3 = p_3,\quad X_4 = ex^3p_2-x^2p_3. \end{aligned}\right\} \tag{32.27}$$

As the two group structures are dissimilar, the spaces defined by these two metrics are essentially different.

Simply-transitive G_4 Groups of Motion

Two simply-transitive groups of identical structure and the same number of variables are *similar* ([147], p. 78); hence if a V_4 space admits a simply-transitive G_4 group of motions, any four operators can be taken provided they satisfy the structure equations.

† The metric (32.26) also admits another operator and hence a G_5 group: cf. (33.1).

It is thus immaterial whether we define the fourth operator starting from a G_3 subgroup acting on V_3 or from a G_3 acting on $\mathring{V}_3$.

We adopt the first approach, assuming moreover a semi-geodesic system of coordinates in V_4 (§ 7) in which

$$ds^2 = e_4\, dx^{4^2} + g_{ij}\, dx^i\, dx^j \qquad (i, j \neq 4), \quad e_4 = \pm 1.$$

As $\underset{4}{\xi}^4 \neq 0$ for a simply-transitive group, the Killing equations show that $\partial_4 \underset{4}{\xi}^4 = 0$. Using these relations and the operators of a subgroup G_3 acting on V_3 (§ 31), we find the fourth operator from the structure equations (10.17) and (10.19); some integration constants can be reduced to zero or ± 1 by means of allowed transformations.

Let us consider a G_4 I group containing a G_3 II subgroup.

The structure equations lead to the set (32.1) but α takes now values 1 to 4:

$$\underset{4}{\xi}^\alpha = \underset{4}{\xi}^\alpha(x^1, x^2, x^3, x^4), \quad \partial_4 \underset{4}{\xi}^4 = 0, \quad \underset{4}{\xi}^4 \neq 0.$$

Integrating this set shows that

$$\left.\begin{aligned} \underset{4}{\xi}^1 &= (c-1)x^1 + A(x^4), \\ \underset{4}{\xi}^2 &= cx^2 - C(x^4)x^1 + B(x^4), \\ \underset{4}{\xi}^3 = x^3 + C(x^4), \quad \underset{4}{\xi}^4 &= D \neq 0, \quad D = \text{const.} \end{aligned}\right\} \tag{32.28}$$

The Killing equations show that

$$\left.\begin{aligned} g_{11}\partial_4 \underset{4}{\xi}^1 + g_{12}\partial_4 \underset{4}{\xi}^2 + g_{13}\partial_4 \underset{4}{\xi}^3 &= 0, \\ g_{12}\partial_4 \underset{4}{\xi}^1 + g_{22}\partial_4 \underset{4}{\xi}^2 + g_{23}\partial_4 \underset{4}{\xi}^3 &= 0, \\ g_{13}\partial_4 \underset{4}{\xi}^1 + g_{23}\partial_4 \underset{4}{\xi}^2 + g_{33}\partial_4 \underset{4}{\xi}^3 &= 0. \end{aligned}\right\} \tag{32.29}$$

The determinant $|g_{ij}| \neq 0$ $(i, j = 1, 2, 3)$ for otherwise V_3 (x^1, x^2, x^3) would be an isotropic manifold; therefore

$$\partial_4 \underset{4}{\xi}^i = 0 \qquad (i = 1, 2, 3),$$

i.e. A, B and C are constant.

The three G_3 II operators and the space V_4 admitting this group of motions, remain invariant under the transformations

$$\left.\begin{aligned} &x^{1'} = x^1 + \alpha, \quad x^{2'} = x^2 - \gamma x^1 + \beta, \\ &x^{3'} = x^3 + \gamma, \quad x^{4'} = \varepsilon x^4 + \delta, \\ &\alpha, \beta, \gamma, \delta = \text{const}, \quad \varepsilon = \pm 1. \end{aligned}\right\} \tag{32.30}$$

Applying these reduces the fourth operator to the form

$$X_4 = (c-1)x^1p_1+cx^2p_2+x^3p_3+Dp_4, \tag{32.31}$$

if $c \neq 1$, and to the form

$$X_4 = Ap_1+x^2p_2+x^3p_3+Dp_4; \quad A, D = \text{const}, \tag{32.32}$$

if $c = 1$.

Applying the transformation $x^{1'} = x^1$, $x^{2'} = x^2$, $x^{3'} = x^3$, $x^4 = x^4/D$ to the operator (32.31) replaces D by unity. This transformation also affects the metric of the V_4 under consideration; we obtain

$$g_{4'4'} = D^2g_{44} = e_4D^2 = \pm D^2. \tag{32.33}$$

The remaining components $g_{\alpha\beta}$ (α, $\beta \neq 4$) do not change; the metric matrix is therefore

G_4 I, $c \neq 1$

$$(g_{\alpha\beta}) = \begin{pmatrix} k_{11}e^{-2(c-1)x^4} & k_{12}e^{-(2c-1)x^4} & k_{12}x^1e^{-(2c-1)x^4}+k_{13}e^{-cx^4} & 0 \\ k_{12}e^{-(2c-1)x^4} & k_{22}e^{-2cx^4} & k_{22}x^1e^{-2cx^4}+k_{23}e^{-(c+1)x^4} & 0 \\ k_{12}x^1e^{-(2c-1)x^4} & k_{22}x^1e^{-2cx^4} & k_{22}x^{1^2}e^{-2cx^4} & \\ \quad +k_{13}e^{-cx^4} & \quad +k_{23}e^{-(c+1)x^4} & \quad +2k_{23}x^1e^{-(c+1)x^4} & \\ & & \quad +k_{33}e^{-2x^4} & 0 \\ 0 & 0 & 0 & \pm D^2 \end{pmatrix}, \tag{32.34}$$

where D and k_{ij} are constants;

$$X_1 = p_2, \quad X_2 = p_3, \quad X_3 = -p_1+x^3p_2,$$
$$X_4 = (c-1)x^1p_1+cx^2p_2+x^3p_3+p_4.$$

A similar investigation for the operators (32.32) leads to the metric

G_4 I, $c = 1$

$$\left.\begin{aligned} & g_{11} = k_{11}, \quad g_{12} = k_{12}e^{-x^4}, \\ & g_{13} = e^{-x^4}[k_{12}(x^1-ax^4)+k_{13}], \\ & g_{23} = e^{-2x^4}[k_{22}(x^1-\alpha x^4)+k_{23}], \quad g_{22} = k_{22}e^{-2x^4}, \\ & g_{33} = e^{-2x^4}[k_{22}(x^1-\alpha x^4)^2+2k_{23}(x^1-\alpha x^4)+k_{33}], \\ & g_{i4} = 0 \quad (i = 1, 2, 3), \quad g_{44} = \pm D^2; \\ & k_{ij}, \ \alpha, \ D = \text{const}; \end{aligned}\right\} \tag{32.35}$$

$$X_1 = p_2, \quad X_2 = p_3, \quad X_3 = -p_1+x^3p_2,$$
$$X_4 = \alpha p_1+x^2p_2+x^3p_3+p_4.$$

Groups G_4 II and G_4 III also contain a G_3 II subgroup. Repeating the analysis leads to the following spaces:

G_4 II

$$\left.\begin{aligned}
g_{11} &= e^{-2x^4}(k_{11}+2k_{13}x^4+k_{33}x^{4^2}),\\
g_{12} &= e^{-3x^4}(k_{12}+k_{23}x^4),\\
g_{13} &= e^{-3x^4}x^1(k_{12}+k_{23}x^4)+e^{-2x^4}(k_{13}+k_{33}x^4),\\
g_{22} &= k_{22}e^{-4x^4},\quad g_{23} = x^1e^{-4x^4}k_{22}+k_{23}e^{-3x^4},\\
g_{33} &= x^{1^2}k_{22}e^{-4x^4}+2x^1k_{23}e^{-3x^4}+k_{33}e^{-2x^4},\\
g_{i4} &= 0\ (i = 1, 2, 3),\quad g_{44} = \pm D^2;\quad k_{ij},\ D = \text{const};
\end{aligned}\right\}\qquad(32.36)$$

$$X_1 = p_2,\quad X_2 = p_3,\quad X_3 = -p_1+x^3p_2,$$
$$X_4 = x^1p_1+\tfrac{1}{2}(4x^2+x^{1^2})p_2+(x^3-x^1)p_3+p_4.$$

G_4 III

$$\left.\begin{aligned}
g_{11} &= e^{-qx^4}[c_1+c_2(q\cos\omega x^4+\omega\sin\omega x^4)\\
&\quad +c_3(q\sin\omega x^4-\omega\cos\omega x^4)],\\
g_{12} &= e^{-(3/2)qx^4}\left[-c_4\sin\frac{\omega}{2}x^4+c_5\cos\frac{\omega}{2}x^4\right]\frac{\omega}{2}-\frac{1}{2}qP,\\
g_{13} &= x^1g_{12}+e^{-qx^4}\left(-\frac{q}{2}c_1-2c_2\cos\omega x^4-2c_3\sin\omega x^4\right),\\
g_{22} &= c_6e^{-2qx^4},\quad g_{23} = x^1c_6e^{-2qx^4}+P,\\
g_{33} &= x^{1^2}c_6e^{-2qx^4}+2Px^1+e^{-qx^4}[c_1+c_2(q\cos\omega x^4\\
&\quad -\omega\sin\omega x^4)+c_3\,(q\sin\omega x^4+\omega\cos\omega x^4)],\\
g_{i4} &= 0\ \ (i = 1, 2, 3),\quad g_{44} = \pm D^2;\\
&\qquad D,\ c_\nu(\nu = 1, \ldots, 6) = \text{const},
\end{aligned}\right\}\qquad(32.37)$$

where $\omega = \sqrt{4-q^2}$, $q^2 < 4$, q is a structure constant, and

$$P = e^{-\frac{3}{2}qx^4}\left(c_4\cos\frac{\omega}{2}x^4+c_5\sin\frac{\omega}{2}x^4\right);$$

$$X_1 = p_2,\quad X_2 = p_3,\quad X_3 = -p_1+x^3p_2,$$
$$X_4 = (qx^1-x^3)p_1+[qx^2+\tfrac{1}{2}(x^{3^2}-x^{1^2})]p_2+x^1p_3+p_4.$$

G_4 IV contains a G_3 III subgroup; as with G_4 I, two cases can be distinguished:

(1) G_4 IV, $\alpha \neq 0$

$$\left.\begin{aligned}&g_{11} = k_{11}, \quad g_{12} = k_{12}e^{x^1-\alpha x^4}, \quad g_{13} = k_{13}e^{-x^4},\\&g_{22} = k_{22}e^{2x^1-2\alpha x^4}, \quad g_{23} = k_{23}e^{-(1+\alpha)x^4+x^1},\\&g_{33} = k_{33}e^{-2x^4}, \quad g_{i4} = 0 \quad (i = 1, 2, 3), \; g_{44} = \pm D^2,\\&D, \alpha, k_{ij} = \text{const};\end{aligned}\right\} \qquad (32.38)$$

$$X_1 = p_3, \quad X_2 = p_2, \quad X_3 = -p_1+x^2p_2,$$
$$X_4 = \alpha p_1+x^3p_3+p_4.$$

(2) G_4 IV

$$\left.\begin{aligned}&g_{11} = k_{11}+2k_{12}x^4+k_{22}x^{4^2},\\&g_{12} = e^{x^1}(k_{12}+k_{22}x^4), \quad g_{22} = k_{22}e^{2x^1},\\&g_{13} = e^{-x^4}(k_{13}+k_{23}x^4), \quad g_{23} = k_{23}e^{x^1-x^4},\\&g_{33} = k_{33}e^{-2x^4},\\&g_{i4} = 0 \; (i = 1, 2, 3), \quad g_{44} = \pm D^2; \quad D, k_{ij} = \text{const};\end{aligned}\right\} \qquad (32.39)$$

$$X_1 = p_3, \quad X_2 = p_2, \quad X_3 = -p_1+x^2p_2.$$
$$X_4 = e^{-x^1}p_2+x^3p_3+p_4.$$

However, these two groups are similar since they have the same structure.

For G_4 V, we start from a G_3 V subgroup; a similar method leads to the following metric:

G_4 *V*

$$\left.\begin{aligned}&g_{11} = k_{11}, \quad g_{12} = e^{x^1-\alpha x^4}(k_{12}\cos x^4+k_{13}\sin x^4),\\&g_{13} = e^{x^1-\alpha x^4}(k_{12}\sin x^4-k_{13}\cos x^4),\\&g_{22} = e^{2x^1-2\alpha x^4}(k_{33}-k_{22}\sin 2x^4+k_{23}\cos 2x^4),\\&g_{23} = e^{2x^1-2\alpha x^4}(k_{22}\cos 2x^4+k_{23}\sin 2x^4),\\&g_{33} = e^{2x^1-2\alpha x^4}(k_{33}+k_{22}\sin 2x^4-k_{23}\cos 2x^4),\\&g_{i4} = 0 \quad (i = 1, 2, 3), \quad g_{44} = \pm D^2, \quad k_{ij}, \quad D = \text{const};\end{aligned}\right\} \qquad (32.40)$$

$$X_1 = p_2, \quad X_2 = p_3, \quad X_3 = -p_1+x^2p_2+x^3p_3,$$
$$X_4 = \alpha p_1-x^3p_2+x^2p_3+p_4, \quad \alpha = \text{const}.$$

The analysis is simpler for G_4 containing an Abelian G_3 subgroup, as the first three operators can always be taken as

$$X_1 = p_2, \quad X_2 = p_3, \quad X_3 = -p_1,$$

and $X_4 = \underset{4}{\xi}^{\alpha}\, \partial_{\alpha}$; we obtain

$$\partial_{\beta}\underset{4}{\xi}^4 = 0 \qquad (\beta = 1, 2, 3, 4).$$

The classification of these V_4 leads to the spaces

G_4 VI$_1$

$$\left.\begin{aligned} &g_{11} = k_{11}e^{-2lx^4}, \quad g_{12} = k_{12}e^{-(l+\varepsilon)x^4}, \\ &g_{13} = k_{13}e^{-(k+l)x^4}, \\ &g_{22} = k_{22}e^{-2\varepsilon x^4}, \quad g_{23} = k_{23}e^{-(k+\varepsilon)x^4}, \quad g_{33} = k_{33}e^{-2kx^4}, \\ &g_{i4} = 0 \quad (i = 1, 2, 3), \quad g_{44} = \pm D^2, \quad k_{ij}, D = \text{const}, \end{aligned}\right\} \tag{32.41}$$

where k and l are structure constants, $\varepsilon = 0$ or 1:

$$X_1 = p_2, \quad X_2 = p_3, \quad X_3 = -p_1,$$
$$X_4 = lx^1p_1 + \varepsilon x^2p_2 + kx^3p_3 + p_4.$$

G_4 VI$_2$

$$\left.\begin{aligned} &g_{11} = k_{11}e^{-2lx^4}, \\ &g_{12} = e^{-(k+l)x^4}(k_{13}\cos x^4 - k_{12}\sin x^4), \\ &g_{13} = e^{-(k+l)x^4}(k_{12}\cos x^4 + k_{13}\sin x^4), \\ &g_{22} = e^{-2kx^4}(k_{22} - k_{23}\cos 2x^4 - k_{33}\sin 2x^4), \\ &g_{23} = e^{-2kx^4}(-k_{23}\sin 2x^4 + k_{33}\cos 2x^4), \\ &g_{33} = e^{-2kx^4}(k_{22} + k_{23}\cos 2x^4 + k_{33}\sin 2x^4), \\ &g_{i4} = 0 \quad (i = 1, 2, 3), \quad g_{44} = \pm D^2; \quad k_{ij}, D = \text{const}, \end{aligned}\right\} \tag{32.42}$$

where k and l are structure constants;

$$X_1 = p_2, \quad X_2 = p_3, \quad X_3 = -p_1,$$
$$X_4 = lx^1p_1 + (kx^2 - x^3)p_2 + (kx^3 + x^2)p_3 + p_4.$$

G_4 VI$_3$, $\varepsilon = 0$ or 1, k is any real number

$$\left.\begin{aligned} &g_{11} = k_{11}e^{-2\varepsilon x^4}, \quad g_{12} = (k_{12} - k_{13}x^4)e^{-(\varepsilon+k)x^4}, \\ &g_{13} = k_{13}e^{-(\varepsilon+k)x^4}, \\ &g_{22} = (k_{22} - 2k_{23}x^4 + k_{33}x^{4^2})e^{-2kx^4}, \\ &g_{23} = (k_{23} - k_{33}x^4)e^{-2kx^4}, \quad g_{33} = k_{33}e^{-2kx^4}, \\ &g_{i4} = 0 \quad (i = 1, 2, 3), \quad g_{44} = \pm D^2; \quad k_{ij}, D = \text{const}; \end{aligned}\right\} \tag{32.43}$$

$$X_1 = p_2, \quad X_2 = p_3, \quad X_3 = -p_1,$$
$$X_4 = \varepsilon x^1p_1 + kx^2p_2 + (kx^3 + x^2)p_3 + p_4.$$

For G_4 VI_4, two cases can be distinguished:

(1) G_4 VI_4, $k \neq \varepsilon$ $(\varepsilon = 0 \text{ or } 1)$

$$\left.\begin{aligned}
g_{11} &= k_{11}e^{-2\varepsilon x^4}, \quad g_{13} = k_{13}e^{-(k+\varepsilon)x^4} + \frac{k_{11}}{(k-\varepsilon)}\, e^{-2\varepsilon x^4}, \\
g_{12} &= (k_{12} - k_{13}x^4)e^{-(k+\varepsilon)x^4} - \frac{k_{11}}{(k-\varepsilon)^2}\, e^{-2\varepsilon x^4}, \\
g_{22} &= \frac{2(k_{13}x^4 - k_{12})}{(k-\varepsilon)^2}\, e^{-(k+\varepsilon)x^4} + \frac{k_{11}}{(k-\varepsilon)^4}\, e^{-2\varepsilon x^4} \\
&\quad + (k_{23} - k_{33}x^4)e^{-2kx^4}, \\
g_{23} &= \left(\frac{k_{12}}{k-\varepsilon} - \frac{k_{13}}{(k-\varepsilon)^2} - \frac{k_{13}x^4}{k-\varepsilon}\right)e^{-(k+\varepsilon)x^4} \\
&\quad - \frac{k_{11}}{(k-\varepsilon)^3}\, e^{-2\varepsilon x^4} + (k_{23} - k_{33}x^4)e^{-2kx^4}, \\
g_{33} &= k_{33}e^{-2kx^4} + \frac{k_{11}}{(k-\varepsilon)^2}\, e^{-2\varepsilon x^4} + \frac{2k_{13}}{k-\varepsilon}\, e^{-(k+\varepsilon)x^4}, \\
g_{i4} &= 0 \quad (i = 1, 2, 3), \quad g_{44} = \pm D_2; \quad k_{ij}, D = \text{const};
\end{aligned}\right\} \qquad (32.44)$$

$$X_1 = p_2, \quad X_2 = p_3, \quad X_3 = -p_1,$$
$$X_4 = (\varepsilon x^1 - x^3)p_1 + kx^2p_2 + (kx^3 + x^2)p_3 + p_4.$$

(2) G_4 VI_4, $k = \varepsilon = 0$ or 1

$$\left.\begin{aligned}
g_{11} &= k_{11}e^{-2\varepsilon x^4}, \\
g_{12} &= (k_{12} - k_{13}x^4 - \tfrac{1}{2}k_{11}x^{4^2})e^{-2\varepsilon x^4}, \\
g_{13} &= (k_{13} - k_{11}x^4)e^{-2\varepsilon x^4}, \\
g_{22} &= [k_{22} - 2k_{23}x^4 + (k_{33} - k_{12})x^{4^2} \\
&\qquad + k_{13}x^{4^3} + \tfrac{1}{4}k_{11}x^{4^4}]e^{-2\varepsilon x^4}, \\
g_{23} &= [k_{23} + (k_{12} - k_{33})x^4 - \tfrac{3}{2}k_{13}x^{4^2} - \tfrac{1}{2}k_{11}x^{4^3}]e^{-2\varepsilon x^4}, \\
g_{33} &= (k_{33} + 2k_{13}x^4 + k_{11}x^{4^2})e^{-2\varepsilon x^4}, \\
g_{i4} &= 0 \quad (i = 1, 2, 3), \quad g_{44} = \pm D^2; \quad k_{ij}, D = \text{const};
\end{aligned}\right\} \qquad (32.45)$$

$$X_1 = p_2, \quad X_2 = p_3, \quad X_3 = -p_1,$$
$$X_4 = (\varepsilon x^1 - x^3)p_1 + \varepsilon x^2p_2 + (\varepsilon x^3 + x^2)p_3 + p_4.$$

Finally, the analysis for insoluble G_4 VII, VIII containing insoluble G_3 VIII and IX respectively, leads to the two spaces

G_4 VII

$$\left.\begin{aligned}
&g_{11} = b_{11}, \quad g_{12} = P,\\
&g_{13} = -Px^2+b_{11}x^1+b_{13}, \quad g_{22} = Q,\\
&g_{23} = -Qx^2+R,\\
&g_{33} = Qx^{2^2}-2Rx^2+P,\\
&g_{i4} = 0 \quad (i = 1, 2, 3), \quad g_{44} = \pm D^2,\\
&\quad P = b_{11}x^{1^2}+2b_{13}x^1+b_{33},\\
&\quad Q = b_{11}x^{1^4}+4b_{13}x^{1^3}+2(b_{12}+2b_{33})x^{1^2}+4b_{23}x^1+b_{22},\\
&\quad R = b_{11}x^{1^3}+3b_{13}x^{1^2}+(b_{12}+2b_{33})x^1+b_{23}.
\end{aligned}\right\} \tag{32.46}$$

where

$$b_{11} = k_{11}, \quad b_{12} = k_{11}x^{4^2}-2\varepsilon k_{13}x^4+k_{12}, \quad b_{13} = -\varepsilon k_{11}x^4+k_{13},$$

$$b_{22} = k_{11}x^{4^4}-4\varepsilon k_{12}x^{4^3}+2(k_{12}+2k_{33})x^{4^2}-4\varepsilon k_{23}x^4+k_{22},$$

$$b_{23} = -\varepsilon k_{11}x^{4^3}+3k_{13}x^{4^2}-\varepsilon(k_{12}+2k_{33})x^4+k_{23},$$

$$b_{33} = k_{11}x^{4^2}-2\varepsilon k_{13}x^4+k_{33}; \quad k_{ij}, D = \text{const}, \quad \varepsilon = \pm 1;$$

$$X_1 = e^{-x^3}(p_1-x^{2^2}p_2-2x^2p_3), \quad X_2 = p_3,$$

$$X_3 = e^{x^3}p_2, \quad X_4 = p_1+\varepsilon p_4.$$

G_4 VIII

$$\left.\begin{aligned}
&g_{11} = \tfrac{1}{2}k_{11}-k_{12}\sin 2x^3-k_{22}\cos 2x^3,\\
&g_{13} = k_{13}\cos x^3-k_{23}\sin x^3,\\
&g_{12} = \cos x^1\,(k_{13}\cos x^3-k_{23}\sin x^3)\\
&\qquad\qquad + \sin x^1\,(k_{12}\cos 2x^3-k_{22}\sin 2x^3),\\
&g_{22} = k_{33}\cos^2 x^1+\sin 2x^1\,(k_{23}\cos x^3+k_{13}\sin x^3)\\
&\qquad\qquad +\sin^2 x^1\,(k_{22}\sin 2x^3+k_{22}\cos 2x^3+\tfrac{1}{2}k_{11}),\\
&g_{23} = k_{33}\cos x^1+\sin x^1\,(k_{23}\cos x^3+k_{13}\sin x^3),\\
&g_{33} = k_{33},\\
&g_{i4} = 0 \quad (i = 1, 2, 3), \quad g_{44} = \pm D^2, \quad k_{ij}, \ D-\text{const};
\end{aligned}\right\} \tag{32.47}$$

$$X_1 = p_2, \quad X_2 = \cos x^2 p_1-\cot x^1 \sin x^2 p_2+\frac{\sin x^2}{\sin x^1}\,p_3,$$

$$X_3 = \partial_2 X_2, \quad X_4 = p_4.$$

This completes the classification of general gravitational fields admitting *intransitive* and *transitive* four-parameter groups of motions.

§ 33. Gravitational Fields Admitting Groups of Motions G_r $(r \geqslant 5)$

We first investigate V_4 with a five-parameter group of motions.

According to a theorem by Fubini, corrected by Yegorov ([230], pp. 187–9) any G_r group of motions $(3 < r \leqslant 5)$ in a V_4 admits a G_{r-1} subgroup; the V_4 with a G_5 group of motions are therefore contained among the V_4 which admit a G_4 (§ 32). It is convenient to consider the two following cases separately:

(1) the G_5 contains an intransitive G_4 subgroup;

(2) the G_5 contains a transitive G_4 subgroup.

Case (1). Suppose that the fifth operator $X_5 = \underset{5}{\xi}{}^{\alpha} p_{\alpha}$. There are three possibilities: (α) $\underset{5}{\xi}{}^4 = 0$; ($\beta$) $\underset{5}{\xi}{}^4 = \lambda$, a constant; ($\gamma$) $\underset{5}{\xi}{}^4 = \underset{5}{\xi}{}^4(x)$.

G_5 A. If $\underset{5}{\xi}{}^4 = 0$, the matrix $(\underset{i}{\xi}{}^{\alpha})(\alpha = 1, \ldots, 4; i = 1, \ldots, 5)$ has rank 3; hence the G_5 group is *intransitive* and its surfaces of transitivity are three-dimensional V_3 or $\overset{*}{V}_2$ manifolds.

The V_4 with these groups is determined by integrating the Killing equations for the appropriate metrics; for G_5 with *non-isotropic* surfaces of transitivity we either find that the original G_4 is the complete group in the space or that the V_4 admits a G_6 group of motions. This does not apply to G_4 with isotropic surfaces of transitivity $\overset{*}{V}_3$; hence V_4 which admit a G_5 acting on $\overset{*}{V}_3$ do exist, and the group is complete on this manifold (cf. below).

G_5 B. When $\underset{5}{\xi}{}^4 = \lambda$, a constant $\neq 0$, the matrix $(\underset{i}{\xi}{}^{\alpha})$ has rank 4, and the G_5 group is transitive. This group thus contains a $G_4\{X_1, X_2, X_3, X_4\}$ subgroup, and for G_3 acting on V_2 or $\overset{*}{V}_2$, it also contains a $G_4\{X_1, X_2, X_3, X_5\}$ group which is intransitive.

G_5 C. This type of G_5 arises in investigating an intransitive G_4 group if $\underset{5}{\xi}{}^4 = \underset{5}{\xi}{}^4(x)$ is a function of the coordinates x^1, x^2, x^3, x^4.

We now consider case (2) where there are again three possible forms for operator $X_5 = \underset{5}{\xi}{}^{\alpha} p_{\alpha}$.

G_5 D. Corresponding to a transitive G_4 and $\underset{5}{\xi}{}^4 = 0$; the rank of $(\underset{i}{\xi}{}^{\alpha})$ is 4, and the G_5 is transitive. We now show that in this case the G_5 must also contain an intransitive G_4. It must be possible to express the commutators of the operators $\{X_1, X_2, X_3, X_5\}$ as a linear combination of these same operators, i.e. they must all have $\underset{i}{\xi}{}^4 = 0$.

It is easy to show that the combination of a transitive G_4 with $\underset{5}{\xi}{}^4 = \lambda$, a constant, gives the same G_5, i.e. if we take the fifth operator as $Y_5 = X_5 - \lambda X_4$ ($\underset{4}{\xi}^4 = 1$) we find that $\underset{5}{\eta}^4 = 0$ and $Y_5 = \underset{5}{\eta}^\alpha p_\alpha$, which is the case already considered.

G_5 E. This type of G_5 arises in investigating transitive G_4 when $\underset{5}{\xi}^4 = \underset{5}{\xi}^4(x)$.

We thus consider V_4 spaces admitting G_5 groups of motion of types A, B, C, D and E.

1. G_5 A. As pointed out above, we need only consider G_5 acting on $\overset{*}{V}_3$. Kruchkovich has shown that there is only one class of V_4 admitting a complete G_5 group on $\overset{*}{V}_3$ ([260], pp. 215–20); the corresponding metric in a special system of coordinates is

$$ds^2 = 2\,dx^1\,dx^4 + a_{22}dx^{2^2} + 2\,a_{23}\,dx^2\,dx^3 + a_{33}\,dx^{3^2}, \quad a_{ij}(x^4). \tag{33.1}$$

Operators:

$$X_i = p_i \quad (i = 1, 2, 3), \quad X_4 = x^2 p_1 + \varphi p_2 + \psi p_3,$$
$$X_5 = x^3 p_1 + \psi p_2 + \nu p_3,$$

where

$$\varphi = -\int \frac{a_{33}}{\Delta}\,dx^4, \quad \psi = \int \frac{a_{23}}{\Delta}\,dx^4, \quad \nu = -\int \frac{a_{22}}{\Delta}\,dx^4, \quad \Delta = a_{22}a_{33} - a_{23}^2.$$

2. G_5 B. By definition this group contains an intransitive subgroup G_4 $\{X_1, X_2, X_3, X_4\}$; if the $G_3\{X_1, X_2, X_3\}$ acts on V_2 or $\overset{*}{V}_2$, the group also admits an intransitive $G_4\{X_1, X_2, X_3, X_5\}$ and the fifth operator is subject to the condition $\underset{5}{\xi}^4 = \lambda$, a constant. The V_4 admitting a G_5 B can be determined by using the results of § 32. The condition $\underset{5}{\xi}^4 = \lambda$, a constant, enables us to integrate the Killing equations directly to obtain the G_5 group and the required V_4.

We begin with the G_4 acting on V_3; if we choose a semi-geodesic system of coordinates in V_4, the Killing equations (for $\underset{5}{\xi}^4 = \lambda$, a constant) show that $\partial_4 \underset{5}{\xi}^\alpha = 0$. Moreover, since $\underset{5}{\xi}^4 =$ constant and $\partial_4 \underset{\alpha}{\xi}^i = 0$ ($\alpha = 1, \ldots 4$; $i = 1, 2, 3$) the commutators $[X_i X_5]$ must be linear combinations of the first four operators, i.e.

$$[X_i X_5] = a_i X_1 + b_i X_2 + c_i X_3 + d_i X_4. \tag{33.2}$$

Putting i equal to 1, 2 and 3 in (33.2) provides expressions for $\partial_1 \underset{5}{\xi}^\alpha$, $\partial_2 \underset{5}{\xi}^\alpha$, $\partial_3 \underset{5}{\xi}^\alpha$, which enables us to integrate the Killing equations for the given V_4.

We apply this method to the structure of type G_4 I. The Killing equations for this space show that

$$\partial_1 \underset{5}{\xi}^3 = 0,\ a_{13}\partial_2 \underset{5}{\xi}^3 + a_{22}\partial_1 \underset{5}{\xi}^2 = 0,\ \lambda\partial_4 a_{13} + a_{13}\partial_3 \underset{5}{\xi}^3 + a_{13}\partial_1 \underset{5}{\xi}^1 + x^1 a_{22}\partial_1 \underset{5}{\xi}^2 = 0,$$

$$\lambda\partial_4 a_{22} + 2a_{22}(\partial_2 \underset{5}{\xi}^2 + x^1\partial_2 \underset{5}{\xi}^3) = 0, \tag{33.3}$$

$$\underset{5}{\xi}^1 a_{22} + \lambda x^1\partial_4 a_{22} + a_{22}(\partial_3 \underset{5}{\xi}^2 + x^1\partial_3 \underset{5}{\xi}^3 + x^1\partial_2 \underset{5}{\xi}^2 + x^{1^2}\partial_2 \underset{5}{\xi}^3) + a_{13}\partial_2 \underset{5}{\xi}^1 = 0,$$

$$2x^1 a_{22}\underset{5}{\xi}^1 + \lambda x^{1^2}\partial_4 a_{22} + 2a_{22}(x^1\partial_3 \underset{5}{\xi}^2 + x^{1^2}\partial_3 \underset{5}{\xi}^3) + 2a_{13}\partial_3 \underset{5}{\xi}^1 = 0.$$

But according to (33.2) with $i = 1, 2, 3$

$$\partial_2 \underset{5}{\xi}^1 = -c_1 - d_1 x^1, \quad \partial_2 \underset{5}{\xi}^2 = a_1 + c_1 x^3, \quad \partial_2 \underset{5}{\xi}^3 = b_1 + d_1 x^3,$$

$$\partial_3 \underset{5}{\xi}^1 = -c_2 - d_2 x^1, \quad \partial_3 \underset{5}{\xi}^2 = a_2 + c_2 x^3, \quad \partial_3 \underset{5}{\xi}^3 = b_2 + d_2 x^3, \tag{33.4}$$

$$-\partial_1 \underset{5}{\xi}^1 + x^3\partial_2 \underset{5}{\xi}^1 = -c_3 - d_3 x^1, \quad -\partial_1 \underset{5}{\xi}^2 + x^3\partial_2 \underset{5}{\xi}^2 = a_3 + c_3 x^3 + \underset{5}{\xi}^3,$$

$$-\partial_1 \underset{5}{\xi}^3 + x^3\partial_2 \underset{5}{\xi}^3 = b_3 + d_3 x^3.$$

Equations (33.3) and (33.4) show that $d_1 = d_2 = d_3 = 0$, i.e. $\{X_1, X_2, X_3, X_5\}$ form a group and hence the required G_5 belongs to the type $G_5 D$, i.e. it contains an intransitive $G_4\{X_1, X_2, X_3, X_4\}$ and a transitive $G_4\{X_1, X_2, X_3, X_5\}$; this type of space is considered below.

A similar result is obtained for all the other spaces in § 32 except those with $G_3\{X_1, X_2, X_3\}$ acting on V_2 or $\overset{*}{V}_2$; there are then two intransitive subgroups, $G_4\{X_1, X_2, X_3, X_4\}$ and $G_4\{X_1, X_2, X_3, X_5\}$, which by definition correspond to case G_5 B. The fifth operator is found as before. We obtain the following V_4 admitting a G_5 B from (32.14)–(32.16):

(1) $$ds^2 = \left(2\frac{\varepsilon k}{\lambda}e^{\lambda x^4} + \mu\right)dx^{1^2} + e^{\lambda x^4 + kx^1}(2dx^1\,dx^2 + e^{-x^1}dx^{3^2}) + dx^{4^2},$$

$$\varepsilon = 0, \pm 1, \lambda = \alpha(k+1) + \beta(\nu - k) \neq 0;\ \alpha, \beta, k, \mu = \text{const}, \tag{33.5}$$

$$X_1 = p_2,\ X_2 = p_3,\ X_3 = x^3 p_2 - e^{x^1}p_3,$$

$$X_4 = -\frac{2}{k-1}p_1 + \frac{2k}{k-1}x^2 p_2 + x^3 p_3,$$

$$X_5 = (\alpha - \beta)p_1 + [x^2(\alpha + \beta) + \varepsilon e^{-kx^1}]p_2 + \alpha x^3 p_3 + p_4.$$

(2) $$ds^2 = (-2\varepsilon e^{2x^4} + \mu)\,dx^{1^2} + e^{2x^4 - 2x^1}(2dx^1\,dx^2 + dx^{3^2}) + dx^{4^2},$$

$$\varepsilon = 0, \pm 1,\ \mu - \text{const},\ X_1 = p_2, X_2 = p_3,\ X_3 = x^2 p_2 - x^1 p_3, \tag{33.6}$$

$$X_4 = p_1 + 2x^2 p_2 + x^3 p_3, \quad X_5 = p_1 + \varepsilon e^{2x^1}p_2 + p_4.$$

(3) $ds^2 = (-q\varepsilon e^{-2x^4}+\mu)\,dx^{1^2}+e^{qx^1-2x^4}\left(2\,dx^1\,dx^2+\frac{4}{\omega^2}\cos^2\frac{\omega x^1}{2}\,dx^{3^2}\right)$

$$+dx^{4^2},\quad \omega=\sqrt{4-q^2},\quad \varepsilon=0,\pm 1,\quad \mu=\text{const},\qquad (33.7)$$

$$X_1=p_2,\quad X_2=p_3,\quad X_3=x^3p_2-\frac{\omega}{2}\tan\frac{\omega x^1}{2}\,p_3,$$

$$X_4=-p_1+\left(qx^2+\frac{x^{3^2}}{2}\right)p_2-x^3\left(\frac{\omega}{2}\tan\frac{\omega x^1}{2}-\frac{q}{2}\right)p_3,$$

$$X_5=(\varepsilon e^{-qx^1}+2x^2)p_2+x^3p_3+p_4.$$

(4) $ds^2=\left(2\frac{\varepsilon}{\lambda}\,e^{+\lambda x^4}+\mu\right)dx^{1^2}+e^{\lambda x^4}(2dx^1\,dx^2+e^{-x^1}\,dx^{3^2})+dx^{4^2},$

$$\lambda=\alpha+\beta\neq 0,\ \alpha,\beta,\mu=\text{const},\ \varepsilon=0,\pm 1,\qquad (33.8)$$

$$X_1=p_2,\ X_2=p_3,\ X_3=x^3p_2-e^{x^1}p_3,\ X_4=2p_1+x^3p_3,$$

$$X_5=(\alpha-\beta)p_1+[(\alpha+\beta)x^2+\varepsilon x^1]p_2+\alpha x^3p_3+p_4.$$

3. We now consider G_5 C. According to the second theorem of Lie

$$[X_iX_5]=\lambda_i^kX_k,\ \lambda_i^k - \text{const}\ (i=1,\ldots,4;\ k=1,\ldots,5).\qquad (33.9)$$

We first consider V_4 admitting a G_4 acting on V_3. Noting that $\underset{i}{\xi}^4=0$ $(i=1,\ldots,4)$, (33.9) shows that

$$\underset{i}{\xi}^\alpha\frac{\partial\underset{5}{\xi}^4}{\partial x^\alpha}=\lambda_i^5\underset{5}{\xi}^4\qquad(\alpha=1,2,3).\qquad (33.10)$$

Moreover, $\partial_4\underset{5}{\xi}^4=0$ in a semi-geodesic system of coordinates. If the G_4 has a G_3 subgroup $\{X_1, X_2, X_3\}$ acting on V_3, i.e. $|\underset{\beta}{\xi}^\alpha|\neq 0$ $(\alpha,\beta=1,2,3)$ and $\lambda_\alpha^5=0$ $(\alpha=1,2,3)$, then (33.10) for $i=1,2$, and $\partial_4\underset{5}{\xi}^4=0$ show that $\underset{5}{\xi}^4=$ constant, which is the case G_5 B considered above. Hence V_4 admitting a G_5 C only occurs when the G_3 subgroup acts on V_2 or $\check{V}_2$, or when at least one of the constants λ_α^5 $(\alpha=1,2,3)$ is non-zero. It is easy to see that the spaces (32.3), (32.4), (32.8) and (32.10) do not admit a G_5 C, for the Jacobi relations give $\lambda_1^5=\lambda_2^5=\lambda_3^5=0$, and the G_3 subgroups of these G_5 have $|\underset{\beta}{\xi}^\alpha|\neq 0$. The remaining spaces can admit a G_5 C; these V_4's are determined by the same method as for G_5 B, including the Jacobi relations.

Thus, for example, the Jacobi relations for the G_4 V (32.6) show that

$$\lambda_1^3=\lambda_1^4=\lambda_1^5=\lambda_2^3=\lambda_2^4=\lambda_2^5=0$$

and (33.9) shows that

$$\underset{5}{\xi}^1 = A(x^4)\, e^{-\lambda_3^5(\lambda_1^1+1)x^1}, \quad \underset{5}{\xi}^2 = \lambda_2^2 x^2 - \lambda_1^2 x^3 - \lambda_4^2, \tag{33.11}$$

$$\underset{5}{\xi}^3 = \lambda_4^1 + \lambda_1^1 x^3 - \lambda_2^1 x^2, \quad \underset{5}{\xi}^4 = De^{-\lambda_3^5 x^1}, \quad D = \text{const.}$$

It is easy to see that the operators $\{X_1, X_2, X_3, X_5\}$ form a simply-transitive G_4 group. A non-singular coordinate transformation therefore exists which converts this G_4 to one of the forms determined in § 32. After this transformation, the group $G_4\{X_1, X_2, X_3, X_4\}$ remains either intransitive, corresponding to a G_5 D, or transitive, corresponding to a G_5 E: both of these are considered below. The same result holds for G_4 IV, VI.

The spaces (32.14)–(32.16) (with $G_3\{X_1, X_2, X_3\}$ subgroups acting on $\mathring{V}_2$) cannot be reduced to G_5 D or G_5 E, but the method outlined at the beginning of the present section 3 can be used.

There are three possibilities for (32.14):

(1) $$ds^2 = \left[k_{44}\left(\frac{1-k}{1+k}\right) - 2k_2k_{12}e^{-(k+1)x^4} + k_{11}e^{-2x^4}\right] dx^{1^2}$$
$$+ k_{12}e^{kx^1-(k+1)x^4}(2dx^1\,dx^2 + e^{-x^1}dx^{3^2}) + k_{44}\,dx^{4^2}, \tag{33.12}$$

where

$$k_{ij},\ k,\ k_2 = \text{const}, \qquad k \neq \pm 1.$$

Operators:

$$X_1 = p_2, \quad X_2 = p_3, \quad X_3 = x^3p_2 - e^{x^1}p_3,$$
$$X_4 = -\frac{2}{k-1}p_1 + \frac{2k}{k-1}x^2p_2 + x^3p_3,$$
$$X_5 = e^{x^1}(p_1 + x^3p_3 + p_4)$$
$$+ \left[-\frac{x^{3^2}}{2} + e^{-(k+1]x^1}\left(-\frac{k_{44}}{k_{12}(k+1)}\, e^{(k+1)x^4} + k_2\right)\right] p_2.$$

(2) $$ds^2 = \left[k_{44}\left(\frac{1+k}{1-k}\right) - 2k_2k_{12}e^{-(k-1)x^4} + k_{11}e^{2x^4}\right] dx^{1^2}$$
$$+ k_{12}e^{kx^1-(k-1)x^4}(2dx^1\,dx^2 + e^{-x^1}\,dx^{3^2}) + k_{44}\,dx^{4^2}, \tag{33.13}$$

where

$$k_{ij},\ k,\ k_2 - \text{const}, \quad k \neq \pm 1.$$

Operators:

$$X_1 = p_2, \quad X_2 = p_3, \quad X_3 = x^3p_2 - e^{x^1}p_3,$$
$$X_4 = -\frac{2}{k-1}p_1 + \frac{2k}{k-1}x^2p_2 + x^3p_3,$$
$$X_5 = e^{-x^1}(p_1+p_4) + \ldots + e^{-(k+1)x^1}\left(\frac{k_{44}}{k_{12}(k-1)}e^{(k-1)x^4} + k_2\right)p_2.$$

(3) $ds^2 = (l-2kx^4)e^{2x^4}dx^{1^2} + e^{2x^4-x^1}(2dx^1\,dx^2 + e^{-x^1}dx^{3^2}) + dx^{4^2},$
$$l,\ k \text{ — const}, \tag{33.14}$$
$$X_1 = p_2, \quad X_2 = p_3, \quad X_3 = x^3p_2 - e^{x^1}p_3,$$
$$X_4 = p_1 + x^2p_2 + x^3p_3, \quad X_5 = e^{-x^1}(p_1+p_4) + (kx^1 - \tfrac{1}{2}e^{-2x^4})p_2.$$

The spaces for (32.15) and (32.16) admit a six-parameter group and are therefore not considered here.

4. We now consider the fourth type—G_5 D.

By definition a G_5 D group contains a transitive $G_4\{X_1, X_2, X_3, X_4\}$ and an intransitive $G_4\{X_1, X_2, X_3, X_5\}$ sharing a common $G_3\{X_1, X_2, X_3\}$ subgroup. This suggests an algorithm for finding V_4 admitting a G_5 D.

We take a transitive $G_4\{X_1, X_2, X_3, X_4\}$ and add a fourth operator of an intransitive G_4 group with the same $G_3\{X_2, X_2, X_3\}$ subgroup as the original G_4. We then integrate the Killing equations for the additional operator and the original V_4.

This method is used to investigate insoluble G_4 VII, VIII.

A transitive G_4 VII contains a G_3 VIII $\{X_1, X_2, X_3\}$ subgroup and has the operators

$$X_1 = e^{-x^3}(p_1 - x^{2^2}p_2 - 2x^2p_3), \quad X_2 = p_3, \quad X_3 = e^{x^3}p_2,$$
$$X_4 = p_1 + \varepsilon p_4, \quad \varepsilon = \pm 1. \tag{33.15}$$

The only intransitive G_4 containing the same G_3 VIII subgroup is the G_4 VII with operators

(a) $X_1 = e^{-x^3}(p_1 - x^{2^2}p_2 - 2x^2p_3)$, $X_2 = p_3$, $X_3 = e^{x^3}p_2$, $X_4 = p_1$ (acting on V_3);

(b) $X_1 = p_2$, $X_2 = x^2p_2 + p_3$, $X_3 = (x^{2^2} + \varepsilon e^{2x^3})p_2 + 2x^2p_3 - e^{x^3}p_1$, $X_4 = p_1$ (acting on V_3^*).

Hence the additional operator must be $X = p_1$, which forms a group with the operators (33.15).

Thus we finally have the two spaces

1. $$ds^2 = 2k_{12}(dx^1\,dx^2 - x^2dx^1\,dx^3) + k_{22}(dx^2 - x^2\,dx^3)^2 - \tfrac{1}{2}k_{12}\,dx^{3^2} + k_{44}\,dx^{4^2},\quad k_{ij} = \text{const.} \tag{33.16}$$

Operators:

$$X_\nu = p_\nu \quad (\nu = 1, 3, 4), \quad X_2 = e^{x^3}p_2,$$
$$X_5 = e^{-x^3}(p_1 - x^{2^2}p_2 - 2x^2p_3).$$

2. $$ds^2 = 2dx^1\,dx^4 + 2\varepsilon e^{-x^3}\,dx^2\,dx^4 + k_{22}(e^{-2x^3}\,dx^{2^2} - \varepsilon\,dx^{3^2}),\quad k_{22} = \text{const} \quad (\varepsilon = \pm 1). \tag{33.17}$$

Operators:

$$X_1 = p_2, \quad X_2 = x^2p_2 + p_3,$$
$$X_3 = (x^{2^2} + \varepsilon e^{2x^3})p_2 + 2x^2p_3 - e^{x^3}p_1, \quad X_4 = p_1, \quad X_5 = p_4.$$

Repeating the analysis for G_4 I–VIII we obtain the following spaces admitting a G_5 D:

(1) $$ds^2 = k_{11}x^4e^{-2\varepsilon x^4} + k_{12}e^{-2\varepsilon x^4}(2dx^1\,dx^2 + dx^{3^2}) + k_{44}\,dx^{4^2},\quad k_{ij} = \text{const} \quad (\varepsilon = 0, 1). \tag{33.18}$$

Operators:

$$X_i = p_i \quad (i = 1, 2, 3), \quad X_4 = -x^3p_2 + x^1p_3,$$
$$X_5 = \varepsilon x^1p_1 + (\varepsilon x^2 + x^1)p_2 + \varepsilon x^3p_3 + p_4.$$

(2) $$ds^2 = k_{11}(dx^{1^2} + \sin^2 x^1\,dx^{2^2}) + k_{33}(\cos x^1\,dx^2 + dx^3)^2 + k_{44}\,dx^{4^2},\quad k_{ij} = \text{const.} \tag{33.19}$$

Operators:

$$X_1 = p_2, \quad X_2 = \cos x^2p_1 - \cot x^1 \sin x^2p_2 + \frac{\sin x^2}{\sin x^1}p_3,$$
$$X_3 = \partial_2X_2, \quad X_4 = p_3, \quad X_5 = p_4.$$

(3) $$ds^2 = 2dx^1\,dx^4 + k_{22}(\cos^2 x^3\,dx^{2^2} + dx^{3^2}) + 2\sin x^3\,dx^2\,dx^4,\quad k_{22} = \text{const.} \tag{33.20}$$

Operators:

$$X_1 = p_2, \quad X_2 = \sec x^3 \cos x^2p_1 - \tan x^3 \cos x^2p_2 + \sin x^2p_3,$$
$$X_3 = \partial_2X_2, \quad X_4 = p_1, \quad X_5 = p_4.$$

(4) $$ds^2 = 2k_{13}e^{-cx^4}\,dx^1\,dx^3 + k_{22}e^{-2cx^4}(dx^2 + x^1\,dx^3)^2 + k_{44}\,dx^{4^2},\quad c, k_{ij} = \text{const.} \tag{33.21}$$

Operators:

$$X_1 = p_2, \quad X_2 = p_3, \quad X_3 = -p_1 + x^3 p_2,$$
$$X_4 = -x^1 p_1 + x^3 p_3, \quad X_5 = c(x^1 p_1 + x^2 p_2) + p_4.$$

(5) $$ds^2 = k_{11} e^{-2x^4}(dx^{1^2} + dx^{3^2}) + k_{22} e^{-4x^4}(dx^2 + x^1\, dx^3)^2 + k_{44}\, dx^{4^2},$$
$$k_{ij} = \text{const.} \qquad (33.22)$$

Operators:

$$X_1 = p_2, \quad X_2 = p_3, \quad X_3 = -p_1 + x^3 p_2,$$
$$X_4 = -x^3 p_1 + \tfrac{1}{2}(x^{3^2} - x^{1^2}) p_2 + x^1 p_3, \quad X_5 = x^1 p_1 + 2x^2 p_2 + x^3 p_3 + p_4.$$

(6) $$ds^2 = k_{11}(dx^{1^2} + dx^{3^2}) + k_{22}(dx^2 + x^1\, dx^3)^2 + k_{44}\, dx^{4^2}, \qquad (33.23)$$
$$k_{ij} = \text{const.}$$

Operators:

$$X_1 = p_2, \quad X_2 = p_3, \quad X_3 = -p_1 + x^3 p_2,$$
$$X_4 = -x^3 p_1 + \tfrac{1}{2}(x^{3^2} - x^{1^2}) p_2 + x^1 p_3, \quad X_5 = p_4.$$

(7) $$ds^2 = k_{11}\, dx^{1^2} + 2k_{23} e^{-(1+\alpha)x^4 + x^1}\, dx^2\, dx^3 + k_{44}\, dx^{4^2}; \quad k_{ij},\ \alpha = \text{const.} \qquad (33.24)$$

Operators:

$$X_1 = p_2, \quad X_2 = p_3, \quad X_3 = -p_1 + x^2 p_2,$$
$$X_4 = -p_1 + x^3 p_3, \quad X_5 = (1+\alpha) p_1 + p_4.$$

(8) $$ds^2 = k_{11} dx^{1^2} + k_{22} e^{-2\alpha x^4 + 2x^1}(dx^{2^2} + dx^{3^2}) + k_{44}\, dx^{4^2};$$
$$k_{ij},\ \alpha = \text{const.} \qquad (33.25)$$

Operators:

$$X_1 = p_2, \quad X_2 = p_3, \quad X_3 = -p_1 + x^2 p_2 + x^3 p_3,$$
$$X_4 = -x^3 p_2 + x^2 p_3, \quad X_5 = \alpha p_1 + p_4.$$

(9) $$ds^2 = 2dx^1\, dx^4 + k_{22} e^{-x^1 + \alpha x^4}(dx^{2^2} + dx^{3^2}), \quad k_{22} = \text{const.} \qquad (33.26)$$

Operators:

$$X_1 = p_2, \quad X_2 = p_3, \quad X_3 = 2p_1 + x^2 p_2 + x^3 p_3,$$
$$X_4 = -x^3 p_2 + x^2 p_3, \quad X_5 = \alpha p_1 + p_4.$$

(10) $$ds^2 = 2dx^1\, dx^4 + k_{23} e^{-x^1 + \alpha x^4}\, dx^2\, dx^3, \quad k_{23} = \text{const.} \qquad (33.27)$$

Operators:

$$X_1 = p_2, \quad X_2 = p_3, \quad X_3 = p_1 + x^3 p_3,$$
$$X_4 = p_1 + x^2 p_2, \quad X_5 = \alpha p_1 + p_4.$$

(11) $$ds^2 = k_{11} e^{-2lx^4}\, dx^{1^2} + 2k_{23} e^{-(k+\varepsilon)x^4}\, dx^2\, dx^3 + k_{44}\, dx^{4^2},$$
$$k_{ij}, \quad k,\ l = \text{const}\ (\varepsilon = 0, 1). \qquad (33.28)$$

Operators:

$$X_1 = p_2, \quad X_2 = p_3, \quad X_3 = -p_1, \quad X_4 = x^2p_2 - x^3p_3,$$
$$X_5 = lx^1p_1 + \varepsilon x^2p_2 + kx^3p_3 = p_4.$$

(12) $$ds^2 = k_{12}e^{-2kx^4}(2dx^1\,dx^2 + dx^{3^2}) + k_{22}e^{-2\varepsilon x^4}\,dx^{2^2} + k_{44}\,dx^{4^2},$$
$$k_{ij},\ k,\ l = \text{const} \quad (2k = l + \varepsilon, \quad \varepsilon = 0, 1). \qquad (33.29)$$

Operators:

$$X_1 = p_2, \quad X_2 = p_3, \quad X_3 = -p_1, \quad X_4 = -x^3p_1 + x^2p_3,$$
$$X_5 = lx^1p_1 + \varepsilon x^2p_2 + kx^3p_3 + p_4.$$

(13) $$ds^2 = k_{11}e^{-x^4}\,dx^{1^2} + k_{12}e^{-2x^4}(2\,dx^1\,dx^2 + dx^{3^2}) + k_{44}\,dx^{4^2}, \quad k_{ij} = \text{const}. \qquad (33.30)$$

Operators:

$$X_1 = p_2, \quad X_2 = p_3, \quad X_3 = -p_1, \quad X_4 = -x^1p_3 + x^3p_2,$$
$$X_5 = x^1p_1 + 3x^2p_2 + 2x^3p_3 + p_4.$$

(14) $$ds^2 = k_{11}e^{-2lx^4}\,dx^{1^2} + k_{22}e^{-2kx^4}(dx^{2^2} + dx^{3^2}) + k_{44}\,dx^{4^2}, \qquad (33.31)$$
$$k_{ij},\ l,\ k = \text{const}.$$

Operators:

$$X_1 = p_2, \quad X_2 = p_3, \quad X_3 = -p_1, \quad X_4 = -x^3p_2 + x^2p_3,$$
$$X_5 = lx^1p_1 + k(x^2p_2 + x^3p_3) + p_4.$$

We now come to the last case.

5. If a V_4 admits a G_5 E then by definition it admits a simply-transitive G_4 group $G_4\{X_1, X_2, X_3, X_4\}$; hence the required V_4 are contained among those admitting simply-transitive groups. Suppose that $X_5 = \underset{5}{\xi^\alpha} p_\alpha$ ($\alpha = 1, \ldots, 4$) is the fifth operator of the G_5 E. According to Lie's second theorem

$$[X_iX_5] = \lambda_i^k X_k \quad (i = 1, \ldots, 4; \quad k = 1, \ldots, 5), \quad \lambda_i^k = \text{const}. \qquad (33.32)$$

The fifth operator can be determined as follows:

(1) we equate the coefficients of the X_i in the Jacobi identities

$$[[X_iX_j]X_5] + [[X_jX_5]X_i] + [[X_5X_i]X_j] = 0$$

and determine all possible constants λ_i^k;

(2) we determine $\underset{5}{\xi^\alpha}$ from the structure equations, after reduction to a simpler form by linear substitutions, and integrate the Killing equations for the given V_4; we thus find the operators and the required metric. Thus for example the Jacobi relations for G_4 V show that $\lambda_1^3 = \lambda_1^4 = \lambda_1^5 = \lambda_2^3 = \lambda_2^4 = \lambda_2^5 = 0$; (33.32) with $i = 1, \ldots, 4$, provides the following

equations for $\underset{5}{\xi}^4$:

$$\partial_2 \underset{5}{\xi}^4 = \partial_3 \underset{5}{\xi}^4 = 0, \quad -\partial_1 \underset{5}{\xi}^4 = \lambda_3^4 + \lambda_3^5 \underset{5}{\xi}^4, \quad \alpha \partial_1 \underset{5}{\xi}^4 = \lambda_4^4 + \lambda_4^5 \underset{5}{\xi}^4. \quad (33.33)$$

The analysis must now consider the four possible cases

$$(1)\ \ \lambda_3^5 = \lambda_4^5 = 0; \quad (2)\ \ \lambda_3^5 = 0,\ \ \lambda_4^5 \neq 0; \quad (3)\ \ \lambda_3^5 \neq 0,\ \ \lambda_4^5 \neq 0;$$
$$(4)\ \ \lambda_3^5 \neq 0,\ \ \lambda_4^5 = 0.$$

The Jacobi relations for (1) and (2) give $\lambda_3^4 = \lambda_4^4 = 0$, i.e. $\underset{5}{\xi}^4 =$ constant (since $\partial_4 \underset{5}{\xi}^4 = 0$ for a semi-geodesic system of coordinates); this kind of group has already been considered.

We obtain $\lambda_4^5 = -\alpha\lambda_3^5$ for cases (3) and (4). Taking the Jacobi relations into account, the linear substitution

$$Y_1 = \lambda_4^1 X_1 + \lambda_4^2 X_2, \quad Y_2 = -\lambda_4^2 X_1 + \lambda_4^1 X_2, \quad Y_3 = X_3, \quad Y_4 = X_4,$$
$$Y_5 = -\frac{1}{(\alpha\lambda_3^5)^2+1}(-\lambda_4^2 X_1 + \lambda_4^1 X_2) - \frac{\alpha\lambda_3^5}{1+(\alpha\lambda_3^5)^2}(\lambda_4^1 X_1 + \lambda_4^2 X_2)$$
$$-\lambda_1^1 X_3 - \lambda_1^2 X_4 + X_5 \quad (33.34)$$

reduces the structure (33.32) to

$$[Y_1 Y_5] = [Y_2 Y_5] = 0, \quad [Y_3 Y_5] = \lambda_3^5 Y_5, \ [Y_4 Y_5] = -\alpha\lambda_3^5 Y_5. \quad (33.35)$$

The Killing equations show, however, that the metric for this fifth operator is singular.

A similar analysis for the remaining spaces leads to the following V_4 admitting a G_5 of type E:

$$(1)\quad ds^2 = k_{11} e^{x^4}\, dx^{1^2} + \lambda k_{11} e^{-x^4} (dx^2 + x^1\, dx^3)^2 - \lambda k_{44} e^{-2x^4}\, dx^{3^2} + k_{44}\, dx^{4^2}. \quad (33.36)$$

Operators:

$$X_1 = p_2, \quad X_2 = p_3, \quad X_3 = -p_1 + x^3 p_2,$$
$$X_4 = -\tfrac{1}{2}x^1 p_1 + \tfrac{1}{2}x^2 p_2 + x^3 p_3 + p_4,$$
$$X_5 = -(x^2 + x^1 x^3) p_1 + \left(x^2 x^3 - \frac{1}{\lambda} x^1 e^{-2x^4}\right) p_2$$
$$+\left(x^{3^2} + \frac{1}{\lambda} e^{2x^4}\right) p_3 + 2x^3 p_4, \quad k_{11},\ k_{44},\ \ \lambda = \text{const},$$

but

$$|g| = \lambda^2 k_{11}^2 k_{44}^2 e^{-2x^4} > 0.$$

(2) $$ds^2 = k_{22}(2\lambda e^{-(1/2)x^4}\, dx^1\, dx^3 + e^{-x^4}(dx^2 + x^1\, dx^3)^2 - \tfrac{1}{4}\lambda^2\, dx^{4^2}) + k_{33}e^{-2x^4}\, dx^{3^2}. \tag{33.37}$$

Operators:

$$X_1 = p_2, \quad X_2 = p_3, \quad X_3 = -p_1 + x^3 p_2,$$
$$X_4 = -\tfrac{1}{2}x^1 p_1 + \tfrac{1}{2}x^2 p_2 + x^3 p_3 + p_4,$$
$$X_5 = (-x^2 - x^1 x^3 + \lambda e^{-(1/2)x^4})p_1 + x^2 x^3 p_2 + x^{3^2} p_3 + 2x^3 p_4;$$
$$k_{22},\ k_{33},\ \lambda = \text{const}\ (|g| = \tfrac{1}{4}\lambda^4 k_{22}^4 e^{-2x^4} > 0).$$

(3) $$ds^2 = k_{11}\, dx^{1^2} + \lambda\varepsilon k_{11} e^{-x^4}\, dx^1\, dx^3 + k_{22} e^{\lambda x^4 + 2x^1}\, dx^{2^2} + \lambda^2 k_{11}\, dx^{4^2}. \tag{33.38}$$

Operators:

$$X_1 = p_1, \quad X_2 = p_3, \quad X_3 = -p_1 + x^2 p_2,$$
$$X_4 = -\lambda p_1 + x^3 p_3 + p_4, \quad X_5 = e^{-(1/\lambda)x^1}(-\lambda p_1 + \varepsilon e^{x^4} p_3 + p_4);$$
$$k_{11},\ k_{22},\ \lambda = \text{const}\ (\varepsilon = \pm 1, \quad |g| = -\lambda^4 k_{11}^3 k_{22} e^{(2\lambda - 2)^{x^4} + 2x^1},$$
when $$k_{11}k_{22} > 0 \quad |g| < 0).$$

(4) $$ds^2 = k_{11}\, dx^{1^2} + 2\varepsilon k_{44} e^{-x^4}\, dx^1\, dx^2 + 2\lambda\mu\, dx^1\, dx^3 + \lambda^2\mu\, dx^{3^2} + k_{44}\, dx^{4^2}. \tag{33.39}$$

Operators:

$$X_1 = p_2, \quad X_2 = p_3, \quad X_3 = -p_1, \quad X_4 = x^2 p_2 + p_4,$$
$$X_5 = e^{-x^1}\left(-p_1 + \varepsilon e^{x^4} p_2 + \frac{1}{\lambda} p_3 + p_4\right);$$
$$k_{11},\ k_{44},\ \lambda,\ \mu = \text{const}\ (\varepsilon = \pm 1, \quad \mu = k_{11} - k_{44}, \quad \mu \neq 0,$$
$$|g| = k_{44}^3 \lambda^2 \mu, \quad \text{when} \quad k_{44}\mu < 0 \quad |g| < 0).$$

This completes the classification of general gravitational fields admitting five-parameter groups of motion. Though some of these V_4 do not correspond to real gravitational fields, they have been automatically derived together with the required spaces, but such spaces of high mobility are of interest to geometers. These V_4 are characterized by $|g| > 0$. Some of the constants of integration in the various metrics can, of course, be reduced to 0 or 1 by means of elementary transformations.

We now consider general gravitational fields admitting higher order groups of motion.

As all V_4 with signature $(---+)$ which admit a G_r group with $r > 7$ are spaces of *constant curvature* ([230]), the only cases of $r > 5$ we need consider are (1) V_4 with G_7, (2) V_4 with intransitive G_6, (3) V_4 with transitive G_6.

The first and second cases have been fully treated by Yegorov [230] and Kruchkovich [303], who derived the following theorems.

THEOREM 1. Any Riemann space admitting an intransitive group of motions $G_{n(n-1)/2}$ is a Kagan sub-projective space with one of the following metrics:

$$\text{(A)}\quad ds^2 = e^{-2\mu(x^1)} \sum_{i=1}^{n} e_i (dx^i)^2, \qquad e_i = \pm 1,$$

$$\text{(B)}\quad ds^2 = e^{-2\mu(R)} \sum_{i=1}^{n} e_i (dx^i)^2, \qquad e_i = \pm 1, \qquad R = \sqrt{\sum_{i=1}^{n} e_i (x^i)^2},$$

$$\text{(C)}\quad ds^2 = e^{-2\mu(x^1)} \left[2\, dx^1\, dx^2 + \sum_{k=3}^{n} e_k (dx^k)^2 \right], \qquad e_k = \pm 1. \tag{33.40}$$

The group acts on a *non-isotropic* hypersurface in (A) and (B) and on an *isotropic* hypersurface in (C).

THEOREM 2. If a V_4 is not an Einstein space and admits a G_r with $r > \frac{1}{2}(n-1)(n-2)+3$, it is a Kagan sub-projective space; when $n = 4$ the G_r are G_7 and the corresponding metric either

$$ds^2 = dx^{1^2} + k\, ds_1^2(x^2, x^3, x^4) \tag{33.41}$$

or

$$ds^2 = e^{-2\mu(x^1)} \left[2\, dx^1\, dx^2 + \sum_{k=3}^{n} e_k (dx^k)^2 \right]. \tag{33.42}$$

k is a constant in (33.41) and ds_1^2 the metric of a space of constant curvature. The function $\mu(x^1)$ in (33.42) is given by

$$\mu' = \frac{Ax^1 + B}{Ax^{1^2} + Cx^1 + D}, \qquad A, B, C, D = \text{const.}$$

Kaigorodov has shown that Einstein spaces with $n = 4$ and a Minkowski signature do not admit G_r with $r > 6$ (ignoring spaces of constant curvature) [505]. Thus theorem 2 is true for all V_4 with a Minkowski signature.

Only V_4 with a transitive G_6 remains to be considered.

We use the following method; we first find the structure equations for the G_6 groups admitted by V_4 spaces, and then integrate the Killing and structure equations to determine the metrics of the required spaces.

We start from the Weyl tensor:

$$C_{ijkl} = R_{ijkl} + \frac{1}{2}(g_{ik}R_{jl} - g_{il}R_{jk} + g_{jl}R_{ik} - g_{jk}R_{il}) + \frac{R}{6} g_{ijkl},$$

where R_{ijkl} is the curvature tensor, R_{ik} the Ricci tensor and R the scalar curvature of the space. According to theorem 3 of § 20 there are only three types of gravitational field $\boldsymbol{T}_i$; the Weyl tensor has three canonical forms in some non-holonomic orthonormal tetrad, one for each type (20.14), (20.16) and (20.18) $\omega = 0$. The relation between the Weyl and curvature tensors and the theorem on the order of the groups of motions admitted by V_n (§ 10) lead to the following theorem:

In order that a V_n should admit a group of motions G_r it is necessary that the rank of the set of equations

$$\underset{L}{\delta} C_{ijkl} = 0, \quad \underset{L}{\delta} C_{ijkl}, m = 0, \ldots \tag{33.43}$$

(in the unknowns ξ^k and $\xi_{k,l}$) should be less than or equal to $\frac{1}{2}n(n+1)-r$.

The method of finding the required G_6 group structures is completely analogous to that used to find the G_r group structures admitted by an Einstein space (cf. §§ 23, 24) since the canonical matrices $C_{\alpha\beta}$ and $R_{\alpha\beta}$ have identical forms.

Let us consider $\boldsymbol{T}_1$ spaces. The matrix for the first relation in (33.43) is identical with (23.5) except that $C_{ijkl,\,m}$ goes into the right-hand column.

If the rank of this matrix is less than 4, all fourth order minor determinants vanish, and hence $\alpha_i = \beta_i = 0$, so that $C_{ijkl} = 0$, corresponding to a conformally flat space. The metrics of the required spaces can therefore be found by using Levine's results which provide a complete summary of the G_r groups of motion admitted by conformally flat V_4 of arbitrary signature [115]. The required V_4 have the following metrics:

$$\begin{gathered} ds^2 = \frac{1}{R^2}\varphi\left(\frac{x^4}{R}\right)(-dx^{1^2}-dx^{2^2}-dx^{3^2}+dx^{4^2}), \\ R = -(x^1)^2-(x^2)^2-(x^3)^2+(x^4)^2, \ X_1 = 2x^1(x^k p_k)+Rp_1, \\ X_2 = 2x^2(x^k p_k)+Rp_2, \ X_3 = 2x^3(x^k p_k)+Rp_3, \quad X_4 = -x^1p_2+x^2p_1, \\ X_5 = -x^1p_3+x^3p_1, \quad X_6 = -x^2p_3+x^3p_2. \end{gathered} \tag{33.44}$$

$$\begin{gathered} ds^2 = \frac{1}{x^3}\varphi\left(\frac{x^4}{x^3}\right)(-dx^{1^2}-dx^{2^2}-dx^{3^2}+dx^{4^2}), \\ X_1 = p_1, \quad X_2 = p_2, \quad X_3 = 2x^1(x^k p_k)+Rp_1, \\ X_4 = 2x^2(x^k p_k)+Rp_2 \quad X_5 = -x^1p_2+x^2p_1, \\ X_6 = x^k p_k \qquad (k = 1, \ldots, 4). \end{gathered} \tag{33.45}$$

Suppose that the rank of the set (33.43) is 4. As all fifth order minor determinants vanish we obtain

$$\alpha_2 = \alpha_3, \quad \beta_2 = \beta_3, \quad C_{aa, m} = C_{23, m} = 0 \quad (a = 1, 2, 3).$$

Taking ξ^l in an orthonormal tetrad in the form

$$\underset{k}{\xi^l} \doteq \delta^l_k, \quad \underset{5}{\xi^l} \doteq \underset{6}{\xi^l} \doteq 0 \qquad (k, l = 1, \ldots, 4)$$

and noting that $\underset{L}{\delta} C_{ijkl} = 0$ and that the group is transitive, we obtain

$$\underset{5}{\xi}_{k,l} \doteq \begin{cases} 1, & kl = 14 \\ 0, & kl \neq 14, \end{cases} \qquad \underset{6}{\xi}_{k,l} \doteq \begin{cases} 1, & kl = 23 \\ 0, & kl \neq 23, \end{cases}$$

$$\underset{p}{\xi}_{k,l} \doteq \begin{cases} 0, & kl = 14, 23 \\ a_p, & kl \neq 14, 23. \end{cases} \qquad (33.46)$$

The structure constants are determined by considering relations (24.5); this set is heavily overdetermined and easily solved. We find that the required G_6 group is a direct product of two subgroups

$$G_6 = \bar{G}_3 \{X_1, X_4, X_5\} \times \boldsymbol{G}_3\{X_2, X_3, X_6\},$$

where $\bar{G}_3$ has the structure

$$[X_1X_4] = aX_5: \quad [X_1X_5] = -X_4: \quad [X_4X_5] = -X_1, \quad a = \text{const},$$

and $\boldsymbol{G}_3$ the structure

$$[X_2X_3] = bX_6,$$
$$[X_2X_6] = X_3, \quad [X_3X_6] = -X_2, \quad b = \text{const}.$$

The required space is clearly reducible:

$$ds^2 = C_1 ds_1^2(x^1, x^2) + C_2 ds_2^2(x^3, x^4); \quad C_1, C_2 = \text{const}, \qquad (33.47)$$

where ds_1^2, ds_2^2 are metrics of constant (possibly zero) curvature. Thus the classes of space (33.47) have the metrics

(1) $$ds^2 = C_1(dx^{1^2} + e_1 dx^{2^2}) + C_2(dx^{3^2} + e_2 \cos^2 x^3 \, dx^{4^2}),$$
$$e_1 = \pm 1, \quad e_2 = \pm 1, \quad C_1 e_1 C_2 e_2 < 0. \qquad (33.48)$$

(2) $$ds^2 = C_1(dx^{1^2} + e_1 dx^{2^2}) + C_2(dx^{3^2} + e_2 \cosh^2 x^3 \, dx^{4^2}),$$
$$e_1 = \pm 1, \quad e_2 = \pm 1, \quad C_1 e_1 C_2 e_2 < 0. \qquad (33.49)$$

(3) $$ds^3 = C_1(dx^{1^2} + e_1 \cos^2 x^1 \, dx^{2^2}) + C_2(dx^{3^2} + e_2 \cos^2 x^3 \, dx^{4^2}),$$
$$e_1 = \pm 1, \quad e_2 = \pm 1, \quad C_1 e_1 C_2 e_2 < 0. \qquad (33.50)$$

(4) $$ds^2 = C_1(dx^{1^2} + e_1 \cosh^2 x^1 \, dx^{2^2}) + C_2(dx^{3^2} + e_2 \cos^2 x^3 \, dx^{4^2}),$$
$$e_1 = \pm 1, \quad e_2 = \pm 1, \quad C_1 e_1 C_2 e_2 < 0. \qquad (33.51)$$

$$(5) \quad ds^2 = C_1(dx^{1^2}+e_1\cosh^2 x^1\, dx^{2^2})+C_2(dx^{3^2}+e_2\cosh^2 x^3\, dx^{4^2}),$$
$$e_1 = \pm 1, \quad e_2 = \pm 1, \quad C_1e_1C_2e_2 < 0. \tag{33.52}$$

We now consider type $\boldsymbol{T}_2$. The appropriate method is completely analogous to that used to determine the general case of Einstein space $R_{ij} = \varkappa g_{ij}(\varkappa \neq 0)$ of maximum mobility, the only difference being that the curvature tensor components in (24.8) are unknown, so that there are more undetermined parameters among the required structure constants. The required G_6 structure is

$$\begin{gathered}
[X_iX_j] = 0 \quad (i, j = 1, 2, 3), \quad [X_1X_4] = 0, \quad [X_2X_4] = X_1, \quad [X_3X_4] = 0,\\
[X_1X_5] = 0, \quad [X_2X_5] = 0, \quad [X_3X_5] = X_1, \quad [X_4X_5] = 0, \quad [X_1X_6] = bX_1,\\
[X_2X_6] = cX_5+dX_4+eX_3, \quad [X_3X_6] = X_5+bX_3+cX_4,\\
[X_4X_6] = -X_2-cX_3+bX_4, \quad [X_3X_6] = -cX_2-eX_4+fX_3,\\
b, c, d, e, f = \text{const.}
\end{gathered} \tag{33.53}$$

The first four operators give a G_4 VI_2 structure ($k = \varepsilon = 0$). But the results of § 32 show that the required V_4's belong to the class of spaces with the metric

$$ds^2 = 2dx^1\, dx^4 + A(x^4)\, dx^{2^2} + 2B(x^4)\, dx^3\, dx^2 + C(x^4)\, dx^{3^2}, \tag{33.54}$$

and the subgroup $\{X_1, \ldots, X_5\}$ acts on a $\overset{*}{V}_3$ with operators

$$X_i = p_i \quad (i = 1, 2, 3), \quad X_4 = x^2p_1+\varphi(x^4)p_2+\psi(x^4)p_3,$$
$$X_5 = x^3p_1+\psi(x^4)p_2+\theta(x^4)p_3,$$

where

$$A = -\Delta\theta', \quad B = \Delta\psi', \quad C = -\Delta\varphi', \ \Delta = AC-B^2.$$

the prime denoting differentiation with respect to x^4.

The structure and Killing equations for the sixth operator show that

$$\begin{aligned}
X_6 = {} & (bx^1+cx^2x^3-\tfrac{1}{2}\, d\cdot x^{2^2}+\tfrac{1}{2}x^{3^2})p_1+[x^3(c\varphi+\psi)+x^2(c\psi-d\cdot\varphi)]\,p_2\\
& +[x^3(b+c\psi+\theta)+x^2(c\theta-d\cdot\psi+e)]p_3+(p-bx^4)p_4,
\end{aligned} \tag{33.55}$$
$$b, c, d, e, p = \text{const},$$

where the functions φ, ψ and θ satisfy the set of equations

$$\begin{aligned}
\varphi' &= \frac{1}{\omega}\,(1-b\varphi+\psi^2-d\cdot\varphi^2+2c\varphi\psi),\\
\psi' &= \frac{1}{\omega}\,(c-e\varphi+c\psi^2-d\cdot\varphi\psi+\theta\psi+c\theta\varphi),\\
\theta' &= \frac{1}{\omega}\,(2c\psi\theta-d\cdot\psi^2+e\psi+\theta^2+b\theta+e\psi-f),
\end{aligned} \tag{33.56}$$

where

$$\omega = p - bx^4.$$

Finally, in the case of the third type $\boldsymbol{T}_3$, the matrix of the set of equations $\underset{L}{\delta} C_{ijkl} = 0$ has a non-zero sixth order minor determinant, so that this type can only admit a group of order $r \leqslant 4$.

Note that the problem of similarity does not arise here, i.e. the spaces all correspond to different structures.

This concludes the classification of real general gravitation fields by groups of motions.

EXAMPLES

1. The space defined by the metric

$$ds^2 = 2dx^1\, dx^4 + e^{-x^1}(dx^{2^2} \pm dx^{3^2})$$

is a *sub-projective* space. It alone admits a G_6 group on V_3 ([260], p. 219).

2. The space defined by the metric

$$ds^2 = 2dx^1\, dx^4 + x^{4^2}\, dx^{2^2} + 2dx^2\, dx^3$$

admits a transitive G_8 group of motions. This V_4 is a symmetrical, empty Einstein ($R_{\alpha\beta} = 0$) doubly projective space which is fully harmonic and has two absolutely parallel vector fields. It cannot correspond to any real gravitational field since its metric at a point has signature $(--++)$.

CHAPTER 6

CONFORMAL MAPPING OF EINSTEIN SPACES

§ 34. Conformal Mapping of Riemann Spaces

If two Riemann spaces V_n and $\overset{*}{V}_n$ possess metric tensors $g_{\alpha\beta}(x)$ and $\overset{*}{g}_{\alpha\beta}(x)$ respectively such that

$$\overset{*}{g}_{\alpha\beta}(x) = e^{2\sigma} g_{\alpha\beta}(x), \quad \sigma = \sigma(x), \tag{34.1}$$

the correspondence between V_n and $\overset{*}{V}_n$ is said to be conformal; two manifolds satisfying eqn. (34.1) are described as *conformal*. The geometrical significance of this definition is as follows: if (34.1) holds, the lengths of the vectors with components dx^α at points in V_n and $\overset{*}{V}_n$ with the same coordinates, differ by a factor depending only on the particular point; formula (2.4) shows that the angles between two corresponding directions at corresponding points are equal, i.e. the *mapping* of V_n on to $\overset{*}{V}_n$ is *conformal*.

Using (34.1), we can easily find the corresponding relations between the covariant metric tensors, the Riemann connectivities, and curvature tensors, etc.

The definition of $g^{\alpha\beta}$ and eqn. (34.1) shows that

$$\overset{*}{g}{}^{\alpha\beta} = e^{-2\sigma} g^{\alpha\beta}. \tag{34.2}$$

The formulae (3.1) and (3.2), defining the Christoffel symbols of the first and second kinds, and eqns. (34.1) and (34.2) show that

$$\overset{*}{\Gamma}_{\alpha,\,\beta\gamma} = e^{2\sigma}(\Gamma_{\alpha,\,\beta\gamma} + g_{\alpha\beta}\sigma_{,\,\gamma} + g_{\alpha\gamma}\sigma_{,\,\beta} - g_{\beta\gamma}\sigma_{,\,\alpha}), \tag{34.3}$$

$$\overset{*}{\Gamma}{}^{\alpha}_{\beta\gamma} = \Gamma^{\alpha}_{\beta\gamma} + \delta^{\alpha}_{\beta}\sigma_{,\,\gamma} + \delta^{\alpha}_{\gamma}\sigma_{,\,\beta} - g_{\beta\gamma}g^{\alpha\tau}\sigma_{,\,\tau}, \tag{34.4}$$

where $\sigma_\alpha \equiv \dfrac{\partial\sigma}{\partial x^\alpha} \equiv \sigma_{,\,\alpha}$ ([77], p. 89). The relation between the curvature tensors of the spaces V_n and $\overset{*}{V}_n$ found by replacing $g_{\alpha\beta}$ and $\Gamma_{\sigma,\,\alpha\beta}$ in (5.7) with $\overset{*}{g}_{\alpha\beta}$ and $\overset{*}{\Gamma}_{\sigma,\,\alpha\beta}$ by means of (34.1), (34.2) and (34.3), and expressing the right-hand side in terms of σ and $R_{\sigma\beta\gamma\delta}$. This leads to the expression

$$e^{-2\sigma}\overset{*}{R}_{\alpha\beta\gamma\delta} = R_{\alpha\beta\gamma\delta}+g_{\alpha\delta}\sigma_{\beta\gamma}+g_{\beta\gamma}\sigma_{\alpha\delta}-g_{\alpha\gamma}\sigma_{\beta\delta} -g_{\beta\delta}\sigma_{\alpha\gamma}+\Delta_1\sigma(g_{\alpha\delta}g_{\beta\gamma}-g_{\alpha\gamma}g_{\beta\delta}), \tag{34.5}$$

using the compact notation

$$\sigma_{\alpha\beta} = \sigma_{,\alpha\beta}-\sigma_{,\alpha}\,\sigma_{,\beta} \quad (\sigma_\alpha = \sigma_{,\alpha}),$$

and $\Delta_1\sigma$ the first Beltrami operator defined by formula (3.18). Contracting (34.5) over the indices α, δ by means of the tensor $\overset{*}{g}{}^{\alpha\delta}$ given by (34.2), provides the following relation between the Ricci tensors,

$$\overset{*}{R}_{\beta\gamma} \equiv \overset{*}{g}{}^{\alpha\delta}\overset{*}{R}_{\alpha\beta\gamma\delta} = R_{\beta\gamma}+(n-2)\sigma_{\beta\gamma}+[\Delta_2\sigma+(n-2)\Delta_1\sigma]g_{\beta\gamma}, \tag{34.6}$$

where $\Delta_2\sigma$ is the second Beltrami operator [cf. (3.18)]. Finally, a second contraction gives the relation between the scalar curvatures of V_n and $\overset{*}{V}_n$;

$$\overset{*}{R} \equiv \overset{*}{g}{}^{\beta\gamma}\overset{*}{R}_{\beta\gamma} = e^{-2\sigma}[R+2(n-1)\Delta_2\sigma+(n-1)(n-2)\Delta_1\sigma]. \tag{34.7}$$

The integral coefficients on the right-hand side of the equation shows that the cases $n=1$ and $n=2$ need special consideration; both cases, however, are trivial. The case $n=1$ is clearly without interest, and $n=2$ can also be excluded, for any two binary quadratic forms can be reduced to $\lambda(e_1\,dx^{1^2}+e_2\,dx^{2^2})$, e_1, $e_2 = \pm 1$, so that any two V_2 with the same signatures are conformal; a conformal correspondence is defined by means of real coefficients. We therefore assume that $n>2$.

We can rewrite (34.7) in the form

$$\overset{*}{g}_{\alpha\beta}\overset{*}{R} = [R+2(n-1)\Delta_2\sigma+(n-1)(n-2)\Delta_1\sigma]g_{\alpha\beta} \tag{34.8}$$

by means of (34.1). Eliminating the second Beltrami operator $\Delta_2\sigma$ between (34.6) and (34.8) shows that

$$\sigma_{\alpha\beta} = \frac{1}{n-2}(\overset{*}{R}_{\alpha\beta}-R_{\alpha\beta}) -\frac{1}{2(n-1)(n-2)}(\overset{*}{R}\overset{*}{g}_{\alpha\beta}-Rg_{\alpha\beta})-\frac{1}{2}\Delta_1\sigma g_{\alpha\beta}. \tag{34.9}$$

Equation (34.2), and raising one index in (34.5), provide the following relation between the mixed components of the curvature tensors:

$$\overset{*}{R}{}^{\alpha}{}_{\beta\gamma\delta} = R^{\alpha}{}_{\beta\gamma\delta}+\delta^{\alpha}_{\delta}\sigma_{\beta\gamma}-\delta^{\alpha}_{\gamma}\sigma_{\beta\delta}+g^{\alpha\nu}(g_{\beta\gamma}\sigma_{\nu\delta}-g_{\beta\delta}\sigma_{\nu\gamma})+\Delta_1\sigma(\delta^{\alpha}_{\delta}g_{\beta\gamma}-\delta^{\alpha}_{\gamma}g_{\beta\delta}). \tag{34.10}$$

We now introduce the geometric object

$$C^{\alpha}{}_{\beta\gamma\delta} \equiv R^{\alpha}{}_{\beta\gamma\delta}+\frac{1}{n-2}(\delta^{\alpha}_{\gamma}R_{\beta\delta}-\delta^{\alpha}_{\delta}R_{\beta\gamma}+R^{\alpha}_{\gamma}g_{\beta\delta}-R^{\alpha}_{\delta}g_{\beta\gamma})$$

$$+\frac{R}{(n-1)(n-2)}(\delta^{\alpha}_{\delta}g_{\beta\gamma}-\delta^{\alpha}_{\gamma}g_{\beta\delta}), \quad (34.11)$$

which is a tensor in virtue of its construction; it was called the *conformal curvature* tensor by Weyl, who first introduced it in analysing conformal Riemann spaces ([49]). It is easy to show that the conformal curvature tensor in any V_3 is zero (cf. example 3, § 5).

Eliminating the tensor $\sigma_{\alpha\beta}$ in (34.10) by means of (34.9), provides the following characteristic relation for conformal Riemann manifolds.

$$\overset{*}{C}{}^{\alpha}{}_{\beta\gamma\delta} = C^{\alpha}{}_{\beta\gamma\delta}. \quad (34.12)$$

EXAMPLES

1. Show that if

$$R_{\beta\gamma\delta} \equiv R_{\beta\gamma,\,\delta}-R_{\beta\delta,\gamma}+\frac{1}{2(n-1)}(g_{\beta\delta}R_{,\gamma}-g_{\beta\gamma}R_{,\,\delta}),$$

$$R^{\sigma}{}_{\gamma\delta} = g^{\sigma\alpha}R_{\alpha\gamma\delta},$$

the following relation holds for the covariant derivative of the conformal curvature tensor:

$$C^{\alpha}{}_{\beta(\gamma\delta,\,\sigma)} = \frac{1}{n-2}(R_{\beta[\delta\sigma}\delta^{\alpha}_{\gamma]}+g_{\beta[\delta}R^{\alpha}{}_{\gamma\sigma]}). \quad (34.13)$$

2. Show also that

$$R^{\sigma}{}_{\sigma\alpha} = 0, \quad C^{\sigma}{}_{\alpha\beta\gamma,\,\sigma} = \frac{n-3}{n-2}R_{\alpha\beta\gamma}. \quad (34.14)$$

3. In order that a V_3 should map conformally on an S_3 it is necessary and sufficient that the tensor

$$R_{\alpha\beta\gamma} = 0 \qquad \text{([58], p. 80).} \quad (34.15)$$

4. In order that a V_n $(n > 3)$ should be conformal with an S_n, it is necessary and sufficient that

$$C_{\alpha\beta\gamma\delta} = 0 \qquad \text{([58], p. 80).} \quad (34.16)$$

5. Show that when $n > 2$, the most general conformal mapping of a Euclidean space on itself can be represented as the product of inversions with respect to a hypersphere, motions and similarity transformations ([14], pp. 375–6).

§ 35. Conformal Mapping of Riemann Spaces on Einstein Spaces

The problem of the conformal mapping of Riemann spaces on Einstein spaces is a natural extension of the problem of conformally flat Riemann manifolds. First raised by Brinkmann, who provided the basic solution [70], [71], [73], it has since been considered by other authors [113], [114], [129]–[133], [161], [174], [179].

We now assume that $\overset{*}{V}_n$ is an Einstein space and that $n > 3$.

Hence

$$\overset{*}{R}_{\alpha\beta} = \frac{\overset{*}{R}}{n}\overset{*}{g}_{\alpha\beta}, \quad \overset{*}{R} = \text{const.} \tag{35.1}$$

We introduce the tensor

$$\omega_{\alpha\beta} = \sigma_{,\alpha\beta} - \sigma_{,\alpha}\sigma_{,\beta} + \tfrac{1}{2}\Delta_1\sigma g_{\alpha\beta} \tag{35.2}$$

and the scalar

$$\omega = g^{\alpha\beta}\omega_{\alpha\beta}.$$

Equation (34.6) can now be restated in the form

$$\overset{*}{R}_{\alpha\beta} = R_{\alpha\beta} + (n-2)\omega_{\alpha\beta} + \omega g_{\alpha\beta} \tag{35.3}$$

and (34.7) in the form

$$e^{2\sigma}\overset{*}{R} = R + 2(n-1)\omega.$$

As the Ricci tensor of an Einstein space satisfied condition (35.1), eqn. (35.3) reduces to

$$\omega_{\alpha\beta} + \frac{1}{n-2}\left(R_{\alpha\beta} - \frac{R}{2(n-1)}g_{\alpha\beta}\right) = \frac{\overset{*}{R}e^{2\sigma}}{2n(n-1)}g_{\alpha\beta}. \tag{35.4}$$

The required differential equations assume a more convenient form, if we denote the second term on the left-hand side (35.4) by $-\tau_{\alpha\beta}$ so that

$$\tau_{\alpha\beta} = -\frac{1}{n-2}\left(R_{\alpha\beta} - \frac{R}{2(n-1)}g_{\alpha\beta}\right).$$

Using (35.2) reduces eqn. (35.4) to the form

$$\sigma_{,\alpha\beta} - \sigma_{,\alpha}\sigma_{,\beta} + \nu g_{\alpha\beta} = \tau_{\alpha\beta}, \tag{35.5}$$

where

$$\nu = \tfrac{1}{2}\Delta_1\sigma - \frac{\overset{*}{R}}{2n(n-1)}e^{2\sigma}, \quad \overset{*}{R} = \text{const.} \tag{35.6}$$

This set of equations assumes a more convenient form if the derivative of the scalar ν is expressed as

$$\nu_{,\alpha} = \nu\sigma_{,\alpha}+\sigma_{,\varrho}\tau^{\varrho}_{\alpha} \tag{35.7}$$

by means of (35.5) and (35.6). If the scalars σ and ν satisfy eqns. (35.5) and (35.7), they also satisfy (35.6) for some particular constant R. In (35.5) and (35.6) $\sigma_{,\alpha}$ is by definition a *gradient* vector. But the set of equations (35.5), (35.7) can be replaced by

$$\sigma_{\alpha,\beta}-\sigma_{\alpha}\sigma_{\beta}+\nu g_{\alpha\beta} = \tau_{\alpha\beta}, \tag{35.8}$$

$$\nu_{,\alpha} = \nu\sigma_{\alpha}+\sigma_{\beta}\tau^{\beta}_{\alpha}, \tag{35.9}$$

where σ_{α} must now be a gradient vector field and the scalar ν is arbitrary. Clearly, any vector σ_{α} satisfying (35.8) must also satisfy the condition

$$\sigma_{\alpha,\beta}-\sigma_{\beta,\alpha} = 0$$

for a gradient vector, so that we can put $\sigma_{\alpha} = \sigma_{,\alpha}$.

Thus ν and σ satisfy (35.5) and (35.7); σ is determined by given components $\sigma_{,\alpha}$ to within a trivial constant. V_n can be mapped conformally on $\overset{*}{V}_n$ only if the set of equations (35.8) and (35.9) are compatible. A necessary condition that (35.8) be integrable is obtained by again differentiating the set covariantly, alternating over the indices of differentiation and applying the Ricci identities (5.13). Using (35.8) and (35.9), we obtain

$$\sigma^{\varrho}C_{\varrho\alpha\beta\gamma} = S_{\alpha\beta\gamma}, \tag{35.10}$$

where $C_{\varrho\alpha\beta\gamma}$ is the conformal curvature tensor defined by (34.11) which can also be written in the form

$$C_{\alpha\beta\gamma\delta} = R_{\alpha\beta\gamma\delta}+g_{\alpha\delta}\tau_{\beta\gamma}-g_{\alpha\gamma}\tau_{\beta\delta}+g_{\beta\gamma}\tau_{\alpha\delta}-g_{\beta\delta}\tau_{\alpha\gamma}, \tag{35.11}$$

and

$$S_{\alpha\beta\gamma} = \tau_{\alpha\beta,\gamma}-\tau_{\alpha\gamma,\beta}, \quad \sigma^{\alpha} = g^{\alpha\varrho}\sigma_{\varrho}. \tag{35.12}$$

A consistent set (35.8) and (35.9) has a most general solution when the integrability condition (35.10) is satisfied identically for any σ^{ϱ}. However, this implies that the coefficients of the σ^{ϱ} are identically zero; $C_{\alpha\beta\gamma\delta} = 0$ and therefore $S_{\alpha\beta\gamma} = 0$, i.e. the $\overset{*}{V}_n$ are *conformally flat* (cf. examples 3, 4, § 34). We neglect this trivial case and assume that (35.10) is not satisfied identically; we must now also consider the set of equations obtained by differentiating (35.10) covariantly. This process of successive differentiation of the compatibility conditions must be continued until it leads either to a contradiction, or to a set of conditions satisfied identically by virtue of the previous sets.

In the second stage in this process, we differentiate (35.10) covariantly and simplify the resulting expressions by means of the original equations (35.8) and (35.9). The second set of conditions is

$$\sigma^{\varrho}(C_{\varrho\alpha\beta\gamma,\,\delta}+g_{\varrho\delta}S_{\alpha\beta\gamma})-\nu C_{\delta\alpha\beta\gamma} = S_{\alpha\beta\gamma,\,\delta}-\tau_{\varrho\delta}C^{\varrho}{}_{\alpha\beta\gamma}. \tag{35.13}$$

As $\tau_{\alpha\beta}$, $S_{\alpha\beta\gamma}$, $C_{\alpha\beta\gamma\delta}$ are independent of the unknowns σ_α and ν, successively differentiating (35.13) and eliminating $\sigma_{\alpha,\,\beta}$ and $\nu_{,\,\alpha}$ by means of the original equations must lead to

$$\sigma^{\varrho}A_{\varrho}-\nu B = T, \tag{35.14}$$

where A_ϱ, B and T are tensors of order $q+4$, $q+3$, $q+3$, respectively; then $q = 0$, 1 corresponds to (35.10) and (35.13) respectively. It is not difficult to find recurrence relations connecting the $(q+1)$th compatibility conditions with the first q. These relations can be obtained by again differentiating (35.14). If the required $(q+1)$th set has coefficients $\boldsymbol{A}^{\varrho}$, $\boldsymbol{B}$ and $\boldsymbol{T}$, then

$$\left.\begin{aligned} \boldsymbol{A}^{\varrho} &= A^{\varrho}{}_{,\,\pi}+\delta^{\varrho}_{\pi}T-B\tau^{\varrho}_{\pi}, \\ \boldsymbol{B} &= B_{,\,\pi}+A_{\pi}, \\ \boldsymbol{T} &= T_{,\,\pi}-A^{\varrho}\tau_{\varrho\pi}. \end{aligned}\right\} \tag{35.15}$$

We thus have the following theorem: in order that a V_n should map *conformally on an Einstein space* $\overset{*}{V}_n$, *the complete sequence of compatibility conditions* (35.14) *must form a consistent set of equations for* σ_α *and* ν.

In general, a V_n is not conformal with an Einstein space; the theorem just proved gives the necessary conditions for conformal correspondence. In order to obtain a set of necessary and *sufficient* conditions, we denote the qth compatibility conditions (35.14) by S_q and consider the complete sequence of conditions

$$S_0,\ S_1,\ \ldots,\ S_p,\ \ldots \tag{35.16}$$

If, at some point, this set of equations for the unknowns σ_α and ν is no longer consistent, the corresponding V_n cannot be conformally mapped on an Einstein space. Hence the V_n which can be so mapped correspond to the case when on reaching some number p, any solution for $S_0, \ldots, S_p$ also satisfies S_{p+1}. We first prove the following theorem: *if the sequence* (35.16) *as far as* p, *has* r *linearly independent solutions which all satisfy* S_{p+1}, *the set* (35.8) *and* (35.9) *is compatible and its solution contains* $(r-1)$ *arbitrary constants.*

Let us assume that σ_α and ν satisfy equations S_q and S_{q+1}. When the coefficients (35.15) are introduced after differentiating S_q, we obtain

$$A^\varrho(\sigma_{\varrho,\,\alpha}-\sigma_\varrho\sigma_\alpha+\nu g_{\varrho\alpha}-\tau_{\varrho\alpha})-B(\nu_{,\,\alpha}-\nu\sigma_\alpha-\sigma_\varrho\tau^\varrho_\alpha)+(\sigma_\varrho \boldsymbol{A}^\varrho-\nu\boldsymbol{B}-\boldsymbol{T}) = 0.$$

S_{q+1} shows that

$$A^\varrho(\sigma_{\varrho,\,\alpha}-\sigma_\varrho\sigma_\alpha+\nu g_{\varrho\alpha}-\tau_{\varrho\alpha})-B(\nu_{,\,\alpha}-\nu\sigma_\alpha-\sigma_\varrho\tau^\varrho_\alpha) = 0. \tag{35.17}$$

Denoting by $\boldsymbol{S}_q$ the homogenous equation corresponding to S_q,

$$A^\varrho\sigma_\varrho-\nu B = 0,$$

we see that

$$\sigma_{\varrho,\,\alpha}-\sigma_\varrho\sigma_\alpha+\nu g_{\varrho\alpha}-\tau_{\varrho\alpha}, \quad \nu_{,\,\alpha}-\nu\sigma_\alpha-\sigma_\varrho\tau^\varrho_\alpha$$

is a solution of $\boldsymbol{S}_q$ for any α.

We first assume that the set (35.16) has a unique solution, i.e. $r = 1$. If this solution is σ_α, ν it must also satisfy S_{p+1}.

The homogeneous set of equations $\boldsymbol{S}_0, \boldsymbol{S}_1, \ldots, \boldsymbol{S}_p$ is then incompatible and the result obtained above shows that σ_α and ν satisfy (35.8) and (35.9), i.e. the theorem is true for $r = 1$.

Now suppose that $\sigma^{(\varrho)}_\alpha$, $\nu^{(\varrho)}$ ($\varrho = 1, \ldots, r-1$, $r>1$) is a complete set of linearly independent solutions of the homogeneous sequence $\boldsymbol{S}_0 \ldots, \boldsymbol{S}_p$. Let $\overset{*}{\sigma}_\alpha$, $\overset{*}{\nu}$ be any particular solution of the inhomogeneous sequence (35.16). It follows from (35.17) that

$$\overset{*}{\sigma}_{\alpha,\,\beta}-\overset{*}{\sigma}_\alpha\overset{*}{\sigma}_\beta+\overset{*}{\nu}g_{\alpha\beta}-\tau_{\alpha\beta} = \sum_\varrho \mu^{(\varrho)}_\beta\sigma^{(\varrho)}_\alpha, \tag{35.18}$$

$$\overset{*}{\nu}_{,\,\beta}-\overset{*}{\nu}\overset{*}{\sigma}_\beta-\overset{*}{\sigma}_\lambda\tau^\lambda_\beta = \sum_\varrho \mu^{(\varrho)}_\beta\nu^{(\varrho)}. \tag{35.19}$$

If $\overset{*}{\sigma}_\alpha+\sigma^{(\varrho)}_\alpha$ and $\overset{*}{\nu}+\nu^{(\varrho)}$ are substituted for $\overset{*}{\sigma}_\alpha$, $\overset{*}{\nu}$, in these equations, and (35.18) and (35.19) subtracted from the resulting equations, we obtain

$$\sigma^{(\varrho)}_{\alpha,\,\beta}-\overset{*}{\sigma}_\alpha\sigma^{(\varrho)}_\beta+\nu^{(\varrho)}g_{\alpha\beta} = \sum_\sigma \chi^{(\sigma\varrho)}_\beta\sigma^{(\sigma)}_\alpha, \tag{35.20}$$

$$\nu^{(\varrho)}_{,\,\beta}-\sigma^{(\varrho)}_\beta\overset{*}{\nu}-\sigma^{(\varrho)}_\lambda\tau^\lambda_\beta = \sum_\sigma \chi^{(\sigma\varrho)}_\beta\nu^{(\sigma)}, \tag{35.21}$$

where $\sigma, p = 1, \ldots, r-1$. The general solution of inhomogeneous sequence (35.16) is of the form

$$\sigma_\alpha = \overset{*}{\sigma}_\alpha+\sum_\varrho b^{(\varrho)}\sigma^{(\varrho)}_\alpha, \quad \nu = \overset{*}{\nu}+\sum_\varrho b^{(\varrho)}\nu^{(\varrho)} \quad (\varrho = 1, \ldots, r-1). \tag{35.22}$$

We determine the functions $b^{(\varrho)}$ so that σ_α and ν satisfy equations (35.8) and (35.9). Substituting from (35.22) into (35.8) and (35.9) shows that the $b^{(\varrho)}$ must satisfy the equations

$$b^{(\varrho)}_{,\alpha}+\mu^{(\varrho)}_{\alpha}+\sum_{\sigma} b^{(\sigma)}\chi^{(\sigma\varrho)}_{\alpha}-b^{(\varrho)}\sigma_{\alpha} = 0, \tag{35.23}$$

where σ_α is given by (35.22). The proof that the system (35.23) is fully integrable can be based on the four equations (35.18)–(35.21). This completes the proof of the theorem. Thus *in order that a* V_n *map conformally on an Einstein space* $\overset{*}{V}_n$, *and that this mapping be carried out by means of* σ, ν, *depending on* $r-1$ *arbitrary parameters, it is necessary and sufficient that an integer p exists for which the sequence* (35.16) *has r linearly independent solutions, all satisfying the next set* S_{p+1} *in the sequence of compatibility conditions* [71].

As noted above in deriving (35.8) and (35.9), the scalar σ is defined to within an arbitrary constant. We can use this to further specify $\overset{*}{R}$, the scalar curvature of the Einstein space. When $r = 1$, the above theorem shows that there is no freedom in choosing $\overset{*}{R}$. If $r > 1$, however, we can reduce $\overset{*}{R}$ to some specified value.

We now prove the following theorem; *when* $r>1$ *a mapping can always be defined for which* $\overset{*}{R} = 0$*: to find a mapping for which* $\overset{*}{R} \neq 0$, *it is sufficient to impose the condition*

$$g^{\alpha\beta}\xi_\alpha\eta_\beta \neq 0,$$

where ξ_α, $\overset{1}{\nu}$ *and* η_α, $\overset{2}{\nu}$ *are some two (possibly identical) solutions of the homogeneous sequence* $\boldsymbol{S}_0, \boldsymbol{S}_1, \ldots, \boldsymbol{S}_p$.

Now $\overset{*}{\sigma}_\alpha = \overset{*}{\sigma}_{,\alpha}$, $\overset{*}{\nu}$ and $\overset{*}{\sigma}^{(\varrho)}_{,\alpha} = \overset{*}{\sigma}_{,\alpha}+\sigma^{(\varrho)}_{\alpha}$, $\overset{*}{\nu}^{(\varrho)} = \overset{*}{\nu}+\nu^{(\varrho)}$ satisfy eqns. (35.8), 35.9) so that $\overset{*}{\sigma}$ and $\overset{*}{\sigma}^{(\varrho)}$ define conformal mappings with scalar curvatures R and $R^{(\varrho)}$ respectively. We put

$$\sigma^{(\varrho)} = \overset{*}{\sigma}^{(\varrho)}-\overset{*}{\sigma}.$$

We now determine the $b^{(\varrho)}$ in equations of the form (35.22) so that σ_α and ν satisfy (35.8), (35.9). Denoting the corresponding scalar curvature by $\overset{*}{R}$, we easily obtain

$$\sum_{\sigma,\varrho} b^{(\sigma)}b^{(\varrho)}\Delta(\sigma^{(\varrho)}, \sigma^{(\sigma)})+\sum_{\sigma} D^{(\sigma)}b^{(\sigma)}+\frac{R}{n(n-1)}e^{2\overset{*}{\sigma}} = \frac{\overset{*}{R}}{n(n-1)}e^{2\sigma},$$

where

$$\Delta(\sigma^{(\varrho)}, \sigma^{(\sigma)}) = g^{\alpha\beta}\sigma^{(\varrho)}_{,\alpha}\sigma^{(\sigma)}_{,\beta},$$

$$D^{(\sigma)} = -\Delta_1\sigma^{(\sigma)}+\frac{e^{2\overset{*}{\sigma}}}{n(n-1)}[R-R^{(\sigma)}e^{2\sigma^{(\sigma)}}].$$

This equation shows that by choosing appropriate values for the $b^{(\varrho)}$ we can assign a specified value to $\mathring{R}$, providing that

$$\Delta(\sigma^{(\varrho)}, \sigma^{(\sigma)}) \neq 0 \quad (\sigma, \varrho = 1, \ldots, r-1),$$
$$R = R^{(\varrho)} \neq 0 \quad (\varrho = 1, 2, \ldots, r-1).$$

This proves the theorem.

The following lemma is important in the particular case $n = 4$: *when* $n = 4$ *the equation* $\boldsymbol{S}_0$, *i.e.*

$$\sigma^{\alpha} C_{\alpha\beta\gamma\delta} = 0, \tag{35.24}$$

shows that either

$$C_{\alpha\beta\gamma\delta} = 0, \tag{35.25}$$

or

$$g^{\alpha\beta}\sigma_{\alpha}\sigma_{\beta} = 0. \tag{35.26}$$

The following proof was suggested by Schouten.

Equation (35.24) means that the tensor $C_{\alpha\beta\gamma\delta}$ lies in the three-dimensional space orthogonal to the vector σ^{α}. If this vector is non-isotropic, (35.26) is not true, and the hypersurface containing the tensor $C_{\alpha\beta\gamma\delta}$ is non-isotropic (cf. § 7), i.e. it is a Riemannian V_3 with a non-degenerate metric.

Any tensor P_{ijkl} in V_3 possessing the algebraic properties of the curvature tensor can be represented in the form

$$P_{ijkl} = g_{ik}Q_{jl} - g_{il}Q_{jk} + g_{jl}Q_{ik} - g_{jk}Q_{il},$$

where Q_{ij} is a symmetric tensor (cf. example 8, § 7). If, moreover, $g^{il}P_{ijkl} = 0$, the tensor $P_{ijkl} = 0$. As the conformal curvature tensor possesses all the properties stated for P_{ijkl}, $C_{\alpha\beta\gamma\delta} = 0$.

When $n = 4$ we can therefore distinguish two cases:

$$(\alpha) \quad \Delta_1\sigma \equiv g^{\alpha\beta}\sigma_{,\alpha}\sigma_{,\beta} \neq 0$$

when the mapping is *non-isotropic*, and

$$(\beta) \quad \Delta_1\sigma \equiv 0,$$

corresponding to an *isotropic* mapping.

If the metric of the V_4 is definite, and the mapping is carried out with real functions, case (β) is impossible. Both cases are possible in an indefinite metric, and require separate treatment.

EXAMPLES

1. Show that the system (35.23) is fully integrable.

2. In order that a V_4 which is not conformally flat should map conformally and non-isotropically on an Einstein space, it is necessary and sufficient that the equation S_0 has a solution (necessarily unique) which together with some scalar ν (also unique) can satisfy S_1 and S_2 (Brinkmann [71], p. 278).

3. Prove that a non-conformally flat V_4 which is not an Einstein space and for which $S_{\alpha\beta\gamma} = 0$ (cf. (35.10) and (35.12)) cannot be mapped non-isotropically on to an Einstein space ([71], p. 278).

§ 36. Conformal Mapping of Einstein Spaces on Einstein Spaces; Non-isotropic Case

If both $\overset{*}{V}_n$ and V_n are Einstein spaces

$$R_{\alpha\beta} = \frac{R}{n} g_{\alpha\beta}, \quad R = \text{const.} \tag{36.1}$$

The simplest case of this kind of mapping corresponds to

$$\sigma_{,\alpha} = 0, \quad \nu = -\frac{R}{2n(n-1)} e^{2\sigma}, \tag{36.2}$$

which clearly satisfies equations S_q. This mapping corresponds geometrically to simply changing the scales along the parametric curves, and is therefore *trivial*.

It was shown in § 35 that non-trivial mappings are possible in which a space with a non-zero scalar curvature is mapped on to a space of zero scalar curvature, and vice versa. Denoting spaces with non-zero and zero scalar curvatures by G_n and $\overset{0}{G}_n$ respectively and applying the theorems derived in § 35, we see that if a $G_n(\overset{0}{G}_n)$ can be mapped on a $\overset{0}{G}_n(G_n)$ in more than one fundamentally different way it can also be mapped non-trivially on to $G_n(\overset{0}{G}_n)$. Moreover, if a G_n can be mapped non-trivially on a G_n it can also be mapped on a $\overset{0}{G}_n$. Finally, if a $\overset{0}{G}_n$ admits a non-trivial mapping on a $\overset{0}{G}_n$ it can also be mapped on a G_n, except perhaps when the following relation holds between the two functions σ, σ, defining two non-trivial mappings

$$g^{\alpha\beta}\sigma_{,\alpha}\sigma_{,\beta} = 0.$$

The following substitutions are convenient for the subsequent analysis:

$$\sigma = -\ln\psi, \quad \nu\psi = N - \tfrac{1}{2}K\psi, \quad K = \frac{R}{n(n-1)}. \tag{36.3}$$

As

$$\overset{*}{R}_{\alpha\beta} = \frac{1}{n}\overset{*}{R}\overset{*}{g}_{\alpha\beta}, \tag{36.4}$$

and

$$\overset{*}{g}_{\alpha\beta} = \frac{1}{\psi^2}g_{\alpha\beta}, \tag{36.5}$$

replacing all the $\overset{*}{V}_n$ tensors in (36.4) by the corresponding tensors in V_n and ψ shows that $\overset{*}{V}_n$ is an Einstein space only when

$$\psi_{,\alpha\beta} = Ng_{\alpha\beta}, \tag{36.6}$$

$$2N\psi = K\psi^2 + \Delta_1\psi - \overset{*}{K}, \tag{36.7}$$

where

$$\Delta_1\psi = g^{\alpha\beta}\psi_{,\alpha}\psi_{,\beta}, \quad \overset{*}{K} = \frac{\overset{*}{R}}{n(n-1)}.$$

We note that N is a linear function of ψ; for differentiating (36.7) covariantly, we obtain

$$N_{,\alpha}\psi + N\psi_{,\alpha} = K\psi\psi_{,\alpha} + g^{\sigma\tau}\psi_{,\sigma}\psi_{,\tau\alpha}$$

and using (36.6), we find that $N_{,\alpha} = K\psi_{,\alpha}$, so that

$$N = K\psi + D, \quad D = \text{const.}$$

Substituting for N in (36.6), (36.7), we obtain

$$\psi_{,\alpha\beta} = (K\psi + D)g_{\alpha\beta}, \tag{36.8}$$

$$\Delta_1\psi = K\psi^2 + 2D\psi + \overset{*}{K}. \tag{36.9}$$

Equation (36.9) is a direct consequence of (36.8) for

$$[\Delta_1\psi - (K\psi^2 + 2D\psi)]_{,\alpha} = 2g^{\sigma\tau}\psi_{,\sigma}\psi_{,\tau\alpha} - 2K\psi\psi_{,\alpha} - 2D\psi_{,\alpha} = 0,$$

and since $\overset{*}{K}$ is constant, (36.9) must be true.

If σ is replaced by ψ, the definition of an *isotropic* mapping given in the previous section becomes

$$\Delta_1\psi \equiv 0.$$

Equation (36.9) shows that a non-trivial mapping is isotropic only when

$$K = \overset{*}{K} = D = 0. \tag{36.10}$$

Hence *an isotropic mapping of one Einstein space on to another is only possible when both scalar curvatures are zero.*

Suppose that $\psi^{(1)}$ and $\psi^{(2)}$ are two solutions of the set (36.8) with corresponding constants $D^{(1)}$ and $D^{(2)}$ respectively; putting

$$\psi = a\psi^{(1)} + b\psi^{(2)} + c,$$

where a, b and c are constants, we find

$$\psi_{\alpha\beta} = (K\psi + D)g_{\alpha\beta},$$

where

$$D = aD^{(1)} + bD^{(2)} - Kc,$$

i.e. (36.8) behaves as a linear set. The sequence of solutions $\psi^{(1)}, \psi^{(2)}, \ldots, \psi^{(p)}$ is *independent* if there are no relations of the type

$$\sum_{r=1}^{p} c_r \psi^{(r)} + d = 0; \quad c_r, d = \text{const} \neq 0.$$

Conformal mappings corresponding to independent solutions are themselves said to be *independent*. Hence the most general Einstein space $\overset{*}{V}_n$ conformal to an Einstein space V_n with the metric ds^2 is

$$\overset{*}{ds}{}^2 = \frac{1}{(c_1\psi^{(1)} + \ldots + c_r\psi^{(r)} + d)^2}\, ds^2.$$

We now consider the case of isotropic mapping when

$$\Delta_1 \psi \neq 0. \tag{36.11}$$

As the hypersurface ψ = constant is non-isotropic it can be used as a coordinate surface. We introduce a non-isotropic and *quasi-semigeodesic* system of coordinates; this differs from a semi-geodesic system by taking some arbitrary function of the arc length and not the arc length as the parameter along the coordinate lines. Clearly, such a system can always be constructed if (36.11) holds, and $x^{n'} = \psi\,(x^1, \ldots, x^n)$. The components of the metric tensor in this reference system are

$$g_{i'n'} = g^{i'n'} = 0, \quad \partial_{i'} g_{n'n'} = 0 \quad (i' \neq n') \tag{36.12}$$

(cf. § 7). Using the notation

$$g^{n'n'} = \frac{1}{g_{n'n'}} \equiv f,$$

equation (36.8) restated in the new system of coordinates (cf. example 1, § 36) shows that

$$f = Kx^{n^2} + 2Dx^n + B, \quad B = \text{const}, \tag{36.13}$$

$$g_{ij} = f\check{g}_{ij}, \quad \check{g}_{ij} = \check{g}_{ij}(x^1, x^2, \ldots, x^{n-1}) \quad (i, j \neq n). \tag{36.14}$$

Hence the metric of the V_n is

$$ds^2 = \frac{1}{f}\, dx^{n^2} + f\, d\sigma^2, \quad d\sigma^2 = \check{g}_{ij}\, dx^i\, dx^j, \tag{36.15}$$

in this system of coordinates, where $d\sigma^2$ is therefore the metric of a non-isotropic $\check{V}_{n-1}$.

To confirm that the metric (36.15) defines an Einstein space, we need only show that the following conditions are satisfied:

$$R_{\alpha\beta} = (n-1)Kg_{\alpha\beta}. \tag{36.16}$$

The components of the Ricci tensor for the metric (36.15) are

$$R_{ij} = \check{R}_{ij} + [\tfrac{1}{2}ff'' + \tfrac{1}{4}(n-2)f'^2]\check{g}_{ij},$$

$$R_{in} = 0, \quad R_{nn} = \tfrac{1}{2}(n-1)\frac{f''}{f},$$

where the prime denotes differentiation with respect to x^n. Therefore (36.16) is only valid when

$$\check{R}_{ij} = [(n-1)Kf - \tfrac{1}{2}ff'' - \tfrac{1}{4}(n-2)f'^2]\check{g}_{ij}, \quad f'' = 2K. \tag{36.17}$$

Substituting for f from (36.13), we obtain

$$\check{R}_{ij} = (n-2)\check{K}\check{g}_{ij} = \frac{\check{R}}{n-1}\check{g}_{ij}, \tag{36.18}$$

where

$$\check{K} = \frac{\check{R}}{(n-1)(n-2)} = BK - D^2. \tag{36.19}$$

We thus have the following theorem ([73], p. 125): *an Einstein space can be mapped conformally and non-isotropically on to another Einstein space only when its metric has the form* (36.15) *in a special system of coordinates; f is defined by formula* (36.13); *$d\sigma^2$ defines an Einstein space $\check{V}_{n-1}$ with the scalar curvature* (36.19).

If $\check{V}_{n-1}$ is a space of constant curvature with $n > 3$, so is V_n. For if $\check{V}_{n-1}$ is an S_{n-1} and f is some function of x^n, the tensor of conformal curvature is

$$C_{ijkl} = f\check{C}_{ijkl} \qquad (i, j, k, l = 1, \ldots, n-1),$$
$$C_{nijn} = 0, \quad C_{nijk} = 0$$

(§ 34), i.e. $C_{\alpha\beta\gamma\delta} = 0$. But when $n > 3$, this is the necessary and sufficient condition that V_n should be conformally flat (cf. example 4, § 34). Every conformally flat Einstein space is a space of constant curvature. In particular, when $n = 4$ the metric $d\sigma^2$ defines an S_3, since every three-dimensional Einstein space is a space of constant curvature (cf. § 13). Therefore *any Einstein space V_4 which can be mapped non-isotropically on to another is a space of constant curvature S_4.*

Examples

1. Show that if two Einstein spaces with non-zero scalar curvatures admit a non-isotropic conformal mapping on to each other they are isometric, or can be made isometric by a change of scale.

2. Prove the analogous theorem when both scalar curvatures are zero.

3. Show that if *only one* of the two scalar curvatures is zero, the spaces cannot be isometric ([73], p. 126).

4. The metric (36.15) admits the substitution $x^n \to x^{n'} + a$. Show that $K' = Ka^2 + 2Da + B$ ([73], p. 126).

5. Prove that an Einstein space admitting a conformal non-isotropic mapping on another Einstein space can also admit an isotropic mapping on still another Einstein space, but only if the $\check{V}_{n-1}$ space admits a non-trivial mapping on to an $(n-1)$-dimensional Einstein space ([73], p. 130).

§ 37. Conformal Mapping of Einstein Spaces; Isotropic Case

We now consider Einstein spaces which admit isotropic conformal mapping on some other Einstein space. According to (36.8), (36.9) and (36.10), such a mapping is only possible when the function ψ satisfies the set of equations

$$\psi_{,\alpha\beta} = 0, \tag{37.1}$$
$$\Delta_1 \psi = 0. \tag{37.2}$$

Geometrically, eqn. (37.2) implies that the surfaces ψ = constant are isotropic. Although we cannot introduce an analogous non-isotropic semi-geodesic system of coordinates based on hypersurfaces ψ = constant as in the previous section, we can introduce an isotropic semi-geodesic system (cf. § 7) in which

$$g^{in} = 0 \quad (i = 1, \ldots, n-2), \quad g^{nn} = 0, \quad g^{(n-1)n} = 1 \tag{37.3}$$

and hence

$$g_{i(n-1)} = 0, \quad g_{(n-1)(n-1)} = 0, \quad g_{n(n-1)} = 1.$$

The matrices $(g_{\alpha\beta})$ and $(g^{\alpha\beta})$ have the following forms

$$\left(\begin{array}{ccc|cc} & & & 0 & m_1 \\ & & & \cdot & \cdot \\ & g_{ij} & & \cdot & \cdot \\ & & & \cdot & \cdot \\ & & & 0 & m_{n-2} \\ \hline 0 & \ldots & 0 & 0 & 1 \\ m_1 & \ldots & m_{n-2} & 1 & m_n \end{array}\right), \quad \left(\begin{array}{ccc|cc} & & & -m^1 & 0 \\ & & & \cdot & \cdot \\ & g^{ij} & & \cdot & \cdot \\ & & & \cdot & \cdot \\ & & & -m^{n-2} & 0 \\ \hline -m_1 & \ldots & -m^{n-2} & \Omega & 1 \\ 0 & \ldots & 0 & 1 & 0 \end{array}\right)$$

$$(i, j = 1, \ldots, n-2), \tag{37.4}$$

where

$$m^i = g^{ij} m_j, \quad \Omega = -m_n + m^i m_i. \tag{37.5}$$

The metric of the V_n in this system of coordinates is therefore

$$ds^2 = g_{ij}\, dx^i\, dx^j + (2dx^{n-1} + 2m_i\, dx^i + m_n\, dx^n) dx^n. \tag{37.6}$$

As shown in § 7, this coordinate system is only appropriate to V_n with indefinite metrics; the metric must be indefinite for the isotropic mappings we are considering (cf. § 35). But (37.1) shows that in the present case the gradient vector $\psi_{,\alpha}$ is also covariantly constant. As

$$\psi_{,\alpha\beta} \equiv \partial_{\alpha\beta}\psi - \Gamma^\sigma_{\alpha\beta}\psi_{,\sigma} = 0, \tag{37.7}$$

and ψ plays the role of $x^{n'}$ in the new system of coordinates, we obtain

$$\Gamma^n_{\alpha\beta} = 0, \quad \partial_{n-1} g_{\alpha\beta} = 0 \quad (\alpha, \beta = 1, \ldots, n), \tag{37.8}$$

i.e. the $g_{\alpha\beta}$ are functions of $x^1, \ldots, x^{n-2}, x^n$ only in the new coordinate system.

It is necessary to check under what conditions a space with the metric (37.6) is an Einstein space.

We consider a $\check{V}_{n-2}$ space with metric

$$d\sigma^2 = g_{ij}\, dx^i\, dx^j \tag{37.9}$$

so that the metric of the V_n can be written as

$$ds^2 = d\sigma^2 + (2dx^{n-1} + 2m_i\, dx^i + m_n\, dx^n)\, dx^n, \tag{37.10}$$

and we denote all quantities belonging to $\check{V}_{n-2}$ by the symbol '$\check{\ }$'.

If we put

$$G_{ij} = \partial_n g_{ij}, \quad G^i_j = g^{ik}G_{kj}, \quad G^{ij} = g^{ik}G^j_k, \quad G = G^i_i, \tag{37.11}$$

$$F_{ij} = G_{ij}+m_{i,\,j}-m_{j,\,i} \quad F^i_j = g^{ik}F_{kj}, \tag{37.12}$$

$$M_i = \partial_n m_i, \quad M^i = g^{ik}M_k, \quad M = \partial_n m_n \tag{37.13}$$

the non-zero components of the Ricci tensor take the form

$$R_{ij} = \check{R}_{ij}, \tag{37.14}$$

$$R_{in} = \tfrac{1}{2}(G_{,i}-F^k_{i,k}), \tag{37.15}$$

$$R_{nn} = \tfrac{1}{2}(\check{\Delta}_2 m_n+\partial_n G+\tfrac{1}{2}F^k_i F^i_k-2M^i_{,i}), \tag{37.16}$$

where

$$\check{\Delta}_2 m_n \equiv g^{ik}m_{n,\,ik}.$$

As $R_{(n-1)n} = 0$ *and* $g_{(n-1)n} = 1$, *it is obvious that the required* V_n *can only be an Einstein space of zero scalar curvature. It follows from* (37.14) *that the* $\check{V}_{n-2}$ *must also be an Einstein space of zero scalar curvature.*

Moreover, (37.15) and (37.16) show that

$$G_{,i}-F^k_{i,\,k} = 0, \tag{37.17}$$

$$\check{\Delta}_2 m_n+\partial_n G+\tfrac{1}{2}F^k_i F^i_k-2M^i_{,i} = 0. \tag{37.18}$$

Suppose that a given $\check{V}_{n-2}$ is an Einstein space of zero scalar curvature. We now show that functions m_i and m_n of $x^1, \ldots, x^{n-2}, x^n$ can be chosen so that the required V_n is also an Einstein space. We denote all terms in (37.17) independent of $m_{i,\,jk}$ by f_i, so that

$$g^{kl}(m_{i,\,kl}-m_{l,\,ik})= f_i, \tag{37.19}$$

where

$$f_i = G^k_{i,\,k}-G_{,\,i},$$

or, raising the index

$$[g^{ji}g^{kl}(\partial_l m_i-\partial_i m_l)]_{,\,k} = f^j,$$

where

$$f^j = g^{jk}f_k.$$

As the tensor inside the square brackets is antisymmetric, the divergence can be obtained from example 4, § 3 and inserted in the last equation, giving

$$\partial_k[\sqrt{g}\, g^{ji}g^{kl}(\partial_l m_i-\partial_i m_l)] = \sqrt{g} f^j. \tag{37.20}$$

A necessary condition for eqns. (37.20) to be integrable is found by non-covariant differentiation with respect to x^i and contraction over the indices i and j; we obtain

$$\partial_j(\sqrt{g}f^j) = 0. \tag{37.21}$$

We now show that this is also a sufficient condition for (37.20) to be integrable. According to example 3, § 3,

$$\frac{1}{\sqrt{g}}\partial_i(\sqrt{g}f^i) = f^i_{,i} = G^{ij}{}_{,ij} - g^{ij}G_{,ij} = \partial_n R,$$

but as $R = 0$ and $\sqrt{g} \neq 0$,

$$\partial_i(\sqrt{g}f^i) = 0,$$

i.e. if (37.21) is true the integrability conditions for (37.20) are satisfied. Hence we can find m_i so as to satisfy (37.20). We determine m_n by substituting these values of m_i into (37.18). V_n is now an Einstein space. In the particular case of metric

$$ds^2 = g_{ij}\,dx^i\,dx^j + 2dx^{n-1}\,dx^n \quad (i, j = 1, \ldots, n-2),$$

eqns. (37.18) and (37.19) reduce to

$$G_{,i} - G^k{}_{i,k} = 0, \quad \partial_n G + \tfrac{1}{2}G^i{}_k G^k{}_i = 0. \tag{37.22}$$

As, moreover, it can be shown that the non-zero components of the curvature tensor are

$$\left.\begin{aligned} R^i{}_{kjl} &= \check{R}^i{}_{kjl}, \quad R^i{}_{njk} = \tfrac{1}{2}(G^i{}_{k,j} - G^i{}_{j,k}),\\ R^{n-1}{}_{jkl} &= \tfrac{1}{2}(G_{jk,l} - G_{jl,k}),\\ R^i{}_{nnj} &= \tfrac{1}{2}(G^i{}_j + \tfrac{1}{2}G^i{}_k G^k{}_j),\\ R^{n-1}{}_{ijn} &= \tfrac{1}{2}(\partial_n G_{ij} - \tfrac{1}{2}G^k{}_i G_{kj}),\end{aligned}\right\} \tag{37.23}$$

it is obvious that in general *even if* $\check{V}_{n-2}$ *is a flat space,* V_n *is not necessarily flat* (cf. example 2, § 37).

We now consider the case $n = 4$ which is of particular interest. As any $\check{V}_{n-2}$ admitting the type of mapping we are considering has been shown to be an Einstein space of zero scalar curvature, and as any V_2 with Ricci tensor $R_{\alpha\beta} = 0$ is a flat space, the linear element of the required V_n reduces to

$$ds^2 = g_{11}\,dx^{1^2} + 2g_{12}\,dx^1\,dx^2 + g_{22}\,dx^{2^2} + (2dx^3 + 2a\,dx^1 + 2b\,dx^2 + m\,dx^4)\,dx^4, \tag{37.24}$$

where the form

$$d\sigma^2 = g_{ij}\,dx^i\,dx^j \qquad (i, j = 1, 2)$$

defines a flat space with signature $(++)$ or $(+-)$ for all values of x^4, neglecting possible multiplication by -1; a, b and m are functions of x^1, x^2, x^4. Given this metric $d\sigma^2$ defining a flat space, we integrate eqns. (37.17) and (37.18) and obtain the general form of four-dimensional Einstein spaces admitting an isotropic conformal mapping on some other Einstein space.

Examples

1. Show that a coordinate transformation of the form

$$x^{i'} = x^{i'}(x^1, \ldots, x^{n-2}, x^n) \qquad (i = 1, \ldots, n-2),$$
$$x^{n-1'} = x^{n-1}+z(x^1, x^{n-2}, x^n), \quad x^{n'} = x^n$$

does not alter the metric (37.6). Simplify this metric by means of the arbitrary functions $x^{i'}(i = 1, \ldots, n-2)$, z.

2. Show that the metric

$$ds^2 = 2dx^{n-1}\, dx^n+(ax^n+b)^2 g_{ij}\, dx^i\, dx^j \qquad (i, j = 1, \ldots, n-2),$$

where a and b are constants and $g_{ij} = g_{ij}(x^1, \ldots, x^{n-2})$ defines an $(n-2)$-dimensional Einstein space, admits an isotropic conformal mapping on some Einstein space $\overset{*}{V}_n$. Find the metric of the $\overset{*}{V}_n$. Consider the particular cases which arise when $a = 0$ and $b = 0$ ([73], p. 136).

3. When does the V_4 metric

$$ds^2 = 2dx^1\, dx^2+2dx^3\, dx^4+2f(x^1, x^4)\, dx^1\, dx^4+2g(x^2, x^4)\, dx^2\, dx^4$$

define an Einstein space and admit an isotropic conformal mapping on another Einstein space $\overset{*}{V}_4$? Determine the metric of the $\overset{*}{V}_4$.

4. Show that if an Einstein space V_4 which is not a space of constant curvature, admits a conformal mapping independent of $x^4 \equiv \psi$ on to another Einstein space, this mapping is necessarily isotropic.

5. When does a V_4 with the metric

$$ds^2 = 2dx^1\, dx^2+2dx^3\, dx^4+2f(x^1, x^4)\, dx^1\, dx^4$$

satisfy the conditions of example 4?

6. The V_4 of example 5 can be embedded as a hypersurface in a five-dimensional Euclidean space [76].

7. The V_4 of example 3 can be embedded as a five-dimensional surface in a seven-dimensional Euclidean space.

8. Show that V_4 with metric

$$ds^2 = 4[dx^1\, dx^2+dx^3\, dx^4+(x^1\, dx^1+x^3\, dx^3)^2]$$

is an Einstein space and can be embedded in a five-dimensional pseudo-Euclidean space with metric

$$ds^2 = dy^{1^2} - dy^{2^2} + dy^{3^2} - dy^{4^2} + dy^{5^2}$$

by means of transformation

$$y^1 = x^1 + x^2, \quad y^2 = x^1 - x^2, \quad y^3 = x^3 + x^4,$$
$$y^4 = x^3 - x^4, \quad y^5 = x^{1^2} + x^{3^2}.$$

9. Show that the metric

$$ds^2 = \frac{e_1\, ds^{1^2} + \ldots + e_p\, dx^{p^2}}{\left[1 + \frac{n-2}{p-1}\frac{K_0}{4}(e_1 x^{1^2} + e_p\, dx^{p^2})\right]^2}$$
$$+ \frac{e_{p+1}\, dx^{p+1^2} + \ldots + e_n\, dx^{n^2}}{\left[1 + \frac{n-2}{n-p-1}\frac{K_0}{4}(e_{p+1} x^{p+1^2} + \ldots + e_n x^{n^2})\right]^2}, \quad e_k = \pm 1$$

defines an Einstein space which can be embedded in an S_{n+1}, by means of the equations

$$e_1 z^{1^2} + \ldots + e_{p+1} z^{p+1^2} = \frac{p-1}{n-2}\frac{1}{K_0},$$

$$e_{p+2} z^{p+2^2} + \ldots + e_{n+2} z^{n+2^2} = \frac{n-p-1}{n-2}\frac{1}{K_0},$$

where $e = \pm 1$, and S_{n+1} is defined by the equation

$$e_1 z^{1^2} + \ldots + e_{n+2} z^{n+2^2} = \frac{1}{K_0}.$$

10. Consider the particular case of example 9 when

$$ds^2 = \frac{dx^{1^2} + dx^{2^2}}{\left[1 + \frac{K}{4}(x^{1^2} + x^{2^2})\right]^2} + \frac{dx^{3^2} + dx^{4^2}}{\left[1 + \frac{K}{4}(x^{3^2} + x^{4^2})\right]^2}$$

([76], pp. 103–4).

CHAPTER 7

GEODESIC MAPPING OF GRAVITATIONAL FIELDS

SUPPOSE we are given a gravitational field determined by the Einstein equations

$$R_{\alpha\beta}-\frac{R}{2}g_{\alpha\beta}=\lambda T_{\alpha\beta}, \tag{38.0}$$

where $T_{\alpha\beta}$ might include a cosmological term as well as the energy-momentum tensor corresponding to a real distribution of matter. One method of investigating the structure of the field is to study the behaviour of *test* particles. The behaviour of such particles does not, however, determine a unique metric, for the trajectories are geodesics (isotropic or non-isotropic depending on the rest mass of the particles), and geodesics do not determine the metric uniquely. In other words, the law of inertia for particles determines the structure of the space-time to a considerable extent but not in general uniquely. In principle the same test particle behaviour is possible in gravitational fields with different energy-momentum tensors and different functions defining the state of the material system; one field is then said to *simulate* another.

In order to formulate the mathematical problem correctly, we assume that the metrics of two gravitational fields, $g_{ij}(x)$ and $\bar{g}_{ij}(x)$, are defined in a common system of coordinates $\{x^\alpha\}$. The problem of the geodesic mapping of gravitational fields *is to determine the general form of two Riemann metrics, $g_{ij}(x)$ and $\bar{g}_{ij}(x)$ in a common system of coordinates such that the geodesics of $g_{ij}(x)$ are also geodesics of $\bar{g}_{ij}(x)$ and vice versa.* We assume that (1) $n=4$ $(i, j=1,\ldots, 4)$; (2) the signature of both metrics is $(---+)$; (3) the problem is to be solved in the C^r class of functions, $r\leqslant 4$.

This chapter gives a general solution of the problem in closed form and a full account of possible simulating fields.

§ 38. Historical Survey

A general solution of the problem for spaces with positive-definite metrics was given by Levi-Civita [8a]. His results served as the starting point for Fubini's and Solodovnikov's work on the geodesic mapping of Riemann spaces from the group point of view (projective transformations in V_n). Fubini considered the case $n = 3$ [18a], and Solodovnikov all V_n with definite metrics [256 a]. Levi-Civita showed that if ds^2 and $d\bar{s}^2$ are the positive-definite metrics of two Riemann spaces with common geodesics and if the roots

$$\underset{\varrho_1}{\lambda}, \quad \underset{\varrho_2}{\lambda}, \ldots, \underset{\varrho_i}{\lambda}, \ldots, \quad \underset{\varrho_{n-m+1}}{\lambda} \tag{38.1}$$

of the equation

$$|\bar{g}_{ij} - \lambda g_{ij}| = 0 \tag{38.2}$$

have multiplicity $\varrho_1 - \varrho_0$, $\varrho_2 - \varrho_1, \ldots, \varrho_i - \varrho_{i-1}, \ldots$, respectively, where $\varrho_0 = 0$, $\varrho_{n-m+1} = n$, there is a system of coordinates in which

$$ds^2 = \sum_{i=1}^{n-m+1} \prod_{j=1}^{n-m+1}{}' |\varphi_{\varrho_j} - \varphi_{\varrho_i}| \sum_{r,s=\varrho_{i-1}+1}^{\varrho_i} K_{rs}\, dx^r\, dx^s,$$

$$d\bar{s}^2 = \frac{1}{\varphi_{\varrho_1} \cdots \varphi_{\varrho_{n-m+1}}} \sum_{i=1}^{n-m+1} \frac{1}{\varphi_{\varrho_i}} \prod_{j=1}^{n-m+1}{}' |\varphi_{\varrho_j} - \varphi_{\varrho_i}| \sum_{r,s=\varrho_{i-1}+1}^{\varrho_i} K_{rs}\, dx^r\, dx^s, \tag{38.3}$$

where φ_{ϱ_i} is a function of x^{ϱ_i} only, if $\underset{\varrho_i}{\lambda}$ is a simple root, but constant otherwise; K_{rs} is a function of the variables $x^{\varrho_{i-1}+1}, x^{\varrho_{i-1}+2}, \ldots, x^{\varrho_i}$ but unity if $\underset{\varrho_i}{\lambda}$ is a simple root. The roots of eqn. (38.2) are real for definite quadratic forms, and the elementary divisors of the λ matrix

$$(\bar{g}_{ij} - \lambda g_{ij}) \tag{38.4}$$

linear. However, neither fact is in general true for Riemann spaces with indefinite metrics as in the general theory of relativity, and the problem is considerably more complicated.

Papers left by Shirokov show that he first solved the problem in 1934 for indefinite metrics with $n = 2$.

A classification of geodesically corresponding three-dimensional Riemann spaces and the appropriate canonical forms have been given by Petrov [165a]. His paper contains a method for solving the problem in n-dimensions, and the particular solution for one type of Riemann space; Levi-Civita's result is generalized to cover linear elementary divisors of the λ-matrix (38.4) even when some of the roots of eqn. (38.2) are complex.

Some interesting special geodesic mappings of Riemann spaces with $n > 3$ are considered by Sinyukov in [219], [256], [274].

We have already given a classification of four-dimensional Riemann manifolds admitting geodesic mappings and the corresponding canonical forms for spaces with a Minkowski signature $(- - - +)$. These results must therefore include the canonical metrics for geodesically corresponding gravitational fields of the most general form, derived from the Einstein equations (38.0) for an arbitrary energy-momentum tensor T_{ij}. V_4 with different signatures, and V_n with metric signatures $n-2$ or $-(n-2)$ can easily be dealt with by the same method, but having no physical interest are not considered.

The problem of the geodesic mapping of Einstein spaces ($T_{ij} = \sigma g_{ij}$, $n = 4$, Minkowski signature) has been solved by Petrov [438]; either the curvatures of the spaces are constant or their metrics are trivially related by $\bar{g}_{ij} = \nu g_{ij}$, $\nu =$ constant. A general solution of the problem for V_4 with Minkowski signature has been obtained by Golikov [347].

§39. Algebraic Classification of the Possible Cases

One way of subdividing the problem is to consider the algebraic structure of the two metrics

$$ds^2 = g_{ij}\, dx^i\, dx^j \quad \text{and} \quad d\bar{s}^2 = \bar{g}_{ij}\, dx^i\, dx^j \tag{39.1}$$

of the Riemann manifolds V_4 and $\bar{V}_4$ referred to a common system of coordinates x^i; Minkowski signatures are assumed.

The algebraic structure of the pair of quadratic forms is determined by the invariant elements of the pair and these (the principal directions and invariant surfaces of the tensor $\bar{g}_{ij}$) depend on the elementary divisors of the λ-matrix (38.4) ([91], p. 188).

Hence the possible cases can be classified by first enumerating all the possible Segre characteristics of the λ-matrix with the corresponding invariant elements. This classification is an algebraic problem involving a local study of the manifolds V_4 and $\bar{V}_4$.

The following Segre characteristics are possible when $n = 4$:

I. The λ-matrix has a simple Segre characteristic [1111]; the elementary divisors are linear, the roots of eqn. (38.2) real and distinct. If any root is repeated, the corresponding Segre characteristics are II [(11)11], III [(111)1], IV [(1111)], V [(11)(11)], round brackets indicating identical roots; in all the four cases, the elementary divisors are linear and the roots of eqn. (38.2) real, even when multiple.

These five cases are characteristic of definite forms (39.1) and the canonical metrics of the corresponding spaces with common geodesics are given by (38.3), except that the signature is now $(- - - +)$.

VI. $[11\bar{1}\bar{1}]$. The λ-matrix has four different elementary divisors but the roots of (38.2) are complex conjugates. Complex conjugate bases are denoted by a bar.

VII. $[(11)(\bar{1}\bar{1})]$. The roots of (38.2) are complex conjugate and multiple.

The last two cases can only occur with indefinite metrics g_{ij}, but as they clearly cannot correspond to a Minkowski signature, they are not considered further. It should be noted, however, that the corresponding canonical forms of metrics with common geodesics can be obtained automatically from the results of § 41.

VIII. $[111\bar{1}]$. Two of the roots of (38.2) are real and two complex-conjugate. In real space, $\bar{g}_{ij}$ has two principal directions and one two-dimensional invariant surface.

If we do not limit the analysis to real space and choose the four mutually orthogonal principal directions of the tensor $\bar{g}_{ij}$ as coordinate directions at each point, we obtain a non-holonomic reference system in general. Denoting the components of g_{ij} and $\bar{g}_{ij}$ in this coordinate system by $\overset{0}{g}_{ij}$ and $\overset{0}{\bar{g}}_{ij}$ respectively, we can put

$$(\overset{0}{g}_{ij}) = \begin{pmatrix} \underset{1}{e} & & & \\ & \underset{2}{e} & & \\ & & \underset{3}{e} & \\ & & & \underset{4}{e} \end{pmatrix}, \qquad (\overset{0}{\bar{g}}_{ij}) = \begin{pmatrix} \underset{1}{e}\underset{1}{\lambda} & & & \\ & \underset{2}{e}\underset{2}{\lambda} & & \\ & & \underset{3}{e}(\alpha+i\beta) & \\ & & & \underset{4}{e}(\alpha-i\beta) \end{pmatrix} \tag{39.2}$$

where $\underset{1}{\lambda}, \underset{2}{\lambda}, \alpha \pm i\beta$ are the roots of (38.2) and $\underset{1}{e} = \underset{2}{e} = \underset{3}{e} = -\underset{4}{e} = -1$.

IX. $[(11)1\bar{1}]$. Two of the roots of (38.2) are complex conjugates and the other is real and multiple.

In real space, $\bar{g}_{ij}$ has two principal directions (the two-dimensional cluster defined by these directions consists entirely of principal directions) and one invariant two-dimensional surface. In complex space, however, using a non-holonomic reference system based on the principal directions of the tensor $\bar{g}_{ij}$, we obtain (39.2) with $\underset{1}{\lambda} = \underset{2}{\lambda} = \lambda$.

A non-holonomic reference system is used in all the remaining cases, and all the matrices $(\overset{0}{g}_{ij})$ can therefore be reduced to the diagonal form.

X. [112]. There are three principal directions, one of which is isotropic, and one two-dimensional invariant surface containing the isotropic principal direction. The roots of (38.2) are real. By choosing an appropriate coordinate system, we can reduce the quadratic forms (39.1) at each point to

$$(\overset{0}{g}_{ij}) = \begin{pmatrix} \underset{1}{e} & & & \\ & \underset{2}{e} & & \\ & & 0 & \underset{3}{e} \\ & & \underset{3}{e} & 0 \end{pmatrix}, \quad (\overset{0}{\bar{g}}_{ij}) = \begin{pmatrix} \underset{1}{e}\,\underset{1}{\lambda} & 0 & 0 & 0 \\ 0 & \underset{2}{e}\,\underset{2}{\lambda} & 0 & 0 \\ 0 & 0 & 0 & \underset{3}{e}\,\underset{3}{\lambda} \\ 0 & 0 & \underset{3}{e}\,\underset{3}{\lambda} & \underset{3}{e} \end{pmatrix}, \quad \underset{1}{e} = \underset{2}{e} = \underset{3}{e} = -1$$

(cf. § 9), where $\underset{1}{\lambda}$, $\underset{2}{\lambda}$ and $\underset{3}{\lambda}$ are the real roots of (38.2), $\underset{3}{\lambda}$ corresponding to the isotropic principal direction of the tensor $\bar{g}_{ij}$. By using these equations to investigate how the vector function $\bar{g}^i_j u_i$ transforms the coordinate vectors, we can easily verify the exact nature of the invariant elements. Applying the transformation

$$x^1 = y^1, \quad x^2 = y^2, \quad x^3 = \tfrac{1}{2}(y^3 - y^4), \quad x^4 = y^3 + y^4,$$

we find

$$(\overset{0}{g}_{ij}) = \begin{pmatrix} \underset{1}{e} & & & \\ & \underset{2}{e} & & \\ & & \underset{3}{e} & \\ & & & -\underset{3}{e} \end{pmatrix}, \quad (\overset{0}{\bar{g}}_{ij}) = \begin{pmatrix} \underset{1}{e}\,\underset{1}{\lambda} & 0 & 0 & 0 \\ 0 & \underset{2}{e}\,\underset{2}{\lambda} & 0 & 0 \\ 0 & 0 & \underset{3}{e}(\underset{3}{\lambda}+1) & \underset{3}{e} \\ 0 & 0 & \underset{3}{e} & -\underset{3}{e}(\underset{3}{\lambda}-1) \end{pmatrix}. \qquad (39.3)$$

XI. [1$\bar{1}$2]. Equation (38.2) has a pair of complex conjugate roots. In real space, $\bar{g}_{ij}$ has one principal isotropic direction and two two-dimensional invariant surfaces, one containing the principal isotropic direction. In complex space on the other hand, and in a non-holonomic orthonormal tetrad based on the invariant elements of the pair of forms (39.1), we obtain (39.3) with $\underset{1}{e} = \underset{2}{e} = e$, $\underset{1}{\lambda} = \alpha + i\beta$, $\underset{2}{\lambda} = \alpha - i\beta$.

As this case clearly cannot correspond to a Minkowski signature, it will not be considered further; the solution can easily be obtained, however, from case X by the method described by Petrov [165a].

XII. [(11)2]. Both roots of eqn. (38.2) are multiple; there are only three elementary divisors. The tensor $\bar{g}_{ij}$ has one isotropic and two non-isotropic principal directions; the cluster defined by the non-isotropic principle directions consists entirely of principal directions. A two-dimensional invariant surface contains the isotropic direction. $(\overset{0}{g}_{ij})$ and $(\overset{0}{\bar{g}}_{ij})$ are given by eqns. (39.3) with $\underset{1}{\lambda} = \underset{2}{\lambda}$.

XIII. [1(12)]. The λ-matrix (38.4) has three elementary divisors and one root of (38.2) is multiple. In contrast to case X, the cluster defined by the isotropic and one of the non-isotropic principal directions consists entirely of principal directions. We must put $\underset{2}{\lambda} = \underset{3}{\lambda}$ in (39.3).

XIV. [(112)]. The root of (38.2) is multiple. There are one isotropic and two non-isotropic principal directions, and the three-dimensional cluster defined by these consists entirely of principal directions. We must put $\underset{1}{\lambda} = \underset{2}{\lambda} = \underset{3}{\lambda}$ in (39.3).

XV. [13]. There are two principal directions, one of which is isotropic. In addition, there is one two-dimensional and one three-dimensional invariant surface passing through the isotropic principal direction. The metrics at any point can be written (cf. § 9)

$$(\overset{0}{g}_{ij}) = \begin{pmatrix} \underset{1}{e} & 0 & 0 & 0 \\ 0 & 0 & 0 & \underset{2}{e} \\ 0 & 0 & \underset{2}{e} & 0 \\ 0 & \underset{2}{e} & 0 & 0 \end{pmatrix}, \quad (\overset{0}{\bar{g}}_{ij}) = \begin{pmatrix} \underset{1}{e}\underset{1}{\lambda} & 0 & 0 & 0 \\ 0 & 0 & 0 & \underset{2}{e}\underset{2}{\lambda} \\ 0 & 0 & \underset{2}{e}\underset{2}{\lambda} & \underset{2}{e} \\ 0 & \underset{2}{e}\underset{2}{\lambda} & \underset{2}{e} & 0 \end{pmatrix}, \quad \underset{1}{e} = \underset{2}{e} = -1,$$

where $\underset{1}{\lambda}$ and $\underset{2}{\lambda}$ are the roots of equation (38.2), $\underset{2}{\lambda}$ corresponding to the isotropic principal direction. The transformation

$$x' = y', \quad x^2 = \tfrac{1}{2}(y^2 - y^4), \quad x^3 = y^3, \quad x^4 = y^2 + y^4$$

reduces these matrices to the form

$$(\overset{0}{g}_{ij}) = \begin{pmatrix} \underset{1}{e} & & & \\ & \underset{2}{e} & & \\ & & \underset{2}{e} & \\ & & & -\underset{2}{e} \end{pmatrix}, \quad (\overset{0}{\bar{g}}_{ij}) = \begin{pmatrix} \underset{1}{e}\underset{1}{\lambda} & 0 & 0 & 0 \\ 0 & \underset{2}{e}\underset{2}{\lambda} & \underset{2}{e} & 0 \\ 0 & \underset{2}{e} & \underset{2}{e}\underset{2}{\lambda} & \underset{2}{e} \\ 0 & 0 & \underset{2}{e} & -\underset{2}{e}\underset{2}{\lambda} \end{pmatrix}. \tag{39.4}$$

XVI. [(13)]. There are two elementary divisors. The root of eqn. (38.2) is multiple.

The tensor $\bar{g}_{ij}$ has one isotropic and one non-isotropic principal direction; in contrast to case XV, the cluster defined by these directions consists entirely of principle directions; $\underset{1}{\lambda} = \underset{2}{\lambda}$ in (39.4).

The remaining possible Segre characteristics for $n = 4$, i.e. [22], [2$\bar{2}$], [(22)] and [4] cannot correspond to gravitational fields since their signatures are zero. Thus we need only consider cases VIII–X and XII–XVI; we can in each case construct a holonomic reference system in which to determine the tensors g_{ij} and $\bar{g}_{ij}$.

§40. The Invariant Equations for $\bar{g}_{ij}$ in a Non-holonomic Orthonormal Tetrad

Eisenhart ([77], p. 135) shows that for two Riemann spaces to have identical geodesic equations in a common system of coordinates, it is necessary and sufficient that their metric tensors satisfy the equations

$$2\mu\bar{g}_{ij,\,k}+2\bar{g}_{ij}\mu_{,\,k}+\bar{g}_{ik}\mu_{,\,j}+\bar{g}_{jk}\mu_{,\,i} = 0, \tag{40.1}$$

where

$$\mu = C\left(\frac{g}{\bar{g}}\right)^{1/5}, \quad C = \text{const.}$$

g and $\bar{g}$ are $|g_{ij}|$ and $|\bar{g}_{ij}|$ respectively and the commas denote covariant derivatives with respect to the metric g_{ij}. We choose an orthonormal tetrad $\underset{m}{i}^k$ at a given point and reduce the form $ds^2 = g_{ij}\, dx^i dx^j$ to $ds^2 = \sum\limits_{i}^{m} \underset{i}{e}\, (dx^i)^2$ and the tensors g_{ij} and $\bar{g}_{ij}$ to

$$g_{ij} = \sum_{m} \underset{m}{e}\, \underset{m}{i}_i\, \underset{m}{i}_j, \quad \bar{g}_{ij} = \sum_{r.s} \underset{r}{e}\, \underset{s}{e}\, \overset{0}{\bar{g}}_{rs} \underset{r}{i}_i \underset{s}{i}_j. \tag{40.2}$$

Equation (40.1) is then reduced to the invariant form (Petrov [165a])

$$2\mu\left\{\underset{r}{X}\overset{0}{\bar{g}}_{pq}+\sum_{s}\underset{s}{e}\overset{0}{\bar{g}}_{sq}\gamma_{spr}+\sum_{t}\underset{t}{e}\overset{0}{\bar{g}}_{tp}\gamma_{tqr}\right\}$$
$$+2\overset{0}{\bar{g}}_{pq}\underset{r}{X}\mu+\overset{0}{\bar{g}}_{pr}\underset{q}{X}\mu+\overset{0}{\bar{g}}_{qr}\underset{p}{X}\mu = 0, \tag{40.3}$$

where $\underset{r}{X}f = \dfrac{\partial f}{\partial x^i}\underset{r}{i}^i$, $\underset{5}{e} = \pm 1$, the $\overset{0}{\bar{g}}_{st}$ are given by the matrices (39.3) and (39.4), and the Ricci rotation coefficients, the invariants γ_{spq}, are defined by the equations

$$\gamma_{enk} = \underset{e}{i}_{i,\,j}\,\underset{n}{i^i}\,\underset{k}{i^j}, \quad \gamma_{nke} = -\gamma_{kne}, \quad \gamma_{nne} = 0. \tag{40.4}$$

On substituting from (39.3), (39.4) for $\overset{0}{\bar{g}}_{\alpha\beta}$ into (40.3), we obtain forty equations covering all possible values of the indices p, q, r, $(n = 4)$ for all types of Segre characteristic:

1. (X) [112].

$$\gamma_{123} = \gamma_{124} = \gamma_{132} = \gamma_{142} = \gamma_{231} = \gamma_{241} = 0;$$

$$\gamma_{134} = \gamma_{143} = -\frac{\underset{3}{e}\,\underset{1}{\lambda}}{2(\underset{3}{\lambda}-\underset{1}{\lambda})^2}\underset{1}{X}\ln\mu;$$

$$\gamma_{234} = \gamma_{243} = -\frac{\underset{3}{e}\,\underset{2}{\lambda}}{2(\underset{3}{\lambda}-\underset{2}{\lambda})^2}\underset{2}{X}\ln\mu;$$

$$\gamma_{341} = -\tfrac{1}{2}\underset{3}{e}\underset{1}{X}\ln\mu; \quad \gamma_{342} = -\tfrac{1}{2}\underset{3}{e}\underset{2}{X}\ln\mu;$$

$$\gamma_{121} = +\frac{\underset{1}{e}\,\underset{1}{\lambda}}{2(\underset{2}{\lambda}-\underset{1}{\lambda})}\underset{2}{X}\ln\mu; \quad \gamma_{212} = \frac{\underset{2}{e}\,\underset{2}{\lambda}}{2(\underset{1}{\lambda}-\underset{2}{\lambda})}\underset{1}{X}\ln\mu; \tag{40.5}$$

$$\gamma_{131} = \gamma_{141} = \frac{\underset{1}{e}\,\underset{1}{\lambda}}{2(\underset{3}{\lambda}-\underset{1}{\lambda})}\underset{3}{X}\ln\mu; \quad \gamma_{343} = \underset{3}{e}\,\frac{-\underset{3}{\lambda}-4}{4}\,\underset{3}{X}\ln\mu;$$

$$\gamma_{232} = \gamma_{242} = \frac{\underset{2}{e}\,\underset{2}{\lambda}}{2(\underset{3}{\lambda}-\underset{2}{\lambda})}\underset{3}{X}\ln\mu; \quad \gamma_{434} = \underset{3}{e}\,\frac{-\underset{3}{\lambda}+4}{4}\,\underset{3}{X}\ln\mu;$$

$$\gamma_{313} = -\underset{3}{e}\,\frac{\underset{3}{\lambda}^2-\underset{1}{\lambda}\,\underset{3}{\lambda}-\underset{1}{\lambda}}{2(\underset{3}{\lambda}-\underset{1}{\lambda})^2}\underset{1}{X}\ln\mu; \quad \gamma_{414} = \underset{3}{e}\,\frac{\underset{3}{\lambda}^2-\underset{1}{\lambda}\,\underset{3}{\lambda}+\underset{1}{\lambda}}{2(\underset{3}{\lambda}-\underset{1}{\lambda})^2}\underset{1}{X}\ln\mu;$$

$$\gamma_{323} = -\underset{3}{e}\,\frac{\underset{3}{\lambda}^2-\underset{2}{\lambda}\,\underset{3}{\lambda}-\underset{2}{\lambda}}{2(\underset{3}{\lambda}-\underset{2}{\lambda})^2}\underset{2}{X}\ln\mu; \quad \lambda_{424} = \underset{3}{e}\,\frac{\underset{3}{\lambda}^2-\underset{2}{\lambda}\,\underset{3}{\lambda}+\underset{2}{\lambda}}{2(\underset{3}{\lambda}-\underset{2}{\lambda})^2}\underset{2}{X}\ln\mu;$$

$$\underset{1}{X}(\underset{1}{\lambda}\mu^2) = 0; \quad \underset{p}{X}(\underset{1}{\lambda}\mu) = 0; \quad p = 2, 3, 4; \quad \underset{1}{X}(\underset{3}{\lambda}\mu) = 0; \quad \underset{2}{X}(\underset{3}{\lambda}\mu) = 0.$$

$$\underset{2}{X}(\underset{2}{\lambda}\mu^2) = 0; \quad \underset{p}{X}(\underset{2}{\lambda}\mu) = 0; \quad p = 1, 3, 4; \quad \underset{3}{X}(\underset{3}{\lambda}^2\mu^3) = 0; \quad \underset{4}{X}(\underset{3}{\lambda}^2\mu^3) = 0.$$

$$\underset{3}{X}\mu = \underset{4}{X}\mu.$$

2. (XII) [(11)2].

This case differs from (X) only in that $\underset{1}{\lambda} = \underset{2}{\lambda}$; so we have

$$\gamma_{132} = \gamma_{142} = \gamma_{134} = \gamma_{143} = \gamma_{133} = \gamma_{144} = \gamma_{231} = \gamma_{241}$$
$$= \gamma_{234} = \gamma_{243} = \gamma_{233} = \gamma_{244} = \gamma_{341} = \gamma_{342} = 0;$$

$$\gamma_{131} = \gamma_{141} = \frac{\underset{1}{e}\underset{1}{\lambda}}{2(\underset{3}{\lambda}-\underset{1}{\lambda})}\underset{3}{X}\ln\mu; \quad \gamma_{343} = \underset{3}{e}\,\frac{-\underset{3}{\lambda}-4}{4}\underset{3}{X}\ln\mu;$$

$$\gamma_{232} = \gamma_{242} = \frac{\underset{2}{e}\underset{1}{\lambda}}{2(\underset{3}{\lambda}-\underset{1}{\lambda})}\underset{3}{X}\ln\mu; \quad \gamma_{434} = \underset{3}{e}\,\frac{-\underset{3}{\lambda}+4}{4}\underset{3}{X}\ln\mu; \tag{40.6}$$

$$\underset{p}{X}(\underset{1}{\lambda}\mu) = 0; \quad p = 1, 2, 3, 4; \quad \underset{1}{X}\mu = \underset{2}{X}\mu = 0; \quad \underset{3}{X}\mu = \underset{4}{X}\mu;$$
$$\underset{1}{X}(\underset{3}{\lambda}\mu) = \underset{2}{X}(\underset{3}{\lambda}\mu) = 0; \quad \underset{3}{X}(\underset{3}{\lambda}^2\mu^3) = \underset{4}{X}(\underset{3}{\lambda}^2\mu^3) = 0.$$

3. (XIII) [1(12)].

In this case $\underset{2}{\lambda} = \underset{3}{\lambda} = \underset{4}{\lambda}$, so that

$$\gamma_{123} = \gamma_{124} = \gamma_{132} = \gamma_{142} = \gamma_{121} = \gamma_{131} = \gamma_{141} = \gamma_{342} = \gamma_{343} = \gamma_{344} = 0;$$
$$\gamma_{321} = \gamma_{421}; \quad \gamma_{324} = \gamma_{424}; \quad \gamma_{323} = \gamma_{423}; \quad \gamma_{322} = \gamma_{422};$$

$$\gamma_{134} = \gamma_{143} = -\underset{3}{e}\,\frac{\underset{1}{\lambda}}{2(\underset{2}{\lambda}-\underset{1}{\lambda})^2}\underset{1}{X}\ln\mu; \quad \gamma_{431} = \frac{\underset{3}{e}}{2}\underset{1}{X}\ln\mu;$$

$$\gamma_{313} = \underset{3}{e}\,\frac{\underset{1}{\lambda}+\underset{1}{\lambda}\underset{2}{\lambda}-\underset{2}{\lambda}^2}{2(\underset{1}{\lambda}-\underset{2}{\lambda})^2}\underset{1}{X}\ln\mu; \quad \gamma_{122} = \frac{\underset{2}{e}\underset{2}{\lambda}}{2(\underset{2}{\lambda}-\underset{1}{\lambda})}\underset{1}{X}\ln\mu; \tag{40.7}$$

$$\gamma_{414} = \underset{3}{e}\,\frac{\underset{1}{\lambda}-\underset{1}{\lambda}\underset{2}{\lambda}+\underset{2}{\lambda}^2}{2(\underset{1}{\lambda}-\underset{2}{\lambda})^2}\underset{1}{X}\ln\mu; \quad \underset{1}{X}(\underset{1}{\lambda}\mu^2) = 0;$$

$$\underset{p}{X}(\underset{1}{\lambda}\mu) = 0; \quad p = 2, 3, 4; \quad \underset{q}{X}(\underset{2}{\lambda}\mu) = 0; \quad q = 1, 2, 3, 4;$$
$$\underset{p}{X}(\mu) = 0; \quad p = 2, 3, 4.$$

4. (XIV) [(112)].

This case differs from (XIII) in that $\lambda_1 = \lambda_2$.

$$\gamma_{434} = \gamma_{343} = \gamma_{341} = \gamma_{342} = 0;$$
$$\gamma_{313} = \gamma_{413}; \quad \gamma_{314} = \gamma_{414}; \quad \gamma_{131} = \gamma_{141}; \quad \gamma_{313} = \gamma_{413}; \tag{40.8}$$

$$\gamma_{323} = \gamma_{423}; \quad \gamma_{324} = \gamma_{424}; \quad \gamma_{232} = \gamma_{242}; \quad \gamma_{323} = \gamma_{423};$$

$$\gamma_{321} = \gamma_{421}; \quad \gamma_{312} = \gamma_{412};$$

$$\underset{p}{X}(\lambda\mu) = 0; \quad \underset{p}{X}\mu = 0; \quad p = 1,\ 2,\ 3,\ 4.$$

5. (XV) [13].

$$\gamma_{123} = \gamma_{132} = \gamma_{134} = \gamma_{143} = -\underset{2}{e}\frac{\underset{1}{\lambda}}{2(\underset{1}{\lambda}-\underset{2}{\lambda})^2}\underset{1}{X}\ln\mu;$$

$$\gamma_{321} = \gamma_{341} = \gamma_{311} = 0;$$

$$\gamma_{124} = \gamma_{142} = -\underset{2}{e}\frac{\underset{1}{\lambda}}{2(\underset{1}{\lambda}-\underset{2}{\lambda})^3}\underset{1}{X}\ln\mu; \quad \gamma_{421} = \underset{2}{e}\underset{1}{X}\ln\mu;$$

$$\gamma_{212} = \underset{2}{e}\frac{\underset{1}{\lambda}+\underset{2}{\lambda}(\underset{1}{\lambda}-\underset{2}{\lambda})^2}{2(\underset{1}{\lambda}-\underset{2}{\lambda})^3}\underset{1}{X}\ln\mu; \quad \gamma_{313} = \underset{2}{e}\frac{\underset{2}{\lambda}}{2(\underset{1}{\lambda}-\underset{3}{\lambda})}\underset{1}{X}\ln\mu;$$

$$\gamma_{414} = \underset{2}{e}\frac{\underset{1}{\lambda}-\underset{2}{\lambda}(\underset{1}{\lambda}-\underset{2}{\lambda})^2}{2(\underset{1}{\lambda}-\underset{2}{\lambda})^3}\underset{1}{X}\ln\mu; \tag{40.9}$$

$$\gamma_{121} = \gamma_{141} = -\underset{1}{e}\frac{\underset{1}{\lambda}}{2(\underset{1}{\lambda}-\underset{2}{\lambda})}\underset{2}{X}\ln\mu;$$

$$\gamma_{342} = -\gamma_{324} = \frac{\underset{2}{e}}{2\mu}\underset{2}{X}(\underset{2}{\lambda}\mu); \quad \gamma_{343} = \gamma_{323} = -\frac{\underset{2}{e}}{2}\underset{2}{X}\ln\mu;$$

$$\gamma_{424} = \gamma_{422} = \tfrac{3}{2}\underset{2}{e}\underset{2}{X}\ln\mu; \quad \gamma_{423} = \tfrac{1}{2}\underset{2}{e}\,\underset{2}{\lambda}\,\underset{2}{X}\ln\mu;$$

$$\gamma_{344} = -\gamma_{322} = \frac{\underset{2}{e}}{2\mu^2}\underset{2}{X}(\underset{2}{\lambda}\mu^2);$$

$$\underset{1}{X}(\underset{1}{\lambda}\mu^2) = 0; \quad \underset{p}{X}(\underset{1}{\lambda}\mu) = 0; \quad p = 2,\ 3,\ 4, \quad \underset{2}{X}\mu = \underset{4}{X}\mu;$$

$$\underset{1}{X}(\underset{2}{\lambda}\mu) = 0; \quad \underset{2}{X}(\underset{2}{\lambda}^3\mu^4) = \underset{4}{X}(\underset{2}{\lambda}^3\mu^4) = 0; \quad \underset{3}{X}\mu = \underset{3}{X}\underset{2}{\lambda} = 0.$$

6. (XVI) [(13)].

This differs from case 5 in that $\underset{1}{\lambda} = \underset{2}{\lambda} = \lambda$. As moreover $|\underset{r}{i}^k| \neq 0$, the last equations (40.9) show that

$$\frac{\partial\mu}{\partial x^i} = 0; \quad \frac{\partial(\lambda\mu)}{\partial x^i} = 0; \quad i = 1,\ 2,\ 3,\ 4,$$

i.e.

$$\mu = C_1; \quad \lambda = C_2.$$

Hence in this case (40.9) reduces to

$$\begin{aligned}\gamma_{313} &= \gamma_{421} = \gamma_{314} = \gamma_{312} = \gamma_{321} = \gamma_{341} = \gamma_{342} = \gamma_{324} \\ &= \gamma_{343} = \gamma_{323} = \gamma_{242} = \gamma_{232} = \gamma_{131} = \gamma_{424} = \gamma_{434} = \gamma_{423} = 0.\end{aligned} \tag{40.10}$$

$$\gamma_{213} = \gamma_{413}; \quad \gamma_{211} = \gamma_{411}; \quad \gamma_{212} = \gamma_{412}; \quad \gamma_{214} = \gamma_{414};$$
$$\mu = C_1; \quad \lambda = C_2,$$

where C_1 and C_2 are constants.

Although we can select real non-holonomic reference systems for types (VIII) and (IX), and obtain the corresponding equations, it proves more convenient to carry out the investigation in complex space and then transform back to real space.

This is the approach used below.

§41. The Canonical Forms of the Metrics of V_4 and $\bar{V}_4$ in a Holonomic Coordinate System

1. Cases (VIII) and (IX). As the elementary divisors of the λ-matrix are linear, Levi-Civita's result applies in complex space; thus in case VIII,

$$\begin{aligned}ds^2 = {} & \underset{1}{e}(\varphi_2-\varphi_1)(\varphi_3-\varphi_1)(\varphi_4-\varphi_1)(dx')^2+e_2(\varphi_1-\varphi_2)(\varphi_3-\varphi_2)(\varphi_4-\varphi_2)(dx^2)^2 \\ & +\underset{3}{e}(\varphi_1-\varphi_3)(\varphi_2-\varphi_3)(\varphi_4-\varphi_3)(dx^3)^2 \\ & +\underset{3}{e}(\varphi_1-\varphi_4)(\varphi_2-\varphi_4)(\varphi_3-\varphi_4)(dx^4)^2,\end{aligned}$$

$$\begin{aligned}d\bar{s}^2 = \frac{1}{\varphi_1\varphi_2\varphi_3\varphi_4}\Bigg\{ & \frac{\underset{1}{e}}{\varphi_1}(\varphi_2-\varphi_1)(\varphi_3-\varphi_1)(\varphi_4-\varphi_1)(dx^1)^2 \\ & +\frac{\underset{2}{e}}{\varphi_2}(\varphi_1-\varphi_2)(\varphi_3-\varphi_2)(\varphi_4-\varphi_2)(dx^2)^2 \\ & +\frac{\underset{3}{e}}{\varphi_3}(\varphi_1-\varphi_3)(\varphi_2-\varphi_3)(\varphi_4-\varphi_3)(dx^3)^2 \\ & +\frac{\underset{3}{e}}{\varphi_4}(\varphi_1-\varphi_4)(\varphi_2-\varphi_4)(\varphi_3-\varphi_4)(dx^4)^2\Bigg\};\end{aligned}$$

where $\varphi_1 = \varphi_1(x^1)$ and $\varphi_2 = \varphi_2(x^2)$ are real functions and $\varphi_3 = \varphi_3(x^3)$ and $\varphi_4 = \varphi_4(x^4)$ are complex conjugate functions of complex variables.

In case (IX)

$$ds^2 = (\varphi_3 - C)(\varphi_4 - C)\{g_{pq}\,dx^p\,dx^q\} + \underset{3}{e}(C-\varphi_3)(\varphi_4-\varphi_3)(dx^3)^2 + \underset{3}{e}(C-\varphi_4)(\varphi_3-\varphi_4)(dx^4)^2,$$

$$d\bar{s}^2 = \frac{1}{C\varphi_3\varphi_4}\left\{\frac{1}{C}(\varphi_3-C)(\varphi_4-C)(g_{pq}\,dx^p\,dx^q) + \frac{\underset{3}{e}}{\varphi_3}(C-\varphi_3)(\varphi_4-\varphi_3)(dx^3)^2 + \frac{\underset{3}{e}}{\varphi_4}(C-\varphi_4)(\varphi_3-\varphi_4)(dx^4)^2\right\},$$

where g_{pq} $(p, q = 1, 2)$ are real functions of the two variables x^1 and x^2, C is constant, φ_3 and φ_4 are complex conjugate functions of the complex variables x^3 and x^4 respectively.

We return to real space, by introducing the transformation

$$x^1 = y^1, \quad x^2 = y^2, \quad x^3 = y^3 + iy^4, \quad x^4 = y^3 - iy^4.$$

In the new system of coordinates

$$\varphi_1 = \varphi_1(y^1), \quad \varphi_2 = \varphi_2(y^2), \quad \varphi_3 = \sigma + i\tau, \quad \varphi_4 = \sigma - i\tau, \; g_{pq} = g_{pq}(y^1, y^2)$$

where σ and τ are conjugate harmonic functions of the real variables y^3 and y^4. Hence the metrics are as follows:

In case (VIII)

$$ds^2 = |\varphi_1-\varphi_2|\{\underset{1}{e}[(\sigma-\varphi_1)^2+\tau^2](dy^1)^2 + \underset{2}{e}[(\sigma-\varphi_2)^2+\tau^2](dy^2)^2\} + \underset{3}{e}\,4\tau^2(2\sigma-\varphi_1-\varphi_2)[(dy^3)^2-(dy^4)^2] + \underset{3}{e}\,8\tau[(\varphi_1-\sigma)(\varphi_2-\sigma)-\tau^2]\,dx^3\,dy^4.$$

$$d\bar{s}^2 = \frac{1}{\varphi_1\varphi_2(\sigma^2+\tau^2)}\Bigg(|\varphi_1-\varphi_2|\left\{\underset{1}{e}\frac{1}{\varphi_1}[(\sigma-\varphi_1)^2+\tau^2](dy^1)^2 + \underset{2}{e}\frac{1}{\varphi_2}[(\sigma-\varphi_2)^2+\tau^2](dy^2)^2\right\} + \underset{3}{e}\frac{4\tau^2}{\sigma^2+\tau^2}(\sigma^2+\tau^2-\varphi_1\varphi_2)[(dy^3)^2-(dy^4)^2] + \underset{3}{e}\frac{8\tau}{\sigma^2+\tau^2}[\sigma(\varphi_1-\sigma)(\varphi_2-\sigma)-\tau^2(\varphi_1+\varphi_2-\sigma)]\,dy^3\,dy^4\Bigg).$$

(41.1)

In case (IX)

$$ds^2 = [(\sigma-c)^2+\tau^2]g_{pq}\,dy^p\,dy^q+\underset{3}{e}\,4\tau\{-\tau[(dy^3)^2-(dy^4)^2]+2(c-\sigma)\,dy^3\,dy^4\}.$$

$$d\bar{s}^2 = \frac{1}{c(\sigma^2+\tau^2)}\left\{\frac{1}{c}[(\sigma-c)^2+\tau^2]g_{pq}\,dy^p\,dy^q\right.$$
$$\left.-\underset{3}{e}\frac{4c\tau^2}{\sigma^2+\tau^2}[(dy^3)^2-(dy^4)^2]+\underset{3}{e}\frac{8\tau}{\sigma^2+\tau^2}(c\sigma-\sigma^2-\tau^2)\,dy^3\,dy^4\right\}. \qquad (41.2)$$

2. (X) Segre characteristic [112].

We discuss this case in somewhat greater detail to illustrate the method.

Let us consider the following set of equations for the unknown function $f(x^1, \ldots, x^4)$:

$$\underset{s}{X}(f) = \underset{s}{i}^k\frac{\partial f}{\partial x^k} = 0; \qquad (s = m+1, \ldots, 4), \qquad (41.3)$$

where the $\underset{s}{i}^k$ are the vectors of the orthonormal tetrad for which (39.3) and (40.5) are valid. The necessary and sufficient conditions for this set to be fully integrable is that all commutators of the operators of the system should be linear combinations of the operators $\underset{s}{X}(f)$ $(s = m+1, \ldots, 4)$, [147], i.e.

$$[\underset{s}{X}\underset{t}{X}](f) = \underset{s}{X}\underset{t}{X}(f)-\underset{t}{X}\underset{s}{X}(f) = \sum_{e=1}^{4}\underset{e}{e}(\gamma_{est}-\gamma_{ets})\underset{e}{X}(f),$$

hence the coefficients of $\underset{1}{X}(f), \ldots, \underset{m}{X}(f)$ must be zero.

There are thus m independent solutions $f^1, \ldots, f^m$ of (41.3).

Using (39.3), we obtain all the commutators of the operators $\underset{1}{X}(f)$, $\underset{2}{X}(f)$, $\underset{3}{X}(f)$ and $\underset{4}{X}(f)$:

$$\begin{aligned}
[\underset{1}{X}\underset{2}{X}](f) &= \underset{1}{e}\gamma_{121}\underset{1}{X}(f)+\underset{2}{e}\gamma_{212}\underset{2}{X}(f);\\
[\underset{1}{X}\underset{3}{X}](f) &= -\underset{1}{e}\gamma_{131}\underset{1}{X}(f)+\underset{3}{e}\gamma_{313}\underset{3}{X}(f)-\underset{3}{e}(\gamma_{413}-\gamma_{431})\underset{4}{X}(f);\\
[\underset{1}{X}\underset{4}{X}](f) &= -\underset{1}{e}\gamma_{131}\underset{1}{X}(f)+\underset{3}{e}(\gamma_{413}+\gamma_{431})\underset{3}{X}(f)-\underset{3}{e}\gamma_{414}\underset{4}{X}(f);\\
[\underset{2}{X}\underset{3}{X}](f) &= -\underset{2}{e}\gamma_{232}\underset{2}{X}(f)+\underset{3}{e}\gamma_{323}\underset{3}{X}(f)-\underset{3}{e}(\gamma_{432}-\gamma_{423})\underset{4}{X}(f);\\
[\underset{2}{X}\underset{4}{X}](f) &= -\underset{2}{e}\gamma_{232}\underset{2}{X}(f)+\underset{3}{e}(\gamma_{423}+\gamma_{432})\underset{3}{X}(f)-\underset{3}{e}\gamma_{424}\underset{4}{X}(f);\\
[\underset{3}{X}\underset{4}{X}](f) &= -\underset{3}{e}\gamma_{343}\underset{3}{X}(f)-\underset{3}{e}\gamma_{434}\underset{4}{X}(f).
\end{aligned} \qquad (41.4)$$

Hence the set $\underset{2}{X}(f) = \underset{3}{X}(f) = \underset{4}{X}(f) = 0$ is fully integrable and has a single solution, denoted by f^1; the set $\underset{1}{X}(f) = \underset{3}{X}(f) = \underset{4}{X}(f) = 0$ is also fully integrable, with solution denoted by f^2. Noting that $\gamma_{313}-\gamma_{413} = \gamma_{413}-\gamma_{414}$ and $\gamma_{323}-\gamma_{423} = \gamma_{423}-\gamma_{424}$, we can easily see that the set $\underset{1}{X}(f) = \underset{2}{X}(f) = (\underset{3}{X}-\underset{4}{X})(f) = 0$ is also fully integrable, with solution φ^4. Finally the set $\underset{1}{X}(f) = \underset{2}{X}(f) = 0$ is integrable, with two independent solutions one of which can be taken as f^4; we denote the other by f^3.

On introducing the transformation

$$x^{k'} = f^k(x^1, \ldots, x^4),$$

the vector components of the orthonormal tetrad in the new system of coordinates, $\underset{r}{i}^k$ (the prime may be omitted), become

$$\underset{1}{i}^k = P_1(x)\,\delta^k_1; \quad \underset{2}{i}^k = P_2(x)\,\delta^k_2; \quad \underset{3}{i}^k - \underset{4}{i}^k = P_3(x)\,\delta^k_3;$$
$$\underset{3}{i}^1 = \underset{3}{i}^2 = \underset{4}{i}^1 = \underset{4}{i}^2 = 0. \tag{41.5}$$

The allowed transformations which do not alter the components of the orthonormal tetrad vectors are

$$x^{1'} = f^1(x^1), \quad x^{2'} = f^2(x^2), \quad x^{3'} = f^3(x^3, x^4), \quad x^{4'} = f^4(x^4). \tag{41.6}$$

Noting (41.5) and that $|\underset{s}{i}^k| \neq 0$, we find from the last thirteen equations of (40.5) that

$$\mu = C\varphi_1\varphi_2\varphi_4{}^2; \quad \underset{1}{\lambda} = \frac{1}{\varphi_1{}^2\varphi_2\varphi_4{}^2};$$
$$\underset{2}{\lambda} = \frac{1}{\varphi_1\varphi_2{}^2\varphi_4{}^2}; \quad \underset{3}{\lambda} = \frac{1}{\varphi_1\varphi_2\varphi_4{}^3}, \tag{41.7}$$

where C is constant, and $\varphi_1(x^1)$, $\varphi_2(x^2)$ and $\varphi_4(x^4)$ are arbitrary functions. As

$$[\underset{s}{X}\underset{t}{X}](f) = \underset{s}{i}^k\partial_k\underset{t}{i}^e\partial_e f - \underset{t}{i}^k\partial_k\underset{s}{i}^e\partial_e f, \quad \text{where} \quad \partial_k f = \frac{\partial f}{\partial x^k},$$

and as f in (41.4) is arbitrary, we can equate coefficients for the various derivatives of f. We obtain the set of equations

$$\underset{1}{i}^1\partial_1\underset{2}{i}^2 = e\gamma_{212}\underset{2}{i}^2; \quad \underset{2}{i}^2\partial_2\underset{1}{i}^1 = e\gamma_{121}\underset{1}{i}^1;$$
$$\underset{3}{i}^3\partial_3\underset{1}{i}^1 + \underset{3}{i}^4\partial_4\underset{1}{i}^1 = e\gamma_{131}\underset{1}{i}^1; \quad \underset{1}{i}^1\partial_1\underset{3}{i}^3 = e\gamma_{313}\underset{3}{i}^3 - e(\gamma_{413}-\gamma_{431})\underset{4}{i}^3;$$
$$\underset{1}{i}^1\partial_1\underset{3}{i}^4 = e(\gamma_{313}-\gamma_{413}+\gamma_{431})\underset{3}{i}^4; \tag{41.8}$$

$$
\begin{aligned}
&\underset{4}{i^3}\partial_3\underset{1}{i^1}+\underset{4}{i^4}\partial_4\underset{1}{i^1} = \underset{1}{e}\gamma_{131}\underset{1}{i^1}; \quad \underset{1}{i^1}\partial_1\underset{4}{i^3} = \underset{3}{e}(\gamma_{413}+\gamma_{431})\underset{3}{i^3}-\underset{3}{e}\gamma_{414}\underset{4}{i^3};\\
&\underset{3}{i^3}\partial_3\underset{2}{i^2}+\underset{3}{i^4}\partial_4\underset{2}{i^2} = \underset{2}{e}\gamma_{232}\underset{2}{i^2}; \quad \underset{2}{i^2}\partial_2\underset{3}{i^4} = \underset{3}{e}(\gamma_{323}-\gamma_{423}+\gamma_{432})\underset{3}{i^4};\\
&\underset{2}{i^2}\partial_2\underset{3}{i^3} = \underset{3}{e}\gamma_{323}\underset{3}{i^3}-\underset{3}{e}(\gamma_{423}-\gamma_{432})\underset{4}{i^3}; \qquad\qquad (41.8)\\
&\underset{4}{i^3}\partial_3\underset{2}{i^2}+\underset{4}{i^4}\partial_4\underset{2}{i^2} = \underset{2}{e}\gamma_{232}\underset{2}{i^2}; \quad \underset{2}{i^2}\partial_2\underset{4}{i^3} = \underset{3}{e}(\gamma_{423}+\gamma_{432})\underset{3}{i^3}-\underset{3}{e}\gamma_{424}\underset{4}{i^3};\\
&\underset{3}{i^3}\partial_3\underset{4}{i^3}-\underset{4}{i^3}\partial_3\underset{3}{i^3}+\underset{3}{i^4}\partial_4(\underset{4}{i^3}-\underset{3}{i^3}) = -\underset{3}{e}\gamma_{343}\underset{3}{i^3}-\underset{3}{e}\gamma_{434}\underset{4}{i^3};\\
&(\underset{3}{i^3}-\underset{4}{i^3})\partial_3\underset{3}{i^4} = -\underset{3}{e}(\gamma_{343}+\gamma_{434})\underset{3}{i^4}.
\end{aligned}
$$

Integrating equations 1, 8, 11 and 2, 3, 6 of this set and using (40.5) and (41.7), we obtain

$$
\underset{1}{i^1} = \frac{\underset{1}{\theta}(x^1)}{\sqrt{|\varphi_1-\varphi_2|\,|\varphi_4-\varphi_1|^2}}; \quad \underset{2}{i^2} = \frac{\underset{2}{\theta}(x^2)}{\sqrt{|\varphi_1-\varphi_2|\,|\varphi_4-\varphi_2|^2}}.
$$

Equations 4, 7 and 10, 12 show that

$$
\underset{3}{i^3}-\underset{4}{i^3} = \frac{\underset{3}{\theta}(x^3, x^4)}{\sqrt{|\varphi_1\varphi_2(\varphi_1-\varphi_4)(\varphi_2-\varphi_4)|}},
$$

where $\underset{1}{\theta}(x^1)$, $\underset{2}{\theta}(x^2)$, $\underset{3}{\theta}(x^3, x^4)$, $\varphi_1(x^1)$, $\varphi_2(x^2)$ and $\varphi_4(x^4)$ are arbitrary functions. Noting (41.6) we introduce the transformation

$$
x^{1'} = \int\frac{dx^1}{\underset{1}{\theta}}, \quad x^{2'} = \int\frac{dx^2}{\underset{2}{\theta}}, \quad x^{3'} = \int\frac{dx^3}{\underset{3}{\theta}}, \quad x^{4'} = x^4
$$

leading to

$$
\underset{1}{i^k} = \frac{1}{\sqrt{|\varphi_1-\varphi_2|\,|\varphi_4-\varphi_1|^2}}\,\delta^k_1; \quad \underset{2}{i^k} = \frac{1}{\sqrt{|\varphi_1-\varphi_2|\,|\varphi_4-\varphi_2|^2}}\,\delta^k_2; \qquad (41.9)
$$

$$
\underset{3}{i^3}-\underset{4}{i^3} = \frac{1}{\sqrt{|\varphi_1\varphi_2(\varphi_1-\varphi_4)(\varphi_2-\varphi_4)|}}\,\delta^k_3; \quad \underset{3}{i^1} = \underset{3}{i^2} = \underset{4}{i^1} = \underset{4}{i^2} = 0
$$

omitting the primes. The transformations which do not alter the components of the orthonormal tetrad are

$$
x^{1'} = x^1+\text{const}, \quad x^{2'} = x^2+\text{const}, \quad x^{3'} = x^3+f^3(x^4), \quad x^{4'} = f^4(x^4). \qquad (41.10)
$$

Integrating the remaining equations of (41.8), including 4 and 10, we obtain

$$
\underset{3}{i^4} = \underset{4}{i^4} = \frac{\varphi_4^5\sqrt{|\varphi_1\varphi_2|}\ \text{sign}\,(\varphi_1\varphi_2)}{[x^3+\underset{4}{\theta}\,(x^4)]\delta_4\varphi_4\sqrt{|\varphi_1-\varphi_4|\,|\varphi_2-\varphi_4|}}
$$

$$\underset{4}{i}^3 = -\frac{1}{2\sqrt{|\varphi_1\varphi_2(\varphi_1-\varphi_4)(\varphi_2-\varphi_4)|}}$$
$$-\frac{\varphi_4^{4}(\varphi_1+\varphi_2-2\varphi_4)\sqrt{|\varphi_1\varphi_2|}\operatorname{sign}(\varphi_1\varphi_2)}{2(\varphi_1-\varphi_4)(\varphi_2-\varphi_4)\sqrt{|\varphi_1-\varphi_4|\,|\varphi_1-\varphi_4|}}$$
$$-\frac{4\varphi_4^{3}\sqrt{|\varphi_1\varphi_2|}\operatorname{sign}(\varphi_1\varphi_2)}{2\sqrt{|\varphi_1-\varphi_4|\,|\varphi_2-\varphi_4|}}+\frac{\underset{5}{\theta}(x^4)\sqrt{|\varphi_1\varphi_2|}\,\varphi_4^{5}}{2\partial_4\varphi_4(x^3+\underset{4}{\theta})\sqrt{|\varphi_1-\varphi_4|\,|\varphi_2-\varphi_4|}}.$$

On using the transformation

$$x^{1'} = x^1,\quad x^{2'} = x^2,\quad x^{3'} = x^3+\tfrac{1}{2}\operatorname{sign}(\varphi_1\varphi_2)\int \underset{5}{\theta}(x^4)dx^4$$
$$x^{4'} = \varphi_4(x^4),$$

we obtain (41.9) with $\varphi_4(x^4) = x^4$, omitting the primes, and

$$\underset{3}{i}^4 = \underset{4}{i}^4 = -\frac{(x^4)^5\sqrt{|\varphi_1\varphi_2|}\operatorname{sign}(\varphi_1\varphi_2)}{[x^3+\theta(x^4)]\sqrt{|\varphi_1-x^4|\,|\varphi_2-x^4|}}$$

$$\underset{4}{i}^3 = -\frac{1}{2\sqrt{|\varphi_1\varphi_2(\varphi_1-x^4)(\varphi_2-x^4)|}}-\frac{2(x^4)^3\sqrt{|\varphi_1\varphi_2|}\operatorname{sign}(\varphi_1\varphi_2)}{\sqrt{|\varphi_1-x^4|\,|\varphi_2-x^4|}}$$
$$-\frac{(x^4)^4(\varphi_1+\varphi_2-2\varphi_4)\sqrt{|\varphi_1\varphi_2|}\operatorname{sign}(\varphi_1\varphi_2)}{2(\varphi_1-x^4)(\varphi_2-x^4)\sqrt{|\varphi_1-x^4|\,|\varphi_2-x^4|}};\tag{41.11}$$

and the allowed transformation is simply the identity transformation plus a possible shift of the origin.

As

$$g^{ij} = \sum_{k=1}^{4} \underset{k}{e}\,\underset{k}{i}^i\,\underset{k}{i}^j$$

(41.9) and (41.11) show that the contravariant components g^{ij} of the metric tensor are

$$g^{11} = \underset{1}{e}\frac{1}{|\varphi_2-\varphi_1|\,|x^4-\varphi_1|^2};\quad g^{12} = g^{13} = g^{14} = 0;$$

$$g^{22} = \underset{2}{e}\frac{1}{|\varphi_1-\varphi_2|\,|x^4-\varphi_2|^2};\quad g^{23} = g^{24} = 0;$$

$$g^{33} = -\frac{\underset{3}{e}\operatorname{sign}(\varphi_1\varphi_2)}{|\varphi_1-x^4|\,|\varphi_2-x^4|}\left\{4(x^4)^3+\frac{(x^4)^4(\varphi_1+\varphi_2-2x^4)}{(\varphi_1-x^4)(\varphi_2-x^4)}\right\};$$

$$g^{34} = -\underset{3}{e}\frac{(x_4)^4\operatorname{sign}(\varphi_1\varphi_2)}{(x^3+\theta)|\varphi_1-x^4|\,|\varphi_2-x^4|};\quad g^{44} = 0.$$

The covariant tensor g_{ij} is then obtained as the inverse of g^{ij}:

$$g_{11} = \underset{1}{e}\,|\varphi_1-\varphi_2|\,|x^4-\varphi_1|^2; \quad g_{12} = g_{13} = g_{14} = 0;$$

$$g_{22} = \underset{2}{e}\,|\varphi_1-\varphi_2|\,|x^4-\varphi_2|^2; \quad g_{23} = g_{24} = 0;$$

$$g_{33} = 0; \quad g_{34} = -\underset{3}{e}\,\frac{x^3+\theta}{(x^4)^4}\,|x^4-\varphi_1|\,|x^4-\varphi_2|\,\operatorname{sign}(\varphi_1\varphi_2); \tag{41.12}$$

$$g_{44} = \underset{3}{e}\,\frac{(x^3+\theta)^2}{(x^4)^8}\,|x^4-\varphi_1|\,|x^4-\varphi_2|$$

$$\times\left\{4\,(x^4)^3+\frac{(x^4)^4(\varphi_1+\varphi_2-2x^4)}{(\varphi_1-x^4)\,(\varphi_2-x^4)}\right\}\,x\,\operatorname{sign}\,(\varphi_1\varphi_2).$$

Lowering the indices of the contravariant components of the orthonormal tetrad vectors defined by eqns. (41.9) and (41.11) by means of the tensor g_{ij}, we determine the components of the tensor $\bar{g}_{ij}$ from the second formula (40.2):

$$\bar{g}_{11} = \underset{1}{\lambda} g_{11}; \quad \bar{g}_{12} = \bar{g}_{13} = \bar{g}_{14} = 0; \quad \underset{1}{\lambda} = \frac{1}{\varphi_1{}^2\varphi_2(x^4)^2}; \tag{41.13}$$

$$\bar{g}_{22} = \underset{2}{\lambda} g_{22}; \quad \bar{g}_{23} = \bar{g}_{24} = \bar{g}_{33} = 0; \quad \underset{2}{\lambda} = \frac{1}{\varphi_1\varphi_2{}^2(x^4)^2};$$

$$\bar{g}_{34} = \underset{3}{\lambda} g_{34}; \qquad \underset{3}{\lambda} = \frac{1}{\varphi_1\varphi_2(x^4)^3};$$

$$\bar{g}_{44} = \underset{3}{\lambda} g_{44} - g_{34}\,\frac{x^3+\theta}{(x^4)^4\varphi_1\varphi_2}.$$

In eqns. (40.12), (40.13), $\varphi_1(x^1)$, $\varphi_2(x^2)$ and $\theta(x^4)$ are arbitrary functions. Direct substitution confirsm that the tensors (41.12) and (41.13) satisfy eqns. (40.1).

A special case, peculiar to Segre characteristic [112] arises when the roots of the equation $|\bar{g}_{ij}-\lambda g_{ij}| = 0$ are *independent of* x^4. If the above method is used, a general holonomic system of coordinates can be found in which

$$\left.\begin{aligned}
&\underset{1}{i}^k = \frac{1}{\sqrt{|\varphi_1-\varphi_2|}}\,\delta_1^k; \quad \underset{2}{i}^k = \frac{1}{\sqrt{|\varphi_1-\varphi_2|}}\,\delta_2^k; \quad \underset{3}{i}^1 = \underset{4}{i}^1 = \underset{3}{i}^2 = \underset{4}{i}^2 = 0;\\
&\underset{3}{i}^4 = \underset{4}{i}^4 = \sqrt{\frac{|\varphi_1\cdot\varphi_2|}{|\varphi_1-\alpha|\,|\varphi_2-\alpha|}}; \quad \underset{3}{i}^3 = \underset{4}{i}^3 + \frac{1}{\sqrt{|\varphi_1\cdot\varphi_2(\varphi_1-\alpha)(\varphi_2-\alpha)|}};\\
&\underset{4}{i}^3 = -\frac{1}{2}\,\frac{1}{\sqrt{|\varphi_1\varphi_2(\varphi_1-\alpha)(\varphi_2-\alpha)|}}\left[1+\frac{\alpha^4(\varphi_1+\varphi_2-2\alpha)\varphi_1\varphi_2}{(\varphi_1-\alpha)(\varphi_2-\alpha)}\right];\\
&\mu = C\varphi_1\varphi_2\alpha^2; \quad \underset{1}{\lambda} = \frac{1}{\varphi_1^2\varphi_2\alpha^2}; \quad \underset{2}{\lambda} = \frac{1}{\varphi_1\varphi_2^2\alpha^2}; \quad \underset{3}{\lambda} = \frac{1}{\varphi_1\varphi_2\alpha^3},
\end{aligned}\right\}$$

where C and α are non-zero constants, and $\varphi_1(x^1)$ and $\varphi_2(x^2)$ are arbitrary non-zero functions. The non-vanishing components of the metric tensors of these spaces are

$$\left.\begin{aligned}
&g_{11} = e_1|\varphi_1-\varphi_2|; \quad g_{22} = e_2|\varphi_1-\varphi_2|; \quad g_{34} = e_3|\varphi_1-\alpha|\,|\varphi_2-\alpha|;\\
&g_{44} = e_3\alpha^4(\varphi_1+\varphi_2-2\alpha)\ \mathrm{sign}\,[\varphi_1\varphi_2(\varphi_1-\alpha)(\varphi_2-\alpha)];\\
&\bar{g}_{11} = \underset{1}{\lambda} g_{11}; \quad \bar{g}_{22} = \underset{2}{\lambda} g_{22}; \quad \bar{g}_{34} = \underset{3}{\lambda} g_{34};\\
&\bar{g}_{44} = \underset{3}{\lambda} g_{34} + e_3\frac{|\varphi_1-\alpha|\,|\varphi_2-\alpha|}{|\varphi_1|\,|\varphi_2|},
\end{aligned}\right\}$$

where $\qquad e_1 = e_2 = -1; \quad e_3 = \pm 1.$

Thus, *the metric tensors of geodesically corresponding Riemann spaces with Segre characteristic* [112] *can be reduced to the forms* (41.12) *and* (41.13); it follows directly from the method that this representation is unique.

3. (XII). Segre characteristic [(11)2] As the analysis is practically identical for the remaining types, we simply give the results. A holonomic system of coordinates can be chosen for two Riemann spaces V_4 and $\bar{V}_4$ with common geodesics and Segre characteristic [(11)2], in which

$$\begin{gathered}
\underset{1}{i}^4 = \underset{2}{i}^4 = \underset{1}{i}^3 = 0, \quad \underset{3}{i}^1 = \underset{3}{i}^2 = \underset{4}{i}^1 = \underset{4}{i}^2 = 0,\\
\underset{4}{i}^3 = -\underset{3}{i}^3 = \frac{1}{2x^{4^2}}, \quad \underset{4}{i}^4 = \underset{3}{i}^4 = \frac{x^{4^2}}{x^3+\theta(x^4)},\\
\mu = cx^{4^2}, \quad \underset{1}{\lambda} = \frac{1}{c_1x^{4^2}}, \quad \underset{2}{\lambda} = \frac{1}{x^{4^3}},
\end{gathered} \tag{41.14}$$

where c and c_1 are constants, and $\theta(x^4)$ is an arbitrary function.

The metrics of these spaces take the form

$$ds^2 = (c_1-x^4)^2 g_{pq}\, dx^p\, dx^q - 2e_3(x^3+\theta)\, dx^3 dx^4,$$
$$d\bar{s}^2 = \underset{1}{\lambda}(c_1-x^4)^2 g_{pq}\, dx^p\, dx^q - e_3 \underset{3}{\lambda}(x^3+\theta)\Big[2\, dx^3\, dx^4 - \frac{x^3+\theta}{x^4}\, dx^{4^2}\Big], \qquad (41.15)$$

where $g_{pq} = g_{pq}(x^1, x^2)$ $(p, q = 1, 2)$ and $\underset{1}{\lambda}$ and $\underset{3}{\lambda}$ are defined by equs. (41.14). *The representation of the metrics in this form is unique to within transformations in a V_2 with metric $d\sigma^2 = g_{pq} dx^p\, dx^q$.*

As in the previous case, when the roots $\underset{i}{\lambda}$ *are independent of* x^4 there is a holonomic system of coordinates in which

$$\underset{1}{i}^3 = \underset{1}{i}^4 = \underset{2}{i}^3 = \underset{2}{i}^4 = \underset{3}{i}^1 = \underset{3}{i}^2 = \underset{4}{i}^1 = \underset{4}{i}^2 = 0;$$
$$\underset{3}{i}^3 = -\underset{4}{i}^3 = -1; \quad \underset{3}{i}^4 = \underset{4}{i}^4 = 1.$$

In this coordinate system the metrics of V_4 and $\bar{V}_4$ are

$$\left.\begin{aligned} ds^2 &= g_{pq}\, dx^p\, dx^q - 4e_3\, dx^3\, dx^4 \\ d\bar{s}^2 &= \underset{1}{\lambda} g_{pq}\, dx^p\, dx^q - 4e_3 \underset{2}{\lambda}\, dx^3\, dx^4 + 16 e_3\, dx_4^2 \end{aligned}\right\},$$

where $g_{pq} = g_{pq}(x^1, x^2)$, $p, q = 1, 2$ and $\underset{1}{\lambda}$ and $\underset{2}{\lambda}$ are unequal constants. The system of coordinates is determined to within a transformation in a V_2 with metric $d\sigma^2 = g_{pq} dx^p\, dx^q$.

4. For (XIII) with Segre characteristic [1(12)] we can choose a holonomic reference system such that

$$\underset{1}{i}^\alpha = \delta_1^\alpha, \quad \underset{2}{i}^1 = 0, \quad \underset{2}{i}^2 = \frac{1}{\sqrt{|\varphi_1 - e_2|}}, \quad \underset{2}{i}^4 = 0,$$
$$\underset{3}{i}^1 = \underset{4}{i}^1 = 0, \; \underset{3}{i}^2 = \underset{4}{i}^2, \; \underset{3}{i}^3 - \underset{4}{i}^3 = [|\varphi_1||\varphi_1 - c_2|]^{-1/2}, \; \underset{3}{i}^4 = \underset{4}{i}^4 = |\varphi_1|^{1/2}|\varphi_1 - c_2|^{-1/2},$$
$$\mu = c_1\varphi_1, \quad \underset{1}{\lambda} = \frac{1}{\varphi_1^2}, \quad \underset{2}{\lambda} = \frac{1}{\varphi_1 c_2}$$

where c_1 and c_2 are constants and $\varphi_1(x^1)$ is an arbitrary function.

The metrics of spaces V_4 and $\bar{V}_4$ with common geodesics in this system of coordinates are

$$ds^2 = e_1 dx^{1^2} + |\varphi_1 - c_2| \left\{ e_2\, dx^{2^2} + 2e_3\, dx^3\, dx^4 + e_3 \left[\frac{c_2^2 \operatorname{sign} \varphi_1}{\varphi_1 - c_2} - R \right] dx^{4^2} \right\},$$
$$d\bar{s}^2 = e_1 \underset{1}{\lambda}\, dx^{1^2} + \underset{2}{\lambda} |\varphi_1 - c_2| \left\{ e_2\, dx^{2^2} + 2e_3\, dx^3\, dx^4 + e_3 \left[\frac{c_2 \varphi_1}{\varphi_1 - c_2} - R \right] dx^{4^2} \right\}$$
$$(41.16)$$

where $R = R(x^2, x^4)$ is an arbitrary function. This representation of the metrics is unique to within the transformation $x^{k'} = x^k + \delta_3^k f(x^4)$.

5. (XIV). If the λ-matrix has a Segre characteristic [(112)], the differential equations for the geodesics of the two spaces V_4 and $\bar{V}_4$ can only be identical in a common system of coordinates, if the Riemann connectivities $\Gamma^{\alpha}_{\beta\gamma}$ are identical.

As $|\underset{r}{i}^k| \neq 0$, the last equations of (40.8) show that

$$\mu = c_1 = \text{const}, \quad \lambda = c_2 = \text{const},$$

and (40.1) shows that the tensor $\bar{g}_{ij}$ is covariantly constant with respect to the metric g_{ij}. Hence the metrics of the spaces V_4 and $\bar{V}_4$ can be represented by the forms

$$\begin{aligned} ds^2 &= g_{pq}\, dx^p\, dx^q + 2e_3\, dx^3\, dx^4; \qquad (p, q = 1, 2, 4) \\ d\bar{s}^2 &= \lambda\, ds^2 + e_3\, d{x^4}^2 \end{aligned} \tag{41.17}$$

to within the transformation

$$x^{1'} = f^1(x^1, x^2, x^4), \quad x^{2'} = f^2(x^1, x^2, x^4), \quad x^{3'} = x^3 + f^3(x^1, x^2, x^4)$$
$$x^{4'} = x^4 + \text{const}.$$

$g_{pq} = g_{pq}(x^1, x^2, x^4)$ and are otherwise arbitrary functions. The components of the vectors of the orthonormal tetrad corresponding to (40.8) must be such that

$$\underset{3}{i}^1 = \underset{4}{i}^1, \quad \underset{3}{i}^2 = \underset{4}{i}^2, \quad \underset{3}{i}^3 - \underset{4}{i}^3 = 1, \quad \underset{3}{i}^4 = \underset{4}{i}^4 = 1, \quad \underset{1}{i}^4 = \underset{2}{i}^4 = 0.$$

6. (XV). Segre characteristic [13]. As in case 2, we can choose a unique holonomic system of reference in which

$$\begin{aligned}
&\underset{1}{i}^{\alpha} = \delta_1^{\alpha}\, |\varphi_1 - x^4|^{-3/2}, \quad \underset{3}{i}^1 = \underset{3}{i}^4 = 0, \quad \underset{3}{i}^3 = |\varphi_1 - x^4|^{-1/2}, \\
&\underset{3}{i}^2 = -\tfrac{1}{2}{x^4}^5\, |\varphi_1 - x^4|^{-3/2}\, \text{sign}\, [\varphi_1(\varphi_1 - x^4)], \quad \underset{2}{i}^1 = \underset{4}{i}^1 = 0, \\
&\underset{2}{i}^2 = \tfrac{1}{2}[\varphi_1^2\, |\varphi_1 - x^4|]^{-1/2}\{1 + \varphi_1^2 {x^4}^9 [(\varphi_1 - x^4)^{-1} + \tfrac{3}{4}x^4(\varphi_1 - x^4)^{-2}] \\
&\qquad - \tfrac{9}{2}\varphi_1^2 {x^4}^4 x^2 (x^3 + \theta)^{-1}\, \text{sign}\, \varphi_1\}, \\
&\underset{4}{i}^2 = \underset{2}{i}^2 - |\varphi_1|^{-1}\, |\varphi_1 - x^4|^{-1/2}, \quad \underset{2}{i}^4 = \underset{4}{i}^4 = -\tfrac{1}{2}\varphi_1 {x^4}^5 (x^3 + \theta)^{-1}\, |\varphi_1 - x^4|^{-1/2}, \\
&\underset{2}{i}^3 = \underset{4}{i}^3 = -\tfrac{1}{2}\varphi_1\, |\varphi_1 - x^4|^{-1/2} \left\{ \frac{{x^4}^5}{\varphi_1 - x^4} + \frac{3{x^4}^4 x^3 + 2x^2\, \text{sign}\, \varphi_1}{2(x^3 + \theta)} \right\}, \\
&\mu = c\varphi_1 {x^4}^3, \quad \underset{1}{\lambda} = \frac{1}{\varphi_1^2 {x^4}^3}, \quad \underset{2}{\lambda} = \frac{1}{\varphi_1 {x^4}^4},
\end{aligned} \tag{41.18}$$

where c = constant, and $\varphi_1\,(x^1)$ and $\theta\,(x^4)$ are arbitrary functions.

The metrics of two spaces V_4 *and* $\bar{V}_4$ *with common geodesics in this system of coordinates are of the form*

$$
\begin{aligned}
g_{11} &= e_1 |\varphi_1 - x^4|^3, \quad g_{12} = g_{13} = g_{14} = g_{22} = g_{23} = 0, \\
g_{24} &= -e_2 2 {x^4}^{-5}(x^3+\theta) |\varphi_1 - x^4| \operatorname{sign} \varphi_1, \quad g_{33} = e_2 |\varphi_1 - x^4|, \\
g_{34} &= -e_2\{2(x^3+\theta) \operatorname{sign} (\varphi_1 - x^4) + \tfrac{1}{2}{x^4}^{-5} |\varphi_1 - x^4| (3{x^4}^4 x^3 + 2x^2 \operatorname{sign} \varphi_1)\}, \\
g_{44} &= e_2 {x^4}^{-6} |\varphi_1 - x^4| \{\tfrac{1}{4}{x^4}^{-4}[3{x^4}^4 x^3 + 2x^2 \operatorname{sign} \varphi_1]^2 \\
&\quad + 2(x^3+\theta)(\varphi_1 - x^4)^{-1}[{x^4}^5(x^3 - 2\theta) + x^2(g\varphi_1 - 7x^4) \operatorname{sign} \varphi_1]\}, \qquad (41.19) \\
\bar{g}_{11} &= \underset{1}{\lambda} g_{11}, \quad \bar{g}_{12} = \bar{g}_{13} = \bar{g}_{14} = \bar{g}_{22} = \bar{g}_{23} = 0, \\
\bar{g}_{24} &= \underset{2}{\lambda} g_{24}, \quad \bar{g}_{33} = \underset{2}{\lambda} g_{33}, \quad \bar{g}_{34} = \underset{2}{\lambda} g_{34} + |\varphi_1|^{-1} g_{24}, \\
\bar{g}_{44} &= \underset{2}{\lambda} g_{44} + 2e_2[|\varphi_1|\, |\varphi_1 - x^4|]^{-1} g_{24}[g_{34} - 2(x^3+\theta) \operatorname{sign} (\varphi_1 - x^4)].
\end{aligned}
$$

This is easily confirmed by direct substitution from 41.19 into (40.1). If, however, the roots $\underset{i}{\lambda}$ are independent of x^4, this same method gives a unique holonomic system of coordinates in which

$$
\begin{aligned}
&\underset{1}{i}^k = \delta_1^k; \quad \underset{2}{i}^1 = \underset{3}{i}^1 = \underset{4}{i}^1 = \underset{3}{i}^4 = 0; \\
&\underset{3}{i}^3 = \frac{1}{\sqrt{|\varphi_1 - \alpha|}}; \quad \underset{3}{i}^2 = -\frac{1}{2} \frac{\alpha^5 \operatorname{sign} \varphi_1}{(\varphi_1 - \alpha)\sqrt{|\varphi_1 - \alpha|}}; \\
&\underset{2}{i}^2 = \frac{1}{2} \frac{1}{|\varphi_1| \sqrt{|\varphi_1 - \alpha|}} \left[1 + \frac{\varphi_1^2 \alpha^9}{\varphi_1 - \alpha}\left(1 + \frac{3\alpha}{4(\varphi_1 - \alpha)}\right)\right]; \\
&\underset{4}{i}^2 = \underset{2}{i}^2 - \frac{1}{|\varphi_1| \sqrt{|\varphi_1 - \alpha|}}; \quad \underset{2}{i}^3 = \underset{4}{i}^3 = -\frac{1}{2} \frac{\varphi_1 \cdot \alpha^5}{(\varphi_1 - \alpha)\sqrt{|\varphi_1 - \alpha|}}; \\
&\underset{2}{i}^4 = \underset{4}{i}^4 = \frac{|\varphi_1|}{\sqrt{|\varphi_1 - \alpha|}},
\end{aligned}
$$

where α is a constant and $\varphi(x^1)$ is an arbitrary function of x^1. In this coordinate system the metrics of V_4 and $\bar{V}_4$ are of the form

$$
\begin{aligned}
ds^2 &= e_1 d{x^1}^2 + e_2 |\varphi_1 - \alpha| (2dx^2\, dx^4 + d{x^3}^2) + 2e_2 \alpha^5 \operatorname{sign} [\varphi_1(\varphi_1 - \alpha)]\, dx^3\, dx^4 \\
&\quad - e_2 \alpha^9 \operatorname{sign} (\varphi_1 - \alpha)\, d{x^4}^2; \\
d\bar{s}^2 &= e_1 \underset{1}{\lambda}\, d{x^1}^2 + e_2 \underset{2}{\lambda} |\varphi_1 - \alpha| (2dx^2 d{x^4}^2 + d{x^3}^2) + 2e_2 \operatorname{sign} [\varphi_1(\varphi_1 - \alpha)]\, dx^3\, dx^4; \\
\underset{1}{\lambda} &= \frac{1}{\varphi_1^2 \alpha^3}; \quad \underset{2}{\lambda} = \frac{1}{\varphi_1 \alpha^4}; \quad \mu = C\varphi_1 \alpha^3,
\end{aligned}
$$

where C is a constant,

7. (XVI). *When the Segre characteristic of the pair of forms* (40.1) *is* [(13)], *the condition that the corresponding spaces have the same geodesics reduces to the condition that the Riemann connectivities be identical.* As in case 5, the tensor $\bar{g}_{ij}$ is covariantly constant with g_{ij} to within the transformation $x^{k'} = x^k + \delta^k_2 f(x^4)$ $(k = 1, 2, 3, 4)$. It is possible to determine a holonomic coordinate system in which the metrics of the spaces V_4 and $\bar{V}_4$ are

$$
\begin{aligned}
ds^2 &= e_1\, dx^{1^2} + 2e_2\, dx^2\, dx^4 + e_2\, dx^{3^2} + \theta dx^{4^2}, \\
d\bar{s}^2 &= \lambda ds^2 + 2e_2\, dx^3\, dx^4,
\end{aligned}
\tag{41.20}
$$

where $\lambda =$ constant and $\theta(x^1, x^4)$ is an arbitrary function. The components of the vectors of the orthonormal tetrad in which eqn. (40.10) holds satisfy the relations

$$
\underset{1}{i}^1 = 1, \quad \underset{1}{i}^3 = \underset{1}{i}^4 = 0, \quad \underset{3}{i}^\alpha = \delta^\alpha_3, \quad \underset{2}{i}^1 = \underset{4}{i}^1, \quad \underset{2}{i}^2 - \underset{4}{i}^2 = 1,
$$

$$
\underset{2}{i}^3 = \underset{4}{i}^3 = 0, \quad \underset{2}{i}^4 = \underset{4}{i}^4 = 1.
$$

We have thus obtained exact solutions for all gravitational fields with common geodesics. Although these solutions contain many arbitrary functions, the conditions for common geodesics to exist (or simulation to occur) are still rather severe. If one of the spaces is an Einstein space, these conditions are even more severe; this case merits special attention.

It is most important to note that the above analysis is based only on the algebraic structure of $\bar{g}_{ij}$, and does not presuppose any particular energy-momentum tensor. We can thus choose g_{ij} and $\bar{g}_{ij}$ corresponding to some particular T_{ij} and $\bar{T}_{ij}$ so that the λ-matrix $(\bar{g}_{ij} - \lambda g_{ij})$ belongs to one of the possible types that map fields of different physical origins on each other with common geodesics (projective transformation of fields). We can in this way, for example, construct an electromagnetic model for a special type of gravitational field.

Finally, we consider the case where V_4 is an Einstein space and $\bar{V}_4$ is arbitrary.

§ 42. The Projective Mapping of Einstein Spaces

Eisenhart [77] shows that if a V_n with metric

$$
ds^2 = g_{\alpha\beta}\, dx^\alpha\, dx^\beta \tag{42.1}
$$

is subjected to the infinitesimal transformation

$$
\overset{*}{x}^\alpha = x^\alpha + \xi^\alpha\, \delta t \tag{42.2}
$$

the metric tensor of the new $\dot{V}_n$ is

$$\dot{g}_{\alpha\beta} = g_{\alpha\beta} + h_{\alpha\beta}\,\delta t \tag{42.3}$$

where

$$h_{\alpha\beta} = \xi_{\alpha,\,\beta} + \xi_{\beta,\,\alpha}, \tag{42.4}$$

the comma denoting a covariant derivative with respect to the metric (42.1).

If the transformation (42.2) is projective, the metrics $g_{\alpha\beta}$ and $\dot{g}_{\alpha\beta}$ have common geodesics; in order that the vector field $\xi^\alpha(x)$ should correspond to a projective transformation [256a], it is necessary and sufficient that there be a function $\psi(x^1, \ldots, x^n)$ satisfying the equations

$$h_{\alpha\beta,\,\gamma} = 2g_{\alpha\beta}\psi_{,\,\gamma} + g_{\alpha\gamma}\psi_{,\,\beta} + g_{\beta\gamma}\psi_{,\,\alpha}. \tag{42.5}$$

The integrability conditions for these equations are

$$h_{\alpha\sigma}R^{\sigma}{}_{\beta\gamma\delta} + h_{\sigma\beta}R^{\sigma}{}_{\alpha\gamma\delta} = g_{\alpha\gamma}\psi_{,\,\beta\delta} - g_{\alpha\delta}\psi_{,\,\beta\gamma} + g_{\beta\gamma}\psi_{,\,\alpha\delta} - g_{\beta\delta}\psi_{,\,\alpha\gamma}. \tag{42.6}$$

If the V_4 with signature $(- - - +)$ at a point is an Einstein space,

$$R_{\alpha\beta} = \varkappa g_{\alpha\beta} \qquad (\alpha, \beta = 1, 2, 3, 4)$$

with $\varkappa =$ constant. As, according to § 19, there are only three types of Einstein space with distinct algebraic structures of the curvature tensor, we naturally consider each type separately.

For the first type (cf. § 19) $(R_{ab}) = \begin{pmatrix} M & N \\ N & -M \end{pmatrix}$ and the matrices M and N are

$$M = \begin{pmatrix} \alpha_1 & & \\ & \alpha_2 & \\ & & \alpha_3 \end{pmatrix}, \quad N = \begin{pmatrix} \beta_1 & & \\ & \beta_2 & \\ & & \beta_3 \end{pmatrix} \tag{42.7}$$

where

$$\sum_1^3 \alpha_i = -\varkappa, \quad \sum_1^3 \beta_i = 0$$

or introducing the *principal curvatures*

$$k_s \equiv \alpha_s + i\beta_s, \quad \sum_1^3 k_i = -\varkappa.$$

Noting that in the non-holonomic orthonormal tetrad in which (42.7) hold the metric tensor is

$$(\overset{0}{g}_{\alpha\varrho}) = \begin{pmatrix} -1 & & & \\ & -1 & & \\ & & -1 & \\ & & & 1 \end{pmatrix}$$

and using it to raise and lower indices, the set of integrability conditions (42.6) for all possible α, β reduce to

$$k_i(h_{jj}-h_{ll}) = \psi_{,jj}-\psi_{,ll}, \quad k_i(h_{ii}+h_{44}) = \psi_{,ii}+\psi_{,44}, \tag{42.8}$$

$$k_i h_{jl} = \psi_{,jl}, \quad k_i h_{i4} = \psi_{,i4}, \tag{42.9}$$

$$(k_i-k_l)(h_{ij}-{}_ih_{l4}) = (k_i-k_l)(h_{l4}-{}_ih_{ij}) = 0, \tag{42.10}$$

where $i, j, l = 1, 2, 3$ and $i \neq j \neq 1$. Equation (42.10) shows that if at least one $h_{\alpha\beta} \neq 0$ with $\alpha \neq \beta$, then $k_1 = k_2 = k_3 = -\varkappa/3$. This means that the Segre characteristic of the λ-matrix $(R_{ab}-\lambda g_{ab})$ $(a, b = 1, \ldots, 6)$ is of the linear type [(111 111)]; changing from collective to ordinary indices, we obtain

$$R_{\alpha\beta\gamma\delta} = \varkappa/3(g_{\alpha\gamma}g_{\beta\delta}-g_{\alpha\delta}g_{\beta\gamma}),$$

i.e. the V_4 has constant curvature.

But V_4 spaces of constant curvature admit an $n(n+2)$ parameter group of infinitesimal projective transformations, and this group is locally isomorphic with the group of projective transformations of a flat space [256a]. Beltrami's theorem [1a] shows that $\mathring{V}_4$ must also be a space of constant curvature.

If however the $h_{\alpha\beta}$ vanish $(\alpha \neq \beta)$, eqns. (42.10) are all satisfied. Eliminating the $\psi_{,\alpha\alpha}$ $(\alpha = 1, 2, 3, 4)$ from (42.8), we obtain the three equations

$$(k_1-k_2)\,(h_{33}-h_{11}-h_{22}-h_{44}) = 0$$
$$(k_1-k_3)\,(h_{22}-h_{11}-h_{33}-h_{44}) = 0$$
$$(k_2-k_3)\,(h_{11}-h_{22}-h_{33}-h_{44}) = 0.$$

Three possibilities arise in solving these equations: (a) all the k_i are different, (b) two of the k_i are equal, and (c) all the k_i are equal.

In case (a)

$$h_{11} = h_{22} = h_{33} = -h_{44}$$

and $h_{\alpha\beta} = \nu g_{\alpha\beta}$. Equations (42.8) can be written

$$g_{\alpha\beta}\nu_{,\gamma} = 2g_{\alpha\beta}\psi_{,\gamma}+g_{\alpha\gamma}\psi_{,\beta}+g_{\beta\gamma}\psi_{,\alpha}. \tag{42.11}$$

In an orthonormal tetrad for $\alpha = \beta \neq \gamma$ we have

$$e_\alpha\nu_{,\gamma} = 2e_\alpha\psi_{,\gamma} \quad \text{where} \quad e_\alpha = \pm 1.$$

Substituting $\nu_{,\gamma} = 2\psi_{,\gamma}$ into (42.11), we obtain

$$g_{\alpha\gamma}\psi_{,\beta}+g_{\beta\gamma}\psi_{,\alpha} = 0$$

and so for $\beta = \gamma \neq \alpha$ we have $\psi_{,\alpha} = 0$, i.e.

$$\nu_{,\alpha} = \psi_{,\alpha} = 0, \quad \nu = \text{const}, \quad \psi = \text{const}.$$

Their tensor nature ensures that these equations are valid in any system of coordinates. Thus when $k_1 \neq k_2 \neq k_3$

$$\xi_{(\alpha,\beta)} = \tfrac{1}{2}\nu g_{\alpha\beta}, \quad \nu = \text{const.}$$

Hence the one-parameter group consists of similarity transformations, and the metric of the $\overset{*}{V}_4$ having geodesics in common with the given V_4 is $\overset{*}{g}_{\alpha\beta} = \sigma g_{\alpha\beta}$, where $\sigma =$ constant, a trivial case of conformal correspondence.

In case (b) we take $k_1 = k_2 \neq k_3$; the results are essentially the same for the other two subcases ($k_1 = k_3 \neq k_2$ and $k_2 = k_3 \neq k_1$). Hence

$$k_1 - k_2 = 0, \quad k_1 - k_3 \neq 0 \quad k_2 - k_3 \neq 0$$

and there remain the two equations

$$h_{22} - h_{11} - h_{33} - h_{44} = 0, \quad h_{11} - h_{22} - h_{33} - h_{44} = 0.$$

Adding and subtracting, we obtain

$$h_{11} = h_{22}, \quad h_{33} = -h_{44}. \tag{42.12}$$

Hence the Segre characteristics of the λ-matrices $(h_{\alpha\beta} - \lambda g_{\alpha\beta})$ and $(\overset{*}{g}_{\alpha\beta} - \lambda g_{\alpha\beta})$ are of the linear type [(11)(11)]; we can use Solodovnikov's theorem [256a]: if the Segre characteristics of the λ-matrices are [(11) (11)] and $h_{\alpha\beta} \neq \varrho g_{\alpha\beta}$ satisfy eqns. (42.5), then in some system of coordinates

$$ds^2 = g_{pq}\, dx^p\, dx^q + g_{rs}\, dx^r\, dx^s \tag{42.13}$$

$$h_{\alpha\beta}\, dx^\alpha\, dx^\beta = (2e_1 + e_2) g_{pq}\, dx^p\, dx^q + (2e_2 + e_1) g_{rs}\, dx^r\, dx^s \tag{42.14}$$

$$\psi = \tfrac{1}{2}(e_1 + e_2), \tag{42.15}$$

where $p, q = 1, 2, r, s = 3, 4$, e_1 and e_2 are constants; g_{pq} depends only on x^1, x^2, and g_{rs} on x^3, x^4. The proof of this theorem does not depend on the signature but only on the elementary divisors of the matrix $(h_{\alpha\beta} - \lambda g_{\alpha\beta})$ being linear.

Thus V_4 is a *reducible* space, with one of the four possible metrics of reducible Einstein spaces—(14.7), (14.8).

The group of projective transformations for case (b), more general than the group of motions, is the group of affine transformations, i.e. transformations preserving the connectivity [256a]; it consists of similarity transformations acting in both two-dimensional spaces whose topological sum produces the given reducible space.

According to Levi-Civita's fundamental theorem [8a], the $\overset{*}{V}_4$ spaces sharing common geodesics with (14.7) and (14.8) have metrics

$$ds^2 = -a_1(dx^{1^2}+\cos^2\sqrt{-\varkappa x^1}\,dx^{2^2})-a_2(dx^{3^2}-\cos^2\sqrt{-\varkappa x^3}\,dx^{4^2}), \quad (42.16)$$

$$ds^2 = -a_1(dx^{1^2}+\cosh^2\sqrt{\varkappa}\,x^1\,dx^{2^2})-a_2(dx^{3^2}-\cosh^2\sqrt{\varkappa x^3}\,dx^{4^2}), \quad (42.17)$$

etc., where a_1 and a_2 are arbitrary non-zero constants. We note that these spaces are not in general Einstein spaces.

In case (*c*) when $k_1 = k_2 = k_3 = -\varkappa/3$, we conclude that V_4 and $\overset{*}{V}_4$ are both spaces of constant curvature and that the group of projective transformations they admit is isomorphic with the group of projective transformations of a flat space, a Minkowski space in the present case. This completes the study of spaces of the first type.

The matrices M and N for the second type can be written

$$M = \begin{pmatrix} \alpha_1 & & \\ & \alpha_2+1 & \\ & & \alpha_2-1 \end{pmatrix}, \quad N = \begin{pmatrix} \beta_1 & & \\ & \beta_2 & 1 \\ & 1 & \beta_2 \end{pmatrix}, \quad (42.18)$$

$$\alpha_1+2\alpha_2 = -\varkappa, \quad \beta_1+2\beta_2 = 0. \quad (42.19)$$

The integrability conditions (42.6) for all possible indices ($\alpha, \beta, \gamma, \delta = 1, 2, 3, 4$) and in the appropriate non-holonomic orthonormal tetrad for (42.18) reduce to a set of conditions which can easily be solved to give

$$h_{\alpha\beta} = 0 \quad (\alpha\neq\beta), \quad h_{11} = h_{22} = h_{33} = -h_{44}.$$

Hence

$$h_{\alpha\beta} = \nu g_{\alpha\beta}$$

in the orthonormal tetrad. Its tensor nature ensures that this equation is valid in any system of coordinates; that ν is constant follows from the discussion for spaces of the first type [cf. case (a)].

We thus obtain the following theorem: *Einstein spaces of the second type derived by projective transformations wider than the group of motions admit only similarity transformations.* The metric of $\overset{*}{V}_4$ is trivially conformal to the metric $g_{\alpha\beta}$; $\overset{*}{g}_{\alpha\beta} = \sigma g_{\alpha\beta}$, σ = constant.

Spaces of the third type admit only one non-holonomic orthonormal tetrad at each point, for which the matrices M and N are

$$M = \begin{pmatrix} -\varkappa/3 & 1 & 0 \\ 1 & -\varkappa/3 & 0 \\ 0 & 0 & -\varkappa/3 \end{pmatrix}, \quad N = \begin{pmatrix} 0 & 0 & 0 \\ 0 & 0 & -1 \\ 0 & -1 & 0 \end{pmatrix}.$$

The integrability conditions (42.6) for this orthonormal tetrad reduce to

$$h_{\alpha\beta} = 0 \quad (\alpha \neq \beta), \quad h_{11} = h_{22} = h_{33} = -h_{44}.$$

Hence we again have

$$h_{\alpha\beta} = \nu g_{\alpha\beta}, \quad \nu = \text{const},$$

in any holonomic system of coordinates. To sum up, *Einstein spaces V_4 with signature $(- - - +)$ admit a projective mapping on Riemann $\overset{*}{V}_4$ manifolds with the same signature, only when one of the following conditions holds:* (1) *V_4 and $\overset{*}{V}_4$ are spaces of constant curvature;* (2) *$\overset{*}{V}_4$ and V_4 are trivially conformal* ($\overset{*}{g}_{\alpha\beta} = \sigma g_{\alpha\beta}$, $\sigma =$ const); or (3) *V_4 is a reducible space with metrics* (14.7) *or* (14.8) *and $\overset{*}{V}_4$ is defined by the metrics* (42.16), (42.17).

Thus if one of the mapped spaces is an Einstein manifold, the condition that the spaces correspond is so severe that it leads either to conformally trivial mapping of the metrics, or to reducible spaces or to spaces of constant curvature; *an empty space-time cannot be simulated by another V_4—whether empty or not.*

EXAMPLES

1. Find all the static and non-static $\overset{*}{V}_4$ metrics which map geodesically on the reducible spaces (14.7) and (14.8).

2. Show that the $\overset{*}{V}_4$ in example 1 are not in general Einstein spaces; when are they?

3. Prove that an Einstein space V_4 with signature $(- - - +)$ can be mapped on to another Einstein space $\overset{*}{V}_4$ only when (1) the mapping is conformally trivial, $\overset{*}{g}_{\alpha\beta} = \nu g_{\alpha\beta}$, $\nu =$ const; (2) V_4 and $\overset{*}{V}_4$ are both spaces of constant curvature.

Compare with example 2.

4. Generalize the theorem in example 3 for an Einstein Space V_n with signature $(- - - \cdots - +)$.

CHAPTER 8

THE CAUCHY PROBLEM FOR THE EINSTEIN FIELD EQUATIONS

MANY problems analogous to those of classical mechanics can arise in relativistic mechanics, comprising the general theory and the Einstein field equations

$$R_{\alpha\beta} - \tfrac{1}{2} R g_{\alpha\beta} = \lambda T_{\alpha\beta}, \tag{43.0}$$

but only a beginning has been made in solving them ([194], [209], [247], [232], [266], [267]). One exception, however, is the Cauchy problem, the determination of the integrals of the field equations for given initial conditions, which can be precisely formulated and solved. In this chapter we outline the Cauchy problem for the Einstein equations (α); the most complete solution is that of Lichnerowicz ([231], [159], [247]). We must also note Petrovskii's fundamental result for the Cauchy problem applied to systems of partial differential equations ([106], [194]). The correct formulation of the problem for equations (43.0) is discussed in references [189], [226], [209], [247].

The solution of the Cauchy problem essentially depends on the class of functions to which the solutions must belong, and on the conditions imposed at the surface on which the problem is set [194]. Thus for the Einstein equations (43.0), the solution depends on conditions on the components of the metric tensor $g_{\alpha\beta}(x)$, and energy-momentum tensor $T_{\alpha\beta}(x)$, and their derivatives.

§ 43. The Einstein Field Equations

According to the definition introduced in § 1, a function φ is of class C^r if it has continuous partial derivatives up to order r; infinitely differentiable functions and analytic functions are denoted by the symbols C^∞ and C^a respectively.

We now introduce the idea of *a function of class C^2 on a part of a surface* S. Consider a function φ continuous in the neighbourhood of the hypersurface S. Suppose that the coordinate system for V_n is so chosen that the equation of the hypersurface is $x^n = 0$. We assume that φ satisfies the following conditions:

(1) it is of class C^2 in every neighbourhood belonging to the regions $x^n > 0$ and $x^n < 0$;

(2) $\partial_\alpha \varphi$ and $\partial_{\alpha\beta}\varphi$ converge uniformly to definite limits as $x^n \to 0$ without changing sign, but the limits differ depending on whether $x^n > 0$ or $x^n < 0$.

The conventional symbol for the discontinuity of φ at S is $[\varphi]$. In accordance with Hadamard's classical results ([17], pp. 83–87), the discontinuities of the derivatives of φ on S are

$$[\partial_i\varphi] = 0, \quad [\partial_n\varphi] = A, \tag{43.1}$$

$$[\partial_{ij}\varphi] = 0, \quad [\partial_{in}\varphi] = \partial_i A, \quad [\partial_{nn}\varphi] = B \quad (i = 1, \ldots, n-1), \tag{43.2}$$

where A and B are functions defined on S.

Any coordinate transformation of class C^2 in V_n gives a corresponding function φ and hypersurface S defined by the equation $f(x) = 0$ in the new reference system; then (1) φ belongs to class C^2 in every neighbourhood of S satisfying the conditions $f > 0$ and $f < 0$, and (2) the first and second derivatives of φ converge uniformly to definite but generally different limits as $f = +0$ and $f = -0$. We now define a *piece-wise continuous* function of class C^2 in a domain of V_n as one which belongs to class C^2 everywhere except in the neighbourhood of a finite number of surfaces of discontinuity $S_1, S_2, \ldots, S_k$, to all of which conditions (1) and (2) apply.

It should be noted that conditions analogous to these occur in the theory of hydrodynamic waves.

The field equations (43.0) apply to a V_4 space-time with metric

$$ds^2 = g_{\alpha\beta}(x)dx^\alpha dx^\beta, \tag{43.3}$$

defining V_4 as a *differentiable manifold.* To help subsequent discussion the concept of a V_4 must be made more precise. Suppose that we are given an n-dimensional, topologically connected ([205], p. 38) V_n space, each point of which has a neighbourhood *homeomorphic* to a Euclidean sphere (cf. § 1). We consider some *open* domain D in V_n ([205], p. 27) and map this domain on an n-dimensional Euclidean space R_n; this defines a *local system of coor-*

dinates for D, by means of which each point x in D can be associated with a point ξ_r in R_n defined by the n real numbers x^α. *A differentiable manifold of class* C^r is defined as a V_n with which we can connect a set A of local systems of coordinates satisfying the following conditions:

(α) The domains corresponding to A completely cover V_n.

(β) Suppose that D_1 and D_2 are two such domains; consider their intersection $D_1 \cap D_2$. If $x \subset D_1 \cap D_2$, the coordinates of the point x in one local system of coordinates corresponding to D_1 are functions of class C^r (with non-zero Jacobian) of the coordinates of this point in the local system corresponding to D_2.

Axioms (α) and (β) lead to such concepts as the *equivalence* of coverings of a domain, *reflexivity* (self-equivalence), *transitivity* and so on. A detailed exposition of such problems may be found for example in Pontryagin's book [205].

As the space-time manifold in which the field equations (43.0) are to be investigated, we take a differentiable manifold V_4 satisfying (α), (β) and the condition:

(γ) That at the intersection of the domains of two local systems of coordinates, the second derivatives of the functions connecting the coordinate systems are *piece-wise continuous functions of class* C^r.

More precisely, we shall assume that at the intersection of the domains of two coordinate systems, the coordinates of a point in one system are fourfold differentiable functions (with non-zero Jacobian) of the coordinates of the point in the other system; the first and second derivatives are continuous, but the third and fourth may have discontinuities of the Hadamard type.

The metric tensor components $g_{\alpha\beta}(x)$ and their derivatives $\partial_\gamma g_{\alpha\beta}(x)$ are assumed to be of class C^1 and piece-wise continuous of class C^2 respectively.

Taking any hypersurface S passing through the point x^α in V_n, we can construct a flat space R_{n-1} tangent to S at the given point. If the direction of any vector in R_{n-1} is given by $\vec{dx}$ with components dx^α, the tangent flat space is defined by the equation $v_\alpha \, dx^\alpha = 0$, where the v_α are the covariant components of the normal vector to R_{n-1}. The orientation in space-time ascribed to the tangent hypersurface to S depends on the norm of the vector v; when

$$g^{\alpha\beta} v_\alpha v_\beta > 0,$$

for a metric with signature $---+$, the hypersurface is described as oriented in space; when

$$g^{\alpha\beta}v_\alpha v_\beta < 0,$$

as *oriented in time*. These considerations permit us to further specify the hypersurface S on which the initial conditions of the Cauchy problem are given.

The Einstein field equations (43.0) were obtained by assuming that they generalize the Poisson equation for the Newtonian potential, and also satisfy the *conservation law* (cf. § 12). Denoting the left-hand side of (43.0) by $S_{\alpha\beta}$, we have

$$S_{\alpha\beta} = \lambda T_{\alpha\beta}, \tag{43.4}$$

where the energy-momentum tensor $T_{\alpha\beta}$ is completely defined by the distribution and movement of the matter, and $S_{\alpha\beta}$ by the metric of the space-time. When $T_{\alpha\beta} \neq 0$ we have the *interior problem*, when $T_{\alpha\beta} = 0$ (empty space-time) *the exterior problem*. We do not as yet specify the actual structure of $T_{\alpha\beta}$ which depends on the nature of the field, but simply note that this tensor generalizes the density term in Poisson's equation.

The purely geometrical *Einstein tensor* $S_{\alpha\beta}$, satisfies the two conditions:

(1) It is a comitant (a rational function) of $g_{\alpha\beta}$, $\partial_\gamma g_{\alpha\beta}$ and $\partial_\gamma \delta g_{\alpha\beta}$, and is linear in the second derivatives.

(2) It satisfies the conservation law

$$S^\alpha{}_{\beta,\,\alpha} = 0. \tag{43.5}$$

These conditions, introduced as hypotheses, do have a physical basis. As, moreover, they can be obtained directly in deriving equations (43.0) by the variational principle, they must be compatible with this principle. Einstein derived the field equations from (1) and (2) intuitively. Cartan's rigorous analysis ([63]) leads to the theorem: *if* $S_{\alpha\beta}$ *satisfies* (1) *and* (2), *then*

$$S_{\alpha\beta} = h\,\{R_{\alpha\beta} - \tfrac{1}{2}(R+k)g_{\alpha\beta}\}, \tag{43.6}$$

where $R_{\alpha\beta}$ *is the Ricci tensor,* R *the scalar curvature of the space and* h, k *arbitrary constants*. Taking h into the factor λ, we have

$$S_{\alpha\beta} \equiv R_{\alpha\beta} - \tfrac{1}{2}(R+k)\,g_{\alpha\beta} = \lambda T_{\alpha\beta}. \tag{43.7}$$

The so-called *cosmological constant* k, introduced by Einstein in connection with the hypothesis of a static finite universe ([43]), was later abandoned by him on physical grounds, whereupon (43.7) reduces to (43.4). In many problems the presence of this constant affects neither the method of solution nor the results, which can be automatically obtained by letting $R \to R+k$.

For a tensor $S_{\alpha\beta}$ defined by eqn. (43.6), the conservation law (43.5) follows directly from the Bianchi identity (5.16). Contracting this over σ, γ, we obtain

$$-R_{\alpha\beta,\,\delta}+R_{\alpha\delta,\,\beta}+R^{\sigma}{}_{\alpha\delta\beta,\,\sigma} = 0,$$

whence, contracting again over α, β, we find

$$R_{,\,\delta}-2R^{\sigma}{}_{\delta,\,\sigma} = 0.$$

If k is a constant, this identity shows that the conservation law holds for $S_{\alpha\beta}$ independently of whether it is the Einstein tensor or is defined by (43.7).

§ 44. The Exterior Cauchy Problem

We consider a V_4 domain in which the energy-momentum tensor $T_{\alpha\beta}$ can be neglected; eqn. (43.7) shows that

$$R_{\alpha\beta} = \tfrac{1}{2}(R+k)g_{\alpha\beta} = \varkappa g_{\alpha\beta}, \quad \varkappa = \tfrac{1}{2}(R+k). \tag{44.1}$$

Contracting over α and β, we obtain the following relation between the constant k and the scalar curvature of the space

$$R = -2k. \tag{44.2}$$

In the particular case of field equations (43.0), the cosmological constant k is zero, so that $R = 0$; we find that the field equations for empty space are

$$R_{\alpha\beta} = 0. \tag{44.3}$$

The solutions of the Cauchy problem for eqns. (44.1) and (44.3) do not differ sufficiently to be of independent interest but are nevertheless useful when considering the interior problem.

Given some hypersurface S in a V_4 space-time with metric tensor components $g_{\alpha\beta}(x)$, the Cauchy problem is to determine the potentials $g_{\alpha\beta}(x)$ satisfying the field equations (44.1) or (44.3) on and outside S, when they and their derivatives are certain given functions on S. Strictly speaking, (44.1) refers more naturally to the interior problem where $\lambda T_{\alpha\beta} = -(k/2)g_{\alpha\beta}$, but there is no essential difference in the method of solution whether $k = 0$ or not.

We must first define the *Cauchy data*; it is convenient to introduce a special system of coordinates in V_4 so that the hypersurface S is given, at least locally, by the equation $x^4 = 0$, and $g_{44} > 0$. The *index* of any derivative of a potential $g_{\alpha\beta}(x)$ is the number of times the component is differentiated with respect to x^4. The *Cauchy data* consists of the values of the $g_{\alpha\beta}$ on S and of the first partial derivatives of the $g_{\alpha\beta}$ of index 1, i.e. the $\partial_4 g_{\alpha\beta}$. We assume that these data are at least three and four times differentiable respectively.

We now use the field equations to examine the behaviour of the second derivatives of the potentials on S. The second derivatives of index $\leqslant 1$ ($\partial_{4i}g_{\alpha\beta}$, $\partial_{ij}g_{\alpha\beta}$, $i, j = 1, 2, 3$) can be determined directly from the Cauchy data by differentation and they are continuous. The derivatives of index $2\,(\partial_{44}g_{\alpha\beta})$ are determined by the field equations and must be obtained from them; they are in general discontinuous at S.

According to (5.2),

$$\tfrac{1}{2}R^{\sigma}{}_{\alpha\beta\gamma} = \partial_{[\beta}\Gamma^{\sigma}{}_{\gamma]\alpha} + \Gamma^{\tau}_{\alpha[\gamma}\Gamma^{\sigma}_{\beta]\tau}\,.$$

Hence

$$\begin{aligned} R_{\alpha\beta} &= g^{\lambda\nu}(\partial_\alpha \Gamma_{\nu,\,\lambda\beta} - \partial_\lambda \Gamma_{\nu,\,\alpha\beta}) + \omega_{\alpha\beta} \\ &= \tfrac{1}{2}g^{\lambda\nu}(\partial_{\beta\lambda}g_{\alpha\nu} + \partial_{\alpha\nu}g_{\beta\lambda} - \partial_{\lambda\nu}g_{\alpha\beta} - \partial_{\alpha\beta}g_{\lambda\nu}) + \omega_{\alpha\beta}, \end{aligned} \tag{44.4}$$

where the $\omega_{\alpha\beta}$ are rational functions of the Christoffel symbols $\Gamma^{\alpha}_{\beta\gamma}$ and therefore *completely determined by the Cauchy data*. If we take the field equations in the form (44.1), the right-hand side can be included in $\omega_{\alpha\beta}$ since $R = -2k =$ constant, and the resulting expression is still determined by the Cauchy data.

A semi-geodesic system of coordinates can be used, but is not essential; it is sufficient that the hypersurface S be *oriented in space* and given by the equation $x^4 = 0$. In view of the special role of the index 4, we express (44.4) in the form

$$R_{ij} \equiv -\tfrac{1}{2}g^{44}\partial_{44}g_{ij} + \Omega_{ij} = 0, \tag{44.5}$$

$$R_{i4} \equiv \tfrac{1}{2}g^{4j}\partial_{44}g_{ij} + \Omega_{i4} = 0, \tag{44.6}$$

$$R_{44} \equiv -\tfrac{1}{2}g^{ij}\partial_{44}g_{ij} + \Omega_{44} = 0, \tag{44.7}$$

where $\Omega_{\alpha\beta}$ contains $\omega_{\alpha\beta}$ and those second derivatives which can be determined by differentiating the Cauchy data on S. As the hypersurface is oriented in space, $g^{44} \neq 0$ and all the derivatives of the form $\partial_{44}g_{ij}$ can be determined from (44.5). In general there are no derivatives of the form $\partial_{44}g_{4\alpha}$ ($\alpha = 1, 2, 3, 4$) in eqns. (44.5)–(44.7), since all curvature tensor components $R_{44\alpha 4}$ vanish. Nevertheless we must show that *discontinuities in the derivatives* $\partial_{44}g_{4\alpha}$ ($\alpha = 1, 2, 3, 4$) *are not physically significant and therefore not essential to the analysis*. It is sufficient to show that a coordinate transformation exists which does not affect the derivatives $\partial_{44}g_{ij}$ but reduces the $\partial_{44}g_{4\alpha}$ to any specified value.

In the neighbourhood of the hypersurface S, we apply the following coordinate transformation which does not change the coordinates of points on S, since S is given by the equation $x^4 = 0$:

$$x^{\lambda'} = x^{\lambda} + \frac{x^{4^3}}{6}(\varphi^{\lambda}(x^i) + \varepsilon^{\lambda}), \tag{44.8}$$

where it is assumed that $\varepsilon^{\lambda} \to 0$ as $x^4 \to 0$. Clearly

$$\left(\frac{\partial x^{\alpha'}}{\partial x^{\beta}}\right)_S = \delta^{\alpha}_{\beta} \tag{44.9}$$

and

$$(\partial_{\mu}A^{\alpha'}_{4})_S = (\partial_4 A^{\alpha'}_{\beta})_S = (\partial_{i4}A^{\alpha'}_{4})_S = (\partial_{44}A^{\alpha'}_{i})_S = 0, \quad A^{\alpha'}_{\beta} \equiv \frac{\partial x^{\alpha'}}{\partial x^{\beta}}, \tag{44.10}$$

but

$$(\partial_{44}A^{\alpha'}_{4})_S = \varphi^{\alpha}. \tag{44.11}$$

As

$$g_{\alpha\beta} = \frac{\partial x^{\sigma'}}{\partial x^{\alpha}}\frac{\partial x^{\tau'}}{\partial x^{\beta}} g_{\sigma'\tau'},$$

it follows directly that

$$\partial_4 g_{\alpha\beta} = A^{\sigma'}_{\alpha}A^{\tau'}_{\beta}A^{\lambda'}_{4}\,\partial_{\lambda'}g_{\sigma'\tau'} + \partial_4 A^{\sigma'}_{\alpha}A^{\tau'}_{\beta}g_{\sigma'\tau'} + \partial_4 A^{\tau'}_{\beta}A^{\sigma'}_{\alpha}g_{\sigma'\tau'}, \tag{44.12}$$

and (44.10) shows that after the transformation, the Cauchy data on the hypersurface S are defined by

$$g_{\alpha\beta} = g_{\alpha'\beta'}, \quad \partial_4 g_{\alpha\beta} = \partial_{4'}g_{\alpha'\beta'}. \tag{44.13}$$

For the second derivatives, we obtain the expression

$$\partial_{44}g_{\alpha\beta} = A^{\sigma'}_{\alpha}A^{\tau'}_{\beta}A^{\lambda'}_{4}A^{\nu'}_{4}\,\partial_{\lambda'\nu'}g_{\sigma'\tau'} + \partial_{44}A^{\sigma'}_{\alpha}A^{\tau'}_{\beta}g_{\sigma'\tau'} + \partial_{44}A^{\tau'}_{\beta}A^{\sigma'}_{\alpha}g_{\sigma'\tau'} + u_{\alpha\beta}, \tag{44.14}$$

where $u_{\alpha\beta}$ denotes a set of terms containing multiples of the first derivatives of A^{α}_{β}; according to (44.10) these vanish on S so that

$$\partial_{44}g_{ij} = \partial_{4'4'}g_{i'j'},$$
$$\partial_{44}g_{4\alpha} = \partial_{4'4'}g_{4'\alpha'} + g_{\alpha\sigma}\varphi^{\sigma} + g_{4\sigma}\delta^4_{\alpha}\varphi^{\sigma}.$$

Thus while the derivatives $\partial_{44}g_{ij}$ are unchanged, the $\partial_{44}g_{4\alpha}$ can be reduced to a specified form by appropriate choice of the functions φ^{σ} ([231],

§ 14). In particular, we can so choose the $\varphi^{(\alpha)}$ that the $\partial_{44}g_{\alpha 4}$ are continuous on S.

Hence with this reservation in mind, we can say that the second derivatives are *continuous on S*. If in addition, we assume that the Cauchy data possess higher order derivatives, we can extend this result to these derivatives by differentiating (44.10) an appropriate number of times.

We still have to consider eqns. (44.6) and (44.7).

We assume for simplicity that $k = 0$, although, as we have seen, this is not essential, and all the results can easily be extended to $k \neq 0$. Equations (44.6) and (44.7) are of the form $R_{4\alpha} = 0$ ($\alpha = 1, 2, 3, 4$). As

$$S^4{}_i = g^{4j}R_{ij}+g^{44}R_{4i}, \quad S^4{}_4 = R^4{}_4-\tfrac{1}{2}R = \tfrac{1}{2}(g^{44}R_{44}-g^{ij}R_{ij}),$$

this set is equivalent to

$$S^4{}_\alpha = 0 \quad (\alpha = 1, 2, 3, 4), \tag{44.15}$$

which contains no derivatives of index 2 and is therefore a set of conditions imposed on the Cauchy data.

We now prove the theorem that *the Einstein field equations is in involution, i.e. if the potentials* $g_{\alpha\beta}(x)$ *satisfy* (44.5) *and conditions* (44.15) [*or* (44.6) *and* (44.7)] *on the hypersurface S, they also satisfy* (44.5) *in the neighbourhood of S.*

Suppose that the Cauchy data satisfy

$$(S^4{}_\alpha)_S = 0$$

and the corresponding potentials $g_{\alpha\beta}(x)$ satisfy equations (44.5). According to the conservation law (43.5),

$$S^\alpha{}_{\beta,\,\alpha} = 0,$$

or, separating index 4,

$$S^4{}_{\beta,\,4}+\sum_{i=1}^{3} S^i{}_{\beta,\,i} = 0. \tag{44.16}$$

On the other hand, (44.5) shows that

$$\begin{aligned} S^i{}_j &= g^{4i}R_{4j}-\tfrac{1}{2}(g^{44}R_{44}+2g^{4i}R_{4i}),\\ S^i{}_4 &= g^{i\sigma}R_{\sigma 4},\\ S^4{}_i &= g^{44}R_{4i},\\ S^4{}_4 &= \tfrac{1}{2}g^{44}R_{44}. \end{aligned}$$

Hence the set (44.16) can be represented in the form

$$g^{44}\,\partial_4 S^4{}_\beta = M^{i\sigma}{}_\beta\,\partial_i S_\sigma{}^4+N^\sigma{}_\beta S^4{}_\sigma, \tag{44.17}$$

where M and N are of class C^0.

Summarizing, we can say that for any solution of (44.5), the corresponding $S^4{}_\beta$ satisfies the set of four linear homogeneous partial differential equations (44.17) for the derivatives $\partial_4 S^4{}_\beta$. But when the Cauchy data are zero on S, $(S^4_\beta)_S = 0$, and the only solutions of the set are zero, which proves the theorem.

Thus if the Cauchy data satisfy (44.6) and (44.7) on S they also satisfy them in the neighbourhood of S, and these equations serve only to define the Cauchy data whereas the first set of field equations (44.5) determine the change of the potentials in time.

Hence the Cauchy problem for the Einstein field equations falls into two parts ([231], § 15):

(1) *The problem of determining the initial conditions*, i.e. finding Cauchy data which satisfy $S^4{}_\alpha = 0$ on the hypersurface S.

(2) *The problem of integrating with respect to the time* x^4, i.e. integrating the equations $R_{ij} = 0$ $(i, j = 1, 2, 3)$ [or $R_{ij} = \frac{1}{2}(R+k)g_{ij}$, if $k \neq 0$] for Cauchy data derived from the first problem.

The principal difficulty lies in solving the first problem, which depends on the kind of hypersurface S; this has only been solved in closed form by making certain particular assumptions. Fourès has solved the second problem by making simple assumptions about the differentiability of the functions ([159], [247]).

If the solution is restricted to the class of analytic functions (i.e. all the functions appearing in the solution of the second problem are real and analytic), the Cauchy–Kowalewski theorem ([160], § 1; [151], p. 294) for sets of partial differential equations shows that the second problem admits a solution of class C^a which is real and unique to within a transformation of coordinates leaving the coordinates of the hypersurface S and the Cauchy data on S unaltered. It should be noted that in relativistic field theory the $g_{\alpha\beta}(x)$ can be analytic only in certain regions of space-time, for in general, when for example there are gravitating masses present, the potentials must be discontinuous. We must expect the appearance of discontinuities at least in the second derivatives of the $g_{\alpha\beta}(x)$ across hypersurfaces oriented in time separating the moving matter from the regions it "sweeps through" in time. Thus in considering the interior and exterior Cauchy problems, Lichnerowicz ([231], chapter 1) started from the *postulate that the* $\partial_{\sigma\tau} g_{\alpha\beta}$ are discontinuous at S. He determined the Cauchy conditions in closed form only for the case of a minimal hypersurface S, i.e. one with mean curvature $H = 0$, a considerable simplication. The main difficulty is that *solutions must be determined in manifolds whose topology may be complex.*

If the hypersurface S is oriented everywhere in time ($g_{44} < 0$), the foregoing discussion can be repeated *mutatis mutandis* and provides two analogous Cauchy problems.

All these results depend entirely on the possibility of solving the field equations (44.5) explicitly for the derivatives $\partial_{44}g_{ij}$ ($i, j = 1, 2, 3$), which follows from the condition $g^{44} \neq 0$ in the cases given.

If a hypersurface S is so chosen in the local system of coordinates that this condition no longer holds, we obtain fundamentally different results associated with a *characteristic manifold* as in the theory of differential equations.

§45. Freedom Available in Choosing Field Potentials for an Einstein Space

The condition that the solution should be analytic greatly simplifies the Cauchy problem and allows us to apply the classical existence and uniqueness theorems. In spite of its apparently formal nature, this condition warrants further consideration. In empty spaces, for example (cf. Chapter 2), all the well-known solutions of the field equations are defined by potentials belonging to class C^a. Moreover, there is a close connection between the Cauchy problem and that of determining the possible freedom in choosing the field potentials; using the simplest possible coordinate system we can determine what is usually called the "*physical*" *functional freedom in the field*, within the class of analytic functions. Such a qualitative appraisal of the problem is useful for empty or Einstein spaces ($R_{\alpha\beta} = \varkappa g_{\alpha\beta}$) and also for the study of general gravitational fields with $T_{\alpha\beta} \neq \sigma g_{\alpha\beta}$.

Let us consider an Einstein space V_4 for which

$$R_{\alpha\beta} = \varkappa g_{\alpha\beta},$$

and introduce a semi-geodesic reference system (Gaussian coordinates); the coordinate lines x^4 are geodesics which intersect the coordinate hypersurfaces $x^4 =$ constant orthogonally, x^4 being the arc length. The normals to the hypersurfaces are time-like. This kind of coordinate system can be introduced, at least locally, at all points with a Minkowski signature in any V_4.

The metric of the space is

$$ds^2 = \overset{*}{ds}{}^2 + dx^{4^2} \tag{45.1}$$

in this coordinate system where $\overset{*}{ds}{}^2$ is a negative definite form,

$$\overset{*}{ds}{}^2 = g_{ij}(x^1, x^2, x^3, x^4)\, dx^i\, dx^j \qquad (i, j = 1, 2, 3), \tag{45.2}$$

and thus

$$g_{i4} = 0, \quad g_{44} = g^{44} = 1, \tag{45.3}$$

and the determinant of the tensor

$$g \equiv |g_{\alpha\beta}| = \overset{*}{g} \equiv |g_{ij}| \qquad (i, j = 1, 2, 3; \quad \alpha, \beta = 1, 2, 3, 4). \tag{45.4}$$

Hence ds^2 defines a space representing a generalized topological product of a V_3 and a one-dimensional space; the definite form $\overset{*}{ds}{}^2$ defines a manifold on every hypersurface $x^4 =$ constant, homeomorphic with a three-dimensional Euclidean space.

The hypersurfaces $x^4 =$ constant are geodesically parallel, and the *second* fundamental tensor on $x^4 =$ constant is given by

$$h_{ij} = -\tfrac{1}{2}\,\partial_4 g_{ij}, \quad h^{ij} = \tfrac{1}{2}\,\partial_4 g^{ij}. \tag{45.5}$$

It is, however, convenient to modify this tensor by changing the sign

$$\omega_{ij} = -h_{ij}. \tag{45.6}$$

Using the three-dimensional tensor $\overset{*}{g}{}^{ij}$, where

$$(\overset{*}{g}_{ij}) = (g_{ij})_{x^4=c}, \tag{45.7}$$

we obtain

$$\omega^{ij} = \overset{*}{g}{}^{ih}\overset{*}{g}{}^{jk}\omega_{hk} \tag{45.8}$$

and

$$\omega_i^j = \overset{*}{g}{}^{jk}\omega_{ki}. \tag{45.9}$$

Using these, we introduce the three-dimensional invariant

$$H = \omega_i^i,$$

defining the *mean* curvature of the hypersurface, and the invariant

$$T^2 = \omega_i^j\omega_j^i. \tag{45.10}$$

In solving the exterior Cauchy problem in this coordinate system, we write the field equations in two sets, as in the previous section:

$$R_{ij} = \varkappa g_{ij} \qquad (i, j = 1, 2, 3), \tag{45.11}$$

$$R_{\alpha 4} = \varkappa g_{\alpha 4} \qquad (\alpha = 1, 2, 3, 4). \tag{45.12}$$

The use of the term "exterior problem" is formally not quite correct, but the assumption $\varkappa = 0$ in no way affects the subsequent analysis. As in the previous section, eqns. (45.11) and (45.12) define a system in involution and the Cauchy problem reduces to the two problems: (1) of determining the most general Cauchy data satisfying eqns. (45.12) on $x^4 = c$, and (2) of integrating with respect to time x^4, i.e. the integration of eqns. (45.11). The hypersurface S is not a characteristic manifold in the present case.

Problem (1) can now be formulated as follows: functions g_{ij} and ω_{ij} of the independent variables x^1, x^2, x^3, must be determined satisfying the set of equations (45.12). This set can also be represented as two subsets, corresponding to $\alpha = 4$ and $\alpha \neq 4$. When $\alpha = 4$, we obtain

$$H^2 - T^2 - \overset{*}{R} = \varkappa, \tag{45.13}$$

where $\overset{*}{R}$ is the scalar curvature for the three-dimensional manifold defined by the tensor $\overset{*}{g}_{ij}$.

The other three equations can be written

$$(\omega^{ij} - H\overset{*}{g}{}^{ij})_{;j} = 0, \tag{45.14}$$

the semicolon denoting covariant differentiation with respect to the metric $\overset{*}{ds}{}^2$.

We are interested in the maximum freedom in determining the g_{ij} and ω_{ij} from these four equations. To assess this, we set down the equations and show that they are equivalent to a Cauchy–Kowalewski set for four unknown functions.

We solve eqn. (45.13) for a highest order derivative of one of the unknown functions with respect to one of the arguments, for example x^3. T contains no derivatives of the unknown functions with respect to x^i ($i = 1, 2, 3$), but $\overset{*}{R}$ does:

$$\overset{*}{R} = \overset{*}{g}{}^{pq}\overset{*}{g}{}^{st}\{\tfrac{1}{2}(\partial_{pt}\overset{*}{g}_{sq} + \partial_{sq}\overset{*}{g}_{pt} - \partial_{st}\overset{*}{g}_{pq}) + \overset{*}{g}{}^{lm}(\overset{*}{\Gamma}_{m,\,pt}\overset{*}{\Gamma}_{l,\,sq} - \overset{*}{\Gamma}_{m,\,pq}\overset{*}{\Gamma}_{l,\,st})\}. \tag{45.15}$$

The second bracket contains only the first derivatives of $\overset{*}{g}_{ij}$ occurring in the Christoffel symbols $\overset{*}{\Gamma}_{p,\,qr}$ for the three-dimensional Riemann space, but the first bracket contains second derivatives; we are interested in the second derivative $\partial_{33}g_{ij}$. There are clearly no derivatives $\partial_{33}g_{3j}$ ($j = 1, 2, 3$) on the right-hand side, and we therefore choose the derivative $\partial_{33}g_{22}$ for example, and solve the equation for this. The coefficient of this derivative is

$$\overset{*}{g}{}^{23^2} - \overset{*}{g}{}^{22}\overset{*}{g}{}^{33},$$

so that eqn. (45.15) is of the form

$$(\overset{*}{g}^{23^2}-\overset{*}{g}^{22}\overset{*}{g}^{33})\partial_{33}\overset{*}{g}_{22} = M, \tag{45.16}$$

where

$$M = H^2-T^2-\varkappa-\overset{*}{K},$$

and $\overset{*}{K}$ denotes the difference between $\overset{*}{R}$ and the left-hand side of this equation.

We now turn to (45.14); on lowering the index i, we obtain

$$\omega^j{}_{k,\,j}-H_{,\,k} \equiv \partial_j\omega^j{}_k-\overset{*}{\Gamma}{}^p_{kj}\omega^j{}_p+\overset{*}{\Gamma}{}^j_{pj}\omega^p{}_k-\partial_k H = 0.$$

First taking $k = 1$, we consider the mean curvature H. Cancelling similar terms, we obtain

$$\partial_2\omega^2{}_1-\partial_3\omega^3{}_1-\partial_1\omega^2{}_3-\partial_1\omega^3{}_3-\overset{*}{\Gamma}{}^p_{1j}\omega^j{}_p+\overset{*}{\Gamma}{}^j_{pj}\omega^p_1 = 0.$$

In the same way, putting $k = 2, 3$, we obtain

$$\partial_1\omega^1{}_2+\partial_3\omega^3{}_2-\partial_2\omega^1{}_1-\partial_2\omega^2{}_3-\overset{*}{\Gamma}{}^p_{2j}\omega^j{}_p+\overset{*}{\Gamma}{}^j_{pj}\omega^p{}_2 = 0,$$

$$\partial_1\omega^1{}_3+\partial_2\omega^2{}_3-\partial_3\omega^1{}_1-\partial_3\omega^2{}_2-\overset{*}{\Gamma}{}^p_{3j}\omega^j{}_p+\overset{*}{\Gamma}{}^j_{pj}\omega^p{}_3 = 0.$$

Only first order partial derivatives of $\overset{*}{g}_{ij}$ and $\omega^i{}_j$ occur in the last three equations. We now solve them for the derivatives of the three unknown functions ω_{ij} with respect to x^3, after replacing $\omega^i{}_j$ by ω_{ij} where

$$\omega^i{}_j = \overset{*}{g}^{ki}\omega_{kj}.$$

We rewrite the first equation in the form

$$\partial_3\omega^3{}_1 = P_1,$$

where

$$P_1 = P_1(\partial_k\omega^p{}_q,\quad \partial_s\overset{*}{g}^{pr},\quad \partial_s\overset{*}{g}_{pr},\quad \overset{*}{g}_{pr},\quad \omega_{pq}),\qquad k < 3.$$

As

$$\partial_3\omega^3{}_1 = \partial_3(\overset{*}{g}^{3s}\omega_{s1}),$$

the first equation takes the form

$$\overset{*}{g}^{33}\partial_3\omega_{13} = P_1-\omega_{13}\partial_3\overset{*}{g}^{33}-\sum_{s\neq 3}\partial_3(\overset{*}{g}^{3s}\omega_{s1}).$$

Similarly, the second and third equations can be written

$$\overset{*}{g}^{33}\partial_3\omega_{23} = P_2-\omega_{23}\partial_3\overset{*}{g}^{33}-\sum_{s\neq 3}\partial_3(\overset{*}{g}^{3s}\omega_{s2}),$$

$$\overset{*}{g}^{22}\partial_3\omega_{22} = P_3-\omega_{22}\partial_3\overset{*}{g}^{22}-\sum_{s\neq 2}\partial_3(\overset{*}{g}^{2s}\omega_{2s})-\sum\partial_3(\overset{*}{g}^{1s}\omega_{1s}).$$

We can substitute from the first two equations for the derivatives of ω_{13} and ω_{23} on the right-hand side of the last equation. We now impose the condition that the coefficients of these derivatives do not vanish. We can in general assume that the components

$$\overset{*}{g}{}^{22}, \quad \overset{*}{g}{}^{33}, \quad \overset{*}{g}{}^{23^2} - \overset{*}{g}{}^{22}\overset{*}{g}{}^{33}$$

do not vanish on the hypersurface $x^4 = 0$. We thus obtain the following set of equations equivalent to (45.12):

$$\partial_{33}\overset{*}{g}_{22} = \frac{1}{\Delta} M, \quad \Delta \equiv \overset{*}{g}{}^{23^2} - \overset{*}{g}{}^{22}\overset{*}{g}{}^{33},$$

$$\partial_3\omega_{13} = \frac{1}{\overset{*}{g}{}^{33}} \{P_1 - \omega_{13}\partial_3\overset{*}{g}{}^{33} - \sum_{s\neq 3} \partial_3(\overset{*}{g}{}^{3s}\omega_{1s})\},$$

$$\partial^3\omega_{23} = \frac{1}{\overset{*}{g}{}^{33}} \{P_2 - \omega_{23}\partial_3\overset{*}{g}{}^{33} - \sum_{s\neq 3} \partial_3(\overset{*}{g}{}^{3s}\omega_{2s})\},$$

$$\begin{aligned}\partial_3\omega_{22} = \frac{1}{\overset{*}{g}{}^{22}} \{&P_3 - \omega_{22}\partial_3\overset{*}{g}{}^{22} - \partial_3(\overset{*}{g}{}^{21}\omega_{21}) \\ &- \omega_{23}\partial_3\overset{*}{g}{}^{23} - \sum_{s\neq 3} \partial_3(\overset{*}{g}{}^{s1}\omega_{s1}) - \omega_{13}\partial_3\overset{*}{g}{}^{13} \\ &- \frac{\overset{*}{g}{}^{13}}{\overset{*}{g}{}^{33}} [P_1 - \omega_{13}\partial_3\overset{*}{g}{}^{33} - \sum_{s\neq 3} \partial_3(\overset{*}{g}{}^{3s}\omega_{1s})] \\ &- \frac{\overset{*}{g}{}^{23}}{\overset{*}{g}{}^{33}} [P_2 - \omega_{23}\partial_3\overset{*}{g}{}^{33} - \sum_{s\neq 3} \partial_3(\overset{*}{g}{}^{3s}\omega_{2s})]\}.\end{aligned}$$

If we are given the following *completely arbitrary functions of* x^1, x^2, x^3,

$$\overset{*}{g}_{11}, \overset{*}{g}_{12}, \overset{*}{g}_{13}, \overset{*}{g}_{23}, \overset{*}{g}_{33}, \omega_{11}, \omega_{12}, \omega_{33},$$

we are left with a set of four equations for the highest order derivatives of the remaining unknown functions

$$\overset{*}{g}_{22}, \omega_{13}, \omega_{23}, \omega_{22}$$

with respect to x^3. As these derivatives do not appear on the right-hand side, the set is of the Cauchy–Kowalewski type. Assuming that the functions are analytic permits the use of the Cauchy–Kowalewski theorem.

The results are, of course, still valid if the *indices are renumbered.* Hence the four functions are defined to within the possible freedom

$$\begin{gathered}\omega_{13}(x^1, x^2), \quad \omega_{23}(x^1, x^2), \quad \omega_{22}(x^1, x^2), \\ g_{22}(x^1, x^2), \quad \partial_3 g_{22} = f(x^1, x^2)\end{gathered} \tag{45.17}$$

at $x^3 = \overset{0}{x}{}^3$, where these functions are arbitrary and analytic in a domain containing the point $x^i = \overset{0}{x}{}^i$, and take given values at this point as must their derivatives. This defines the degree of freedom with which the Cauchy data are determined in solving problem (1).

We now turn to problem (2), i.e. integrating the field equations for the initial Cauchy data. Equation (45.11) can be written in the form

$$\partial_{44} g_{ij} = F_{ij} \qquad (i, j = 1, 2, 3),$$

where F_{ij} is completely determined by the Cauchy data. This is a Cauchy–Kowalewski set of six equations in the six unkowns g_{ij} and hence only admits a solution in the class of analytic functions with arbitrariness

$$g_{ij} = g_{ij}(x^1, x^2, x^3),$$
$$\partial_4 g_{ij} = \varphi_{ij}(x^1, x^2, x^3) \equiv \omega_{ij}$$

for $x^4 = 0$. These functions are not entirely arbitrary for they satisfy (45.12) on the hypersurface $x^4 = 0$, i.e. they are the Cauchy data determined in the solution of problem (1).

We can now set down the required potentials in a semi-geodesic system of coordinates, separating the *parametric part* which depends on the *arbitrary functions* from the *principal part* containing the definite components.

We expand the field potentials $g_{ij}(x)$ in the power series

$$g_{ij} = \sum_{k_1, \ldots, k_4 = 0}^{\infty} C_{ij,\, k_1 k_2 k_3 k_4} (x^1 - \overset{0}{x}{}^1)^{k_1} (x^2 - \overset{0}{x}{}^2)^{k_2} (x^3 - \overset{0}{x}{}^3)^{k_3} (x^4 - \overset{0}{x}{}^4)^{k_4}, \tag{45.18}$$

where the coefficients $C_{ij,\, k_1 k_2 k_3 k_4}$ are given by

$$C_{ij,\, k_1 k_2 k_3 k_4} = \frac{1}{k_1! k_2! k_3! k_4!} \left(\frac{\partial^{k_1 + k_2 + k_3 + k_4}}{\partial x^{1^{k_1}} \partial x^{2^{k_2}} \partial x^{3^{k_3}} \partial x^{4^{k_4}}} g_{ij} \right)_{x^\alpha = \overset{0}{x}{}^\alpha}.$$

In order to separate out the *parametric* part of this expansion, we must eliminate all terms containing derivatives which occur on the left-hand sides of eqns. (45.11) and (45.12), and terms containing derivatives of these derivatives (the so-called *principal derivatives* of the set of field equations).

Let us consider the potential $g_{22}(x)$. The following derivatives of this unknown function occur on the left-hand sides of (45.11) and (45.12): $\partial_{33} g_{22}$, $\partial_3 \omega_{22} = \partial_{34} g_{22}$, $\partial_{44} g_{22}$. The first derivative entails the elimination of all terms divisible by $(x^3-\overset{0}{x}{}^3)^2$ from the expansion; what remains is the parametric part for the derivative. It must consist of (1) terms not divisible by $(x^3-\overset{0}{x}{}^3)$, which can therefore be obtained by putting $x^3 = \overset{0}{x}{}^3$, giving an arbitrary function of x^1, x^2, x^4; (2) terms containing $(x^3 - \overset{0}{x}{}^3)$ and thus obtained by differentiating (45.18) with respect to x^3 before putting $x^3 = \overset{0}{x}{}^3$, which again defines an arbitrary function of x^1, x^2, x^4. Thus the parametric part for the principal derivative is

$$A(x^1, x^2, x^4)+(x^3-\overset{0}{x}{}^3)B(x^1, x^2, x^4). \tag{45.19}$$

The second principal derivative $\partial_{34}g_{22}$ entails the elimination of all terms divisible by $(x^3 - \overset{0}{x}{}^3)(x^4 - \overset{0}{x}{}^4)$ from the expansion. This means all terms which (1) are not divisible by $(x^3-\overset{0}{x}{}^3)$, i.e. $A(x^1, x^2, x^4)$, and (2) all terms containing $(x^3 - \overset{0}{x}{}^3)$ but not $(x^4 - \overset{0}{x}{}^4)$, obtained by differentiating (45.19) with respect to x^3 and putting $x^4 = \overset{0}{x}{}^4$, which leads to functions of x^1, x^2; hence the expansion for g_{22} has the form

$$A(x^1, x^2, x^4)+(x^3-\overset{0}{x}{}^3)B(x^1, x^2). \tag{45.20}$$

Finally, the third derivative $\partial_{44}g_{22}$ entails the elimination of all terms divisible by $(x^4 - \overset{0}{x}{}^4)^2$. The remaining terms are (1) those not divisible by $(x^4 - \overset{0}{x}{}^4)$, obtained by putting $x^4 = \overset{0}{x}{}^4$ in (45.20), i.e.

$$A(x^1\ x^2)+(x^3-\overset{0}{x}{}^3)B(x^1, x^2),$$

and (2) terms containing $(x^4-\overset{0}{x}{}^4)$ but not $(x^4-\overset{0}{x}{}^4)^2$, obtained by differentiating the series with respect to x^4 and then putting $x^4 = \overset{0}{x}{}^4$, which gives a function of x^1, x^2. Hence the parametric part has the form

$$A(x^1, x^2)+(x^3-\overset{0}{x}{}^3)B(x^1, x^2)+(x^4-\overset{0}{x}{}^4)C(x^1, x^2).$$

We have thus completely determined the arbitrariness in $g_{22}(x)$; the *principal part* of this potential is, in general, a function of the four independent variables x^α ($\alpha = 1, 2, 3, 4$). It is clear that A, B and C are the Cauchy data on $x^4 =$ constant: $g_{22}(x^1, x^2)$, $\partial_3 g_{22}(x^1, x^2)$, $\omega_{22}(x^1, x^2)$.

Thus

$$g_{22} = A(x^1, x^2)+(x^3-\overset{0}{x^3})B(x^1, x^2)$$
$$+(x^4-\overset{0}{x^4})C(x^1, x^2)+\psi_{22}(x^1, x^2, x^3, x^4),$$

where ψ_{22} the principal part of the potential, is completely determined by the field equations if the Cauchy data are specified.

Repeating the analysis for all the potentials and changing the notation for the arbitrary functions, we obtain the following expressions:

$$\left.\begin{aligned}
g_{11} &= f_{11}(x^1, x^2, x^3)+(x^4-\overset{0}{x^4})\,\varphi_{11}(x^1, x^2, x^3)+\psi_{11},\\
g_{12} &= f_{12}(x^1, x^2, x^3)+(x^4-\overset{0}{x^4})\,\varphi_{12}(x^1, x^2, x^3)+\psi_{12},\\
g_{13} &= f_{13}(x^1, x^2, x^3)+(x^4-\overset{0}{x^4})\,\varphi_{13}(x^1, x^2)+\psi_{13},\\
g_{22} &= f_{22}(x^1, x^2)+(x^4-\overset{0}{x^4})\,\varphi_{22}(x^1, x^2)\\
&\quad+(x^3-\overset{0}{x^3})\,\theta_{22}(x^1, x^2)+\psi_{22},\\
g_{23} &= f_{23}(x^1, x^2, x^3)+(x^4-\overset{0}{x^4})\,\varphi_{23}(x^1, x^2)+\psi_{23},\\
g_{33} &= f_{33}(x^1, x^2, x^3)+(x^4-\overset{0}{x^4})\,\varphi_{33}(x^1, x^2, x^3)+\psi_{33},\\
g_{i4} &= 0 \qquad (i = 1, 2, 3), \quad g_{44} = 1,
\end{aligned}\right\} \tag{45.21}$$

where the ψ_{ij} are the principal parts of the potentials. The functions f_{ij}, φ_{ij}, θ_{ij} define the degree of freedom in the general solution for a semi-geodesic system of coordinates.

As the choice of the surface S for which the Cauchy problem is solved is itself arbitrary, there is one more arbitrary function $x^4 = \Phi(x^1, x^2, x^3)$.

We are concerned with *the arbitrariness of the structure of the space irrespective of the choice of the initial hypersurface*. We must therefore note that in the final estimate the total number of arbitrary functions is reduced by one. There still remains some freedom in choosing the coordinate system. Although the coordinate system in the space itself is specified, it is still possible without changing the set of hypersurfaces $x^4 =$ constant in V_4 to introduce a transformation

$$x^{i'} = x^{i'}(x^1, x^2, x^3) \qquad (i = 1, 2, 3)$$

on the hypersurface $x^4 =$ constant. Applying this transformation on the Cauchy hypersurface can reduce three of the potentials $g_{ij}(x^1, x^2, x^3)$ to specified functions in accordance with Riemann's well-known theorem.

This is done by introducing a specialized coordinate system on $x^4 = 0$, for example a semigeodesic system of coordinates in V_3, or by imposing certain other conditions; this automatically reduces the total number of arbitrary *functions by three, each of three arguments.*

Some arbitrariness in the choice of the coordinates on x^4 = constant still remains; this can be written

$$x^{s'} = x^{s'}(x^1, x^2) \qquad (s' = 1, 2),$$

apart from renumbering the coordinates and so the degree of freedom of the potentials on $x^4 = 0$ can be reduced by *two functions of two variables.*

Thus *the arbitrariness in determining the field potentials in a semi-geodesic system of coordinates is set by four functions of three arguments and three functions of two arguments (the Cauchy conditions).*

In this most simple semi-geodesic coordinate system, the final degree of freedom reflects the irreducible residue which, in the general case, represents *the purely gravitational effect*, and as such is conventionally known as the *physical arbitrariness.*

Examples

1. Derive formulae (45.21) for all the potentials, following the pattern for $g_{22}(x)$.

2. Show that the degree of freedom for T_1 spaces is not less than that for T_2 and T_3 spaces and is thus defined by four functions of three arguments and three functions of two arguments.

3. Show that the "physical" arbitrariness for T_2 spaces in a semigeodesic system of coordinates is defined by two functions of three arguments and seven functions of two arguments ([261], p. 237).

4. Show that the "physical"arbitrariness for T_3 is defined by nine functions of two independent variables ([261], p. 237).

§46. Characteristic and Bicharacteristic Manifolds

When the metric of a V_4 is indefinite, there is a cone of isotropic directions

$$g_{\alpha\beta}x^\alpha x^\beta = 0$$

at each point, dividing the domain of real vectors with a positive norm from that of real vectors with a negative norm (cf. §2). Every vector lying in this cone is isotropic. If the hypersurface S touches the corresponding isotropic cone at each point, the tangent plane to S at every point is isotropic and the metric is degenerate (§ 7), and hence the normal to the hypersurface is isotropic. If the equation is $f(x^\alpha) = 0$, then

$$\Delta_1 f = g^{\alpha\beta} f_{,\alpha} f_{,\beta} = 0. \tag{46.1}$$

This equation defines an *isotropic* hypersurface of the V_4 space (denoted as before by $\overset{*}{V}_3$) called the *characteristic manifold* of the field equations. In the special system of coordinates where S is given by the equation $x^4 = 0$, (46.1) reduces to

$$\Delta_1(x^4) = g^{44} = 0.$$

This implies that $\partial_{44} g_{ij}$ cannot be determined explicitly from (44.5); according to Petrovskii ([106], first part) the $\partial_{44} g_{ij}$ may be discontinuous at S and there can be an infinite number of solutions of the field equations for the same Cauchy data; *the isotropic cones are the characteristics of the Einstein field equations. The isotropic $\overset{*}{V}_3$ hypersurfaces tangent to the isotropic cones are characteristic manifolds of the field equations.* In the terminology of geometric optics, the isotropic normal to V_3 represents the *wave vector* tangent to the light ray, f is the *eikonal*, and eqn. (46.1) the *eikonal equation* (Landau and Lifshitz [156], p. 271).

We now show that the characteristic manifolds $\overset{*}{V}_3$ can be obtained from certain curves C_0, known as the *bicharacteristics of the field equations*, and a one-parameter family of hyperplanes defined at all points of the bicharacteristic.

Consider the bilinear form

$$H(x^\alpha, y_\sigma) = \tfrac{1}{2} g^{\alpha\beta} x_\alpha y_\beta$$

and the inhomogeneous partial differential equation

$$2H(x^\alpha, \partial_\sigma f) = C,$$

where C is an arbitrary constant. This equation corresponds to the set of ordinary equations

$$\frac{dx_1}{\dfrac{\partial H}{\partial y_1}} = \frac{dx^2}{\dfrac{\partial H}{\partial y_2}} = \frac{dx^3}{\dfrac{\partial H}{\partial y_3}} = \frac{dx^4}{\dfrac{\partial H}{\partial y_4}} = -\frac{dy_1}{\dfrac{\partial H}{\partial x^1}} = -\frac{dy_2}{\dfrac{\partial H}{dx^2}} = -\frac{dy_3}{\dfrac{\partial H}{\partial x^3}} = \frac{dt}{2H}$$

in the variables x^α, y_σ, admitting a first integral

$$2H(x^\alpha, y_\sigma) = C,$$

where C = constant. We can represent the variables x^α, y_σ as functions of an auxiliary parameter u by means of the canonical system

$$\frac{dx^\alpha}{du} = \frac{\partial H}{\partial y_\alpha}, \quad \frac{\partial y_\alpha}{\partial u} = -\frac{\partial H}{\partial x^\alpha}. \tag{46.2}$$

The right-hand sides are known if the Hamiltonian $H(x^\alpha, y_\sigma)$ is given. Denoting dx^α/du by $\dot{x}^\alpha$, the first set of equations (46.2) takes the form

$$\dot{x}^\alpha = g^{\alpha\beta} y_\beta \tag{46.3}$$

and, conversely,

$$y_\beta = g_{\beta\sigma}\dot{x}^\sigma.$$

This shows that x^α and (46.2) define the extremals of the Lagrangian

$$2L = g_{\alpha\beta}\dot{x}^\alpha\dot{x}^\beta,$$

since by changing from the variables $(x^\alpha, \dot{x}^\beta)$ to (x^α, y_σ) we obtain the classical relation between L and H:

$$H = \dot{x}^\alpha \frac{\partial L}{\partial \dot{x}^\alpha} - L.$$

The solutions of this equation are extremals satisfying the first integral

$$2L = C, \quad C = \text{const.} \tag{46.4}$$

This means that the extremals are also extremals for

$$\sqrt{2L} = \sqrt{g_{\alpha\beta}\dot{x}^\alpha\dot{x}^\beta},$$

if (46.3) is true, i.e. the curves x^α define geodesics of the V_4 space (§ 6). If in particular $C = 0$, the eikonal equations (46.1) have a first integral f = constant, and the $\overset{*}{V}_3$ manifolds can be found from the isotropic geodesics, taking the isotropic hyperplanes touching the isotropic cones along the tangent to the isotropic curve.

We can now see that the bicharacteristics are isotropic curves, for the theory of partial differential equations shows that given a set of $\overset{*}{V}_3$ manifolds touching the elementary cone in a V_4 along the generator g, the tangent to the curve C_0 in V_4 associated with the $\overset{*}{V}_3$ is g. However, not every isotropic curve is a geodesic. The above analysis shows that *the bicharacteristics of the Einstein field equations are isotropic geodesics.*

Several theories of gravitational waves have been propounded by analogy with Hadamard's theory of the propagation of waves, treating the wave fronts as surfaces of discontinuity for the second derivatives of the field potentials $g_{\alpha\beta}(x)$ ([231], § 16; [258]; [288]; [289]).

Before considering the interior Cauchy problem, we must say something about the energy-momentum tensor since the metric of the V_4 space and the field potentials $g_{\alpha\beta}(x)$ depend largely on the physical properties of this tensor.

§ 47. The Energy-momentum Tensor

More careful treatment than it has hitherto received must be given to the problem of the structure of the energy-momentum tensor $T_{\alpha\beta}$ in the Einstein field equations, and in particular that of *the cosmological term* in these equations, which can always be associated with $T_{\alpha\beta}$ and must be examined for the special conditions of the problem and not in general. The energy momentum tensor for a system of non-interacting particles, i.e. $T_{\alpha\beta} = m_0c^2u_\alpha u_\beta$, the case to which Einstein and many others restricted themselves, corresponds to the simplest physical state only. The outlook of the present work calls for a general study of the energy-momentum tensor covering all the various possible cases. The basic task is to provide *the algebraic structure of* $T_{\alpha\beta}$, stating at a given point in space-time (1) *the latent roots*, (2) *the eigenvectors*, and (3) the *Segre characteristic* defined by the elementary divisors. Having defined the structure of $T_{\alpha\beta}$ at a point and written it in tensor form, we obtain the structure of $T_{\alpha\beta}(x)$ in any domain, not reducible to a point, with unchanging *Segre characteristic*. In the solution of any particular problem, the elements defining the structure of $T_{\alpha\beta}$ must be derived from physical considerations; they cannot by their very nature be universal.

We consider the pair of symmetric tensors $T_{\alpha\beta}$ and $g_{\alpha\beta}$ at a point in space-time, and the corresponding λ-matrix

$$(T_{\alpha\beta} - \lambda g_{\alpha\beta}). \qquad (47.1)$$

We assume that the metric of space-time corresponds to the indefinite form $g_{\alpha\beta}\,dx^\alpha\,dx^\beta$ with signature $(- - - +)$. The general theorem of § 9 on the reduction of a pair of forms, one of which is non-degenerate ($|g_{\alpha\beta}| \neq 0$), *over the field of real numbers*, enables us to set down all the possible Segre characteristics for the elementary divisors and the canonical forms for the tensors $T_{\alpha\beta}$ and $g_{\alpha\beta}$ at the given point. By writing these in tensor form, we obtain the expressions for the *tensor fields* $T_{\alpha\beta}(x)$ and $g_{\alpha\beta}(x)$ in a domain. These tensor expressions can be obtained by considering $T_{\alpha\beta}$ and $g_{\alpha\beta}$ in a uniquely defined *quasi-orthonormal tetrad*; the prefix "quasi" implies that some of the vectors of the tetrad may be isotropic (cf. § 9).

The matrix (47.1) can have the following types of Segre characteristic:

(1) [1111], (2) [211], (3) [31], (4) [22], (5) [4].

These can be subdivided into the various cases corresponding to multiple latent roots for the elementary divisors, e.g. [(21)1], when the elementary divisors have the form $(\lambda-\lambda_1)^2$, $(\lambda-\lambda_1)$, $(\lambda-\lambda_2)$; moreover, the latent roots may be *complex* and the eigenvectors *isotropic* or *complex*.

As the signature is $(- - - +)$, the number of possible Segre characteristics is reduced to three. This can most easily be seen from (9.4) and (9.6) which render cases (4) and (5) impossible.

We have already given a detailed treatment of the problem of the structure of a pair of tensors in § 39; we now apply these results to the classification of possible energy-momentum tensor structures by replacing $\bar{g}_{\alpha\beta}$ with $T_{\alpha\beta}$.

Type I. [1111]

(a) *The latent roots are real.*

Depending on the *multiplicity* of the latent roots, we have: (a_1) [1111], (a_2) [(11)11], (a_3) [(111)1], (a_4) [(11)(11)], (a_5) [(1111)]. The canonical forms relative to the orthonormal tetrad formed by the eigenvectors are

$$(g_{\alpha\beta}) = \begin{pmatrix} -1 & & & \\ & -1 & & \\ & & -1 & \\ & & & 1 \end{pmatrix}, \quad (T_{\alpha\beta}) = \begin{pmatrix} -\lambda_1 & & & \\ & -\lambda_2 & & \\ & & -\lambda_3 & \\ & & & \lambda_4 \end{pmatrix} \quad (47.2)$$

where the λ_j are the latent roots corresponding to the elementary divisors. In cases a_2, a_3, a_4, a_5, the corresponding λ_j in (47.2) are equal.

(b) *Some latent roots are complex.*

There are only two subcases: (b_1) [1111], $T_{\alpha\beta}$ has two real and two complex conjugate latent roots, and thus in real space has two real eigenvectors and one invariant 2-plane; (b_2) [(11) 11]—two complex-conjugate and one real latent root, in real space $T_{\alpha\beta}$ has a two-dimensional cluster of eigenvectors and one invariant 2-plane. Equation (9.6) shows that the canonical forms for (b_1) relative to a real quasi-orthonormal tetrad are

$$(g_{\alpha\beta}) = \begin{pmatrix} -1 & & & \\ & -1 & & \\ & & 1 & 1 \\ & & 1 & -1 \end{pmatrix}, \quad (T_{\alpha\beta}) = \begin{pmatrix} -\lambda_1 & & & \\ & -\lambda_2 & & \\ & & \alpha-\beta, & \alpha+\beta \\ & & \alpha+\beta, & -\alpha+\beta \end{pmatrix} \tag{47.3}$$

where λ_1 and λ_2 are the real latent roots and $\alpha \pm i\beta$ the complex conjugate ones. In case (b_2) $\lambda_1 = \lambda_2$. It is not difficult to find a real linear transformation reducing the matrix $(g_{\alpha\beta})$ in (47.3) to the diagonal form. No other cases with Segre characteristic [1111] are admitted by a metric with signature $(- - - +)$.

Type II. [112]

It is easily verified that only real latent roots are possible; depending on their multiplicity we have: (1) [112], (2) [(11)2], (3) [1(12)], (4) [(112)]. In case (1) there is a unique orthonormal tetrad which gives the canonical forms

$$(g_{\alpha\beta}) = \begin{pmatrix} -1 & & & \\ & -1 & & \\ & & -1 & \\ & & & 1 \end{pmatrix}, \quad (T_{\alpha\beta}) = \begin{pmatrix} -\lambda_1 & & & \\ & -\lambda_2 & & \\ & & -\lambda_3-1, & -1 \\ & & -1, & \lambda_3-1 \end{pmatrix}. \tag{47.4}$$

The tensor $T_{\alpha\beta}$ has three eigenvectors, one of which is isotropic, and one invariant 2-plane containing the isotropic eigenvector. In case (2) $\lambda_1 = \lambda_2$, and we thus have a 2-plane of eigenvectors formed by the linear pencil of the two non-isotropic eigenvectors. In case (3) we must put $\lambda_2 = \lambda_3$ and in case (4) $\lambda_1 = \lambda_2 = \lambda_3$; there is a three-dimensional cluster of eigenvectors formed by the three independent eigenvectors.

Type III. [13]

The Minkowski signature admits only real latent roots. There are two possibilities: (1) [13] and (2) [(13)].

In the first case there are two eigenvectors, one non-isotropic corresponding to the simple elementary divisor and the other isotropic. There is a three-dimensional invariant cluster containing the isotropic eigenvector. In case (1) we can always define a unique orthonormal tetrad for which

$$(g_{\alpha\beta}) = \begin{pmatrix} -1 & & & \\ & -1 & & \\ & & -1 & \\ & & & 1 \end{pmatrix}, \quad (T_{\alpha\beta}) = \begin{pmatrix} -\lambda_1 & & & \\ & -\lambda_2 & 1 & 0 \\ & -1 & -\lambda_2 & -1 \\ & 0 & -1 & \lambda_2 \end{pmatrix}. \tag{47.5}$$

In case (2) $\lambda_1 = \lambda_2$; we thus have an invariant 2-plane containing the isotropic and non-isotropic eigenvectors and consisting entirely of eigenvectors. The orthonormal tetrad is therefore no longer unique.

In general, *multiple latent roots imply non-unique orthonormal tetrads* (47.2), (47.3), (47.4), (47.5).

This completes the classification of the possible energy-momentum tensors for real gravitational fields.

Only the simplest physical cases corresponding to the simpler Segre characteristics of $T_{\alpha\beta}$, have as yet been studied. We now consider some typical examples.

1. The energy-momentum tensor of "pure matter", or, as it is sometimes called *the energy-momentum tensor of the mass current* ([188], p. 304) corresponds to the case when $T_{\alpha\beta}$ has a single eigenvector with a non-zero latent root, oriented in time, and interpreted physically as the tangent vector τ^α to the four-dimensional trajectory of the mass current; the latent root $\lambda_4 \equiv \mu_0 c^2$ where $\mu_0(x)$ is the *rest density*. The remaining latent roots $\lambda_i = 0$ ($i = 1$, 2, 3). The vector τ^α together with three vectors $\underset{i}{\xi}^\alpha$ (i = 1, 2, 3) forms an orthonormal tetrad in a non-unique way. We thus have

$$g_{\alpha\beta} = \sum_{\sigma=1}^{4} e_\sigma \underset{\sigma}{\xi}_\alpha \underset{\sigma}{\xi}_\beta, \quad e_4 = -e_1 = -e_2 = -e_3 = 1, \quad \underset{4}{\xi}_\alpha = \tau_\alpha$$

and

$$T_{\alpha\beta} = \mu_0 c^2 \tau_\alpha \tau_\beta. \tag{47.6}$$

As is well known ([188], p. 305), the components of $T_{\alpha\beta}$ in the orthonormal tetrad have the following physical significance: *T_{44} is the energy density of the flow; T_{4i} ($i = 1, 2, 3$) the components of the momentum density along the axes Ox, Oy, Oz, multiplied by c, or the energy flux density along the* ith *axis divided by c; T_{ij} ($i = 1, 2, 3$) is the flux density of the momentum along the* ith *axis across the surface normal to the* jth *axis.*

The Segre characteristic is of the simple type I, in fact $[1\overset{0}{(111)}]$, the 0 over the round bracket showing that the three corresponding latent roots are zero.

2. *The energy-momentum tensor for macroscopic bodies* such as perfect fluids. As the momentum flux across an element $d\sigma$ of surface is just the force acting on $d\sigma$, $T_{\alpha\beta}\, d\sigma^\beta$ is the αth component of this force. As the maximum variation of stress with direction is vanishingly small in solid bodies compared with the stresses entering into the theory of relativity, the con-

ditions are analogous to those in a perfect fluid to a close approximation. We introduce an internal reference system into each element of volume; the pressure acting on an element of surface is perpendicular to the surface and is the same for all directions (Pascal's law).

Hence

$$T_\alpha \, d\sigma^\beta = p \, d\sigma_\alpha,$$

i.e.

$$T_{ij} = p g_{ij} \qquad (i, j = 1, 2, 3).$$

The components of momentum density T_{i4} are all zero; the component T_{44} is the internal energy density ε of the body, so that ε/c^2 is the mass density, i.e. the mass per unit proper volume. Thus in the internal system of reference

$$(T_{\alpha\beta}) = \begin{pmatrix} p & 0 & 0 & 0 \\ 0 & p & 0 & 0 \\ 0 & 0 & p & 0 \\ 0 & 0 & 0 & \varepsilon \end{pmatrix}. \qquad (47.7)$$

Let the velocity of the macroscopic motion of the element of volume be u_α; in the internal system of reference, $u_i = 0$ $(i = 1, 2, 3)$, $u_4 = 1$. Hence, the expression corresponding to the matrix (47.7) is

$$T_{\alpha\beta} = (p+\varepsilon)u_\alpha u_\beta - p g_{\alpha\beta}, \qquad (47.8)$$

which holds in any system of coordinates by virtue of its tensor nature. Equation (47.7) shows that the elementary divisors of the λ-matrix (47.1) are all of the first degree, and that the Segre characteristic is of type I—[(111)1]. Putting the pressure $p = 0$, we obtain (47.6) with $\varepsilon = \mu_0 c^2$. The tensors (47.6) and (47.8) are widely used in various problems of relativistic hydrodynamics ([210]–[212]).

3. *The energy-momentum tensor of the electromagnetic field.* According to the special theory of relativity, the energy-momentum tensor of the electromagnetic field can be derived from an antisymmetric tensor $F_{\alpha\beta} = -F_{\beta\alpha}$, *the electromagnetic field tensor* ([173], pp. 143–6; [156], § 32; [272]; [273]; [304]; [218]; [237]). The actual relation is

$$T_{\alpha\beta} = \tfrac{1}{4} F_{\sigma\tau} F^{\sigma\tau} g_{\alpha\beta} - F_{\alpha\sigma} F^{\sigma}_{\beta}, \qquad (47.9)$$

so that *the "trace" of the energy-momentum tensor of an electromagnetic field is zero:*

$$T^\sigma_\sigma = g^{\alpha\beta} T_{\alpha\beta} = 0. \qquad (47.10)$$

We can find the algebraic structure of the tensor $T_{\alpha\beta}$ (47.9) as follows. The special theory of relativity shows that in a local non-holonomic orthonormal tetrad, the components F_{i4} determine the special components of the electric field, and F_{ij} $(i, j = 1, 2, 3)$ those of the magnetic field. If the electric and magnetic field vectors have components $\{\alpha_1, \alpha_2, \alpha_3\}$ and $\{\beta_1, \beta_2, \beta_3\}$ respectively, then

$$\left.\begin{aligned}\alpha_1 &= F_{14} = -F^{14}, & \beta_1 &= F_{23} = F^{23},\\ \alpha_2 &= F_{24} = -F^{24}, & \beta_2 &= F_{31} = F^{31},\\ \alpha_3 &= F_{34} = -F^{34}, & \beta_3 &= F_{12} = F^{12},\end{aligned}\right\} \tag{47.11}$$

the indices being raised and lowered by means of the Minkowski metric tensor in the given orthonormal tetrad. According to (47.9) and (47.11)

$$(T_{\alpha\beta}) = \begin{pmatrix} \nu & \alpha_3\beta_2-\alpha_2\beta_3 & \alpha_1\beta_3-\alpha_3\beta_1 & \alpha_2\beta_1-\alpha_1\beta_2 \\ \alpha_3\beta_2-\alpha_2\beta_3 & \nu-\alpha_1^2-\beta_1^2 & -\alpha_1\alpha_2-\beta_1\beta_2 & -\alpha_1\alpha_3-\beta_1\beta_3 \\ \alpha_1\beta_3-\alpha_3\beta_1 & -\alpha_1\alpha_2-\beta_1\beta_2 & \nu-\alpha_2^2-\beta_2^2 & -\alpha_2\alpha_3-\beta_2\beta_3 \\ \alpha_2\beta_1-\alpha_1\beta_2 & -\alpha_1\alpha_3-\beta_1\beta_3 & -\alpha_2\alpha_3-\beta_2\beta_3 & \nu-\alpha_3^2-\beta_3^2 \end{pmatrix}, \tag{47.12}$$

where

$$\nu = \tfrac{1}{2}\sum_{i=1}^{3}(\alpha_i^2+\beta_i^2). \tag{47.13}$$

The complete group of Lorentz rotations consists of three spatial and three Lorentz rotations. If we fix some vector $\underset{4}{\bar{e}}$ and its orthogonal non-isotropic hyperplane, we can use the three purely spatial rotations to obtain a special coordinate system on this hyperplane, which has a negative-definite metric, and satisfies any three given conditions. It may easily be seen for example that *in real space* in the new system of coordinates

$$\alpha_3' = \beta_3' = \alpha_1'\alpha_2'+\beta_1'\beta_2' = 0. \tag{47.14}$$

As in general

$$(\alpha_1\alpha_2+\beta_1\beta_2)^2+(\alpha_2\beta_1-\alpha_1\beta_2)^2 = (\alpha_1^2+\beta_1^2)(\alpha_2^2+\beta_2^2) \tag{47.15}$$

for any orthonormal tetrad, if we put

$$\alpha_1^2+\beta_1^2 = a^2, \quad \alpha_2^2+\beta_2^2 = b^2,$$

eqns. (47.14) and (47.15) show that (47.12) in the new system of coordinates (omitting primes) reduces to

$$(T_{\alpha\beta}) = \begin{pmatrix} \dfrac{a^2+b^2}{2} & 0 & 0 & \pm ab \\ 0 & \dfrac{-a^2+b^2}{2} & 0 & 0 \\ 0 & 0 & \dfrac{a^2-b^2}{2} & 0 \\ \pm ab & 0 & 0 & \dfrac{a^2+b^2}{2} \end{pmatrix}, \tag{47.16}$$

whence it follows that the roots of the characteristic equation for (47.1) are $\lambda_1 = \lambda_2 = -\lambda_3 = -\lambda_4 = \lambda = \frac{1}{2}(a^2-b^2)$.

Hence *the energy-momentum tensor of the electromagnetic field has four real characteristic numbers all equal in magnitude but two with opposite signs* (Synge [96]; Ruse [99]). We can extend this treatment by using the elementary divisors of the λ-matrix $(T_{\alpha\beta}-\lambda g_{\alpha\beta})$ (§ 9).

Two cases can be distinguished; (a) $\lambda \neq 0$ and (b) $\lambda = 0$. In the first case, the linear elementary divisors and the equal absolute values of the characteristic numbers of $T_{\alpha\beta}$ suggests the use of (9.4) to obtain the following canonical forms for $(g_{\alpha\beta})$ and $(T_{\alpha\beta})$:

$$(g_{\alpha\beta}) = \begin{pmatrix} -1 & 0 & 0 & 0 \\ 0 & -1 & 0 & 0 \\ 0 & 0 & -1 & 0 \\ 0 & 0 & 0 & +1 \end{pmatrix},$$

$$(T_{\alpha\beta}) = \begin{pmatrix} +\lambda & 0 & 0 & 0 \\ 0 & -\lambda & 0 & 0 \\ 0 & 0 & -\lambda & 0 \\ 0 & 0 & 0 & -\lambda \end{pmatrix}.$$

The Segre characteristic is thus of the simple type [(11) (11)]. These matrices enable $T_{\alpha\beta}$ and $g_{\alpha\beta}$ to be expressed as follows in terms of the eigenvectors and characteristic numbers of $T_{\alpha\beta}$:

$$g_{\alpha\beta} = \sum_{\sigma} e_{\sigma}\underset{\sigma}{\xi}_{\alpha}\underset{\sigma}{\xi}_{\beta}, \quad T_{\alpha\beta} = \sum_{\sigma} e_{\sigma}\underset{\sigma}{\lambda}\underset{\sigma}{\xi}_{\alpha}\underset{\sigma}{\xi}_{\beta}.$$

These relations are valid only for this simple type of characteristic.

We thus obtain

$$T_{\alpha\beta} = \lambda(-\underset{1}{\xi}_{\alpha}\underset{1}{\xi}_{\beta}+\underset{2}{\xi}_{\alpha}\underset{2}{\xi}_{\beta}+\underset{3}{\xi}_{\alpha}\underset{3}{\xi}_{\beta}-\underset{4}{\xi}_{\alpha}\underset{4}{\xi}_{\beta}),$$
$$g_{\alpha\beta} = \sum_{\sigma} e_{\sigma}\underset{\sigma}{\xi}_{\alpha}\underset{\sigma}{\xi}_{\beta}, \tag{47.17}$$

where the unit orthogonal vectors $\underset{2}{\xi}^{\alpha}$ and $\underset{3}{\xi}^{\alpha}$ are both spatial, defined to within a rotation in the surface $\{\underset{2}{\xi}, \underset{3}{\xi}\}$, and $\underset{1}{\xi}^{\alpha}$ and $\underset{4}{\xi}^{\alpha}$ are unit orthogonal vectors defined to within a Lorentz rotation in surface $\{\underset{1}{\xi}, \underset{4}{\xi}\}$, so that $\underset{1}{\xi}$ is spatial and $\underset{4}{\xi}$ temporal.

In the *special case* (b), $\lambda = (a^2-b^2)/2 = 0$ so that $b = \varepsilon a$, $\varepsilon = \pm 1$; determining the elementary divisors of the λ-matrix $(T_{\alpha\beta}-\lambda g_{\alpha\beta})$, where $T_{\alpha\beta}$ is given by (47.16) and $(g_{\alpha\beta})$ is a Minkowski metric, shows that the Segre characteristic is of the non-simple type $[(2\overset{0}{1}1)]$, cf. § 9. The tensor $T_{\alpha\beta}$ has therefore one *isotropic eigenvector*, and the canonical forms of $T_{\alpha\beta}$ and $g_{\alpha\beta}$ are particularly simple if we replace the orthonormal tetrad by a quasi-orthonormal tetrad with the isotropic vector as one coordinate vector. According to (9.4), the general form of the canonical matrices are

$$(g_{\alpha\beta}) = \begin{pmatrix} e_1 & 0 & 0 & 0 \\ 0 & e_2 & 0 & 0 \\ 0 & 0 & 0 & 1 \\ 0 & 0 & 1 & 0 \end{pmatrix}, \quad (T_{\alpha\beta}) = \begin{pmatrix} e_1\underset{1}{\lambda} & 0 & 0 & 0 \\ 0 & e_2\underset{2}{\lambda} & 0 & 0 \\ 0 & 0 & 0 & \underset{3}{\lambda} \\ 0 & 0 & \underset{3}{\lambda} & 1 \end{pmatrix}.$$

But in the present case $\underset{1}{\lambda} = \underset{2}{\lambda} = \underset{3}{\lambda} = 0$, $e_1 = e_2 = -1$, so that

$$(g_{\alpha\beta}) = \begin{pmatrix} -1 & 0 & 0 & 0 \\ 0 & -1 & 0 & 0 \\ 0 & 0 & 0 & 1 \\ 0 & 0 & 1 & 0 \end{pmatrix}, \quad (T_{\alpha\beta}) = \begin{pmatrix} 0 & 0 & 0 & 0 \\ 0 & 0 & 0 & 0 \\ 0 & 0 & 0 & 0 \\ 0 & 0 & 0 & 1 \end{pmatrix}. \tag{47.18}$$

Hence

$$T_{\alpha\beta} = \underset{4}{e}_{\alpha}\underset{4}{e}_{\beta}, \tag{47.19}$$

where the vector $\underset{4}{e}_{\alpha}$ is isotropic as can easily be seen by considering the matrix $(g_{\alpha\beta})$ in (47.18); (47.19) is of course true in any coordinate system.

The isotropic vector $\underset{4}{e}_\alpha$ can be multiplied by any number *without changing its norm*; hence we can introduce any factor which is the square of a scalar on the right-hand side of (47.19). We can easily transform from the quasi-orthonormal to an orthonormal tetrad by means of the substitution

$$x_3 \rightarrow x_3+x_4, \quad x_4 \rightarrow x_3-x_4.$$

The special case of an electromagnetic field provides the first example of an energy-momentum tensor with a non-simple Segre characteristic.

We can also consider energy-momentum tensors corresponding to the superposition of two or more fields.

4. Thus

$$T_{\alpha\beta} = (p+\varepsilon)u_\alpha u_\beta - pg_{\alpha\beta}+\tfrac{1}{4}(F^{\sigma\tau}F_{\sigma\tau})g_{\alpha\beta}-F_{\alpha\sigma}F^\sigma_\beta \tag{47.20}$$

for a perfect fluid and an electromagnetic field, where the non-isotropic vector u_α is the velocity of the particles ([173], p. 153).

5. We can also consider *the energy-momentum tensor for dissipative processes* when deriving the relativistic hydrodynamic equations, allowing for *viscosity and heat conduction* ([193], p. 505). It is obtained by adding an appropriate tensor $\tau_{\alpha\beta}$ to the right-hand side of (47.8), and an additional set of components ν_α to n_α, the particle flux-density vector for the substance. Thus

$$\left.\begin{aligned} T_{\alpha\beta} &= (p+\varepsilon)u_\alpha u_\beta - pg_{\alpha\beta}+\tau_{\alpha\beta}, \\ n_\alpha &= nu_\alpha+\nu_\alpha. \end{aligned}\right\} \tag{47.21}$$

The velocity u^α is so defined that the momentum is zero in the *internal system of reference* of each element of fluid, and the energy is the same function of the other thermodynamic quantities used in the absence of dissipative processes. In the internal system of reference $\tau_{44} = \tau_{4i} = 0$ and $u^i = 0$, so that

$$\tau_{\alpha\beta}u^\beta = 0. \tag{47.22}$$

On account of its tensor nature, this relation is true in any system of coordinates. As the component n^4 of the particle flux-density *4-vector* in the internal system must be the particle number density n,

$$\nu_\alpha n^\alpha = 0. \tag{47.23}$$

The equations of motion of the fluid must follow from

$$\partial_\sigma T^\sigma_\alpha = 0, \quad \partial_\alpha n^\alpha = 0, \tag{47.24}$$

which gives

$$u^\alpha \partial_k T^k_\alpha = -T\partial_\alpha (su^\alpha) + \mu \partial_\alpha \nu^\alpha + u^\alpha \partial_\sigma \tau^\sigma_\alpha,$$

where T in the absolute temperature, s the entropy density and $\mu = (P+\varepsilon-Ts)/n$ is the relativistic chemical potential of the substance. Using (47.22), we obtain

$$\partial_\alpha \left(su^\alpha - \frac{\mu}{T} \nu^\alpha \right) = -\nu^\alpha \partial_\alpha \left(\frac{\mu}{T} \right) - \frac{1}{T} \tau^\sigma_\alpha \partial_\sigma u^\alpha. \tag{47.25}$$

The left-hand side of this equation is the 4-divergence of the entropy flux and hence the expression on the right-hand side is the rate of increase of entropy due to the dissipative processes. The entropy flux density 4-vector is therefore

$$s^a = su^\alpha - \frac{\mu}{T} \nu^\alpha.$$

The tensor $\tau_{\alpha\beta}$ and the vector ν^α must be linear functions of the gradients of velocity and thermodynamic quantities, such that the right-hand side of (47.25) is positive. This condition together with (47.22) and (47.23) uniquely determines $\tau_{\alpha\beta}$ and γ^α as follows:

$$\tau_{\alpha\beta} = \eta \left(\frac{\partial u_\alpha}{\partial x^\beta} + \frac{\partial u_\beta}{\partial x^\alpha} + u_\beta u^\sigma \frac{\partial u_\alpha}{\partial x^\sigma} + u_\alpha u^\sigma \frac{\partial u_\beta}{\partial x^\sigma} \right) - (\zeta - \tfrac{2}{3}\eta) \frac{\partial u_\sigma}{\partial x^\sigma} (g_{\alpha\beta} - u_\alpha u_\beta),$$

$$\nu_\alpha = -\frac{\chi}{c} \left(\frac{nT}{p+\varepsilon} \right)^2 \left[\partial_\alpha \left(\frac{\mu}{T} \right) + u_\alpha u^\sigma \partial_\sigma \left(\frac{\mu}{T} \right) \right],$$

where η and ζ are viscosity coefficients, and χ the thermal conductivity. The equivalent invariant form for the first equation is

$$\tau^{\alpha\beta} = \eta[(g^{\alpha\sigma} - u^\alpha u^\sigma) u^\beta_{;\sigma} + (g^{\beta\sigma} - u^\beta u^\sigma) u^\alpha_{;\sigma}] - (\zeta - \tfrac{2}{3}\eta)(u^\alpha u^\beta - g^{\alpha\beta}) u^\sigma_{;\sigma}. \quad (47.26$$

Other field combinations are possible. It may be noted that all the tensors $T_{\alpha\beta}$ except (47.26) have the simplest type of characteristic according to the classification of (47.2) and (47.3).

A study of $T_{\alpha\beta}$, defining the distribution, movement and properties of matter, is of fundamental importance and leads to a series of interesting physical conclusions as in [273].

§ 48. The Conservation Law for the Energy-momentum Tensor

If the field equations are stated in the form

$$S_{\alpha\beta} = \lambda T_{\alpha\beta} \tag{48.1}$$

the Einstein tensor $S_{\alpha\beta}$ satisfies the identity (43.5), i.e. $S^{\alpha}{}_{\beta,\alpha} = 0$, $\lambda =$ constant, so that $T_{\alpha\beta}$ satisfies the conservation law

$$T^{\alpha}{}_{\beta,\alpha} = 0. \tag{48.2}$$

Let us see what physical conclusions can be deduced from this equation for particular energy-momentum tensors. We first consider $T_{\alpha\beta}$ for the flow of matter and of a perfect fluid defined by (47.6) and (47.8) respectively. Combining these two cases, the following energy-momentum tensor is considered,

$$T_{\alpha\beta} = ru_{\alpha}u_{\beta} - \theta_{\alpha\beta}, \tag{48.3}$$

where u_{α} is a unit velocity vector oriented in time, r the scalar density and $\theta_{\alpha\beta}$ a relative symmetric tensor. Thus

$$g_{\alpha\beta}u^{\alpha}u^{\beta} = 1 \tag{48.4}$$

and

$$u^{\alpha}{}_{,\beta}u_{\alpha} = 0. \tag{48.5}$$

Denoting

$$\theta^{\alpha}{}_{\beta,\alpha} = r\sigma_{\beta}, \tag{48.6}$$

the conservation law (48.2) takes the form

$$(ru^{\alpha}\, u_{\beta})_{,\alpha} = r\sigma_{\beta}.$$

Hence besides (48.4), (48.5), the scalar r and the vector u_{α} also satisfy the relation

$$(ru^{\alpha})_{,\alpha}u_{\beta} + ru^{\alpha}u_{\beta,\alpha} = r\sigma_{\beta}. \tag{48.7}$$

After contraction with u^{β}, (48.4) and (48.5) show that

$$(ru^{\alpha})_{,\alpha} = r\sigma_{\alpha}u^{\alpha}. \tag{48.8}$$

As $r \neq 0$,

$$u^{\alpha}u_{\beta,\alpha} = (g_{\alpha\beta} - u_{\alpha}u_{\beta})\sigma^{\alpha}. \tag{48.9}$$

The last two equations can be provisionally interpreted as follows. We note first that the left-hand side of (48.8) is reminiscent of the well-known expression in the continuity equation. If, furthermore, $u^{\alpha} = dx^{\alpha}/ds$, eqns. (48.9) can be interpreted as a system of differential equations defining a *flow line* for a given σ^{α}.

The conservation law for the mass flow tensor. Equation (48.3) shows that

$$r = \mu_0, \quad \theta_{\alpha\beta} = 0, \quad \sigma_{\beta} = 0,$$

and eqns. (48.8) and (48.9) reduce to

$$(\mu_0 u^\alpha)_{,\alpha} = 0, \tag{48.10}$$

$$u^\alpha u_{\beta,\alpha} = 0. \tag{48.11}$$

Equation (48.10) is the *continuity equation* for the medium, stating *that the divergence of the product of the rest density of the medium and the unit velocity vector is zero*; (48.11) states that *the flow lines are non-isotropic geodesics of* V_4.

If we put

$$r = p+\varepsilon, \quad \theta_{\alpha\beta} = pg_{\alpha\beta}, \quad \sigma_\beta = \frac{1}{p+\varepsilon}(p\delta^\alpha_\beta)_{,\alpha} = \frac{1}{p+\varepsilon}\partial_\beta p,$$

we obtain the energy-momentum tensor (47.8). We furthermore suppose that other considerations, of a thermodynamic nature for example, determine the equation of state connecting the internal energy density and the pressure

$$\varepsilon = \varphi(p).$$

Hence

$$\sigma = \int_{p_0}^{p} \frac{dp}{\varphi(p)+p},$$

and it can be shown that the flow lines are again non-isotropic geodesics, not in the actual V_4 but in $\boldsymbol{V}_4$, another Riemann space conformal to V_4 with metric

$$\boldsymbol{ds}^2 = \omega^2\, ds^2,$$

where

$$\omega = e^{\int_{p_0}^{p} \frac{dp}{p+\varepsilon}}.$$

An analogous result holds whenever σ_β is a gradient field ([231], § 17).

Any further investigation of the conservation law for the energy-momentum tensor of an electromagnetic field is not required (cf. [231], chapter II, section II).

Solving the interior Cauchy problem depends on the physical significance and structure of the energy-momentum tensor; we restrict our attention to the flow of matter, (47.6), and of a perfect fluid, (47.8).

§ 49. The Interior Cauchy Problem for the Flow of Matter

We consider the interior Cauchy problem for the energy-momentum tensor (47.6).

Given a gravitational field on a hypersurface S in terms of the Cauchy data introduced in § 45, we have to determine this field outside S when it satisfies the equation

$$S^{\alpha}_{\beta} = \chi \varrho u^{\alpha} u_{\beta}, \tag{49.1}$$

where $\varrho > 0$ and u_{α} is a unit vector. We suppose that S does not touch the isotropic cone. If a local system of coordinates is chosen so that S has the equation $x^4 = 0$, then $g^{44} \neq 0$. As in the solution of the exterior problem, we can replace (49.1) by two sets of equations. The first consists of the six equations

$$R_{ij} = \chi \varrho (u_i u_j - \tfrac{1}{2} g_{ij}) \qquad (i, j = 1, 2, 3), \tag{49.2}$$

and the second

$$S^{4}_{\alpha} = \chi \varrho u^{4} u_{\alpha}. \tag{49.3}$$

We also have the conditions

$$g^{\sigma\tau} u_{\sigma} u_{\tau} = 1 \tag{49.4}$$

and

$$\varrho > 0. \tag{49.5}$$

The conservation law ensures that every set of solutions $g_{\alpha\beta}$, $u_{\alpha, \varrho}$ of these equations also satisfies the following equations analogous to (48.11), (48.10):

$$u^{\alpha} u_{\beta, \alpha} \equiv u^4 \partial_4 u_{\beta} + \Psi_{\beta} = 0, \tag{49.6}$$

$$(\varrho u^{\alpha})_{, \alpha} \equiv u^4 \partial_4 \varrho + \varrho \partial_4 u^4 + F = 0, \tag{49.7}$$

where Ψ_{β} depends on u_{α}, $\partial_i u_{\alpha}$ and the Cauchy data on S, and F is a function of u_{α}, $\partial_i u_{\alpha}$, ϱ, $\partial_i \varrho$ and the Cauchy data. The $g_{\alpha\beta}$ and $\partial_4 g_{\alpha\beta}$ must be respectively three and two times continuously differentiable on S. Having the Cauchy data at our disposal, we determine the components $S^{4}{}_{\alpha}$ uniquely on S; we can then determine ϱ and u_{α} on S. We use the relation

$$(\chi \varrho u^4)^2 = g^{\sigma\tau} S^{4}{}_{\sigma} S^{4}{}_{\tau}.$$

The right-hand side is essentially positive so that S^{4}_{α} is a time-like vector. Denoting its norm by

$$g^{\sigma\tau} S^{4}{}_{\sigma} S^{4}{}_{\tau} = \Omega^2,$$

we obtain

$$\chi \varrho u^4 = \Omega.$$

The second group of field equations (49.3) then gives

$$u_\alpha = \frac{1}{\Omega} S^4{}_\alpha, \quad u^4 = \frac{1}{\Omega} S^{44}, \quad \chi\varrho = \frac{\Omega^2}{S^{44}}.$$

As $\varrho > 0$

$$S^{44} > 0. \tag{49.8}$$

In order to determine u_α uniquely from the Cauchy data we must somehow fix the sign of Ω.

As in the corresponding exterior problem the first group of field equations (49.2) determines the derivatives $\partial_{44}g_{ij}$ on the hypersurface S. Moreover, (49.8) shows that $u^4 \neq 0$, and hence it is possible to determine $\partial_4 u_\alpha$ and $\partial_4\varrho$ on S from (49.6) and (49.7).

Thus if all the above assumptions are true, *the quantities* u_α, ϱ, $\partial_{44}g_{ij}$, $\partial_4 u_\alpha$, $\partial_4\varrho$ *are continuous and completely determined on the hypersurface S.*

If the Cauchy data are locally differentiable a larger number of times in any particular case, the above result can be extended to the higher derivatives of the set of functions $\{g_{\alpha\beta}, u_\alpha, \varrho\}$ by simply differentiating (49.2), (49.6), (49.7) with respect to x^4.

We complete the analysis by considering the set of functions $\{g_{\alpha\beta}, u_\alpha, \varrho\}$ which satisfy conditions (49.3), (49.4) (49.5) on the hypersurface S, and examining their behaviour in the neighbourhood of S. It follows from (49.6) and (49.7) that

$$(S^\alpha{}_\beta - \lambda T^\alpha{}_\beta)_{,\alpha} = 0.$$

Setting down this set of equations for the solutions (49.2), we obtain

$$g^{44}\partial_4(S^4{}_\alpha - \lambda T^4{}_\alpha) = A^{i\sigma}_\alpha \partial_i(S^4{}_\sigma - \lambda T^4{}_\sigma) + B^\sigma{}_\alpha(S^4{}_\sigma - \lambda T^4{}_\sigma).$$

Repeating the exterior problem analysis up to (44.17), we find that (49.3) is also satisfied in the neighbourhood of S; we have a system in involution.

This solves the problem of determining the Cauchy data, and we are left with the problem of *integrating the set of equations* (49.2), (49.6), (49.7) *in "time"*; the method indicated in § 44 does not impose very stringent conditions on the class of the functions involved.

In the case of functions of class C^a, we can use the classical Cauchy–Kowalewski existence theorem to obtain a solution, which is unique to within a change of coordinates (cf. § 44) conserving the hypersurface *point-wise* and the Cauchy data provided they satisfy the conditions

$$g^{44} \neq 0, \quad \Omega^2 = g^{\sigma\tau} S^4{}_\sigma S^4{}_\tau > 0, \quad S^{44} > 0.$$

More characteristic manifolds are possible in the interior problem than in the corresponding exterior problem. There are two types of hypersurface at which the Cauchy data can be discontinuous:

(1) the hypersurface S touches the elementary cone, the case already considered in the exterior problem when

$$g^{44} = 0$$

corresponding to an isotropic hypersurface;

(2) V_3 manifolds corresponding to the condition

$$\Omega = 0,$$

i.e.

$$u^4 = 0,$$

so that

$$S^4{}_\alpha = 0.$$

Hence the hypersurface S is formed by the trajectories, a case not found in the exterior problem.

If a finite density ϱ is assumed, the case $S^{44} = 0$ reduces to the preceding one.

§ 50. The Interior Cauchy Problem for a Perfect Fluid

The solution of the interior Cauchy problem for a perfect fluid involves a basically similar analysis to that in the previous section, but with one interesting physical difference.

Suppose that the energy-momentum tensor for a perfect fluid is of the form

$$T_{\alpha\beta} = (p+\varepsilon)u_\alpha u_\beta - p g_{\alpha\beta}, \tag{50.1}$$

where ε can be defined by some function $\varphi(p)$ of the pressure. Corresponding to this tensor, there are the two following sets of field equations:

$$R_{ij} = \lambda[(p+\varepsilon)u_i u_j - \tfrac{1}{2}(\varepsilon - p)g_{ij}], \tag{50.2}$$

$$S^4{}_\alpha = \lambda[(p+\varepsilon)u_\alpha u^4 - p\delta^4_\alpha], \tag{50.3}$$

subject to the condition

$$g^{\sigma\tau}u_\sigma u_\tau = 1, \tag{50.4}$$

that u_α is a unit time-like vector. The conservation law for the energy-momentum tensor and the continuity equation, similar to those in the previous section, are as follows:

$$u^{\alpha}u_{\beta,\alpha}-\frac{1}{p+\varepsilon}\,\partial_{\alpha}p(\delta^{\alpha}_{\beta}-u^{\alpha}u_{\beta}) = 0 \tag{50.5}$$

and

$$[(p+\varepsilon)u^{\alpha}]_{,\alpha}-u^{\alpha}\partial_{\alpha}p = 0. \tag{50.6}$$

Assuming provisionally that p is known on the hypersurface S, (50.3) shows that

$$\lambda(p+\varepsilon)u_{\alpha}u^{4} = S^{4}{}_{\alpha}+\lambda p\delta^{4}_{\alpha},$$

and, as u_{α} is a unit vector,

$$[\lambda(p+\varepsilon)u^{4}]^{2} = g^{\lambda\mu}(S^{4}{}_{\lambda}+\lambda p\delta^{4}_{\lambda})(S^{4}{}_{\mu}+\lambda p\delta^{4}_{\mu}).$$

Denoting the positive-definite left-hand side, in general a function of p, by Ω^{2}, we easily obtain

$$\left.\begin{aligned} u_{\alpha} &= \frac{1}{\Omega}\,(S^{4}{}_{\alpha}+\lambda p\delta^{4}_{\alpha}), \quad u^{4} = \frac{1}{\Omega}\,(S^{44}+\lambda pg^{44}), \\ \lambda(p+\varepsilon) &= \frac{\Omega^{2}}{S^{44}+\lambda pg^{44}}. \end{aligned}\right\} \tag{50.7}$$

If $\varepsilon = \varphi(p)$, the third equation is an implicit equation determining the possible values of p. We can then determine u_{α} and ε from (50.7) and then $\partial_{44}g_{ij}$ from (50.2).

$\partial_{4}\varepsilon$ and $\partial_{4}p$ must now be determined from the equation of state. If the index β in (50.5) is raised and put equal to 4, eqns. (50.5) and (50.6) can be written as

$$\begin{aligned} (p+\varepsilon)u^{4}\partial_{4}u^{4}-(g^{44}-u^{4^{2}})\partial_{4}p &= A^{4}, \\ (p+\varepsilon)\partial_{4}u^{4}+u^{4}\varphi'(p)\partial_{4}p &= B, \end{aligned}$$

where A^{4} and B are fully determined on S. This pair of algebraic equations can be satisfied simultaneously for the unknowns $\partial_{4}p$ and $\partial_{4}u^{4}$, only when the determinant of the coefficients on the left-hand side does not vanish, i.e.

$$g^{44}-(1-\varphi')u^{4^{2}} \neq 0. \tag{50.8}$$

If this condition is satisfied, the quantities $\partial_{4}p$ and $\partial_{4}u^{4}$ are determined uniquely on S; the quantities $\partial_{4}u^{i}$ are then determined by the remaining equations of (50.5). Clearly the whole analysis of the previous section can be repeated almost exactly; (50.8) leads to new types of characteristic manifolds.

In contrast to the previous section there are three possible manifolds at which field discontinuities can arise:

(1) Characteristic isotropic $\mathring{V}_3$ manifolds corresponding to the condition $g^{44} = 0$;

(2) $\boldsymbol{V}_3$ manifolds tangential to, or formed by, the "stream lines";

(3) $\overline{V}_3$ manifolds for which

$$g^{44}-(1-\varphi')u^{4^2} = 0.$$

These manifolds satisfy the equation

$$[g^{\sigma\tau}-u^\sigma u^\tau(1-\varphi')]\partial_\sigma f\partial_\tau f = 0.$$

The *pressure gradient* may be discontinuous at these manifolds; this represents the relativistic generalization of the classical wavefront. If the wave fronts are *oriented in time*, they admit an obvious interpretation in relativistic physics ([231], § 19).

For such wavefronts,

$$\Delta_1 f \equiv g^{\sigma\tau}\partial_\sigma f\partial_\tau f \leqslant 0.$$

Equation (50.7) shows that

$$\Delta_1 f = (u^\sigma\partial_\sigma f)^2(1-\varphi'),$$

and hence

$$\varphi' \geqslant 1. \tag{50.9}$$

This result enables us to determine the "propagation velocity" of these hydrodynamic waves. We take two manifolds, $\overline{V}_3^0$ and $\overline{V}_3^\delta$, defined by the equations

$$f(x^\alpha) = 0, \quad f(x^\alpha) = \delta,$$

where δ is infinitesimally small. We draw a stream line through some point on $\overline{V}_3^0$ and find the point of intersection with $\overline{V}_3^\delta$, retaining only quantities of the first order in δ.

The required point on $\overline{V}_3^\delta$ is of the form $x^\alpha+\eta u^\alpha$, where η is the scalar defined by the relation

$$\eta u^\alpha\partial_\alpha f = \delta. \tag{50.10}$$

Let us consider a point $x^\alpha \subset \overline{V}_3^0$, and a unit vector n^α normal to $\overline{V}_3^0$ ($n^\alpha n_\alpha = -1$). The covariant components of $\boldsymbol{n}$ are

$$n_\alpha = \frac{\partial_\alpha f}{\sqrt{-\Delta_1 f}}.$$

The orthogonal trajectory to $\overline{V}_3^0$ at x^α intersects $\overline{V}_3^\delta$ at the point $x^\alpha+\eta n^\alpha$, where η is defined by

$$\eta n^\sigma \partial_\sigma f = \delta. \tag{50.11}$$

Hence

$$\eta = \frac{\delta}{n^\sigma \partial_\sigma f} = \frac{-\delta}{\sqrt{-\Delta_1 f}}. \tag{50.12}$$

Consider the vector $t^\alpha = \eta u^\alpha - \eta n^\alpha$; it clearly lies in the tangent plane to the wavefront. In fact,

$$\eta u^\alpha n_\alpha = \eta \frac{u^\sigma \partial_\sigma f}{\sqrt{-\Delta_1 f}} = \frac{\delta}{\sqrt{-\Delta_1 f}} = -\eta,$$

and thus

$$t^\alpha n_\alpha = (\eta u^\alpha - \eta n^\alpha) n_\alpha = 0.$$

As the norm of t^α

$$t^\alpha t_\alpha = \eta^2 - \eta^2 - 2n\eta u^\alpha n_\alpha = \eta^2 + \eta^2 > 0,$$

he vector is oriented "in time".

Hence $\eta\, u^\alpha$ is the sum of two vectors, one normal to the wave surface and oriented in space, and the other tangential to the surface and oriented in time. The "propagation velocity" v of the wave is defined as the limit of the ratio of these vectors:

$$v = \lim_{\delta \to 0} \left| \frac{\eta}{t^\alpha t_\alpha} \right|.$$

Whence it follows that

$$v^{-2} = \lim_{\delta \to 0} \left(1 + \frac{\eta^2}{\eta^2}\right) = 1 - \frac{\Delta_1 f}{(u^\sigma \partial_\sigma f)^2} = 1-(1-\varphi') = \varphi'.$$

Thus the wave velocity is

$$v = \frac{1}{\sqrt{\varphi'}},$$

subject to condition (50.9). This result, which holds in classical mechanics. generalizes the concept of wave velocity. Moreover, (50.9) shows that $v \leqslant c$, since the velocity of light is taken as unity; this is in agreement with relativistic physic.

The solution of the interior Cauchy problem for distributions and motions of matter defined by more complex energy-momentum tensors can in principle be obtained by the above methods, but the solution and the corresponding characteristic manifolds depend essentially on the form of $T_{\alpha\beta}$. An investigation of electromagnetic fields and superposed electromagnetic and purely material fields can be found in [231].

Problems involving dissipative processes, approximate solutions, and the formulation of the Cauchy problem for the so-called unified field theories not considered here are dealt with in [267], [268], [196], [213].

CHAPTER 9

SPECIAL TYPES OF GRAVITATIONAL FIELD

IN THIS chapter we consider V_4 spaces corresponding to special types of gravitational field defined by conditions of either geometrical or physical nature.

§ 51. Reducible and Conformal-reducible Einstein Spaces

A Riemann V_n space is said to be *reducible*, if its metric has the form

$$ds^2 = \sum_{k=1}^{r} \varphi_k, \tag{51.1}$$

in some holonomic system of coordinates, where φ_k is a quadratic form in the variables $x^{\alpha_{k-1}+1}, \ldots, x^{\alpha_k}$, and $r > 1$.

In relativity theory, we restrict ourselves to the case $n = 4$, and assume that (51.1) has a signature $(- - - +)$ at every point of the domain under investigation. The metric (51.1) is therefore either

(1) $$ds^2 = \varphi_1(x^1) + \varphi_2(x^2, x^3, x^4),$$

or

(2) $$ds^2 = \varphi_1(x^1, x^2) + \varphi_2(x^3, x^4).$$

The first term in case (1) can always be reduced to $\varphi_1 = e_1 dx^{1^2}$ $(e_1 = \pm 1)$ and hence the metric admits a non-isotropic Killing vector $\underset{1}{\xi}{}^{\alpha} = \delta^{\alpha}_{1}$; the form φ_2 can always be reduced to an orthogonal form in the V_3 space of the variables x^2, x^3, x^4, [10] so that

$$ds^2 = e_1\, dx^{1^2} + \sum_{k=2}^{4} e_k H_k^2\, dx^{k^2},$$
$$H_k = H_k(x^2, x^3, x^4), \quad e_k = \pm 1.$$

In the case of reducible Einstein spaces, ($R_{\alpha\beta} = \varkappa g_{\alpha\beta}$) the field equations show that $\varkappa = 0$ and φ_2 corresponds to a three-dimensional Einstein space with zero Ricci tensor $R_{i,j}$ $(i, j = 2, 3, 4)$; hence φ_2 defines a flat space and the required V_4 is a flat manifold (cf. § 13). The metric (1) can therefore be reduced to the Minkowski form.

It is easily shown that case (2) leads only to metrics (14.7) and (14.8) or to the trivial case of a flat space. *Thus the metric of any reducible Einstein space with* $n = 4$ *and signature* $(- - - +)$ *is in a special system of coordinates either a Minkowski metric or* (14.7) *or* (14.8).

We now consider *conformal-reducible* Einstein spaces. A V_n is said to be *conformal-reducible* if its metric has the form

$$ds^2 = H^2 \sum_{k=1}^{r} \varphi_k \tag{51.2}$$

in some system of coordinates, where the φ_k are defined as in (51.1), $r > 1$, and $H = H(x^1, x^2, x^3, x^4)$ is an arbitrary scalar.

We consider those conformal-reducible V_n's with (1) $n = 4$; (2) metric signature $(- - - +)$; (3) metric satisfying the field equations $R_{\alpha\beta} = \varkappa g_{\alpha\beta}$.

Within the general theory of relativity, there is a whole series of solutions of the field equations $R_{\alpha\beta} = \varkappa g_{\alpha\beta}$, found by Schwarzschild, Kottler, Weyl, Levi-Civita, Kasner, Wilson [51a], Delsarte, Narlikar and Karmakar, Einstein and Rosen, etc. (cf. § 14), which are conformal-reducible V_n satisfying the above three conditions, so that a special study of this class of gravitational fields is of undoubted interest.

There are two possible cases when $n = 4$:

$$ds^2 = H^2[\varphi_1(x^1) + \varphi_2(x^2, x^3, x^4)] \tag{51.3}$$

and

$$ds^2 = H^2[\varphi_1(x^1, x^2) + \varphi_2(x^3, x^4)]. \tag{51.4}$$

As in the case (1) of reducible spaces, the metric (51.3) can be reduced by changing the coordinate x^1 and using a three-dimensional transformation $x^i = \psi^i(x^j)$ $(i, j = 2, 3, 4)$ leaving the structure of the form unaltered; we obtain

$$ds^2 = H^2(e_1\, dx^{1^2} + \sum_{i=2}^{4} g_{ii}\, dx^{1^2}), \quad g_{ij} = g_{1i} = 0 \quad (i, j = 2, 3, 4) \tag{51.5}$$

according to Cotton's well-known theorem for V_3 [10]. The field equations for this metric are

$$(\alpha)\ R_{1j} = 0, \quad (\beta)\ R_{ij} = 0, \quad (\gamma)\ R_{ii} = \varkappa H^2 g_{ii}, \quad (\delta)\ R_{11} = e_1 \varkappa H^2.$$

Inserting the Ricci tensor components, we obtain

$$(\alpha)\ 2H_1H_j - HH_{1j} = 0 \quad (\beta)\ 4\alpha_i\alpha_j - 2\alpha\alpha_{ij} + \alpha^2 \overset{\bullet}{R}_{ij} = 0$$

$$(\gamma)\ -e_1(H_1^2 + HH_{11}) - \frac{2HH_{ii}}{g_{ii}} - H\Delta_2 H + H^2 \overset{\bullet}{R}_{ii} - \Delta_1 H + \frac{4}{g_{ii}} H_1^2 = \varkappa H^4$$

$$(\delta)\ 3e_1(H_1^2 - HH_{11}) - H\Delta_2 H - \Delta_1 H = \varkappa H^4,$$

where $\Delta_1 f \equiv g^{ij} f_{,i} f_{,j}$, $\Delta_2 f \equiv g^{ij} f_{,ij}$ are the first and second Beltrami parameters, $H_j \equiv \partial_j H$, $H_{ij} \equiv \partial_{ij} H$, and $\overset{\bullet}{R}_{ij}$ is the Ricci tensor of the three-dimensional space with metric $g_{ij}\, dx^i\, dx^j$.

Integrating (α), we find

$$H^{-1} = \varphi(x^2, x^3, x^4) + \psi(x^1). \tag{51.6}$$

Equations (β), (γ) and (δ) can now be reduced to the equivalent set

$$2\varphi_{ij} + (\varphi + \psi)\overset{\bullet}{R}_{ij} = 0 \tag{51.7}$$

$$-2e_1\psi_{11} + 2g^{ii}\varphi_{ii} + (\varphi + \psi) g^{ii} \overset{\bullet}{R}_{ii} = 0 \tag{51.8}$$

(no summation over i)

$$-3e_1\psi_1^2 + 6e_1\psi_{11}(\varphi + \psi) - 3\Delta_1\varphi - \tfrac{1}{2}\overset{\bullet}{R}(\varphi + \psi)^2 = \varkappa. \tag{51.9}$$

As ψ depends only on x^1, and φ and $\overset{\bullet}{R}_{ij}$ only on x^2, x^3, x^4, (51.7) shows that either

$$(1)\quad \psi \neq \text{const}, \quad \overset{\bullet}{R}_{ij} = 0, \quad \varphi_{ij} = 0$$

or

$$(2)\quad \psi = \text{const}.$$

In the second case, we can without loss of generality put $\psi = 0$. Thus two cases are possible:

$$(1)\quad \overset{\bullet}{R}_{ij} = 0, \quad \varphi_{ij} = 0, \quad \psi \neq \text{const}. \tag{51.10}$$

$$(2)\quad 2\varphi_{ij} + \overset{\bullet}{R}_{ij}\varphi = 0, \quad \psi = 0. \tag{51.11}$$

In the first case (51.8) shows that

$$2g^{ii}\varphi_{ii} + \varphi g^{ii} \overset{\bullet}{R}_{ii} = -g^{ii}\overset{\bullet}{R}_{ii}\psi + 2e_1\psi_{11}$$

(no summation over i). Hence

$$g^{ii}\overset{\bullet}{R}_{ii} = \overset{\bullet}{\varkappa}_i = \text{const}, \quad 2e_1\psi_{11} - \overset{\bullet}{\varkappa}_i\psi = 2g^{ii}\varphi_{ii} + \overset{\bullet}{\varkappa}_i\varphi = \lambda_i = \text{const}, \tag{51.12}$$

so that

$$\dot{\varkappa}_i = \dot{\varkappa}_j = \dot{\varkappa} \qquad (i \neq j)$$

$$\lambda_i = \lambda_j = \lambda \qquad (i \neq j)$$

and

$$\dot{R}_{ii} = \dot{\varkappa} g_{ii}.$$

(51.10) and (51.12) show that $\sum g_{ii}\, dx^{i^2}$ defines a three-dimensional Einstein space, i.e. a space of constant curvature (cf. § 13). Hence the metric $ds^2 \equiv e_1 dx^{1^2} + g_{ij}\, dx^i\, dx^j$ defines a so-called sub-projective Kagan space of the principal kind [451]. But every sub-projective V_4 space is *conformally flat*. Thus ds^2 defines a conformally flat Einstein space and hence a space of constant curvature (§ 13). Thus case (1) leads to a trivial result.

We now consider case (2), i.e. (51.11). φ cannot be a constant, for then the space would be *reducible*.

The field equations then show that

$$\dot{R}_{ij} = -\frac{2\varphi_{ij}}{\varphi}. \tag{51.13}$$

Thus: *apart from spaces of constant curvature and reducible spaces, the problem of determining the metrics of conformal-reducible spaces of the type* (51.3) *is equivalent to that of finding the three-dimensional Riemann spaces (in the variables x^2, x^3, x^4) defined by* (51.13) *and the equation*

$$3\Delta_1\varphi + \frac{\dot{R}}{2}\varphi^2 = -\varkappa, \quad \varphi \neq \text{const.} \tag{51.14}$$

We note that the second Beltrami parameter of the function φ satisfied the equation

$$2\Delta_2\varphi + \dot{R}\varphi = 0$$

or using (51.14)

$$3\Delta_1\varphi - \varphi\Delta_2\varphi + \varkappa = 0$$

and that the integrability conditions for eqns. (51.13) are

$$(\dot{R}^s{}_j g_{ik} - \dot{R}^s{}_k g_{ij})\varphi_s + \tfrac{1}{2}\dot{R}(g_{ij}\varphi_k - g_{ik}\varphi_j) + \varphi \dot{R}_{i[j,\,k]} = 0.$$

Particular cases of conformal-reducible Einstein spaces type (51.4) are given by the solutions obtained by Delsarte, Schwarzschild, Kottler, the cosmological models of Einstein and de Sitter, and gravitational fields with charged particles, etc.

If $p, q = 1, 2$; $a, b = 3, 4$ and $\alpha, \beta = 1, 2, 3, 4$, the coordinate transformation

$$x^p = x^p(x^{q'}), \quad x^a = x^a(x^{b'})$$

gives new metric tensor components

$$g_{p'a'} = 0,\ g_{p'q'} = H^2 A^p_{p'} A^q_{q'} g_{pq},\ g_{a'b'} = H^2 A^a_{a'} A^b_{b'} g_{ab}, \qquad A^\alpha_{\alpha'} \equiv \frac{\partial x^\alpha}{\partial x^{\alpha'}}.$$

Thus without altering the structure of the metric (51.4), we can reduce both φ_1 and φ_2 to diagonal form so that

$$(g_{\alpha\beta}) = \begin{pmatrix} e_1 H^2 \beta^2 & 0 & 0 & 0 \\ 0 & e_2 H^2 & 0 & 0 \\ 0 & 0 & e_3 H^2 \gamma^2 & 0 \\ 0 & 0 & 0 & e_4 H^2 \end{pmatrix}$$

where $\beta = \beta(x^1, x^2)$, $\gamma = \gamma(x^3, x^4)$, $e_i = \pm 1$.

As all curvature tensor components $R_{\alpha\beta\gamma\delta}$ with four different indices vanish owing to the orthogonality of the coordinate system, the field equations reduce to

$$2H_i H_a - HH_{ia} = 0 \qquad (i = 1, 2,\ a = 3, 4), \tag{51.15}$$

$$\left.\begin{aligned} 2H_1H_2 - HH_{12} + \frac{H}{\beta} H_1 \beta_2 &= 0, \\ 2H_3H_4 - HH_{34} + \frac{H}{\gamma} H_3 \gamma_4 &= 0, \end{aligned}\right\} \tag{51.16}$$

$$\frac{e_1}{\beta^2}\left(2\frac{H}{\beta} H_1\beta_2 - 2HH_{11} + H_1^2\right) - e_2\left(\frac{2H}{\beta} H_2\beta_2 + 2HH_{22} - H_2^2\right) \tag{51.17}$$

$$-\frac{e^2H^2}{\beta}\beta_{22} + \frac{e_3}{\gamma^2}\left(\frac{H}{\gamma} H_3\gamma_3 - HH_{33} - H_3^2\right) - e_4\left(\frac{H}{\gamma} H_4\gamma_4 + HH_{44} + H_4^2\right) = \varkappa H^4,$$

$$\frac{e_1}{\beta^2}\left(\frac{H}{\beta} H_1\beta_1 - HH_{11} + 2H_1^2\right) = e_2\left(\frac{H}{\beta} H_2\beta_2 - HH_{22} + 2H_2^2\right), \tag{51.18}$$

$$\frac{e_1}{\beta^2}\left(\frac{H}{\beta} H_1\beta_1 - HH_{11} - H_1^2\right) - e_2\left(\frac{H}{\beta} H_2\beta_2 + HH_{22} + H_2^2\right)$$

$$+\frac{e_3}{\gamma^2}\left(\frac{2H}{\gamma} H_3\gamma_3 - 2HH_{33} + H_3^2\right) - e_4\left(\frac{2H}{\gamma} H_4\gamma_4 + 2HH_{44} - H_4^2\right)$$

$$-\frac{e_4H^2}{\gamma}\gamma_{44} = \varkappa H^4, \tag{51.19}$$

$$\frac{e_3}{\gamma^2}\left(\frac{H}{\gamma} H_3\gamma_3 - HH_{33} + 2H_3^2\right) = e_4\left(\frac{H}{\gamma} H_4\gamma_4 - HH_{44} + 2H_4^2\right). \tag{51.20}$$

Equation (51.15) shows that

$$H^{-1} = \varphi(x^1, x^2)+\psi(x^3, x^4)$$

and (51.16) that

$$\varphi_{12} = \varphi_1 \frac{\beta_2}{\beta}, \quad \psi_{34} = \psi_3 \frac{\gamma_4}{\gamma},$$

i.e.

$$\varphi_1 = \omega(x^1)\beta, \quad \psi_3 = \nu(x^3)\gamma. \tag{51.21}$$

Using linear combinations of the remaining equations, we obtain

$$\frac{e_1}{\beta^2}\left(\varphi_{11}-\frac{\varphi_1\beta_1}{\beta}\right) = e_2\left(\varphi_{22}-\varphi_2\frac{\beta_2}{\beta}\right), \tag{51.22}$$

$$\frac{e_3}{\gamma^2}\left(\psi_{33}-\psi_3\frac{\gamma_3}{\gamma}\right) = e_4\left(\psi_{44}-\psi_4\frac{\gamma_4}{\gamma}\right), \tag{51.23}$$

$$6e_2\varphi_{22}-\frac{6e_1\varphi_1^2}{\beta^2(\varphi+\psi)}-\frac{6e_2\varphi_2^2}{\varphi+\psi}+6e_4\psi_{44}-\frac{6e_3\psi_3^2}{\gamma^2(\varphi+\psi)}-\frac{6e_4\psi_4^2}{\varphi+\psi}$$
$$-e_2\frac{\beta_{22}}{\beta}(\varphi+\psi)-e_4\frac{\gamma_{44}}{\gamma}(\varphi+\psi) = \frac{2\varkappa}{\varphi+\psi}, \tag{51.24}$$

$$2e_2\varphi_{22}-2e_4\varphi_{44}-e_2\frac{\beta_{22}}{\beta}(\varphi+\psi)+e_4\frac{\gamma_{44}}{\gamma}(\varphi+\psi) = 0. \tag{51.25}$$

Thus the set of equations stating the necessary and sufficient conditions for a V_4 of the second type to be a conformal-reducible Einstein space reduces to the *six* equations (51.21)–(51.25) in the four unknowns φ, ψ, β, γ, under the condition $H^{-1} = \varphi + \psi$.

As the functions φ, β depend only on the variables x^1, x^2, and ψ, γ on x^3, x^4, we can use the method of separation of variables in integrating the above set. We consider a point $P(\overset{0}{x}{}^3, \overset{0}{x}{}^4)$ in a binary domain $\{x^3, x^4\}$; using the notation

$$\psi(P) = C_1, \quad \left(-2\psi_{44}+\frac{\gamma_{44}}{\gamma}\psi\right)_P = C_2, \quad \left(\frac{\gamma_{44}}{\gamma}\right)_P = C_3,$$

(51.24) and (51.25) show that

$$2e_2\varphi_{22}+e_2C_2-\frac{e_4\beta_{22}}{\beta}(\varphi+C_1)+e_4C_3\varphi = 0.$$

Taking another point $\bar{P}$ in the same domain we obtain

$$2e_2\varphi_{22}+e_2\bar{C}_2-e_4\frac{\beta_{22}}{\beta}(\varphi+\bar{C}_1)+e_4\bar{C}_3\varphi = 0$$

so that

$$(C_2-\bar{C}_2)e_4+(\bar{C}_1-C_1)e_2\frac{\beta_{22}}{\beta}+e_4(C_3-\bar{C}_3)\varphi = 0.$$

Two possibilities must now be distinguished: (1) the coefficient of β_{22} is zero, i.e. $C_1 = \bar{C}_1$; (2) $C_1 \neq \bar{C}_1$. In the first case $\psi(P) = \psi(\bar{P})$; as points P and $\bar{P}$ are arbitrary $\psi =$ constant. Hence $\varphi \neq$ constant, for otherwise the Einstein spaces would be *reducible;* we finally obtain the two possibilities

$$(1)\quad \frac{\beta_{22}}{\beta} = k\varphi+C,\quad (2)\quad \psi = \dot{C}_2,\quad \frac{\gamma_{44}}{\gamma} = \dot{C}_1, \tag{51.26}$$

where k, C, $\dot{C}_1$, $\dot{C}_2$ are constants. In the second case we can always without loss of generality put $\psi = 0$ (combining ψ with φ).

Clearly on starting from the binary domain $\{x^1, x^2\}$, we obtain

$$(1')\quad \frac{\gamma_{44}}{\gamma} = \bar{k}\psi+\bar{C},\quad (2')\quad \varphi = \dot{\bar{C}}_2,\quad \frac{\beta_{22}}{\beta} = \dot{\bar{C}}_1. \tag{51.27}$$

Combining conditions (51.26) and (51.27) gives in principle four possibilities; however, if we reject the *reducible* spaces given by combination $2 \rightarrow 2'$, we have

$$\begin{aligned} &(1)\quad \frac{\beta_{22}}{\beta} = k\varphi+c, && \frac{\gamma_{44}}{\gamma} = \bar{k}\psi+\bar{c}, \\ &(2)\quad \varphi = 0, && \frac{\beta_{22}}{\beta} = c_1, \\ &(3)\quad \psi = 0, && \frac{\gamma_{44}}{\gamma} = c_1, \end{aligned} \tag{51.28}$$

where, k, $\bar{k}$, c, $\bar{c}$, and c_1 are constants.

We thus have to integrate the set of field equations (51.23)–(51.25) for all three cases; as this task does not present any essential difficulties, we leave the details to the reader, and simply state the final results.

The metric of conformal-reducible Einstein spaces of the type

$$ds^2 = H^2[\Phi_1(x^1, x^2)+\Phi_2(x^3, x^4)]$$

can always be reduced to one of the following forms:

$$\text{I.}\quad ds^2 = \frac{1}{(\varphi+\psi)^2}\left[e_1\varphi_2^2\left(f-e_1e_2\omega'\int\frac{dx^2}{\varphi_2^2}\right)^2 dx^{1^2}+e_2\,dx^{2^2}\right.$$

$$\left.+e_3\psi_4^2\left(\bar{f}-e_3e_4\nu'\int\frac{dx^4}{\psi_4^2}\right)^2 dx^{3^2}+e_4\,dx^{4^2}\right]$$

where

$$\varphi_2^2 = \frac{k}{3}\varphi^3+\frac{c\varphi^2}{2}+e_2t\varphi+e_2m-e_1e_2\omega^2$$

$$\psi_4^2 = \frac{\bar{k}}{3}\psi^3+\frac{\bar{c}k_2}{2}-e_2t\psi-e_1m-e_3e_4\nu^2-\frac{e_4\varkappa}{3}.$$

k, $\bar{k}$, c, $\bar{c}$, t, m are constants, $e_2k = e_4\bar{k}$, $e_2c = -e_4\bar{c}$, $\omega(x^1)$ and $\nu(x^3)$ are arbitrary functions.

$$\text{II.}\quad ds^2 = \frac{1}{(\varphi+\psi)^2}\left\{e_1\left(\frac{\varphi'}{\omega}\right)^2 dx^{1^2}+e_2\,dx^{2^2}\right.$$

$$\left.+e_3\left[\mu\sqrt{-e_4\tau}x^4+\frac{e_3e_4\nu'x^4}{\sqrt{-e_4\tau}}\right]^2 dx^{3^2}+e_4\,dx^{4^2}\right\},$$

$$\psi = \sqrt{-e_4\tau}x^4+f(x^3),\quad f' = \nu\mu\sqrt{-e_4\tau}$$

$$\text{III.}\quad ds^2 = \frac{1}{(\varphi+\psi)^2}\left\{e_1\left(\frac{\varphi'}{\omega}\right)^2 dx^{1^2}+e_2\,dx^{2^2}+e_3\left(\frac{\psi'}{\nu}\right)^2 dx^{3^2}+e_4\,dx^{4^2}\right\},$$

$$e_1\omega^2+e_3\nu^2 = -\frac{\varkappa}{3}.$$

This is a space of constant curvature.

$$\text{IV.}\quad ds^2 = \frac{1}{\psi^2}\left\{e_1\beta^2\,dx^{1^2}+e_2\,dx^{2^2}+e_3\psi_4^2\left(f-e_3e_4\nu'\int\frac{dx^4}{\psi_4^2}\right)^2 dx^{3^2}+e_4\,dx^{4^2}\right\},$$

$$\beta = \mu_1(x^1)\cosh kx^2+\mu_2(x^1)\sinh kx^2,$$

$$\psi_4^2 = n\psi^3-e_2e_4k^2\psi^2-e_3e_4\nu^2-e_4\frac{\varkappa}{3}\quad k, n = \text{constant}.$$

$$\text{V.}\quad ds^2 = \frac{1}{\psi^2}\left[e_1\beta^2\,dx^{1^2}+e_2\,dx^{2^2}+e_3\psi_4{}^2(f-e_3e_4\nu'\int\frac{dx^4}{\psi_4^2}\right)^2 dx^{3^2}+e_4\,dx^{4^2}\right\},$$

$$\beta = \mu_1(x^1)\cos kx^2+\mu_2(x^1)\sin kx^2,$$

$$\psi_4^2 = n\psi^3+e_2e_4k^2\psi^2-e_3e_4\nu^2-e_4\frac{\varkappa}{3},\quad k, n = \text{constant}.$$

$$\text{VI.}\ ds^2 = \frac{1}{\psi^2}\left\{e_1\beta^2\, dx^{1^2} + e_2\, dx^{2^2} + e_3\psi_4{}^2\left(f - e_3 e_4 \nu' \int \frac{dx^4}{\psi_4{}^2}\right)^2 dx^{3^2} + e_4\, dx^{4^2}\right\},$$

$$\beta = \mu_1(x^1)x^2 + \mu_2(x^1),$$

$$\psi_4{}^2 = n\psi^3 - e_3 e_4 \nu^2 - e_4\frac{\varkappa}{3}, \quad n = \text{constant.}$$

These solutions allow for the symmetric role of the functions φ and ψ and the arbitrary $e_\alpha = \pm 1$; solutions corresponding to real gravitational fields are obtained by so choosing e_α that the signature is $(- - - +)$. The choice of e_α is otherwise arbitrary so we can derive both *static* and *non-static* solutions.

Different choices of the arbitrary functions and constants in these metrics give rise to many well-known solutions; Thus, putting $k = 12$, $m = \omega = \nu = \varkappa = 0$, $e_2 = -1$, $e_4 = 1$, in I gives Delsarte's solution.

$$ds^2 = \frac{1}{[\wp(x^2) - \wp(x^4)]^2}\left[-\wp'^2(x^2)\, dx^{1^2} - dx^{2^2} - \wp'^2(x^4)\, dx^{3^2} + dx^{4^2}\right]$$

where $\wp$ is the Weierstrass elliptic function.

Putting $\mu_1 = \nu = \varkappa = 0$, $\mu_2 = 1$, $n = -2a$, $k^2 = 1$, $e_1 = e_2 = e_4 = -1$, $e_3 = 1$, $f = 1$ in V clearly gives Schwarzschild's solution.

The method of solution and the result for $\varkappa = 0$ are due to the author; the generalization for $\varkappa \neq 0$ was obtained in a series of works by Anchikov [507], [508].

Example

Prove that a V_n is reducible only when it admits a constant, symmetric, idempotent tensor field $t_{\alpha\beta}$:

$$t_{\alpha\beta,\gamma} = 0, \quad t_{[\alpha\beta]} = 0, \quad t^\sigma_\alpha t_{\sigma\beta} = t_{\alpha\beta}$$

(Shirokov [108]).

§ 52. Symmetric Gravitational Fields

In this section, the symbol T_i is used to denote both T_i and $\dot{T}_i$ spaces defined in § 18.

A T_i space and its corresponding field are said to be *symmetric* if the curvature tensor of T_i is covariantly constant:

$$R_{\alpha\beta\gamma\delta,\lambda} = 0. \qquad (52.1)$$

The necessary conditions for these equations to be integrable are obtained by again differentiating the equations covariantly and alternating over the indices of differentiation. Then, using the Ricci identities, we obtain

$$R_{\lambda\mu\sigma[\alpha}R^{\sigma}{}_{\beta]\gamma\delta}+R_{\lambda\mu\sigma[\gamma}R^{\sigma}{}_{\delta]\alpha\beta} = 0. \tag{52.2}$$

As these equations contain only curvature tensor components, it is convenient to state them for the canonical non-holonomic orthonormal tetrad in which the $R_{\alpha\beta\gamma\delta}$ have the canonical forms (19.6), (19.12) and (19.17), with metric tensor (11.7); we consider all three types of gravitational field.

For T_1, we can use the curvature tensor components given in (19.6); we have

$$\alpha_1(\alpha_2-\alpha_3)-\beta_1(\beta_2-\beta_3) = \beta_1(\alpha_2-\alpha_3)+\alpha_1(\beta_2-\beta_3) = 0$$

together with three more pairs of equations obtained by cyclic interchange of the indices. By multiplying the second equation of every pair by i before adding to the first, and using the stationary curvatures $k_s = \alpha_s + i\beta_s$ (cf. § 17), we derive the equivalent set of equations

$$k_1(k_2-k_3) = k_2(k_3-k_1) = k_3(k_1-k_2) = 0. \tag{52.3}$$

In addition, (19.7) gives

$$\sum_1^3 k_s = \varkappa. \tag{52.4}$$

If $\varkappa \neq 0$, eqns. (52.3) and (52.4) have only the following solutions (excluding interchange of indices):

$$(\alpha)\quad k_1 = k_2 = k_3 = \frac{\varkappa}{3}, \quad (\beta)\quad k_1 = k_2 = 0, \quad k_3 = \varkappa.$$

Case (α) obviously gives a space of constant curvature and (β), as we now show, a *reducible* space.

In order that a V_n be reducible, it is necessary and sufficient that it should admit a constant, symmetric idempotent tensor field $C_{\alpha\beta}$ (cf. example 1, § 51):

$$(1)\quad C_{\alpha\beta,\gamma} = 0, \quad (2)\quad C_{[\alpha\beta]} = 0, \quad (3)\quad C^{\sigma}_{\alpha}C_{\sigma\beta} = C_{\alpha\beta}.$$

We now show that a tensor satisfying all three conditions does exist for the spaces in case (β); a proof for a non-holonomic orthonormal tetrad will suffice. The set of differential equations (1) are compatible, provided only that the integrability condition

$$C_{\alpha\beta,\,[\gamma\delta]} \equiv R^{\sigma}{}_{\alpha\gamma\delta}C_{\sigma\beta}+R^{\sigma}{}_{\beta\gamma\delta}C_{\alpha\sigma} = 0 \tag{52.5}$$

is satisfied. All further equations obtained by differentiation are *satisfied identically* by virtue of (1) and the symmetry of the space. Applying (52.5) to the orthonormal tetrad (19.6), we obtain

$$C_{12} = C_{34} = 0, \quad C_{11} = C_{22}, \quad C_{33} = -C_{44};$$

applying conditions (2), and the idempotency condition (3) for $C_{\alpha\beta}$ to this tetrad, we obtain

$$C_{ij} = 0 \quad (i \neq j), \quad C_{11}^2+C_{11} = C_{33}^2+C_{33} = 0.$$

Hence if $C_{\alpha\beta}$ for the canonical tetrad is

$$(C_{\alpha\beta}) = \left(\begin{array}{c|c} 0 & 0 \\ \hline 0 & S \end{array}\right), \quad S = \begin{pmatrix} -1 & 0 \\ 0 & 1 \end{pmatrix},$$

all three conditions for a reducible space are satisfied. As moreover $C_{\alpha\beta}$ is a second order tensor, the metric form splits into two *binary forms* [108]; clearly each defines a two-dimensional space of constant curvature—the first with a negative-definite and the second with an indefinite metric. Hence *a symmetric T_1 space has the metric* (14.7) or (14.8) *in a special system of coordinates, or is a space of constant curvature; when $\varkappa = 0$ it is a flat Minkowski space* ([261], p. 242; [287]; [320]).

We now consider a T_2 space with curvature tensor components defined by formulae (19.12) under condition (19.13). Putting $\lambda = 1$, $\mu = 4$, in eqns. (52.2), we obtain $\alpha_1 = \beta_1 = 0$, and hence from (19.13), $\alpha_2 = \varkappa/2$, $\beta_2 = 0$. If we now fix the indices in (52.2) we obtain $\alpha_2 = 0$, $\varkappa = 0$, using the permutation $\begin{pmatrix}\lambda\mu & \alpha\beta & \gamma\delta \\ 24 & 14 & 12\end{pmatrix}$. Thus a *$T_2$ type space can only be symmetric if $\varkappa = 0$, when the geometry is that of an empty space*. But as shown in § 25, there is only one symmetric T_2 with metric satisfying the equations $R_{\alpha\beta} = 0$; this is the T_2 space of maximum mobility with the metric (25.23) and a six-parameter transitive soluble group of motions.

The curvature tensor components for a T_3 space are defined by (19.17); substituting into (52.2) leads to a contradiction for all values of $\varkappa$.

Thus *a symmetric T_1 space is either a space of constant curvature or one of the reducible spaces* (14.7), (14.8)*; when $\varkappa = 0$ a symmetric T_1 is always flat. A symmetric T_2 must have $\varkappa = 0$, and is defined by the metric* (25.23) *in a special system of coordinates. T_3 spaces cannot be symmetric* ([261], p. 243).

Introducing a normal system of coordinates (§ 7), we obtain

$$g_{\alpha\beta} = \overset{0}{g}_{\alpha\beta} - \frac{1}{3}\overset{0}{R}_{\alpha\sigma\beta\tau}x^{\sigma}x^{\tau} - \frac{1}{3!}\overset{0}{R}_{\alpha\sigma\beta\tau,\lambda}\, x^{\sigma}x^{\tau}x^{\lambda} + o(\delta^4).$$

Let us consider a point P of a V_4 space, infinitesimally close to the origin, and the components $g_{\alpha\beta}(P)$ calculated to a certain order of small quantities. The zeroth approximation defines a flat metric, in no way reflecting the effect of the gravitational field; the first order approximation does not alter the picture. The second order approximation is always important if some distribution and movement of matter exists; it is characterized by non-zero curvature tensor components in the orthonormal tetrad, and this first indicates the existence of a field. The third order approximation defines, roughly speaking, the rate of change of the curvature tensor, and this is zero for symmetric spaces so that only higher order corrections remain; we might say that the curvature tensor "changes slowly". In particular, therefore, a "slow change" in the curvature tensor is only possible in T_2 spaces when the space is maximally homogeneous, and can never occur in T_3 spaces. Many investigations have been devoted to symmetric V_n's (cf. § 14).

EXAMPLES

1. Show that a symmetric T_2 has stationary curvatures $k_s \equiv \alpha_s + i\beta_s = 0$.

2. Show that for a symmetric T_2 in a normal system of coordinates (§ 7)

$$g_{\alpha\beta} = \overset{0}{g}_{\alpha\beta} + \frac{1}{2}\sum_{n=1}^{\infty}\frac{(-1)^n 2^{n+2}}{(2n+2)!}\, m_{\alpha}{}^{\sigma_1} m_{\sigma_1}{}^{\sigma_2} \ldots m_{\sigma_{n-1}\beta}$$

where

$$m_{\alpha\beta} = R_{\alpha\lambda\beta\sigma}x^{\lambda}x^{\sigma}$$

the x^{σ} are normal coordinates, and $\overset{0}{g}_{\alpha\beta}$, $\overset{0}{R}_{\alpha\beta\gamma\delta}$ denote values at the origin of the normal coordinates.

Show also that when $\alpha_s = \beta_s = 0$ for the orthonormal tetrad (19.19) we have

$$\overset{0}{R}_{\alpha\beta\gamma\delta} = h_{\alpha\beta}h_{\gamma\delta} - p_{\alpha\beta}p_{\gamma\delta},$$

where

$$(h_{\alpha\varrho}) = \begin{pmatrix} 0 & -1 & 0 & 0 \\ 1 & 0 & 0 & -1 \\ 0 & 0 & 0 & 0 \\ 0 & 1 & 0 & 0 \end{pmatrix}, \quad (p_{\alpha\beta}) = \begin{pmatrix} 0 & 0 & -1 & 0 \\ 0 & 0 & 0 & 0 \\ 1 & 0 & 0 & -1 \\ 0 & 0 & 1 & 0 \end{pmatrix}.$$

3. Using the notation of examples 1 and 2 and putting

$$h_{\alpha\beta}x^{\beta} = y_{\alpha}; \quad p_{\alpha\beta}x^{\beta} \equiv z_{\alpha}$$

show that

$$m_{\alpha}{}^{\sigma_1}m_{\sigma_1}{}^{\sigma_2}\dots m_{\sigma_r\beta} = (-1)^r v^{2r}[y_{\alpha}y_{\beta}+(-1)^{r+1}z_{\alpha}z_{\beta}]; \quad v \equiv i(x^1-x^4)$$

and hence

$$g_{\alpha\beta} = \overset{0}{g}_{\alpha\beta}+Ay_{\alpha}y_{\beta}+Bz_{\alpha}z_{\beta}$$

where, on changing the variable v to $v = x^1 - x^4$,

$$A = \sum_{n=1}^{\infty} \frac{2^{2n+1}}{(2n+2)!} v^{2n-2} = \frac{1}{v^4}(\sinh^2 - v^2),$$

$$B = \sum_{n=1}^{\infty} \frac{(-1)^n 2^{2n+1}}{(2n+2)!} v^{2n-2} = \frac{1}{v^4}(\sin^2 v - v^2),$$

and obtain

$$g_{\alpha\beta} = \overset{0}{g}_{\alpha\beta}+\frac{1}{v^4}[(v^2-\sinh^2 v)y_{\alpha}y_{\beta}+(v^2-\sin^2 v)z_{\alpha}z_{\beta}], \quad v = x^1-x^4.$$

4. Show that a transformation of coordinates can reduce the metric (52.6) to the form

$$ds^2 = -dx^{1^2}-\sinh^2 v\, dx^{2^2}-\sin^2 v\, dx^{3^2}+dx^{4^2}, \quad v = x^1-x^4,$$

and also to

$$ds^2 = 2dx^1\, dx^4-\sinh^2 x^4\, dx^{2^2}-\sin^2 x^4\, dx^{3^2}.$$

§ 53. Static Einstein Spaces

These spaces correspond to gravitational fields admitting a coordinate system with metric tensor components independent of the time x^4, and with both time directions essentially equivalent so that $g_{\alpha 4} = 0$ $(\alpha \neq 4)$. If x^4 does not have to correspond to time, we have a generalization of static spaces defined by the conditions that: (1) the space admits a G_1 group of motions, and (2) that in the system of coordinates with Killing vector $\xi^{\alpha} = \delta^{\alpha}_4$, the space also admits the mirror reflection $x^{4'} = -x^4$, $x^{i'} = x^i$ $(i = 1, 2, 3)$.

In this special systems of coordinates the metric tensor has the form

$$(g_{\alpha\beta}) = \begin{pmatrix} g_{11} & g_{12} & g_{13} & 0 \\ g_{21} & g_{22} & g_{23} & 0 \\ g_{31} & g_{32} & g_{33} & 0 \\ 0 & 0 & 0 & g_{44} \end{pmatrix}, \quad \partial_4 g_{\alpha\beta} = 0, \qquad (53.1)$$

and hence

$$ds^2 = d\sigma^2 + g_{44}\, dx^{4^2},$$

where $d\sigma^2$ is a quadratic form in the x^i $(i=1, 2, 3)$. Let V_4 be a static space with $g_{44} > 0$ and a negative-definite form $d\sigma^2$. Cotton has shown that every V_3 admits a holonomic orthogonal system of coordinates ([10], p. 357). Hence $-d\sigma^2$ can always be represented as a sum of squares by means of a non-singular transformation

$$x^i = f^i(x^{1'}, x^{2'}, x^{3'}) \quad (i = 1, 2, 3), \quad x^4 = x^{4'}, \quad |d_j f^i| \neq 0.$$

As

$$g_{i'4'} = g_{\alpha 4} \frac{\partial x^\alpha}{\partial x^{i'}} = 0 \qquad (i = 1, 2, 3),$$

under such a transformation, *every static V_4 admits a 4-orthogonal system of coordinates* with metric tensor of the form

$$(g_{\alpha\beta}) = \begin{pmatrix} -\underset{1}{H}^2 & 0 & 0 & 0 \\ 0 & -\underset{2}{H}^2 & 0 & 0 \\ 0 & 0 & -\underset{3}{H}^2 & 0 \\ 0 & 0 & 0 & \underset{4}{H}^2 \end{pmatrix}, \tag{53.2}$$

where the Lamé coefficients $\underset{\alpha}{H}$ are functions of the x^i $(i = 1, 2, 3)$. The curvature tensor components for (53.2) are

$$\left.\begin{aligned} R_{i4i4} &= -\underset{4}{H}\left[\underset{4}{H}_{ii} + \underset{i}{H}\left(\underset{4}{H}_i \frac{\underset{i}{H_i}}{\underset{i}{H}^2}\right) - \sum_k^{\bullet} \underset{i}{H}_k\left(\frac{\underset{4}{H_k}}{\underset{k}{H}^2}\right)\right], \\ R_{i4j4} &= -\underset{4}{H}[\underset{4}{H}_{ij} + \underset{4}{H}_i \partial_j \ln \underset{i}{H} + \underset{4}{H}_j \partial_i \ln \underset{j}{H}], \\ R_{i4jk} &= 0, \quad R_{ijij} = \underset{i}{H}\underset{i}{H}_{jj} + \underset{j}{H}\underset{j}{H}_{ii} \\ &\quad + \underset{i}{H}\underset{j}{H}\left(\frac{\underset{i}{H}_l \underset{j}{H}_l}{\underset{l}{H}^2} - \frac{\underset{i}{H}_i \underset{j}{H}_i}{\underset{i}{H}^2} - \frac{\underset{i}{H}_j \underset{j}{H}_j}{\underset{j}{H}^2}\right), \quad l \neq i, j, \\ R_{ijik} &= \underset{i}{H}\underset{i}{H}_{jk} - \underset{i}{H}\left(\frac{\underset{k}{H}_i \underset{i}{H}_k}{\underset{k}{H}} + \frac{\underset{j}{H}_k \underset{i}{H}_j}{\underset{j}{H}}\right), \quad (i \neq j \neq k), \end{aligned}\right\} \tag{53.3}$$

where $i, j = 1, 2, 3$, and the symbol $\sum\limits_k^{\bullet}$ implies that $k \neq i$.

We now investigate the type of static Einstein spaces with the field equations $R_{\alpha\beta} = 0$.

Three of these equations are satisfied identically; the other seven reduce to

$$\left.\begin{aligned} &\underset{j}{H}^2\underset{k}{H}^2 R_{i4i4} + \underset{i}{H}^2\underset{4}{H}^2 R_{jkjk} = 0, \\ &\underset{i}{H}^2 R_{k4j4} + \underset{4}{H}^2 R_{ikkj} = 0, \\ &\underset{i}{H}^2 R_{jkjk} + \underset{j}{H}^2 R_{ikik} + \underset{k}{H}^2 R_{ijij} = 0, \end{aligned}\right\} \tag{53.4}$$

where $i, j, k = 1, 2, 3$ and $i \neq j \neq k$. We first write the λ-matrix

$$A \equiv (R_{ab} - \lambda g_{ab}) \qquad (a, b = 1, \ldots, 6)$$

in bivector space collective indices. Using (53.3), we have

$$A = \begin{pmatrix} P & 0 \\ 0 & Q \end{pmatrix},$$

where

$$P = \begin{pmatrix} R_{1414} + \lambda(\underset{1}{H}\underset{4}{H})^2 & R_{1424} & R_{1434} \\ R_{1424} & R_{2424} + \lambda(\underset{2}{H}\underset{4}{H})^2 & R_{2434} \\ R_{1434} & R_{2434} & R_{3434} + \lambda(\underset{3}{H}\underset{4}{H})^2 \end{pmatrix}$$

and

$$Q = \begin{pmatrix} R_{2323} - \lambda(\underset{2}{H}\underset{3}{H})^2 & R_{2331} & R_{2312} \\ R_{2331} & R_{3131} - \lambda(\underset{1}{H}\underset{3}{H})^2 & R_{3112} \\ R_{2312} & R_{3112} & R_{1212} - \lambda(\underset{1}{H}\underset{2}{H})^2 \end{pmatrix}.$$

Using eqns. (53.4) and applying elementary transformations, we can easily show that

$$P \approx Q$$

so that

$$(R_{ab} - \lambda g_{ab}) \approx \begin{pmatrix} P & 0 \\ 0 & P \end{pmatrix}.$$

Hence we need only consider the elementary divisors of either matrix P or Q. Let us, for example, take Q; dividing the first line by $\underset{2}{H}\underset{3}{H}$, the second by $\underset{1}{H}\underset{3}{H}$, and the third by $\underset{1}{H}\underset{2}{H}$, and then doing the same with the columns, we obtain a matrix equivalent to Q, i.e. with the same elementary divisors, but of the form

$$(A_{MN} - \lambda\delta_{MN}) \qquad (M, N = 1, 2, 3),$$

where

$$A_{MN} \to \frac{R_{ijkl}}{\underset{i\ j\ k\ l}{HHHH}}$$

and δ_{MN} is the Kronecker symbol. We thus have two symmetric matrices, A and the 3×3 unit matrix I. Hence the λ-matrix $(A - \lambda I)$ has a simple Segre characteristic and the elementary divisors have real bases. Thus; *every static Einstein space with signature* $(---+)$ *and* $\varkappa = 0$ *is a space of the first type* T_1 *with real principle curvatures.* It should be noted that, although the field equations cannot be integrated in closed form, there is no difficulty in principle in assessing the arbitrariness involved in the general solution of the static case. Buchdahl and others have considered generalized static spaces with $n > 4$ under the conditions stated above ([215], [248]).

EXAMPLES

1. Show that if the metric of a static V_4 satisfying the equations $R_{\alpha\beta} = 0$ has the form

$$ds^2 = g_{ik}(x^i)\, dx^i\, dx^k + g_{aa}(x^j)\, dx^{a^2} \qquad (a \neq i, j, k),$$

where a is a fixed index subject to summation, the reciprocal metric

$$ds^2 = (g_{aa})^{\frac{2}{n-3}} g_{ik}\, dx^i\, dx^k + (g_{aa})^{-1}\, dx^{a^2}$$

also defines a static solution of these equations [215].

2. Apply the result obtained in example 1 to the Schwarzschild solution and to a flat Minkowski metric.

3. Show that given a static solution of the general field equations $(R_{\alpha\beta} \neq \varkappa g_{\alpha\beta})$, it is possible to construct a reciprocal static solution; the total energies of these fields are equal in magnitude but opposite in sign [248].

4. Investigate the λ-matrix and the type of a four-dimensional generalized *static* Einstein space when x^4 is not a time coordinate.

§ 54. Centro-symmetric Gravitational Fields

It is in general reasonable to suppose that whenever several bodies exist in space these bodies must be in motion, so that no gravitational field can be static. Nevertheless, in many problems it is important to consider static fields, and in particular, static centro-symmetric gravitational fields.

A centro-symmetric field may be caused by any centro-symmetric distribution of matter. If we are dealing with a continuous substance, not only the distribution of matter but also the motion must be centro-symmetric.

Centro-symmetry means that the metric of the space-time must be the same at all points equidistant from the centre of symmetry; it is most conveniently defined in terms of the theory of groups of motion (§§ 10, 24). We preface this definition by the following geometrical discussion ([188], pp. 610–17). We seek static centro-symmetric fields defined by the conditions; (1) x^4 corresponds to time, (2) the negative-definite form $\varphi = \mathbf{g}_{ij}\mathbf{d}\mathbf{x}^i\,\mathbf{d}\mathbf{x}^j$ $(i, j = 1, 2, 3)$ is invariant with respect to all possible orthogonal transformations of the independent variables $\mathbf{x}^1, \mathbf{x}^2, \mathbf{x}^3$, (3) $\mathbf{g}_{ij} = \mathbf{g}_{ij}(x^1, x^2, x^3)$.

We now associate the hypersurfaces $x^4 =$ constant with a series of flat spaces with rectangular coordinates $\mathbf{x}^i$, but, of course, the metric of these spaces differs from that of the hypersurfaces.

We introduce the following coordinates x^α, which are a generalization of spherical polar coordinates in a flat space by putting

$$\mathbf{x}^1 = x^1 \sin x^2 \cos x^3, \quad \mathbf{x}^2 = x^1 \sin x^2 \sin x^3,$$
$$\mathbf{x}^3 = x^1 \cos x^2, \quad \mathbf{x}^4 = x^4.$$

Hence

$$x^1 = r = \sqrt{\mathbf{x}^{1^2}+\mathbf{x}^{2^2}+\mathbf{x}^{3^2}}, \quad x^2 = \frac{\pi}{2}-\theta, \quad x^3 = \varphi,$$

where θ is the latitude and φ the longitude.

The required group of motions defining a centro-symmetric static field is a G_4 group with one evident Killing vector $\underset{4}{\xi^\alpha} = \delta_4^\alpha$; the components of the required tensor must be independent of time. Moreover, the G_4 must contain a subgroup isomorphic with the group of rotations which leave invariant a sphere with centre at the origin of a Cartesian system of coordinates in a Euclidean space. As is well known, such a G_3 is insoluble since it does not contain a two-parameter subgroup; it is therefore of type IX.

Thus the required group in four-dimensional space must also be insoluble. Structures VIII and IX are indistinguishable over the field of complex numbers; but real parameters, as in the present case, require type IX. Clearly, the group must act on two-dimensional manifolds defined by the latitude and longitude, so that the rank of the matrix of the Killing vectors $(\underset{s}{\xi^\alpha})$ ($s = 1, 2, 3$; $\alpha = 1, 2, 3, 4$) must be two; the surfaces of transitivity of the G_3 group are two-dimensional. According to § 24 the group operators can always be taken as

$$\left.\begin{aligned} X_1 &= \cos x^3\, p_2 - \sin x^3 \cot x^2\, p_3, \\ X_2 &= \sin x^3\, p_2 + \cos x^3 \cot x^2\, p_3, \\ X_3 &= -p_3. \end{aligned}\right\} \tag{54.1}$$

As, moreover, the space is static, it must also admit the operator

$$X_4 = p_4. \tag{54.2}$$

We obtain the required metric by integrating the Killing equations for the operators (54.1) and (54.2). We must also take into account those allowed coordinate transformations which leave these operators invariant. The general form of these transformations can be determined by setting down the differential equations corresponding to the invariance of the operators under the transformation, and then finding their general solution; we obtain

$$x^{1'} = \boldsymbol{f}(x^1), \quad x^{2'} = \pm x^2 + \boldsymbol{s}\pi, \quad x^{3'} = x^3 + \boldsymbol{k}\pi, \quad x^{4'} = x^4 + \boldsymbol{\theta}(x^1), \tag{54.3}$$

where $\boldsymbol{f}$ and $\boldsymbol{\theta}$ are arbitrary functions of x^1, $\boldsymbol{s}$ and $\boldsymbol{k}$ are integers and $\boldsymbol{f}' \neq 0$, since the transformation is non-degenerate. The inverse transformation to (54.3) is

$$\left.\begin{aligned} x^1 = f(x^{1'}), \quad x^2 = \pm x^{2'} + s\pi, \quad x^3 = x^{3'} + k\pi, \\ x^4 = x^{4'} + \theta(x^{1'}). \end{aligned}\right\} \tag{54.4}$$

The Killing equations

$$\underset{s}{\xi^\sigma}\partial_\sigma g_{\alpha\beta} + g_{\alpha\sigma}\partial_\beta \underset{s}{\xi^\sigma} + g_{\beta\sigma}\partial_\alpha \underset{s}{\xi^\sigma} = 0$$

for the operators for $s = 3, 4$ reduce to

$$g_{\alpha\beta} = g_{\alpha\beta}(x^1, x^2).$$

For $s = 1, 2$ we obtain the two sets of equations

$$\partial_2 g_{\alpha\beta} - g_{3\alpha} \cot x^2 \delta^3_\beta - g_{3\beta} \cot x^2 \delta^3_\alpha = 0, \tag{54.5}$$

$$g_{2\alpha}\delta^3_\beta + g_{2\beta}\delta^3_\alpha - \frac{1}{\sin^2 x^2}(g_{3\alpha}\delta^2_\beta + g_{3\beta}\delta^2_\alpha) = 0. \tag{54.6}$$

Putting $\alpha, \beta = 1, 2, 3, 4$ in (54.6), we obtain

$$g_{12} = g_{13} = g_{23} = g_{24} = g_{34} = 0, \quad g_{33} = g_{22} \sin^2 x^2. \tag{54.7}$$

Substituting in eqns. (54.5), we obtain

$$\left.\begin{aligned}\partial_2 g_{11} = 0, \quad \partial_2 g_{14} = 0, \quad \partial_2 g_{22} = 0,\\ \partial_2 g_{33} - 2g_{33} \cot x^2 = 0, \quad \partial_2 g_{44} = 0.\end{aligned}\right\} \tag{54.8}$$

Hence $g_{11}, g_{22}, g_{33}, g_{44}$ depend only on x^1 and

$$g_{33} = g_{22} \sin^2 x^2.$$

In the new system of coordinates resulting from allowed transformations (54.4), we obtain

$$g_{1'4'} = g_{14}f' + g_{44}\theta', \quad g_{2'2'} = g_{22}(x^1),$$

as well as (54.7) and (54.8). If we impose the conditions

$$g_{1'4'} = 0, \quad g_{22}(x^1) = -x^{1^2}$$

we arrive at a set of equations which are compatible *in real space*.

Thus choosing f and θ as integrals of this set, we obtain the required metric in the form

$$ds^2 = -e^{\alpha}\, dx^{1^2} - x^{1^2}(dx^{2^2} + \sin^2 x^2\, dx^{3^2}) + e^{\beta}\, dx^{4^2} \tag{54.9}$$

for the required metric signature, where α, β are functions of x^1.

After evaluating the Ricci tensor components, we can write the field equations in the form

$$\left.\begin{aligned}\varkappa T_1^1 &= e^{-\alpha}\left(\frac{\beta'}{x^1} + \frac{1}{x^{1^2}}\right) - \frac{1}{x^{1^2}},\\ \varkappa T_2^2 = \varkappa T_3^3 &= \frac{1}{2}e^{-\alpha}\left(\beta'' - \frac{\alpha'\beta'}{2} + \frac{\beta' - \alpha'}{x^1} - \frac{(\beta')^2}{2}\right),\\ \varkappa T_4^4 &= e^{-\alpha}\left(\frac{1}{x^{1^2}} - \frac{\alpha'}{x^1}\right) - \frac{1}{x^{1^2}}.\end{aligned}\right\} \tag{54.10}$$

The other energy-momentum tensor components vanish identically.

In *empty* space where $T_\beta^\alpha = 0$, this set is easily integrated to give $\beta = -\alpha$ and

$$e^{-\alpha} = 1 + \frac{\gamma}{x^1}, \quad \gamma = \text{const},$$

the well-known Schwarzschild solution.

As x^1 tends to infinity the metric tends to that of a flat space. If Newton's law of gravitation is to be obeyed at large distances where the field is weak, the constant γ can be expressed in terms of the total mass of the body producing the field. As is well known,

$$g_{44} = 1+\frac{2\varphi}{c^2}$$

in a weak field ([156], p. 272) where the potential φ is given by its Newtonian expression

$$\varphi = -\frac{kM}{r} \equiv -\frac{kM}{x^1},$$

where M is the total mass of the body ([156], p. 303). Hence

$$\gamma = -\frac{2kM}{c^2}.$$

Such an element of length defines a gravitational field in empty space produced by any centro-symmetric distribution of matter, either static or dynamic. If a field in empty space is centro-symmetric, it need not be static, and Birkhoff's assertion to the contrary is incorrect [66].

The above metric is important in the solution of a number of problems and can be obtained in many different ways ([156], pp. 304–7; [188], §§ 129–30; [225], § 57). It is of fundamental importance in calculating the motion of the perihelion of Mercury, and the bending of light rays. The problem was first solved by Einstein in 1915, using the method of successive approximations [35a]. Then Schwarzschild and, independently of him, Droste gave a rigorous solution for the field of a material point ([37], [38]). A considerable simplification was introduced by Weyl who used the variational principle and Cartesian coordinates ([42]).

Attention has been recently paid to the fact that Schwarzschild's metric can be deduced from Lichnerowicz's postulates ([231], pp. 5–6), when the components of the tensor $g_{\alpha\beta}$ are functions of class C^1. If an additional condition is not imposed, Birkhoff's theorem about the statical character of the spherically symmetric field does not follow (Petrov [506]; Hamoui [509]).

Examples

1. Show that the metric

$$ds^2 = -dx^{1^2}-dx^{2^2}-dx^{3^2}-\alpha(x^1\,dx^1+x^2\,dx^2+x^3\,dx^3)^2+\beta\,dx^{4^2}$$

defines a centro-symmetric static distribution of matter. Integrate the equations in empty space and show that

$$\alpha = \frac{2m}{r^2(r-2m)}, \quad \beta = 1-\frac{2m}{r}, \quad r = \sqrt{x^{1^2}+x^{2^2}+x^{3^2}}.$$

2. Obtain the Schwarzschild solution from (54.10).

3. Show that the metric

$$ds^2 = -dx^{1^2} - dx^{2^2} - dx^{3^2} - \frac{(x^1dx^1 + x^2dx^2 + x^3dx^3)^2}{a^2 - r^2} + \left(\frac{3h - h_0}{2hh_0}\right)^2 dx^{4^2},$$

where

$$a = r_0\left(\frac{r_0}{2m}\right)^{1/2}, \quad m = \frac{kM}{c^2}, \quad \frac{1}{h^2} = 1 - \frac{2mr^2}{r_0 3}, \quad \frac{1}{h_0^2} = 1 - \frac{2m}{r_0}$$

can be interpreted as the metric of a field inside an incompressible fluid sphere of radius r_0 and mass M. The energy-momentum tensor is given in the form

$$T_{\alpha\beta} = \left(\mu_0 + \frac{p}{c^2}\right) u_\alpha u_\beta - p g_{\alpha\beta},$$

where μ_0 is the rest density and p is the pressure (Schwarzschild [37a]; Weyl [69]).

4. Calculate the purely "spatial", i.e. Ricci, tensor on the hypersurface x^4 = constant, and the curvature tensor for the metric (54.9).

5. Show that the action function in a centro-symmetric field has the form

$$S = -E_0 t + M\varphi + \int e^{-\frac{\nu}{2}} \sqrt{\frac{E_0^2}{c^2} e^{-\nu} - m^2c^2 - \frac{M^2}{x^{1^2}}}\, dx^1,$$

where E_0 is the constant energy, M the angular momentum, $e^\nu = 1 - \dfrac{2km'}{c^2x^1}$, m' the mass of the body producing the field, m the mass of a test particle and

$$\varphi = \int \frac{\dfrac{M}{x^{1^2}}\, dx^1}{\sqrt{\dfrac{E_0^2}{c^2} - m^2c^2e^\nu - \dfrac{M^2}{x^{1^2}} e^\nu}}$$

(Landau and Lifshitz [156], pp. 312–13).

§ 55. Gravitational Fields with Axial Symmetry

The investigation of gravitational fields with axial symmetry was started by Weyl ([42], [51]) and Levy-Civita ([47], [50]) and has since been continued by many others. The definition of these fields has undergone some changes and still varies with the author.

A study of this type of gravitational field and an analysis of the concept of axial symmetry are to be found, for example, in Lewis [88], van Stockum [100], Marder [269], etc.

Perhaps the most general definition of a field with axial symmetry which is also invariant is as follows:

(α) the field is *constant*, i.e. $g_{\alpha\beta}$ is independent of x^4 in a special system of coordinates; this means that the metric admits a Killing operator of a one-parameter group of motions with a Killing vector norm

$$g_{\alpha\beta}\underset{1}{\xi^\alpha}\underset{1}{\xi^\beta} > 0;$$

(β) the metric admits another non-isotropic operator with a Killing vector norm

$$g_{\alpha\beta}\underset{2}{\xi^\alpha}\underset{2}{\xi^\beta} < 0.$$

These two operators define an *Abelian* G_2 group of motions

$$[X_1X_2] = 0. \tag{55.1}$$

By imposing particular additional conditions we can construct all the fields with axial symmetry considered in the literature.

By choosing the coordinate system so that the Killing vectors assume their simplest form, we can determine one of the possible metric forms; this choice of coordinates cannot, of course, be unique.

The rank of the matrix corresponding to G_2, i.e. $(\underset{s}{\xi^\alpha})$ $(s = 1, 2;\ \alpha = 1, 2, 3, 4)$ is two; the trajectories of two one-parameter groups cannot coincide. A coordinate system can always be introduced into a V_4 so that the Killing vector for the first operator will be

$$\underset{1}{\xi^\alpha} = \delta^\alpha_4,$$

and any transformation

$$x^{i'} = \varphi^i, \quad x^{4'} = x^4+\varphi^4 \quad (i = 1, 2, 3), \tag{55.2}$$

where $\varphi^\alpha = \varphi^\alpha(x^1, x^2, x^3)$ leaves this unchanged. The structure equations (55.1) show that

$$\underset{2}{\xi^\alpha} = \underset{2}{\xi^\alpha}(x^1, x^2, x^3).$$

Using the allowed transformations (55.2), we can further specify the coordinate system so that though the vector $\underset{1}{\xi}^{\alpha}$ remains unaltered, $\underset{2}{\xi}^{\alpha}$ takes a simpler form. In particular, since the group is Abelian $\underset{2}{\xi}^{\alpha}$ could be reduced to the form $\underset{2}{\xi}^{\alpha} = \delta^{\alpha}_{2}$ ([147], p. 49). But as we are mainly interested in the Weyl and Levi-Civita solution we choose a different coordinate system. Let the coordinate curve x^3 be taken along the axis of symmetry so that $g_{\alpha\beta}$ must in general depend on x^3. In some infinitesimal domain V_3 at some instant of time ($x^4 = t =$ constant) there is symmetry about the tangent l to the curve x^3. Hence the required group operator must define a rotation about l and is therefore

$$\underset{2}{\xi}^{\alpha} = x^2\delta^{\alpha}_{1} - x^1\delta^{\alpha}_{2}. \tag{55.3}$$

We can also obtain the same result by starting from the allowed transformations (55.2). In the new coordinate system the vector $\underset{2}{\xi}^{\alpha}$ becomes

$$\underset{2}{\xi}^{i'} = \varphi^{i}_{j}\underset{2}{\xi}^{j}, \quad \underset{2}{\xi}^{4'} = \varphi^{4}_{j}\underset{2}{\xi}^{j} + \underset{2}{\xi}^{4} \qquad (i = 1, 2, 3).$$

If we require that

$$\underset{2}{\xi}^{1'} = \varphi^2, \quad \underset{2}{\xi}^{2'} = -\varphi^1, \quad \underset{2}{\xi}^{3'} = 0, \quad \underset{2}{\xi}^{4'} = 0, \tag{55.4}$$

we obtain a set of four differential equations for the four unknowns $\varphi^{\alpha}(\alpha = 1, 2, 3, 4)$. As $\underset{2}{\xi}^{i}$ cannot define the same trajectories as $\underset{1}{\xi}^{\alpha}$, at least one of the components $\underset{2}{\xi}^{i}$ $(i = 1, 2, 3)$ does not vanish identically in some domain A. Thus (55.4) can be put into the form

$$\varphi^{\alpha}_{s} = F^{\alpha}_{s}(\varphi^{\alpha}_{k}, \varphi^{\alpha}, x), \quad k \neq s.$$

If, moreover, we assume that the right-hand sides are analytic functions in A, these equations constitute a Cauchy–Kowalewski set in the domain A.

Thus it is always possible to introduce a coordinate system in A such that

$$\underset{1}{\xi}^{\alpha} = \delta^{\alpha}_{4}, \quad \underset{2}{\xi}^{\alpha} = x^2\delta^{\alpha}_{1} - x^1\delta^{\alpha}_{2}. \tag{55.5}$$

The allowed transformations are now clearly

$$x^{1'} = f\cos\left[\tan^{-1}\frac{x_1}{x_2} + \varphi\right], \quad x^{2'} = f\sin\left[\tan^{-1}\frac{x_1}{x_2} + \varphi\right],$$

$$x^{3'} = \varphi^3, \quad x^{4'} = \varphi^4, \tag{55.6}$$

where f, φ, φ^3, φ^4 are arbitrary functions of x^3 and $\varrho \equiv (x^{1^2}+x^{2^2})^{1/2}$. Integrating the the Killing equations for the vector $\underset{1}{\xi}^\alpha$, we find that $g_{\alpha\beta} = g_{\alpha\beta}(x^1, x^2, x^3)$; the set of Killing equations for the vector $\underset{2}{\xi}^\alpha$ reduce to

$$x^2\partial_1 g_{33}-x^1\partial_2 g_{33} = x^2\partial_1 g_{34}-x^1\partial_2 g_{34} = x^2\partial_1 g_{44}-x^1\partial_2 g_{44} = 0,$$
$$x^2\,\partial_1 g_{13}-x^1\,\partial_2 g_{13}-g_{23} = x^2\,\partial_1 g_{23}-x^1\,\partial_2 g_{23}+g_{13} = 0,$$
$$x^2\,\partial_1 g_{14}-x^1\,\partial_2 g_{14}-g_{24} = x^2\,\partial_1 g_{24}-x^1\,\partial_2 g_{24}+g_{14}$$
$$= x^2\,\partial_1 g_{11}-x^1\,\partial_2 g_{11}-2g_{12} = x^2\,\partial_1 g_{22}-x^1\,\partial_2 g_{22}+2g_{12}$$
$$= x^2\,\partial_1 g_{12}-x^1\,\partial_2 g_{12}+g_{11}-g_{22} = 0.$$

Hence the components g_{33}, g_{34}, g_{44} are functions of x^3 and ϱ, and

$$g_{13} = A\cos\varphi, \quad g_{23} = A\sin\varphi,$$
$$A = A(\varrho, x^3), \quad \varphi = \tan^{-1}\frac{x_1}{x_2}+B(\varrho, x^3),$$
$$g_{14} = C\cos\omega, \quad g_{24} = C\sin\omega,$$
$$C = C(\varrho, x^3), \quad \omega = \tan^{-1}\frac{x_1}{x_2}+D(\varrho, x^3),$$
$$g_{11} = e^{-\lambda}\left[-1+(1-e^{-\nu})\frac{x^{1^2}}{\varrho^2}\right],$$
$$g_{12} = e^{-\lambda}(1-e^{-\nu})\frac{x^1x^2}{\varrho^2},$$
$$g_{22} = e^{-\lambda}\left[-1+(1-e^{-\nu})\frac{x^{2^2}}{\varrho^2}\right],$$

where A, B, C, D, λ, ν are functions of x^3 and ϱ. As the functions f, φ, φ^3, φ^4 in the allowed transformations (55.6) are arbitrary, we can further specify the coordinate system so that

$$g_{1'3'} = g_{2'3'} = g_{3'4'} = 0,$$
$$g_{4'4'}e^{-\lambda} = 1,$$

where $\overset{*}{\lambda}$ is the value the first factor in $g_{1'1'}$, takes in the new coordinate system. This exhausts all the possible allowed transformations; bearing in mind the required signature, the metric must finally take the form

$$\left.\begin{aligned} g_{pq} &= e^{-\mu}\left[-\delta_{pq}+(1-e^{-\nu})\frac{x^px^q}{\varrho^2}\right] \quad (p, q = 1, 2), \\ g_{31} &= g_{32} = g_{34} = 0, \quad g_{33} = -e^{\lambda-\mu}, \quad g_{44} = e^{\mu}, \end{aligned}\right\} \tag{55.7}$$

where λ, μ, ν are functions of the two variables ϱ and x^3 only. By imposing various additional conditions, we can obtain the different solutions with axial symmetry. The choice of ϱ and x^3 as independent variables has an obvious physical interpretation.

We now consider in more detail the Weyl and Levi-Civita solution obtained from (55.7) by putting $\lambda = \nu$ and requiring that the solution be *static* and not simply *constant*. A solution is static provided that all the metric tensor components in some system of coordinates are independent of $x^4 = t$, and that the metric is invariant under time reversal:

$$x^4 \to -x^4.$$

Under these conditions all the components $g_{4i} = 0$. We thus obtain

$$\left.\begin{aligned} g_{pq} &= e^{-\mu}\left[-\delta_{pq}+(1-e^{-\nu})\frac{x^p x^q}{\varrho^2}\right],\\ g_{13} &= g_{23} = g_{34} = g_{14} = g_{24} = 0,\\ g_{33} &= -e^{\nu-\mu}, \quad g_{44} = e^{\mu}, \end{aligned}\right\} \tag{55.8}$$

where

$$\varrho = \sqrt{x^{1^2}+x^{2^2}},$$

and μ and ν are functions of the variables ϱ and x^3. The Einstein tensor

$$G_{\alpha\beta} \equiv R_{\alpha\beta}-\tfrac{1}{2}Rg_{\alpha\beta}$$

has components

$$\left.\begin{aligned} G_{34} &= 0, \quad G_{4p} = 0,\\ G_{3p} &= \frac{1}{2}\left(-\frac{1}{\varrho}\frac{\partial\nu}{\partial x^3}+\frac{\partial\mu}{\partial\varrho}\frac{\partial\mu}{\partial x^3}\right)\frac{x^p}{\varrho},\\ G_{33} &= \frac{1}{2\varrho}\frac{\partial\nu}{\partial\varrho}-\frac{1}{4}\left[\left(\frac{\partial\mu}{\partial\varrho}\right)^2-\left(\frac{\partial\mu}{\partial x^3}\right)^2\right],\\ G_{pq} &= \left\{-\frac{1}{2\varrho}\frac{\partial\nu}{\partial\varrho}+\frac{1}{4}\left[\left(\frac{\partial\mu}{\partial\varrho}\right)^2-\left(\frac{\partial\mu}{\partial x^3}\right)^2\right]\right\}\frac{x^p x^q}{\varrho^2}\\ &\quad+\left\{-\frac{1}{2}\left(\frac{\partial^2\nu}{\partial\varrho^2}+\frac{\partial^2\nu}{\partial x^{3^2}}\right)\right.\\ &\quad\left.-\frac{1}{4}\left[\left(\frac{\partial\mu}{\partial\varrho}\right)^2+\left(\frac{\partial\mu}{\partial x^3}\right)^2\right]\right\}e^{-\nu}\left(\delta_{pq}-\frac{x^p x^q}{\varrho^2}\right),\\ G_{44} &= e^{2\mu-\nu}\left\{-\left(\frac{\partial^2\mu}{\partial\varrho^2}+\frac{1}{\varrho}\frac{\partial\mu}{\partial\varrho}+\frac{\partial^2\mu}{\partial x^{3^2}}\right)\right.\\ &\quad\left.+\frac{1}{2}\left(\frac{\partial^2\nu}{\partial\varrho^2}+\frac{\partial^2\nu}{\partial x^{3^2}}\right)+\frac{1}{4}\left[\left(\frac{\partial\mu}{\partial\varrho}\right)^2+\left(\frac{\partial\mu}{\partial x^3}\right)^2\right]\right\}. \end{aligned}\right\} \tag{55.9}$$

Using these expressions and introducing the components of the energy-momentum tensor, we obtain the general field equations. In empty space we clearly have

$$\left.\begin{aligned}
&\omega_1 \equiv \frac{\partial^2\mu}{\partial\varrho^2}+\frac{1}{\varrho}\frac{\partial\mu}{\partial\varrho}+\frac{\partial^2\mu}{\partial x^{3^2}} = 0, \quad \omega_3 \equiv \frac{\partial\nu}{\partial x^3}-\varrho\frac{\partial\mu}{\partial\varrho}\frac{\partial\mu}{\partial x^3} = 0,\\
&\omega_2 \equiv \frac{\partial\nu}{\partial\varrho}-\frac{\varrho}{2}\left[\left(\frac{\partial\mu}{\partial\varrho}\right)^2-\left(\frac{\partial\mu}{\partial x^3}\right)^2\right] = 0,\\
&\omega_4 \equiv \frac{\partial^2\nu}{\partial\varrho^2}+\frac{\partial^2\nu}{\partial x^{3^2}}+\frac{1}{2}\left[\left(\frac{\partial\mu}{\partial\varrho}\right)^2+\left(\frac{\partial\mu}{\partial x^3}\right)^2\right] = 0.
\end{aligned}\right\} \tag{55.10}$$

Two of these equations are contracted Bianchi identities; moreover, as can easily be verified,

$$\omega_4 = \frac{\partial\omega_2}{\partial\varrho}+\frac{\partial\omega_3}{\partial x^3}+\varrho\frac{\partial\mu}{\partial\varrho}\omega_1, \tag{55.11}$$

$$\frac{\partial\omega_2}{\partial x^3}-\frac{\partial\omega_3}{\partial\varrho} = \varrho\frac{\partial\mu}{\partial x^3}\omega_1. \tag{55.12}$$

As usual in the analysis of solutions of the field equations, the conditions imposed on the field potentials at infinity must be known. The choice of the conditions at infinity is not, in general, as obvious as is sometimes supposed (cf. § 58). Following Weyl and Levi-Civita, we assume that μ and ν vanish at infinity.

The first equation of (55.8) shows that the g_{pq} have a singularity of the indeterminate type on the x^3 axis, unless $\lim\limits_{\varrho\to 0} \nu = 0$.

The first equation of (55.10) is linear and homogeneous in μ; it is Laplace's equation in cylindrical coordinates for axially symmetric functions. The solutions of this equation, neglecting the case $\mu = 0$, satisfy the boundary conditions at infinity only if there are singularities at finite points. Singularities off the x^3 axis always occur on circles, but reduce to points on the axis itself, corresponding to the form $\sum\limits_i [\varrho^2+(x^3-a_i)^2]^{-1/2}$ or derivatives of such poles with respect to x^3.

Naturally, these solutions must also satisfy the equations $\omega_2 = \omega_3 = 0$ for the function ν. Now if the equation $\omega_1 = 0$ is satisfied in some simply connected domain of the (ϱ, x^3) space $(\varrho \geqslant 0)$, then (55.12) shows that the equations $\omega_2 = \omega_3 = 0$ have solutions, but if singularities exist, the space (ϱ, x^3) is no longer simply connected. Hence the possible singularities must be considered ([152], pp. 208–10).

For a singularity off the x^3 axis, a solution μ of the equation $\omega_1 = 0$ can be chosen with an arbitrary circular singularity, and the integral along a closed contour enclosing this singularity in the (ϱ, x^3) plane

$$\oint \left(\frac{\partial \nu}{\partial \varrho} d\varrho + \frac{\partial \nu}{\partial x^3} dx^3\right) = \oint \left\{\frac{\varrho}{2}\left[\left(\frac{\partial \mu}{\partial \varrho}\right)^2 - \left(\frac{\partial \mu}{\partial x^3}\right)^2\right] d\varrho + \varrho \frac{\partial \mu}{\partial \varrho}\frac{\partial \mu}{\partial x^3} dx^3\right\}, \tag{55.13}$$

does not vanish in general. But if this integral does not vanish, ν is not defined uniquely outside the singularity. Hence a necessary condition for the existence of a solution is that this integral vanishes.

Suppose now that there is a point singularity on the x^3 axis; $\partial \nu/\partial x^3$ is zero on this axis except at the singularity, so that if ν is zero at one point it must be zero along the whole x^3 axis *up to the singularity*.

At the singularity, the integral of the differential

$$d\nu = \frac{\varrho}{2}\left[\left(\frac{\partial \mu}{\partial \varrho}\right)^2 - \left(\frac{\partial \mu}{\partial x^3}\right)^2\right] d\varrho + \varrho \frac{\partial \mu}{\partial \varrho}\frac{\partial \mu}{\partial x^3} dx^3 \tag{55.14}$$

along a small semicircle around the singularity with diameter on the x^3 axis must vanish.

Consider the following typical singularity:

$$\mathring{\mu} = (\varrho^2 + x^{3^2})^{-1/2}.$$

As

$$\frac{\partial \mathring{\mu}}{\partial \varrho} = -\frac{\varrho}{r^3}, \quad \frac{\partial \mathring{\mu}}{\partial x^3} = -\frac{x^3}{r^3}, \quad r = \sqrt{\varrho^2 + x^{3^2}},$$

the differential (55.14) reduces to

$$d\nu = \left(\frac{\varrho}{2}\frac{\varrho^2 - x^{3^2}}{r^6} d\varrho + \frac{\varrho^2 x^3}{r^6} dx^3\right). \tag{55.15}$$

For points on the semicircular contour

$$\varrho = r\cos\varphi, \quad x^3 = r\sin\varphi, \quad r = \text{const},$$
$$d\varrho = -x^3\, d\varphi, \quad dx^3 = \varrho\, d\varphi.$$

From (55.15) we obtain

$$d\nu = \frac{1}{2r^2}\cos\varphi \sin\varphi\, d\varphi, \quad r = \text{const},$$

and integrating between the limits $(-\pi/2, \pi/2)$, we obtain

$$[\nu]_{\varphi=-\pi/2}^{\varphi=\pi/2} = \frac{1}{2r^2}\int_{-\pi/2}^{\pi/2} \cos\varphi \sin\varphi \, d\varphi = \frac{1}{4r^2}[\sin^2\varphi]_{-\pi/2}^{\pi/2} = 0. \quad (55.16)$$

The solution $\overset{*}{\mu}$ is compatible with the conditions of regularity for ν.

A solution with two singularities is of particular interest. The solution at one singular point can be expanded in a power series in ϱ^2 and x^3, allowing for the singularity at the other point; we assume that near the origin the solution μ has the form

$$\overset{**}{\mu} = \frac{a}{r} + \sum_{m,n} a_{m,n}\varrho^{2m}x^{3n}. \quad (55.17)$$

In this expansion, we are interested only in those $a_{m,n}$ contributing to the integral along the semicircle. If the semicircle does not enclose any other singularity besides $r = 0$, the integral cannot depend on r. Thus all the coefficients $a_{m,n}$ which would make the integral depend on r can be omitted. Moreover, the regular part of $\overset{**}{\mu}$ by itself cannot give a non-vanishing integral, so that we need only consider the "cross" products of the singular and regular parts of $\overset{**}{\mu}$. The derivatives of the singular part of $\overset{**}{\mu}$ for a given φ fall off as r^{-2}. They are multiplied by ϱ, and by the coordinate differentials which both increase as r^1, and by derivatives of the regular part of $\overset{**}{\mu}$. The only terms of interest are those depending on φ but not on r; the only such term is $\varrho^0 x^3$. We can therefore take

$$\overset{**}{\mu} = \frac{a}{r} + bx^3 + \dots$$

The required differential is thus

$$d\nu = -\varrho\frac{\partial}{\partial x^3}\left(\frac{a}{r}\right)\frac{\partial}{\partial x^3}(bx^3)\,d\varrho + \varrho\frac{\partial}{\partial\varrho}\left(\frac{a}{r}\right)\frac{\partial}{\partial x^3}(bx^3)\,dx^3 + \dots,$$

and only these terms can contribute to the integral. Hence

$$d\nu = \frac{ab}{r^3}\varrho(x^3\,d\varrho - \varrho\,dx^3) = -ab\cos\varphi\,d\varphi = -ab\,d(\sin\varphi),$$

and the integral between the limits $(-\pi/2, \pi/2)$ does not vanish. Hence in the neighbourhood of one of the singular points, the derivative with respect to x^3 of the regular part of μ must vanish. In other words, *it is impossible* *to have more than one singular point along the x^3 axis.* Physically, this means that the field equations exclude motions of point masses incompatible with the equations of motion ([152], chapter XV).

Example

1. Show that if a solution with axial symmetry admits an Abelian G_3 group of motions, the linear element can be represented in the form

$$ds^2 = e^{2\alpha}\,dt^2 - r^2 e^{-2\alpha}\,d\varphi^2 - e^{2\beta-2\alpha}\,(dr^2+dz^2),$$

where $\alpha = c \ln r + D$, $\beta = c^2 \ln r + \varepsilon$, D, c, ε are constants ([296], pp. 8–9).

§ 56. Harmonic Gravitational Fields

If the x^α denote Cartesian coordinates in an ordinary Euclidean space R_n, the distance of any point from the origin is given by

$$s = \sqrt{g_{\alpha\beta}x^\alpha x^\beta}. \tag{56.1}$$

Differentiating, we obtain

$$(s^p)_{,\alpha} = ps^{p-2}g_{\alpha\beta}x^\beta \tag{56.2}$$

and

$$(s^p)_{,\alpha}{}^{\alpha} \equiv g^{\alpha\beta}(s^p)_{,\alpha\beta} = p(n+p-2)s^{p-2}. \tag{56.3}$$

Hence $s_{,\alpha}{}^{\alpha}$ is a function of s only, and $s^{2-n}\,(n \neq 2)$ is a solution of the Laplace equation

$$v_{,\alpha}{}^{\alpha} = 0. \tag{56.4}$$

In general, such solutions are not valid in an arbitrary V_n. We can therefore distinguish as a special class *those* V_n *which admit harmonic functions depending only on distance*; such spaces are called *harmonic* (or *completely harmonic*). We can also distinguish V_n *harmonic at a given point* $\overset{0}{x}{}^\alpha$; for these the distance from any point to $\overset{0}{x}_\alpha$ satisfies the equation

$$s_{,\alpha}{}^{\alpha} = f(s). \tag{56.5}$$

The special importance of harmonic spaces in the theory of gravitation is explained by the fact that the classical theory of gravitation and electrostatics is based on the laws of Newton and Coulomb, both of which presuppose the possibility of using harmonic functions depending only on distance.

Shirokov was the first to study this class of V_n's and showed in his review of potential theory in V_n, that: (1) in order that a V_n should admit a harmonic function depending only on distance from any point, it is necessary and sufficient that every hypersphere in the space has a constant mean curvature; (2) V_n's in which every hypersphere has a constant mean curvature are Einstein spaces ($R_{\alpha\beta} = \varkappa g_{\alpha\beta}$) ([72], [75], [108]). A hypersphere is a hypersurface with all its points equidistant from a given point – the centre; this is the analogue of a circle on a surface in the Bianchi sense in V_n—a geometrical form applicable to any V_n.

Thus for $n = 3$, harmonic functions only exist in spaces of constant curvature. When $n > 3$ the problem depends on the metric signature of the V_n. The concept of V_n spaces harmonic at a point was introduced by Ruse and Copson [123]; the affine generalization is considered in [185], [186]. Important results concerning harmonic V_n's, V_n's harmonic at a given point, and the connections between these two forms were obtained by Lichnerowicz [139], [197], Walker [128], [141], [142], [145] and others [217], [178], [198].

Because of its physical importance, we give a detailed analysis of four-dimensional spaces with metric signature $(- - - +)$.

We consider a point in a V_n and introduce a corresponding locally geodesic coordinate system. Using the notation

$$\frac{\partial^{p-1}\Gamma^i_{jk_1}}{\partial x^{k_2}\dots\partial x^{k_p}} = \Gamma^i_{jk_1,\,k_2\dots k_p},$$

we obtain, as in § 7,

$$\{\Gamma^i_{j(k,\,lpq)}\}_0 = -\tfrac{3}{5}\{\tfrac{2}{9}R^{ai}{}_{(kl}R_{aq\,|\,j\,|\,p)} + R^i{}_{(pq\,|\,j\,|\,kl)}\}_0. \tag{56.6}$$

We now assume that the V_n admits a harmonic function and is therefore an Einstein manifold. Contracting (56.6) over i, j and noting that $R_{\alpha\beta} = \varkappa g_{\alpha\beta}$, we obtain

$$\{R_{a(k\,|\,b\,|\,l}R^{ab}{}_{qp)}\}_0 = -\tfrac{15}{2}\{\Gamma^a_{a(k,\,lpq)}\}_0. \tag{56.7}$$

Consider the right-hand side of this equation; we first note that (cf. example 1, § 56) the mean curvature of any Bianchi hypersphere must be constant and equal to

$$H = \frac{n-1}{s} + \frac{\partial \ln \sqrt{g}}{\partial x^k}\xi^k = \frac{n-1}{s} + \frac{\partial \ln \sqrt{g}}{\partial s}, \tag{56.8}$$

in the given coordinate system [72]. The required functions $\varphi(s)$ must have the form

$$\varphi = a\int e^{-\int H\,ds}\,ds+C_2. \tag{56.9}$$

Equations (56.8) and (56.9) show that

$$\varphi = C_1\int\frac{ds}{s^{n-1}\sqrt{g}}+C_2,$$

and hence the discriminant of the metric tensor must be a function of s, i.e. $g \equiv |g_{\alpha\beta}| = g(s)$. Thus

$$\Gamma^{\sigma}_{\sigma\alpha} = \tfrac{1}{2}\partial_\alpha \ln g(s)$$

depends only on the distance from the origin. We now differentiate

$$s^2 = g_{\alpha\beta}x^\alpha x^\beta$$

four times and equate corresponding terms in powers of s, noting that $x^\alpha = \xi^\alpha s$, where ξ^α is the unit vector along a geodesic through the origin. As

$$\Gamma^{\sigma}_{\sigma\alpha} \equiv \Gamma_\alpha = \theta_\alpha(s)s_\alpha, \quad s_\alpha = \frac{\partial s}{\partial x^\alpha},$$

$$-\tfrac{15}{2}\{\Gamma^{\sigma}{}_{\sigma\,(k,\,lpq)}\}_0 = \{\lambda g_{(kl}g_{pq)}\}_0$$

and hence

$$R_{a(k\,|\,b\,|\,l}R^{ab}{}_{qp)} = \lambda g_{(kl}g_{pq)} \tag{56.10}$$

at any point of the domain in V_n which admits a geodesic system of coordinates. This condition was first pointed out by Shirokov [72]. Contracting, we find that

$$\lambda = \varkappa^2. \tag{56.11}$$

We now put $n = 4$, and ignoring the problem of different signatures for the moment, we introduce an orthonormal tetrad (imaginary in the case of the signature $---+$) at the given point relative to which

$$g_{\alpha\alpha} = 1, \quad g_{\alpha\beta} = 0 \quad (\alpha \neq \beta). \tag{56.12}$$

In the bivector space corresponding to this tetrad, we have as in § 15,

$$R_{\alpha\beta\gamma\delta} \to R_{ab}, \quad g_{\alpha\gamma}g_{\beta\delta}-g_{\alpha\delta}g_{\beta\gamma} \to g_{ab} \qquad (a, b = 1, \ldots, 6).$$

Then

$$g_{aa} = 1, \quad g_{ab} = 0 \quad (a \neq b)$$

and

$$(R_{ab}) = \begin{pmatrix} M & N \\ N & M \end{pmatrix},$$

where $M \equiv (m_{ij})$, $N \equiv (n_{ij})$ are symmetric 3×3 matrices with *traces*

$$\sum_1^3 m_{ii} = \varkappa, \quad \sum_1^3 n_{ii} = 0,$$

and the bivector indices are numbered as follows: 14→1, 24→2, 34→3, 23→4, 31→5, 12→6. When setting down conditions (56.10) for the bivector space, we must consider the following combinations of the indices k, l, p, q,

1. $k \neq l \neq p \neq q$
2. $k = l \neq p \neq q,$
3. $k = l \neq p = q,$
4. $k = l = p \neq q,$
5. $k = l = p = q,$

thus obtaining five corresponding relations between the m_{ij} and n_{ij}:

$$m_{\alpha\alpha}(n_{\beta\beta}-n_{\gamma\gamma})+m_{\beta\beta}(n_{\gamma\gamma}-n_{\alpha\alpha})+m_{\gamma\gamma}(n_{\alpha\alpha}-n_{\beta\beta}) = 0,$$

$$m_{\alpha\gamma}m_{\beta\gamma}+m_{\alpha\beta}\left(m_{\gamma\gamma}-\frac{\varkappa}{3}\right) = n_{\alpha\gamma}n_{\beta\gamma}+n_{\alpha\beta}n_{\gamma\gamma},$$

$$m_{\gamma\gamma}^2+n_{\gamma\gamma}^2+4(m_{\alpha\beta}^2-n_{\alpha\alpha}n_{\beta\beta})+2(m_{\alpha\alpha}m_{\beta\beta}-n_{\alpha\beta}^2) + (m_{\alpha\gamma}^2+m_{\beta\gamma}^2)+(n_{\alpha\gamma}^2+n_{\beta\gamma}^2) = \varkappa^2,$$

$$m_{\alpha\beta}(m_{\alpha\alpha}-m_{\beta\beta})+m_{\alpha\gamma}m_{\beta\gamma} = n_{\alpha\beta}(n_{\alpha\alpha}-n_{\beta\beta})+n_{\alpha\gamma}n_{\beta\gamma},$$

$$m_{\alpha\alpha}^2+m_{\beta\beta}^2+m_{\gamma\gamma}^2+2(n_{\beta\gamma}^2+n_{\alpha\beta}^2+n_{\alpha\gamma}^2) = \varkappa^2,$$

where $\alpha, \beta, \gamma = 1, 2, 3$, and $\alpha \neq \beta \neq \gamma$.

Examining this algebraic set under all possible conditions shows that

$$\left.\begin{aligned} &n_{\alpha\alpha} = e\left(m_{\alpha\alpha}-\frac{\varkappa}{3}\right), \quad n_{\alpha\beta} = em_{\alpha\beta} \\ &(\alpha, \beta = 1, 2, 3) \quad (\alpha \neq \beta), \quad e = \pm 1. \end{aligned}\right\} \tag{56.13}$$

Thus the λ-matrix $(R_{ab}-\lambda g_{ab})$ $(a, b = 1, \ldots, 6)$ of the bivector space is

$$(R_{ab}-\lambda g_{ab}) = \begin{pmatrix} M & N \\ N & M \end{pmatrix}, \tag{56.14}$$

where

$$M = (m_{\alpha\beta}-\lambda\delta_{\alpha\beta}), \quad N = \left(e\left(m_{\alpha\beta}-\frac{\varkappa}{3}\,\delta_{\alpha\beta}\right)\right). \tag{56.15}$$

The whole discussion has so far been independent of the metric signature of the V_4 space over the field of complex numbers. If we now insist that the gravitational field be real, the metric at all points in the V_4 must reduce to Minkowski's form

$$g_{11} = g_{22} = g_{33} = -g_{44} = -1, \quad g_{\alpha\beta} = 0 \quad (\alpha \neq \beta). \tag{56.16}$$

The transition from (56.12) to (56.16) must be made by the non-degenerate transformation defined by the matrix

$$\left(\frac{\partial x^{\alpha}}{\partial x^{\beta'}}\right)_0 = \begin{pmatrix} i & 0 & 0 & 0 \\ 0 & i & 0 & 0 \\ 0 & 0 & i & 0 \\ 0 & 0 & 0 & 1 \end{pmatrix}.$$

But clearly under this transformation, all the $n_{\alpha\beta}$ are *purely imaginary* and the $m_{\alpha\beta}$ all real. This leads to a contradiction unless $m_{\alpha\beta} = 0$, $m_{\alpha\alpha} = \varkappa/3$, $(\alpha \neq \beta;\ \alpha, \beta = 1, 2, 3)$. Hence a harmonic space V_4 with metric signature $(- - - +)$ corresponds to a λ-matrix of the simple type in the bivector space, with identical elementary divisors and Segre characteristic [(111, 111)]. As

$$R_{ab} = \frac{\varkappa}{3}\, g_{ab} \qquad (a, b = 1, \ldots, 6),$$

in this case,

$$R_{\alpha\beta\gamma\delta} = \frac{\varkappa}{3}\,(g_{\alpha\gamma}g_{\beta\delta} - g_{\alpha\delta}g_{\beta\gamma}),$$

i.e. the V_4 is a *space of constant curvature* S_4. This result can also be obtained from the theorem that any V_n is a conformally flat manifold, if $\overset{0}{R}_{\alpha\beta\gamma\delta} = 0$, $\alpha, \beta, \gamma, \delta \neq 0$ for every orthonormal n-ad in the V_n ([77], § 37). Every conformally flat Einstein space is an S_n. Thus *in the general theory of relativity, a real gravitational field admits harmonic functions dependent on distance, only if the field defines a geometry corresponding to a space with constant curvature.*

A theory of gravitation and electrostatics based on Newton's and Coulomb's laws can only be developed for spaces with constant curvature; in general different laws are required for the geometries defined in the general theory of relativity.

Examples

1. Prove eqns. (56.8) and (56.9) (Shirokov [72]).

2. Show that a harmonic V_n is an S_n if $n = 2, 3$.

3. A harmonic conformally flat V_n is an S_n.

4. A harmonic V_n is an S_n if the metric index is 1 or $n-1$ (i.e. $n-p = 1$ or $n-1$, where p is the number of plus signs in the canonical form of the metric $ds^2 = \sum e_i\, dx^{i^2}$, $e_i = \pm 1$ at the given point in the space ([128], [142a]).

5. Show that the space defined by the metric

$$ds^2 = c(x^2\, dx^1 - x^1\, dx^2)^2 + 2dx^1\, dx^3 + 2dx^2\, dx^4, \quad c = \text{const} \neq 0,$$

is harmonic. Show also that it is an Einstein space ($R_{\alpha\beta} = 0$) with a metric index 2.

6. Every completely harmonic V_{n-1} in an S_n is an S_{n-1} ([217], p. 388).

7. Show that a completely harmonic V_4 with a definite metric tensor is a symmetric V_4 [168a].

8. Show that a V_4 with metric

$$ds^2 = \psi(x, y)\, dx^2 + 2\varphi(x, y)\, dx\, dy + \theta\, (x, y)\, dy^2 + 2dx\, dz + 2dy\, dt$$

is completely harmonic.

§ 57. Spaces Admitting Cylindrical Waves

This section deals with solutions of the field equations in empty space suggested by Einstein and Rosen, in a search for a rigorous cylindrical wave solution. A problem arises in that the approximate solutions found by linearizing the field equations correspond to *plane* gravitational waves, whereas this is not true for the non-linearized equations. Although the problem only leads to a simplified set of field equations which Einstein and Rosen did not solve, they are worth considering as the results obtained in Chapter 4 can be used to indicate whole classes of actual solutions.

It is important to obtain some *invariant* characterization of these spaces, i.e. the type and the group of motions admitted.

Formal considerations led Einstein and Rosen to consider spaces with the metric tensor

$$(g_{\alpha\beta}) = \begin{pmatrix} -A & 0 & 0 & 0 \\ 0 & -B & 0 & 0 \\ 0 & 0 & -C & 0 \\ 0 & 0 & 0 & A \end{pmatrix}, \quad g = -A^2BC; \quad A, B, C > 0, \qquad (57.1)$$

where A, B, C are functions of x^1, x^4 only. By analogy with the theory of electromagnetic waves, this fact served as the basis for the formal construction of the metric. The substitution

$$\alpha = \ln A, \quad \beta = \frac{1}{2}\ln\left(\frac{B}{C}\right), \quad \gamma = \frac{1}{2}\ln(BC)$$

gives five independent field equations [104].

We now use a different method.

Evaluating the curvature tensor components for the metric (57.1) from (5.7), we obtain

$$\left.\begin{aligned}
R_{1414} &= \frac{1}{2}\left(A_{44}-A_{11}-\frac{A_4^2}{A}+\frac{A_1^2}{A}\right),\\
R_{2323} &= \frac{1}{4A}(B_1C_1-B_4C_4),\\
R_{2424} &= \frac{1}{2}\left(B_{44}-\frac{B_4^2}{2B}-\frac{A_1B_1+A_4B_4}{2A}\right),\\
R_{3131} &= \frac{1}{2}\left(C_{11}-\frac{C_1^2}{2C}-\frac{A_1C_1+A_4C_4}{2A}\right),\\
R_{3434} &= \frac{1}{2}\left(C_{44}-\frac{C_4^2}{2C}-\frac{A_1C_1+A_4C_4}{2A}\right),\\
R_{1212} &= \frac{1}{2}\left(B_{11}-\frac{B_1^2}{2B}-\frac{A_1B_1+A_4B_4}{2A}\right),\\
R_{2412} &= \frac{1}{2}\left(-B_{14}+\frac{B_1B_4}{2B}+\frac{A_1B_4+A_4B_1}{2A}\right),\\
R_{3431} &= \frac{1}{2}\left(C_{14}-\frac{C_1C_4}{2C}-\frac{A_1C_4+A_4C_1}{2A}\right).
\end{aligned}\right\} \quad (57.2)$$

All the other components vanish. The subscripts for A, B, C denote partial differentiation, thus

$$A_i \equiv \partial_i A, \quad A_{ij} \equiv \partial_{ij}A, \ \ldots$$

Five of the field equations $R_{\alpha\beta} = 0$ for empty space are satisfied identically, and the remaining five are

$$\left.\begin{aligned}
\frac{1}{A}R_{1414}-\frac{1}{B}R_{1212}-\frac{1}{C}R_{3131} &= 0,\\
\frac{1}{A}(R_{2424}-R_{1212})-\frac{1}{C}R_{2323} &= 0,\\
\frac{1}{A}(R_{3434}-R_{3131})-\frac{1}{B}R_{2323} &= 0,\\
\frac{1}{A}R_{1414}+\frac{1}{B}R_{2424}+\frac{1}{C}R_{3434} &= 0,\\
\frac{1}{B}R_{2412}-\frac{1}{C}R_{3431} &= 0.
\end{aligned}\right\} \tag{57.3}$$

Substituting from (57.2) for the curvature tensor components, we obtain five equations determining the potentials A, B, C, equivalent to the equations of Einstein and Rosen.

The group of motions admitted by the metric (57.1) under conditions (57.3) has two explicit operators, $X = \delta_2^\alpha\partial_\alpha$, $Y = \delta_3^\alpha\partial_\alpha$, defining an Abelian G_2. The Killing equations for this metric admit a solution other than

$$\xi^\alpha = C_1\delta_2^\alpha+C_2\delta_3^\alpha, \quad C_1,\ C_2\text{—const},$$

only when the metric tensor satisfies additional conditions not derivable from the field equations. These are; (1) the metric satisfies the empty space field equations $R_{\alpha\beta} = 0$; (2) the space admits a 4-orthogonal system of coordinates; (3) $g_{11} = -g_{44}$ in such a coordinate system; (4) the space admits an Abelian G_2 acting on non-isotropic two-dimensional surfaces of transitivity with a definite metric; (5) the metric signature is $(---+)$.

The space type is determined by considering the λ-matrix in the bivector space, i.e.

$$(R_{ab}-\lambda g_{ab}) =$$

$$\begin{pmatrix}
R_{1414}+\lambda A^2 & 0 & 0 & 0 & 0 & 0\\
0 & R_{2424}+\lambda AB & 0 & 0 & 0 & R_{2412}\\
0 & 0 & R_{3434}+\lambda AC & 0 & R_{3431} & 0\\
0 & 0 & 0 & R_{2323}-\lambda BC & 0 & 0\\
0 & 0 & R_{3431} & 0 & R_{3131}-\lambda AC & 0\\
0 & R_{2412} & 0 & 0 & 0 & R_{1212}-\lambda AB
\end{pmatrix}.$$

Interchanging columns and rows reduces this matrix to the form

$$\begin{pmatrix} M_1 & 0 & 0 & 0 \\ 0 & M_2 & 0 & 0 \\ 0 & 0 & M_3 & 0 \\ 0 & 0 & 0 & M_4 \end{pmatrix},$$

where M_1 and M_2 are

$$\left(\lambda + \frac{R_{1414}}{A^2}\right), \quad \left(\lambda - \frac{R_{2323}}{BC}\right),$$

respectively, with two linear elementary divisors. The Segre characteristic of the λ-matrix must therefore be either $[111, \overline{111}]$ or $(12, \overline{12})$. Hence the type of space must be either T_1 or T_2. The remaining two submatrices are

$$M_3 = \begin{pmatrix} R_{2424} + \lambda AB & R_{2412} \\ R_{2412} & R_{1212} - \lambda AB \end{pmatrix},$$

$$M_4 = \begin{pmatrix} R_{3434} + \lambda AC & R_{3431} \\ R_{3431} & R_{3131} - \lambda AC \end{pmatrix}.$$

The last equation of (57.3) shows that either both or neither R_{2412} and R_{3431} are zero. If either

$$R_{1212} + R_{2424} \neq \pm 2R_{2412}$$
$$R_{1212} + R_{2424} = \pm 2R_{2412}$$

both submatrices have linear elementary divisors and the space is a T_1. If on the other hand

$$R_{2412} \neq 0,$$

we have two *non-linear* elementary divisors of the form $(\lambda - \lambda_0)^2, (\lambda - \bar{\lambda}_0)^2$, and the space is a T_2. Moreover, it can be shown that

$$\frac{R_{1414}}{A^2} = -\frac{R_{2323}}{BC}; \quad \frac{R_{2424}}{B} = -\frac{R_{3131}}{C}; \quad \frac{R_{1212}}{B} = -\frac{R_{3434}}{C} \tag{57.4}$$

and that the two quadratic equations

$$\left.\begin{aligned} &\lambda^2 + \frac{1}{AB}(R_{2424} - R_{1212})\lambda + \frac{(R_{2412})^2 - R_{2424}R_{1212}}{(AB)^2} = 0, \\ &\lambda^2 + \frac{1}{AC}(R_{3434} - R_{3131})\lambda + \frac{(R_{3431})^2 - R_{3434}R_{3131}}{(AC)^2} = 0, \\ &R_{2412} \neq 0 \end{aligned}\right\} \tag{57.5}$$

have four roots—two repeated complex-conjugate pairs. Conditions (57.4) and (57.5) define all the possible T_2 type spaces satisfying the Einstein–Rosen conditions. Using the classification of T_2 spaces by groups of motion given in § 27, we can derive all the possible integrals of the Einstein–Rosen field equations. One particular integral is the T_2 space of maximum mobility with a G_6 group of motions defined by the metric

$$ds^2 = -dx^{1^2} - \sinh^2 v\, dx^{2^2} - \sin^2 v\, dx^{3^2} + dx^{4^2}, \quad v = x^1 - x^4,$$

and others include spaces satisfying conditions (57.4), (57.5) and admitting a narrower group.

EXAMPLE

1. Show that the Rosen solution of the field equations (cf. § 14) is a particular case of the Einstein–Rosen solution with an Abelian G_2 group of motions and

$$A = e^{2(\gamma-\psi)}, \quad B = x^{1^2} e^{-2\psi}, \quad C = e^{2\psi},$$

where ψ, γ are functions of x^1 and x^4.

§ 58. Spaces and their Boundary Conditions

The problem of the limiting conditions in integrating the gravitational field equations plays a similar role in relativity and classical theories. In the Newtonian theory of gravitation based on action at a distance, the situation is as follows. Consider a number of material bodies acting at some point either inside or outside the bodies (the *interior* or *exterior* problem). This action at the point is described by a potential, the derivatives φ'_x, φ'_y, φ'_z, taken with a negative sign, determining the components of the force. This force depends on the mass density $\varrho(x, y, z)$. In Newtonian theory the potential is determined by Poisson's equation $\nabla^2\varphi = 4\pi\varrho$, or by Laplace's equation $\nabla^2\varphi = 0$ in the exterior problem.

Poisson's equation has the following general solution with a continuous gradient

$$\varphi = \int \frac{\varrho\, dv}{r} + \overset{0}{\varphi},$$

where $\overset{0}{\varphi}$ satisfies the Laplace equation ($\nabla^2\overset{0}{\varphi} = 0$) in infinite space. The *condition at infinity* is important in Newtonian theory so that φ must be a given function at infinity. Thus in classical mechanics $\overset{0}{\varphi}$ is taken as the potential of masses all lying at infinity. For example, $\overset{0}{\varphi}$ = constant is taken as the potential of a homogeneous spherical shell of infinite radius.

In the relativistic theory of gravitation, taken in the wide sense, the situation is essentially the same. The role of the potential φ is taken over by the metric tensor components $g_{\alpha\beta}(x)$; the energy-momentum tensor $T_{\alpha\beta}(x)$ is the analogue of the density ϱ. Instead of Poisson's equation, we now have the general field equations

$$R_{\alpha\beta}-\frac{R}{2}g_{\alpha\beta}=\lambda T_{\alpha\beta},$$

and in regions where the energy-momentum tensor $T_{\alpha\beta}$ vanishes, we have the analogue of Laplace's equation

$$R_{\alpha\beta}=0.$$

In this sense we can say that the $g_{\alpha\beta}(x)$ satisfying these last equations are generalized harmonic functions.

This rather formal analogy is not conclusive but clearly useful in clarifying the problem. As in the classical case, the limiting conditions play as important role in the relativistic theory of gravitation, and must appear in the integration of the field equations [225].

Historically, the problems of boundary conditions was linked primarily with cosmology, and therefore formulated as the problem of the boundary conditions at infinity. Einstein considered the problem more than once and saw only the three following alternatives: (1) the metric *at infinity must tend to that of a flat Minkowski space in a suitable system of coordinates;* (2) *there are no universal boundary conditions but each problem must have its own individual solution;* (3) *an additional cosmological term must be introduced into the general field equations so that the space becomes closed,* thus eliminating the boundary conditions; the universe is closed on itself and is everywhere homogeneous [97].

Hypothesis (1) does not always agree with the relativistic concept of inertia and contradicts certain statistical considerations. Suggestion (2) is no solution to the problem but a refusal to answer it. Case (3) involves a generalization of the field equations not supported by our present knowledge of gravitation.

Although the problem has been much discussed, no definite solution has been obtained. We suggest that the algebric classification of the curvature tensor given in Chapter 3 allows us to formulate a fairly definite and *invariant* solution to the problem.

Hypothesis (1), clearly closely connected with planetary problems usually considered as static, does not apply in general. Let us consider, for example, a T_2 type field. Under certain conditions, this can be regarded as being caused by matter and gravitational waves. According to hypothesis (1) the space defined by the field must tend in the limit to a flat space-time. For this to occur, it is necessary and sufficient that the curvature tensor components of this Riemann space tend to zero for certain values of the argument. But for any T_2 spaces admitting a group of motions, for example, this is easily seen to be impossible for any values of the arguments, if a singular metric is excluded. Thus for the T_2 space of maximum mobility with metric

$$ds^2 = -dx^{1^2} - \sinh^2(x^1 - x^4)\, dx^{2^2} - \sin^2(x^1 - x^4)\, dx^{3^2} + dx^{4^2}$$

one curvature tensor component is $\sinh^2(x^1 - x^4)$, and this tends to zero only when $g \equiv |g_{\alpha\beta}| \to 0$. This point applies even more obviously to T_3 spaces.

One could simply exclude T_2 and T_3 spaces, but such a procedure is hardly scientific unless supported by more weighty arguments. Thus hypothesis (1) is not universal.

At this stage, we note that the boundary conditions must not be formulated for some particular coordinate system, but in an invariant form with an exact physical basis. Moreover, if the boundary conditions for the potentials $g_{\alpha\beta}(x)$ are to describe their behaviour in regions where $T_{\alpha\beta}(x) \to 0$, then in these regions the given field must tend to the field with metric tensor components satisying the empty space equations

$$R_{\alpha\beta} = 0.$$

Moreover, *the type of the space must not change*, for such a change can only be explained physically by the superposition of a new field, and this is excluded by the conditions of the problem.

Finally, as hypothesis (1) agrees well with experimental observations for spaces of the first type, to which all experimentally verified solutions belong, it must be considered valid for T_1 spaces, and taken as a model in formulating the boundary conditions for T_2 and T_3 spaces.

The curvature in T_1 spaces, indicated by the non-zero curvature tensor is, for example, caused by the presence of gravitational masses. When the distance from these masses increases, as on a hypersphere of very large radius for the Schwarzschild metric in polar coordinates, the gravitational effect falls off, the "distortion" of the space decreases, the space becomes more homogeneous and *tends to the T_1 space of maximum mobility* with a G_{10} group of motions, i.e. to a flat space. This is the essence of hypothesis (1). A Lie group of motions has a simple physical interpretation; under all group transformation *the $g_{\alpha\beta}(x)$ transform to the same functions of the new variables*, and thus the *gravitational field is not altered.* It is therefore more natural to approach the problem of boundary conditions from the point of view of groups of motion.

If we assume that T_2 and T_3 spaces owe their origin to the motion of gravitational masses and possibly to the presence of gravitational waves propagated in empty space with the velocity of light, we might argue that it is this second cause which prevents the space from "tending" to a flat manifold as the effect of the masses falls off with the distance. We can, in principle, avoid introducing infinity, for we can imagine some physical phenomenon with its own geometry caused by a "gravitation" unlike that operating between the planets, and occurring only in regions where all other fields can be neglected. This idea was put forward by Lobachevskii in his famous *Introduction* to the book *New Elements of Geometry.*

By analogy with T_1 spaces, it seems most natural to assume that in regions where boundary conditions are important, T_2 and T_3 spaces tend to the corresponding spaces of maximum mobility. We can thus formulate the following *principle for imposing boundary conditions;* if we have a T_i gravitational field ($i = 1, 2, 3$), the metric tensor components $g_{\alpha\beta}(x)$ in regions where $T_{\alpha\beta}(x) \to 0$, differ by arbitrarily small amounts from the corresponding components of the T_i space admitting the maximum possible group of motions.

This formulation is independent of the system of reference and has an *invariant-group* basis. It is not therefore a refusal to find general boundary conditions such as persuaded Einstein against hypothesis (2), but a statement of the fact that only three types of gravitational field exist, *each with its own boundary conditions.*

We can always determine the type of a T_i space and thus the appropriate boundary conditions. Using the results of § 25, we obtain the following scheme:

$T_1 \to$ a flat T_1 space with a G_{10} group of motions,

$T_2 \to$ a T_2 space with a soluble transitive G_6 group of the form (25.8),

$T_3 \to$ a T_3 space with a non-Abelian G_2 group of the form (25.70).

It is obvious that any attempt to impose the wrong type of boundary conditions leads to a contradiction, as there is no physical reason for introducing a singular metric or a change of type.

The principle we have formulated should be important in the study of T_2 and T_3 type spaces (cf. [258], [299]).

In principle, we can approach the problem of bundary conditions for T_2 and T_3 spaces in another way, by considering closed space-time continua, thus eliminating these conditions. This question can only be solved be analysis of the physical significance of these manifolds and, in the final reckoning, by experiment.

REFERENCES

1. B. RIEMANN, Über die Hypothesen, welche der Geometrie zu Grunde liegen, *Abhand. K. Ges. Wiss. Göttingen*, **13,** 133 (1868). Reprinted and edited by H. Weyl, Springer, Berlin (1920).

1a. E. BELTRAMI, Teoria degli spazii di curvatura costante, *Ann. di. Mat.* (Ser. 2) **2,** 232 (1868).

2. E. B. CHRISTOFFEL, Über die Transformation der homogenen Differentialausdrücke zweiten Grades, *Jour. reine und angew. Math. (Crelle)* **70,** 46 (1869).

3. G. RICCI, Sui parametri gli invarianti delle forme quadratiche differenziali, *Ann. di Mat.* (2), **14,** 1 (1886).

4. F. SCHUR, Über den Zusammenhang der Räume constanten Riemann'schen Krümmungmaasses mit den projektiven Räumen. *Math. Ann.* **27,** 537 (1886).

5. S. LIE and F. ENGEL, *Theorie der Transformationsgruppen* Vols. 1 and 2, Teubner, Leipzig (1888) and (1890). Reprinted in 1930.

6. W. KILLING, Über die Grundlagen der Geometrie, *Jour. reine und angew. Math. (Crelle)* **109,** 121 (1892).

7. S. LIE and F. ENGEL, *Theorie der Transformationsgruppen*, Vol. 3, Teubner, Leipzig (1893). Reprinted in 1930.

8. A. P. KOTEL'NIKOV, *Theory of Screws and some Applications to Geometry and Mechanics (Vintovoe schisleniye i nekotorye prilozheniya ego k geometrii i mekhanike)*, Kazan' (1895).

8a. T. LEVI-CIVITA, Sulle trasformazioni della equazioni dinamiche, *Ann. di Mat.* (Ser. 2) **24,** 255 (1896).

9. P. MUTH, *Theorie und Anwendung der Elementartheiler*, Teubner, Leipzig (1899).

10. É. COTTON, Sur les variétés à trois dimensions, *Ann. Fac. Sci. Toulouse* (Ser. 2), **1,** 385, Paris (1899).

11 A. P. KOTEL'NIKOV, *Projective Theory of Vectors (Proyektivnaya teoriya vektorov)*, Kazan' (1899).

12. J. LARMOR, *Aether and Matter*, Cambridge (1900).

13. G. RICCI and T. LEVI-CIVITA, Méthodes du calcul différentiel absolu et leurs applications, *Math. Ann.* **54,** 125 (1901).

14. L. BIANCHI, *Lezioni di geometria differenziale*, vol. 1, Spoerri, Pisa (1902).

15. H. A. LORENTZ, Weiterbildung der Maxwellschen Theorie, Electronentheorie. *Encyklopädie der Math. Wiss.*, Vol. V, Part 2, §§ 64, 65 (1904).

16. G. FUBINI, Sulla teoria degli spazii che ammettono un gruppo conforme, *Atti R. Accad. Sci.* Torino, **38,** 404 (1903).

17. J. HADAMARD, *Leçons sur la propagation des ondes, et les équations de l'hydrodynamique*, Hermann, Paris (1903).

18. G. FUBINI, Sugli spazii che ammettono un gruppo continuo di movimenti, *Ann. di Mat.* (3), **8,** 39 (1903).

18a. G. FUBINI, Sui gruppi di trasformazioni geodetiche, *Memorie R. Acc. Sci.* Torino (Ser. 2), **53,** 261 (1903).

19. H. A. LORENTZ, Electromagnetic phenomena in a system moving with any velocity

smaller than of light, *Proc. Kon. Akad., Wet.* Amsterdam, **6,** 809 (1904). Reprinted in *The Principle of Relativity*, p. 9, Methuen, London (1923).

20. G. Fubini, Sugli spazii a quattro dimensioni che ammettono un gruppo continuo di movimenti, *Ann. di Mat.* (3), **9,** 33 (1904).

21. G. Kowalewski, *Einführung in die Determinantentheorie*, Leipzig (1905).

21a. E. V. Huntingdon, Note on the definitions of abstract groups and fields by sets of independent postulates, *Trans. Amer. Math. Soc.* **6,** 181 (1905).

22. L. E. Dickson, Definitions of a group and a field by independent postulates, *Trans. Amer. Math. Soc.* **6,** 198 (1905).

23. H. Poincaré, Sur la dynamique de l'électron, *Compt. Rend. Acad. Sci.* (Paris), **140,** 1504 (1905).

24. A. Einstein, Zur Elektrodynamik bewegter Körper, *Ann. Physik*, **17,** 891 (1905). English translation in *The Principle of Relativity*, p. 35, Methuen, London (1923).

25. H. Poincaré, Sur la dynamique de l'électron, *Rend. Circolo Mat. Palermo*, **21,** 129 (1906).

26. H. Minkowski, Das Relativitätsprinzip, Lecture before the Math. Ges. Göttingen. (1907), published in *Ann. Physik*, **47,** 927 (1915).

27. L. E. Dickson, Equivalence of Pairs of Bilinear or Quadratic Forms under Rational Transformation, *Trans. Amer. Math Soc.* **10,** 347 (1909).

28. G. Fubini, Sulle rapprezentazioni che conservano le ipersfere *Ann. di Mat.* (3), **16,** 141 (1909).

29. W. von Ignatowsky, Einige allgemeine Bemerkungen zum Relativitätsprinzip, *Phys. Zeit.* **11,** 972 (1910).

30. Ph. Frank and H. Rothe, Über die Transformation der Raumzeitkoordinaten von ruhenden auf bewegte Systeme, *Ann. Physik*, **34,** 825 (1911).

31. W. von Ignatowsky, Eine Bemerkung zum meiner Arbeit: "[29]" *Phys. Zeit.* **12,** 779 (1911).

32. Ph. Frank and H. Rothe, Zur Herleitung der Lorentztransformation, *Phys. Zeit.* **13,** 750 (1912).

33. G. Mie, Grundlagen einer Theorie der Materie, *Ann. Physik*, **37,** 511 (1912); **39,** 1 (1912).

34. A. Einstein, Die formale Grundlage der allgemeinen Relativitätstheorie, *K. Preus. Akad. Wiss. Sitz.* Part 2, 1030 (1914).

35. A. Einstein, Zur allgemeinen Relativitätstheorie, *K. Preus. Akad. Wiss. Sitz.* Part 2, 778, 799 (1915).

35a. A. Einstein, Erklärung der Perihelbewegung des Merkur aus der allgemeinen Relativitätstheorie, *K. Preus. Akad. Wiss. Sitz.* 831 (1915).

36. A. Einstein, Die Grundlage der allgemeinen Relativitätstheorie, *Ann. Physik*, **49,** 769 (1916).

37. K. Schwarzschild, Über das Gravitationsfeld eines Massenpunktes nach der Einsteinschen Theorie, *K. Preus. Akad. Wiss. Sitz.* 189 (1916).

37a. K. Schwarzschild, Über der Gravitationsfeld einer Kugel aus inkompressibler Flüssigkeit nach der Einsteinschen Theorie, *K. Preus. Akad. Wiss. Sitz.* 424 (1916).

38. J. Droste, The field of a single centre in Einstein's theory of Gravitation and the motion of a particle in that field, *Proc. Kon. Akad. Wet.* Amsterdam, **19,** 197 (1916).

39. T. Levi-Civita, Nozione di parallelismo in una varietà qualunque e conseguente specificazione geometrica della curvatura Riemanniana, *Rend. Circ. Mat. di Palermo*, **42,** 173 (1917).

40. F. Klein, Zu Hilberts erster Note über die Grundlagen der Physik, *Göttingen Nachr., Math.-Phys. Klasse*, 469 (1917).

41. D. Hilbert, Die Grundlagen der Physik, 2 Mitteilung, *Göttingen Nachr., Math.-Phys. Klasse*, 395 (1915); 53 (1917).
42. H. Weyl, Zur Gravitationstheorie, *Ann. Physik*, **54**, 117 (1917).
43. A. Einstein, Kosmologische Betrachtungen zur allgemeinen Relativitätstheorie, *K. Preus. Akad. Wiss. Sitz.* 142 (1917). English translation in *The Principle of Relativity*, p. 175, Methuen, London (1923).
44. L. Bianchi, *Lezioni sulla teoria dei gruppi continui finiti di trasformazioni*, Spoerri, Pisa (1918).
45. F. Klein, Über die Differentialgesetze für die Erhaltung von Impuls und Energie in der Einsteinschen Gravitationstheorie, *Göttingen Nachr. Math.-Phys. Klasse*, 171 (1918).
46. F. Kottler, Über die physikalischen Grundlagen der Einsteinschen Gravitationstheorie, *Ann. Physik* (ser. 4), **56**, 401 (1918).
47. T. Levi-Civita, ds^2 Einsteiniani in campi Newtoniani, III, IV, *Atti Accad. Naz. Lincei*, **27**, (2) 183, 220 (1918).
48. J. A. Schouten, Die direkte Analysis zur neueren Relativitätstheorie, *Verh. Kon. Akad. van Weten.* Amsterdam (Ser. 1), **12**, No. 6 (1918).
49. H. Weyl, Reine Infinitesimalgeometrie, *Math. Zeit.* **2**, 384 (1918).
50. T. Levi-Civita, ds^2 Einsteiniani in campi Newtoniani, VIII, IX. *Atti. Accad. Naz. Lincei*, **28** (1), 3, 101 (1919).
51. H. Weyl, Bemerkung über die axialsymmetrischen Lösungen der Einsteisschen Gravitationsgleichungen, *Ann. Physik*, **59**, 185 (1919).
51a. W. Wilson, Space-time Manifolds and corresponding gravitational fields, *Phil. Mag.* (Ser. 6) **40**, 703 (1920).
52. L. P. Eisenhart, Condition that a tensor be the curl of a vector, *Bull. Amer. Math. Soc.* **28**, 425 (1922).
53. E. Goursat, *Leçons sur l'intégration des équations aux dérivées partielles du 1er ordre*, Paris (1921).
54. Th. de Donder, La gravifique einsteinienne, *Ann. de l'Observatoire Roy. de Belgique*, Gauthier-Villars, Paris (1921).
55. J. A. Schouten and D. J. Struik, On some properties of general manifolds relating to Einstein's theory of gravitation, *Amer. Jour. Math.* **43**, 214 (1921).
56. R. Bach, Zur Weylschen Relativitätstheorie und der Weylschen Erweiterung des Krümmungstensorbegriffs, *Math. Zeit.* **9**, 110 (1921).
57. E. Kasner, Geometrical theorems on Einstein's cosmological equations, *Amer. Jour. Math.* **43**, 217 (1921).
58. J. A. Schouten, Über die konforme Abbildung *n*-dimensionaler, Mannigfaltigkeiten mit quadratischer Massbestimmung auf eine Mannigfaltigkeit mit euklidischer Massbestimmung, *Math. Zeit.* **11**, 58 (1921).
59. E. Cartan, Sur une généralisation de la notion de courbure de Riemann et les espaces à torsion, *Compt. Rend. Acad. Sci. (Paris)*, **174**, 593 (1922).
60. E. Fermi, Sopra i fenomeni dei che avvengono in vicinanza di una linea oraria, *Rend. Accad. Lincei (Rome)*, **31** (1), 21, 51 (1922).
61. O. Veblen, Normal coordinates for the geometry of paths, *Proc. Nat. Acad. Sci. U.S.A.* **8**, 192 (1922).
62. R. Bach, Neue Lösungen der Einsteinschen Gravitationsgleichungen, A. Das Feld in der Umgebung eines langsam rotierenden kugelähnlichen Körpers von beliebiger Masse in 1. und 2. Annäherung, *Math. Zeit.* **13**, 119 (1922).
62a. R. Bach and H. Weyl, Neue Lösungen der Einsteinschen Gravitationsgleichungen, B. Explizite Aufstellung statischer axialsymmetrischer Felder, *Math. Zeit.* **13**, 134 (1922).

63. E. CARTAN, Sur les équations de la gravitation d'Einstein, *Jour. Math. Pures et Appl.* (ser. 9), **1**, 141 (1922).
64. V. F. KAGAN, *Fundamentals of the Theory of Determinants (Osnovaniya teorii opredelitelei)*, Ukraine Gos. izdat. Odessa (1922).
65. W. K. CLIFFORD, Preliminary sketch of Biquaternions, *Proc. Lond. Math. Soc.*, **4,** 381 (1873).
66. G. D. BIRKHOFF, *Relativity and Modern Physics*, Harvard Univ. Press, Cambridge, Mass. (1923).
67. O. VEBLEN and T. Y. THOMAS, The geometry of paths, *Trans. Amer. Math. Soc.* **25,** 551 (1923).
68. K. LANCZOS, Ein vereinfachendes Koordinatensystem für die Einsteinschen Gravitationgleichungen, *Phys. Zeit.* **23,** 537 (1922).
69. H. WEYL, *Space, Time, Matter*, Methuen (1922).
70. H. W. BRINKMANN, On Riemann spaces conformal to Einstein spaces, *Proc. Nat. Acad. Sci. U.S.A.* **9,** 172 (1923).
71. H. W. BRINKMANN, Riemann spaces conformal to Einstein spaces, *Math. Ann.* **91,** 269 (1924).
72. P. A. SHIROKOV, On the distinguishing properties of spheres in spaces of constant curvature, *Izv. Kazan. Phys.-Math. Soc.* (Ser. 2), **25** (1925).
73. H. W. BRINKMANN, Einstein spaces which are mapped conformally on each other, *Math. Ann.* **94,** 119 (1925).
74. P. A. SHIROKOV, Constant vector and second order tensor fields in Riemann spaces, *Izv. Kazan Phys.-Math. Soc.* (Ser. 2) **25,** 86 (1925).
75. P. A. SHIROKOV, On functions satisfying the Laplace equation in Riemann three-dimensional spaces and depending only on distance, *Uch. zap. Kazan Gos. Univ.* **85,** 59 (1925).
76. E. KASNER, An algebraic solution of the Einstein equations, *Trans. Amer. Math. Soc.* **27,** 101 (1925).
77. L. P. EISENHART, *Riemannian Geometry*, Princeton Univ. Press (1926).
78. V. HLAVATY, Application des paramètres locaux, *Ann. Soc. Polon. Math.* **5,**19 (1926).
79. V. F. KAGAN, On some systems of numbers arising from Lorentz transformations (in Russian), parts 1 and 2, *Inst. Math. and Mech.*, Moscow State Univ., Moscow (1926).
80. L. P. EISENHART, Non-Riemannian geometry, Colloq. Publ. 8, *Amer. Math. Soc.* (1927).
81. E. CARTAN, La théorie des groupes et la géométrie, *L'enseignement mathématique*, **26,** 200 (1927); also *Oeuvres Complètes*, Part I, Vol. II, p. 841, Gauthier-Villars, Paris (1952).
82. YA. S. DUBNOV, On doubly symmetric orthogonal matrices, *Inst. Math. and Mech.*, Moscow State Univ., Moscow (1927).
83. E. CARTAN, *Leçons sur la géométrie des espaces de Riemann*, Gauthier-Villars, Paris (1928).
84. S. I. VAVILOV, *Experimental Basis of the Theory of Relativity*, G.I.Z. (1928).
85. A. M. LOPSHITS, A vector solution of the problem of doubly symmetric matrices, *Proceedings of the all-Russia Conference of Mathematicians* (1928).
86. O. K. MITTER, On a solution of Einstein's gravitational equations $G_{\mu\nu} = 0$, symmetrical about an axis, *Tohoku Math. Jour.* **34,** 110 (1931).
87. J. A. SCHOUTEN and F. G. VAN DANTZIG, Generelle Feldtheorie, *Zeit. Phys.* **78,** 639 (1932).
88. T. LEWIS, Some special solutions of the equations of axially symmetric gravitational fields, *Proc. Royal Soc.* **136,** 176 (1932).

89. H. W. TURNBULL and A. C. AITKEN, *An Introduction to the Theory of Canonical Matrices*, Blackie, London (1932).
90. M. BÔCHER, *Introduction to Higher Algebra*, MacMillan, New York (1907).
91. P. A. SHIROKOV, *Tensor Calculus Part* 1 *(Tenzornoe ischislenie)* Moscow–Leningrad, Gostekhizdat (1934).
92. T. Y. THOMAS, *The Differential Invariants of Generalized Spaces*, Cambridge Univ. Press (1934).
93. V. HLAVATY, Les courbes de la variété générale à *n* dimensions, *Mém. Sci. Math. Fasc.* **63**, Gauthier-Villars, Paris (1934).
94. J. DELSARTE, Sur les ds^2 d'Einstein à symétrie axiale, *Actualités Scientifiques et Industrielles* No. 157, Hermann, Paris (1934).
94a. J. DELSARTE, Sur les ds^2 binaires et la problème d'Einstein, *Jour. Math. Pures et Appl.* (Liouville) (Ser. 9), **13**, 19 (1934).
95. M. A. LAVRENT'EV and L. A. LYUSTERNIK, *Elements of the Calculus of Variations (Osnovy variatsionnogo ischisleniya)*, vol. 1, part 2, Moscow–Leningrad, United Scientific and Technical Press (1935).
96. J. L. SYNGE, Principal null-directions defined in space-time by an electromagnetic field, Univ. of Toronto Studies, *Appl. Math.* Series No. 1 (1935).
97. See Ref. 43.
98. E. CARTAN, La Géométrie des espaces de Riemann, *Mém. Sci. Math.* IX. Gauthier-Villars, Paris (1925).
99. H. S. RUSE, On the geometry of the electromagnetic field in general relativity, *Proc. London Math. Soc.* (Ser. 2), **41**, 302 (1936).
100. W. J. VAN STOCKUM, The gravitational field of a distribution of particles rotating about an axis of symmetry, *Proc. Roy. Soc. Edinburgh*, **5**, 135 (1936).
101. V. V. STEPANOV, *Kurs differentsial'nykh uravneniĭ*, Moscow–Leningrad, United Scientific and Technical Press (1937). German translation, *Lehrbuch der Differentialgleichungen*, Veb. Deutsch. Ver. Wiss. Berlin (1956).
102. G. A. TIKHOV, Sur la déviation des rayons lumineux dans le champs de gravitation des étoiles, *Comp. Rend. Acad. Sci. U.R.S.S.* **16**, 199 (1937).
103. P. Y. CHOU, Isotropic static solutions of the field equations in Einstein's theory of gravitation, *Amer. Jour. Math.* **59**, 754 (1937).
104. A. EINSTEIN and N. ROSEN, On gravitational waves, *Jour. Franklin Inst.* **223**, 43 (1937).
105. P. A. SHIROKOV, The problem of translations in Riemann spaces, *Izv. Kazan Phys.-Math. Soc.* (Ser. 3) **9** (1937).
106. I. G. PETROVSKII, Über das Cauchysche Problem für Systeme von partiellen Differentialgleichungen, *Mat. Sbor.* (New Ser. 2), **44**, 815 (1937).
107. T. Y. THOMAS, New theorems on Riemann Einstein Spaces, *Mat. sbor.* (New Ser.) **3**, 331 (1938).
108. P. A. SHIROKOV, Symmetric conformal-Euclidean spaces, *Izv. Kazan Phys.-Math. Soc.* (Ser. 3), **11** (1938).
109. A. PIALKOW, Einstein spaces in a space of constant curvature, *Proc. Nat. Acad. Sci. U.S.A.* **24**, 30 (1938).
110. J. A. SCHOUTEN and D. J. STRUIK, *Einführung in die neueren Methoden der Differentialgeometrie*, Vol. 1, Groningen-Batavia, Noordhoff (1935).
111. V. A. FOCK, Sur le mouvement des masses finies d'après la théorie de gravitation Einsteinienne, *J. Physics (Acad. Sci. U.S.S.R)*, **1**, 81 (1939); *Zh. eksp. teor. fiz.* **9**, 375 (1939).
112. F. KLEIN, *Vorlesungen über Höhere Geometrie*, 3rd ed., Springer, Berlin (1926).

113. Y. HAANTJES and W. WRONA, Über konformeuklidische und Einsteinsche Räume gerader Dimension, *Proc. Kon. Ned. Akad. Wet.* **42,** 626 (1939).
114. A. FIALKOW, Totally geodesic Einstein spaces, *Bull. Amer. Math. Soc.* **45,** 423 (1939).
115. J. LEVINE, Groups of motions in conformally flat spaces, II, *Bull. Amer. Math. Soc.* **45,** 766 (1939).
116. H. HÖNL and A. PAPAPETROU, Über die Selbstenergie und das Gravitationsfeld einer elektrischen Punktladung, *Zeit. Phys.* **112,** 65 (1939).
117. H. HÖNL and A. PAPAPETROU, Über die innere Bewegung des Elektrons, I, *Zeit. Phys.* **112,** 512 (1939).
118. A. PAPAPETROU and H. HÖNL, Über die innere Bewegung des Elektrons, II, *Zeit. Phys.* **114,** 478 (1939).
119. A. PAPAPETROU, Drehimpuls- und Schwerpunktsatz in der relativistischen Mechanik, *Praktika Akad. Athenon,* **14,** 540 (1939).
120. N. G. CHEBOTARËV, *The Theory of Lie Groups (Teoriya grupp Li)* Moscow–Leningrad, Gostekhizdat (1940).
121. A. F. BOGORODSKIĬ, *The Nature of the Red Shift in the Spectra of Extra-Galactic Nebulae,* Pulkovo Observatory Circular No. 29, 5 (1940).
122. A. F. BOGORODSKIĬ, *The Kepler Problem in the General Theory of Relativity,* Pulkovo Observatory Circular No. 30, p. 30 (1940).
123. E. T. COPSON and H. S. RUSE, Harmonic Riemannian spaces, *Proc. Roy. Soc. Edinburgh,* **60,** 117 (1940).
124. H. HÖNL and A. PAPAPETROU, Über die innere Bewegung des Elektrons, III, *Zeit. Phys.* **116,** 153 (1940).
125. A. PAPAPETROU, Drehimpuls- und Schwerpunktsatz in der Diracschen Theorie des Elektrons, *Praktika Akad. Athenon,* **15,** 404 (1940).
126. F. R. GANTMAKHER and M. G. KREĬN, *Ostsillyatsionnye matritsy i malye kolebaniya mekhanicheskikh sistem,* Moscow–Leningrad, Gostekhizdat (1941), *Oscillation Matrices and Kernels and Small Oscillations of Mechanical Systems.* English translation of second edition, U.S. Atomic Energy Commission. TR–4481 (1961).
127. H. TAKENO, Equations characterizing various Riemannian spaces treated in cosmologies (in Japanese), *Jour. Sci. Hiroshima Univ.* (ser. A), **12,** 125 (1942).
128. A. G. WALKER, Note on a distance invariant and the calculation of Ruse's invariant, *Proc. Edinburgh Math. Soc.* (2), **7,** 16 (1942).
129. A. FIALKOW, Correction to "totally geodesic Einstein spaces", *Bull. Amer. Math. Soc.* **48,** 167 (1942).
130. S. SASAKI, On a relation between a Riemannian space which is conformal with Einstein spaces and normal conformally connected spaces whose groups of holonomy fix a point or a hypersphere, (in Japanese), *Tensor,* (Old Ser.) **5,** 66 (1942).
131. K. YANO, Conformal and concircular geometries in Einstein spaces, *Proc. Imp. Acad. Tokyo,* **19,** 444 (1943).
132. YUNG-CHOW WONG, Some Einstein spaces with conformally separable fundamental tensors, *Trans. Amer. Math. Soc.* **53,** 157 (1943).
133. YUNG-CHOW WONG, Family of totally umbilical hypersurfaces in an Einstein space, *Ann. Math.* **44,** 271 (1943).
134. A. PAPAPETROU, Les corpuscules à structure multipolaire en relativité restreinte, *Praktika Akad. Athenon,* **18,** 40 (1943).
135. A. PAPAPETROU, La structure intérieure des corpuscules à constitution monobipolaire, *Praktika Akad. Athenon,* **18,** 50 (1943).
136. A. PAPAPETROU, Ondes gravifiques du corpuscule mono-bipolaire, *Praktika. Akad. Athenon,* **18,** 313 (1943).

137. A. Papapetrou, La loi des moments dans un système quelconque de coordonnées, *Praktika Akad. Athenon*, **18**, 317 (1943).
138. A. G. Kurosh, *Group Theory (Teoriya grupp)*, Moscow – Leningrad, Gostekhizdat (1944).
139. A. Lichnerowicz, Sur les espaces Riemanniens complètement harmoniques, *Bull. Soc. Math. France*, **72**, 146 (1944).
140. A. Papapetrou, La Théorie de la gravitation dans la relativité restreinte, *Praktika Akad. Athenon*, **19**, 224 (1944).
141. A. G. Walker, A particular harmonic Riemannian space, *Jour. London Math. Soc.* **20**, 93 (1945).
142. A. G. Walker, On completely harmonic spaces, *Jour. London Math. Soc.* **20**, 159 (1945).
142a. A. Lichnerowicz and A. G. Walker, Sur les espaces harmoniques de type hyperbolique normal, *Comp. Rend. Acad. Sci. (Paris)* **221**, 394 (1945).
143. Yung-Chow Wong, Some theorems on Einstein 4-space, *Duke Math. Jour.* **13**, 601 (1946).
144. V. V. Narlikar and K. R. Karmarkar, On a curious solution of relativistic field equations, *Current Sci.* **15**, 69 (1946).
145. A. G. Walker, Symmetric harmonic spaces, *Jour. London Math. Soc.* **21**, 47 (1946).
146. A. Z. Petrov, One type of Einstein space, *Trudy Kazan. Aviats. Inst.* **7** (1946).
147. L. P. Eisenhart, *Continuous Groups of Transformations*, Princeton Univ. Press (1933).
148. H. Weyl, *The Classical Groups, their Invariants and Representations*, Princeton Univ. Press (1939).
149. W. Pauli, *Relativitätstheorie*, *Encyklopëdie der Math. Wiss.* Vol. 5, Part 2, Section 19 (1921). English translation, *Theory of Relativity*, Pergamon (1958).
150. W. Wrona, Sur les conditions nécessaires et suffisantes pour qu'un espace soit Einsteinien ou conformé-euclidien (cf. the article 'Polish Mathematical Society' in *Usp. Mat. Nauk*, **2**, part 1 (17) (1947).
151. P. K. Rashevskiĭ, *Geometrical Theory of Partial Differential Equations (Geometricheskaya teoriya uravnenii s chastnymi proizvodnymi)* Moscow – Leningrad, Gostekhizdat (1947).
152. P. G. Bergmann, *Introduction to the Theory of Relativity*, Prentice-Hall, New York (1942).
153. V. F. Kagan, *Fundamentals of the Theory of Surfaces in Tensor Presentation (Osnovy teorii poverkhnostei v tensornom izlozhenii)*, part 1, Moscow – Leningrad, Gostekhizdat (1947).
154. A. Papapetrou, A static solution of the equations of the gravitational field for an arbitrary charge distribution, *Proc. Roy. Irish Acad.* **51**, *A*, 191 (1947).
155. A. P. Norden, *Differentsial'naya geometriya*, Moscow, Uchpedgiz (1948). German translation, *Differential geometrie*, Veb. Deutsch. Ver. Wiss. Berlin. Vol. 1 (1956). Vol. II (1957).
156. L. D. Landau and E. M. Lifshitz, *Teoriya Polya*, Moscow – Leningrad, Gostekhizdat (1948); English translation, *The Classical Theory of Fields*, Pergamon (1959).
157. A. Z. Petrov, The curvature of Riemann spaces, *Dokl. Akad. Nauk*, **61**, 211 (1948).
158. B. A. Rozenfel'd, The differential geometry of modes of symmetry, *Dokl. Akad. Nauk*, **59**, 1057 (1948).
159. Mme Fourès and A. Lichnerowicz, Sur un théorème global de réduction des ds^2 statiques généraux d'Einstein, *Comp. Rend. Acad. Sci. (Paris)* **226**, 775 (1948).

160. S. P. Finikov, *The Method of Exterior Cartan Forms in Differential Geometry. (Metod vneshnikh form Kartana v differentsialnoi geometrii)*, Moscow – Leningrad, Gostekhizdat (1948).

161. W. Wrona, Conditions nécessaires et suffisantes qui déterminent les espaces Einsteiniens, conformément euclidiens et de courbure constante, *Ann. Soc. Polon. Math.* **20,** 28 (1948).

162. A. Papapetrou, Einstein's theory of gravitation and flat space, *Proc. Roy. Irish Academy,* **52,** A, 11 (1948).

163. A. Papapetrou, The question of non-singular solutions in the generalized theory of gravitation, *Phys. Rev* **73,** 1105 (1948).

164. N. M. Petrova, The equations of motion and the matter tensor for a system of finite masses in the general theory of relativity, *Zh. eks. tekh. fiz.* **19,** 989 (1949).

165. A. Z. Petrov, The principal axis theorem for a tensor, *Izv. Kazan Phys.-Math. Soc.* (Ser. 3), **14,** 43 (1949).

165a. A. Z. Petrov, The geodesic mapping of Riemann spaces with an indefinite metric, *Uch. zap. Kazan,* **109,** Part 4 (1949).

166. E. Cartan, Sur une classe remarquable d'espace de Riemann, *Bull. Soc. Math. de France,* **54,** 214 (1926); **55,** 114 (1927); see also *Oeuvres Complètes,* Partie 1, Vol. 2, 587, 639, Gauthier-Villars, Paris (1952).

167. A. Einstein and L. Infeld, On the motion of particles in general relativity theory, *Canad. Jour. Math.* **1,** 209 (1949).

168. I. P. Yegorov, Extension of Fubini's theorem on the order of groups of motion of Riemannian spaces, *Dokl. Akad. Nauk,* **66,** 793 (1949).

168a. A. G. Walker, On Lichnerowicz's conjecture for harmonic 4-spaces *Jour. London Math. Soc.* **24,** 21 (1949).

169. B. A. Rozenfel'd and A. A. Abramov, Spaces with affine connection and symmetric spaces, *Usp. Mat. Nauk* (New Ser.) **5,** 72 (1950).

170. A. P. Norden, *Spaces of Affine Connection (Prostranstva affinoi svyaznosti)*, Moscow – Leningrad, Gostekhizdat (1950).

171. A. Z. Petrov, The Simultaneous reduction of a tensor and a bitensor to canonical form, *Uch. zap. Kazan Gos. Univ.* **110,** book 3, (1950).

172. G. B. Gurevich, On some linear transformations of symmetric tensors and polyvectors, *Mat. sbor.* (N.S.) **26,** 463 (1950).

173. A. Lichnerowicz, *Elémentes du calcul tenseuriel,* Paris (1950). English translation, *Elements of Tensor Calculus,* Methuen (1962).

174. G. Y. Rainich, *Mathematics of Relativity,* Wiley, New York (1950).

175. N. H. Kuiper, Einstein spaces and connections, *Proc. Kon. Akad. Wet. Amsterdam,* **53,** I. 1560; II, 1568 (1950).

176. A. Z. Petrov, On spaces defining gravitational fields, *Dokl. Akad. Nauk* **81,** 149 (1951).

177. A. Papapetrou, Equations of motion in general relativity, *Proc. Phys. Soc.* **A 64,** I, 57; II, 302 (1951).

178. A. Z. Petrov, On the existence in a gravitational field of a harmonic function depending only on distance, *Uch. Zap. Kazan. Gos. Univ.* **111,** book 8 (1951).

179. N. H. Kuiper, Sur les propriétés conformes des espaces d'Einstein, Centre Belge de Recherches Math. *Colloque de Géométrie Différentielle,* 165, Louvain (1951), Georges Thone, Liège.

180. A. H. Taub, Empty space-times admitting a three-parameter group of motions, *Ann. Math.* **53,** 472 (1951).

181. A. Z. Petrov, Regular Einstein spaces admitting a transitive group of motions, *Uch. zap. Kazan Gos. Univ.* **112,** book 10, 27 (1952).

182. A. Z. Petrov, Gravitational fields with complex stationary curvatures, *Uch. zap. Kazan Gos. Univ.* **112,** book 10, 35 (1952).
183. A. Z. Petrov, *Gravitational Fields (0 polya tyagoteniya)*, "125 years of the non-Euclidean geometry of Lobachevskii," 179, Moscow–Leningrad, Gostekhizdat (1952).
184. A. P. Shirokov, *The Problem of A-Spaces (K voprosu ob A-prostranstvakh)* "125 years of the non-Euclidean geometry of Lobachevskii," 195, Moscow–Leningrad, Gostekhizdat (1952).
185. H. S. Ruse, Simple harmonic affine spaces of symmetric connection, *Publ. Math. Debrecen*, **2**, 169 (1952).
186. E. M. Patterson, Simply harmonic Riemann extensions, *Jour. London Math. Soc.* **27,** 102 (1952).
187. A. A. Sokolov, Remarks on the quantum theory of a gravitational field, *Vestnik Moscow State Univ.*, Ser. Fiz.-Mat. No. 9, 9 (1952).
188. P. K. Rashevskiĭ, *Riemannian Geometry and Tensor Analysis (Rimanova geometriya i tenzornyi analiz)*, Moscow–Leningrad, Gostekhizdat (1953).
189. F. I. Frankl', Some basic comments on the general theory of relativity, *Usp. Mat. Nauk*, **8,** part 3, (55), 160 (1953).
190. A. D. Aleksandrov, On the meaning of the theory of relativity, *Vestnik Leningrad State Univ.*, Ser. Mat. Fiz. Chim. **8,** 103 (1953).
191. A. D. Aleksandrov and V. V. Ovchinnikov, Remarks on the basis of the theory of relativity, *Vestnik Leningrad State Univ.*, Ser. Mat. Fiz. Chim. **11,** 95, (1953).
192. A. D. Aleksandrov, Some views on the theory of relativity, *Voprosy filosofii*, No. 5 (1953).
193. L. D. Landau and E. M. Lifshitz, *Fluid Mechanics*, Pergamon (1959), English translation of first part of *Mekhanika sploshnykh sred*, Gostekhizdat Moscow (1953).
194. I. G. Petrovskii, *Lectures on Partial Differential Equations*, Interscience Publishers, New York, 1954. English translation of *Lektsii ob uravneniyakh s chastnymi proizvodnymi*, Moscow–Leningrad, Gostekhizdat (1953).
195. F. I. Frankl', Gravitational waves and the motion of gases in strong varying gravitational fields, *Trudy Fiz-Mat. Fak. Kirgiz Gos. Univ.*, part 2, 47 (1953).
196. A. Lichnerowicz, Étude des équations du champ de la théorie unitaire d'Einstein, *Rend. Sem. Mat.-Fis. Milano*, **25,** 121 (1953–4).
197. A. Lichnerowicz, Équations de Laplace et espaces harmoniques, Centre Belge de Recherches Math., *Premier colloque sur les équations aux dérivées partielles*, p. 9, Louvain (1953), Georges Thone, Liège (1954).
198. T. J. Willmore, Mean value theorems in harmonic Riemannian spaces, *Jour. London Math. Soc.* **25,** 54 (1950).
199. B. N. Delone, *Short Exposition of the Proof of the Consistency of Lobachevskii Planimetry (Kratkoe islozhenie dokazatel'stva neprotivorechivosti planimetrii Lobachevskogo)*, Izdat. Akad. Nauk. U.S.S.R., Moscow (1953).
200. G. Vranceanu, Sur les groupes de mouvement d'un espace de Riemann à quatre dimensions, *Acad. Repub. Pop. Romane Stud. Cerc. Mat.* (in Rumanian), **4,** 121 (1953).
201. V. A. Fock, On Frankl's paper "Some Basic Comments on the Theory of Relativity", *Usp. Mat. Nauk*, **9,** part 4 (62), 229 (1954).
202. L. Infeld, Some remarks on the theory of relativity, *Voprosy filosofii.* No. 5 (1954).
203. G. I. Kruchkovich, The classification of three-dimensional Riemannian spaces by groups of motion, *Usp. Mat. Nauk*, **9,** part 1 (59), 3 (1954).

204. P. A. SHIROKOV, Symmetric spaces of the first class, *Uch. zap. Kazan Gos. Univ.* **114,** book 8, 71 (1954).
205. L. S. PONTRYAGIN, *Continuous Groups (Nepreryvnye gruppy)*, 2nd. Ed., Moscow, Gostekhizdat (1954). English translation of lst ed. *Topological Groups*, Princeton Univ. Press (1939).
206. N. ROSEN, Some cylindrical gravitational waves, *Bull Res. Council, Israel*, **3,** 328 (1954).
207. A. Z. PETROV, Classification of spaces defined by gravitational fields, *Uch. zap. Kazan Gos. Univ.* **114,** No. 8, 55 (1954). English translation: Trans. No. 29, Jet Propulsion Lab. California Inst. Tech. Pasadena 1963.
208. A. P. NORDEN, On a complex representation of biplanar space vectors, *Uch. zap. Kazan Gos. Univ.* **114,** book 8, 45 (1954).
209. I. M. KHALATNIKOV, Some problems of relativistic hydrodynamics, *Zh. Eksp. Teor. Fiz.* **27,** 529 (1954).
210. J. R. POUNDER, On relativistically rigid surfaces of revolution, *Comm. Dublin Inst. Adv. Studies* (Ser. A), No. 11 (1954).
211. A. H. TAUB, General relativistic variational principles for perfect fluids, *Phys. Rev.* **94,** 1468 (1954).
212. G. SALZMAN and A. H. TAUB, Born-type rigid motion in relativity, *Phys. Rev.* **95,** 1659 (1954).
213. PHAM MAU QUAN, Mouvements permanents d'un fluide parfait thermodynamique, *Compt. Rend. Acad. Sci. (Paris)* **238,** 324 (1954).
214. E. CLAUSER, Sui fronti d'onda nella teoria unitaria Einsteiniana, *Rend. Inst. Lombardo Sci. Lett.*, Cl. Sci Mat.-Nat. **87,** 473 (1954).
215. H. A. BUCHDAHL, Reciprocal static solutions of the equations $G_{\mu\nu} = 0$, *Quart. Jour. Math.* **5,** 116 (1954).
216. T. TAKASU, Equations of motion of a free particle in the Author's general relativity as a non-holonomic Laguerre geometry realized in the moving three-dimensional Cartesian space, *Proc. Japan Acad.* **30,** 814 (1954).
217. J. A. SCHOUTEN, *Ricci Calculus*, 2nd ed., Springer, Berlin (1954).
218. H. TAKENO, On solutions of electromagnetic equations in non-static spherically symmetric space-times, *Tensor* (New Ser.), **4,** 9 (1954).
219. N. S. SINYUKOV, The geodesic mapping of Riemannian spaces on symmetric Riemannian spaces, *Dokl. Akad. Nauk*, **98,** 21 (1954).
220. A. PAPAPETROU, Eine Theorie des Gravitationsfeldes mit einer Feldfunktion, *Zeit. Phys.* **139,** 518 (1954).
221. A. PAPAPETROU, Eine neue Theorie des Gravitationsfeldes, I, *Math. Nachr.* **12,** 129 (1954).
222. A. PAPAPETROU, Eine neue Theorie des Gravitationsfeldes, II, *Math. Nachr.* **12,** 143 (1954).
223. A. PAPAPETROU and W. URICH, Zur Kohlerschen Formulierung der Gravitationstheorie, *Ann. Physik* (Ser. 6), **14,** 220 (1954).
224. P. K. RASHEVSKIĬ, The theory of spinors, *Usp. mat. nauk*, **10,** part 2, 3 (1955); *Amer. Math. Soc. Translations.* (Ser. 2), **6,** 1 (1957).
225. V. A. FOCK, *Theory of Space, Time and Gravitation*, Pergamon (1959). English translation of *Teoriya prostranstva, vremeni i tyagoteniya*, Moscow, Gostekhizdat (1955).
226. F. I. FRANKL', The correct setting of Cauchy's problem and properties of harmonic coordinates in the general theory of relativity, *Trudy Fiz.-Mat. Fak. Kirgiz. Gos. Univ.*, part 3 (1955).

227. T. Matsumoto, Sur la déduction axiomatique des formules de transformation de Lorentz, *Mem. Coll. Sci. Univ. Kyoto* (Ser. A), Math. **29,** 55 (1955).
228. A. Z. Petrov, Spaces of maximum mobility defined by gravitational fields, *Dokl. Akad. Nauk,* **105,** 905 (1955).
229. B. A. Rosenfeld', *Non-Euclidean Geometries (Neevklidovy geometrii)*, Moscow, Gostekhizdat (1955).
230. I. P. Yegorov, Motions in spaces of affine connectivity (o dvizheniyakh v prostranstvakh affinoi svyaznosti), Doctorate Thesis, Moscow State University (1955).
231. A. Lichnerowicz, *Théories relativistes de la gravitation et de l'électromagnétisme*, Massen & Cie, Paris (1955).
232. E. Sterchi, Su una nuova interpretazione del principio dell' azione potenziale, *Boll. Unione Mat. Ital.* (Ser. III), **10,** 161 (1955).
233. H. Takeno, On conformal transformations in the space-time of relativistic cosmology, *Tensor* (New Ser.), **4,** 141 (1955).
234. G. Kron, A physical interpretation of the Riemann–Christoffel curvature tensor, *Tensor* (New Ser.) **4,** 150 (1955).
235. A. P. Shirokov, A property of covariantly constant affinors, *Dokl. Akad. Nauk,* **102,** 461 (1955).
236. H. Takeno, On groups of conformal transformations in spherically symmetric space-times, *Tensor* (New. Ser.), **5,** 23 (1955).
237. M. Ikeda, On static solutions of Einstein's generalized theory of gravitation, II, *Prog. Th. Phys.* **13,** 265 (1955).
238. H. J. Meister and A. Papapetrou, Die Bewegungsgleichungen in der allgemeinen Relativitätstheorie und die Koordinatenbedingungen, *Bull. Acad. Polon. Sci.*, Cl. III, **3,** 163 (1955).
239. I. P. Yegorov, Riemannian V_4 of non-constant curvature with maximum mobility, *Dokl. Akad. Nauk,* **103,** 9 (1955).
240. A. Papapetrou and W. Urich, Das Pol-Dipol Teilchen im der Gravitationsfeld und elektromagnetischen Feld, *Zeit. Naturf.* **10** A, 109 (1955).
241. S. Bochner, Stationary space-time in general relativity, *Proc. Nat. Acad. Sci.* **41,** 485 (1955).
242. H. Takeno, On the theory of gravitational waves, *Tensor* (New Ser.) **6,** 15 (1956).
243. A. Z. Petrov, The classification of spaces defined by gravitational fields by groups of motion, *Usp. Mat. Nauk,* **11,** Part 4, 181 (1956).
244. A. Z. Petrov, Motions in gravitational fields, *Proceedings of the Third All-Union Mathematical Conference*, Vol. II (1956).
245. V. V. Narlikar, Results of gravitational significance in Riemannian geometry, *Nature,* **177,** 1138 (1956).
246. D. N. Verma and S. R. Roy, Special metric forms and their gravitational significance, *Bull. Calcutta Math. Soc.* **48,** 129 (1956).
247. Mme Y. Fourès-Bruhat, Sur l'intégration des équations de la relativité générale, *Jour. Rat. Mech. and Anal.* **5,** 951 (1956).
248. H. A. Buchdahl, Reciprocal static solutions of the equations of the gravitational field, *Austral. Jour. Phys.* **9,** 13 (1956).
249. K. Just, Strenge Kugelsymmetrische Lösungen Einsteinschen Feldgleıchungen, *Zeit. Phys.* **145,** 235 (1956).
250. J. Géhéniau and R. Debever, Les invariants de courbure de l'espace de Riemann à quatre dimensions, *Bull. Cla. Sci. Acad. Roy. Belgique,* **42,** 114 (1956).
251. F. A. E. Pirani, On the physical significance of the Riemann tensor, *Acta. Phys. Polon.* **15,** 389 (1956).

252. A. TRAUTMAN, Killing's equations and conservation theorems, *Bull. Acad. Polon. Sci.*, Cl. III, **4**, 679 (1956).
253. F. I. FRANKL', The correct setting of Cauchy's problem and properties of harmonic coordinates in the general theory of relativity, *Usp. Mat. Nauk*, **11**, Part 3, 189 (1956).
254. V. A. FOCK, A comment on Frankl's paper "The correct setting of Cauchy's problem and properties of harmonic coordinates in the general theory of relativity", *Usp. Mat. Nauk*, **11**, Part 3, 197 (1956).
255. F. I. MIKHAIL, Solutions of Einstein's field equations for empty space, *A'in Shams Univ. Science Bull.*, No. 1, 125, Cairo (1956).
256. N. S. SINYUKOV, Normal geodesic mappings of Riemann spaces, *Dokl. Akad. Nauk*, **111**, 766 (1956).
256a. A. S. SOLODOVNIKOV, Projective transformation of Riemannian spaces, *Usp. Mat. Nauk* (N.S.) **11**, Part 4, 45 (1956).
257. K. YANO and S. BOCHNER, *Curvature and Betti Numbers*, Princeton (1953).
258. F. A. E. PIRANI, Invariant formulation of gravitational radiation theory, *Phys. Rev.* **105**, 1089 (1957).
259. KH. KERES, On the concept of an inertial system in the general theory of relativity (o ponyatii inertsial'noi sistemy v obshchei teorii otnositel'nosti), collection *Investigations in Theoretical Physics*, Tartu (1957).
260. G. I. KRUCHKOVICH, Motions in Riemann spaces V_4 *Mat. sbor.* **41**, (83), 195 (1957).
261. A. Z. PETROV, Spaces defined by gravitational fields (Prostranstva, opredelyaemye polyami tyagoteniya), Doctorate Thesis, Moscow State University (1957).
262. J. GÉHÉNIAU, Une classification des espaces Einsteiniens, *Compt. Rend. Acad. Sci.* (Paris), **244**, 723 (1957).
263. V. JOSEPH, Physical properties of some empty space-times, *Proc. Cam. Phil. Soc.* **53**, 836 (1957).
264. KH. KERES, Some problems of the general theory of relativity (Nekotorye voprosy obshchei teorii otnositel'nosti), collection *Investigations in Theoretical Physics*, Tartu (1957).
265. Papers from the Conference on the Role of Gravitation in Physics, held at the University of North Carolina, Chapel Hill N.C., Jan., 1957, *Rev. Mod. Phys.* **29**, 351 (1957).
266. PHAM MAU QUAN, Inductions électromagnétiques dans un milieu anisotrope relativiste, *Compt. Rend. Acad. Sci. (Paris)* **245**, 1782 (1957).
267. PHAM MAU QUAN, Inductions électromagnétiques en relativité générale et principe de Fermat, *Arch. Rat. Mech. and Anal.* **1**, 54 (1957).
268. T. AITMURZAYEV, A Method of solving unsteady gas flow equations in general relativity, allowing for dissipative processes, *Dokl. Akad. Nauk*, **113**, 769 (1957).
269. L. MARDER, On the existence of cylindrical gravitational waves, Doctorate Thesis, King's College, London; cf. *Proc. Roy. Soc.* A, **244**, 524 (1958).
270. P. I. PETROV, Second order invariants of the quaternary differential quadratic form, *Dokl. Akad. Nauk*, **113**, 1214 (1957).
271. A. PAPAPETROU, Über periodische nichtsinguläre Lösungen in der allgemeinen Relativitätstheorie, *Ann. Physik* (Ser. 6), **20**, 399 (1957).
272. H. TAKENO, On plane wave solutions of field equations in general relativity, I, *Tensor* (New Ser.), **7**, 97 (1957).
273. H. BONDI, Negative mass in general relativity, *Rev. Mod. Phys.* **29**, 423 (1957).
274. N. S. SINYUKOV, Equidistant Riemannian spaces, *Scientific Annual*, Odessa Univ. (1957).

275. A. Papapetrou, Eine neue Formulierung in der Relativitätstheorie, *Schr. Forschungsinst. Math. Deutschen Akad. d. Wiss. Berlin* **1**, 210 (1957).
276. A. Papapetrou, La problème du mouvement dans la Relativité générale et dans la théorie du champ unifié d'Einstein, *Ann. Inst. Henri Poincaré*, **15**, 173 (1957).
277. H. Treder, Stromladungsdefinition und elektrische Kraft in der einheitlichen Feldtheorie, *Ann. Physik* (Ser. 6), **19**, 369 (1957).
278. H. J. Meister, Die Bewegungsgleichungen in der Theorie des Gravitationsfeldes mit einer Feldfunktion, *Zeit. Phys.* **147**, 531, (1957).
279. A. H. Taub, Singular hypersurfaces in general relativity, *Illinois Jour. Math.* **1**, 370 (1957).
280. A. H. Taub, Approximate solutions of the Einstein equations for isentropic motions of plane-symmetric distributions of perfect fluids, *Phys. Rev.* **107**, 884 (1957).
281. H. Bondi, Plane gravitational waves in general relativity, *Nature*, **179**, 1072 (1957).
282. K. Yano and T. Nagano, Einstein spaces admitting a one-parameter group of conformal transformations, *Ann. Math.* **69**, 451 (1959).
283. F. A. E. Pirani, Tetrad formulation of general relativity theory, *Bull. Acad. Polon. Sci.*, Cl, III, **5**, 143 (1957.)
284. A. Trautman, On the conservation theorems and coordinate systems in general relativity, *Bull. Acad. Polon. Sci.*, Cl. III, **5**, 721 (1957).
285. A. Trautman, Discontinuities of field derivatives and radiation in covariant theories, *Bull. Acad. Polon. Sci.*, Cl. III, **5**, 273 (1957).
286. P. I. Petrov, Correction to the Article "Second Order Invariants of the Quaternary Differential Quadratic Form", *Dokl. Akad. Nauk*, **119**, 846 (1958).
287. A. Z. Petrov, Classification of general gravitational fields, *Izv. Vys. Uch. Zav. Mat. Kazan* No. 6 (7) 226 (1958).
288. A. Trautman, Sur la propagation des discontinuités du tenseur de Riemann, *Compt. Rend. Acad. Sci. (Paris)*, **246**, 1500 (1958).
289. A. Trautman, *Lectures on General Relativity*, pre-print, King's College, London (1958).
290. C. Cattaneo, Sui postulati comuni alla cinematica classica e alla cinematica relativistica, *Atti Accad. Naz. Lincei (8th Ser.) Rend.*, Cl. Sci. fis. mat. e nat., **24**, 526 (1958).
291. L. Bel, Sur la radiation gravitationelle, *Compt. Rend. Acad. Sci. (Paris)*, **247**, 1094 (1958).
292. L. P. Eisenhart, Spaces for which the Ricci scalar R is equal to zero, *Proc. Nat. Acad. Sci. U.S.A.* **44**, 695 (1958).
293. H. Treder, Stosswellen des Gravitationsfeldes, *Ann. Physik* (Ser. 7), **2**, 225 (1958).
294. A. Papapetrou, Über periodische Gravitations- und elektromagnetische Felder in der allgemeinen Relativitätstheorie, *Ann. Physik*, (Ser. 7), **1**, 186 (1958).
295. A. Papapetrou, Über zeitabhängige Lösungen der Feldgleichungen der allgemeinen Relativitätstheorie, *Ann. Physik* (Ser. 7), **2**, 87 (1958).
296. L. Ó Raifeartaigh and J. L. Synge, A property of empty space-time, *Proc. Roy. Soc. A*, **246**, 299 (1958).
297. W. Israel, Discontinuities in spherically symmetric gravitational fields and shells of radiation, *Proc. Roy. Soc. A*, **248**, 404 (1958).
298. J. L. Synge, Whittaker's contributions to the theory of relativity, *Proc. Edinburgh Math. Soc.* **11**, 39 (1958).
299. L. Bel, Étude algébrique d'un certain type de tenseurs de courbure. Le cas 3 de Petrov, *Compt. Rend. Acad. Sci. (Paris)*, **247**, 2096 (1958).
300. A. Trautman and W. Tulczyjew, Gravitation and invariance (in Polish), *Postepy Fizyki*, **9**, 3 (1958).

301. A. TRAUTMAN, Radiation and boundary conditions in the theory of gravitation, *Bull. Acad. Polon. Sci.* **6,** Sér. Sci. Math. Astr. Phys. 407 (1958).
302. A. TRAUTMAN, Boundary conditions at infinity for physical theories, *Bull. Acad. Polon. Sci.* **6,** Sér. Sci. Math. Astr. Phys. 403 (1958).
303. G. I. KRUCHKOVICH, Motions in Kagan sub-projective spaces, *Nauch. Doklad. Vysh. Shkol.* No. 1, 43 (1958).
304. H. TAKENO, On plane wave solutions of the field equations in general relativity, II, *Tensor* (New Ser.), **8,** 59 (1958).
305. V. JOSEPH, On apparent luminosity in general relativity, *Month. Not. Roy. Astr. Soc.* **118,** 631 (1958).
306. I. T. TODOROV, A uniqueness theorem for the wave equation (On the Fock–Frankl' discussion). *Usp. Mat. Nauk,* **13,** Part **2,** 211 (1958).
307. KU CHAO-KHAO, Some types of homogeneous Riemannian spaces, *Dokl. Akad. Nauk,* **122,** 171 (1958).
308. W. KUNDT, Methoden zur Charakterisierung von Lösungen der Einsteinschen Gravitationsfeldgleichungen, Dissertation (pre-print), University, Hamburg (1958).
309. L. BEL, Définition d'une densité d'énergie et d'un état de radiation totale généralisée, *Compt. Rend. Acad. Sci. (Paris)* **246,** 3015 (1958).
310. V. HLAVATÝ, The structure of our space, *Aeronautical Eng. Rev.* **17,** 25 (April, 1958).
311. D. GEISSLER and H. TREDER, Über ebene Wellen in der allgemeinen Relativitätstheorie, *Tensor* (New Ser.), **8,** 165 (1958).
312. PHAM TAN HUANG, La méthode des singularités pour les équations du mouvement en relativité générale et en théorie du champ unifié, *Ann. Schuola norm. super. Pisa,* Sci. fis. e mat. (Ser. 3), **12,** 425 (1958).
313. MME Y. FOURÈS-BRUHAT, La relativité générale, Séminaire de mécanique analytique et de mécanique céleste, held by M. Janet. Fac. Sci. Paris, Ire. année, 1957/58, Paris (1958) (mimeographed).
314. V. A. BRUMBERG, Equations of motion and coordinate conditions in the relativistic N-body problem, *Astr. Zh.* **35,** 893; English translation, *Soviet Astr.* **2,** 839 (1958).
315. YA. I. PUGACHEV, Differential identities and the solubility of the Euler–Lagrange equations, *Trudy Krasnodar Inst. pishchev. prom.,* part 20, 55–57 (1958).
316. A. P. NORDEN, On complex representation of Lorentz-space tensors, *Izv. Vys. Uch. Zav. Mat.,* No. 1 (8) 156 (1959).
317. A. D. ALEKSANDROV, Philosophical content and significance of the theory of relativity, *Voprosy filosofii,* No. 1 (1959).
318. A. P. NORDEN and V. V. VISHNEVSKII, The complex representation of the invariants of a four-dimensional Riemannian space, *Izv. Vys. Uch. Zav. Mat.,* No 2 (9), 176 (1959).
319. T. TAKASU, Ein Seitenstück der Relativitätstheorie als eine erweiterte Laguerresche Geometrie, *Proc. Japan. Acad.* **35,** 65 (1959).
320. A. Z. PETROV, Symmetric gravitational fields, *Izv. Vys. Uch. Zav. Mat.* No. 2 (9) 189 (1959).
321. B. T. VAVILOV, Weak gravitational fields, *Izv. Vys. Uch. Zav. Fiz.* No. **2,** 73 (1959).
322. R. K. SACHS, Propagation laws for null and type III gravitational waves, *Zeit. Phys.* **157,** 462 (1960).
323. L. BEL, Introduction d'un tenseur du quatrième ordre, *Compt. Rend. Acad. Sci. (Paris),* **248,** 1297 (1959).
324. L. BEL, Quelques remarques sur la classification de Petrov. Étude du cas 2, *Compt. Rend. Acad. Sci. (Paris),* **248,** 2561 (1959).

325. V. Hlavatý, The holonomy group, I. The curvature tensor, *Jour. Math. and Mech.* **8,** 285 (1959).
326. V. Hlavatý, The holonomy group, II. The Lie group induced by a tensor, *Jour. Math. and Mech.* **8,** 597 (1959).
327. V. Hlavatý, The holonomy group, III. Metrisable spaces, *Jour. Math. and Mech.* **9,** 89 (1960).
328. V. Hlavatý, The holonomy group, IV. The general L_n with symmetric connection, *Jour. Math. and Mech.* **9** 453 (1960).
329. T. Takasu, Adjusted relativity theory: Applications of extended Euclidean geometry, extended equiform geometry and extended Laguerre geometry to physics, *Yokohama Math. Jour.* **7,** 1 (1959).
330. D. Geissler *et al.*, Die Gravitationstrahlung eines zeitweilig nichtstationären Systems, *Ann. Physik*, (Ser. 7), **2,** 344 (1958).
331. A. Papapetrou and H. Treder, Das Sprungproblem erster Ordnung in der allgemeinen Relativitätstheorie, *Math. Nachr.* **20,** 53 (1959).
332. A. H. Taub, Small motions of spherically symmetric distributions of matter, Les Théories Relativistes de la Gravitation (Colloque International, Royaumont, 1959), Centre National de la Recherche Scientifique, 173 (1962), Paris.
333. R. Arnowitt and S. Deser, Quantum theory of gravitation: general formulation and linearized theory, *Phys. Rev.* **113,** 745 (1959).
334. R. Arnowitt *et al.*, Dynamical structure and definition of energy in general relativity, *Phys. Rev.* **116,** 1322 (1959).
335. P. G. Bergmann, Summary of the 'Colloque International, Royaumont' (1959), Les Théories Relativistes de la Gravitation, Centre National de la Recherche Scientifique, 463 (1962), Paris.
336. D. J. Newman and C. W. Kilmister, A new expression for Einstein's law of gravitation, *Proc. Cam. Phil. Soc.* 55, 139 (1959).
337. A. Z. Petrov *et al.*, Classification of general gravitational fields by groups of motion, I, *Izv. Vys. Uch. Zav. Mat. Kazan.* No. 6, 118 (1959).
338. N. M. Petrova, Conservation laws for a system of rotating bodies in the general theory of relativity, Collection *Investigations of Transfer Processes Problems of the Theory of Relativity*, Alma-Ata, Kazakhsk. Univ., 192 (1959).
339. Z. A. Aitikeyeva and N. M. Petrova, A system of spherically symmetric bodies in the general theory of relativity, Collection *Investigations of Transfer Processes. Problems of the Theory of Relativity*, Alma-Ata, Kazakhsk. Univ., 209 (1959).
340. G. Murzagaliyev, Angular momentum tensors and the moments of a force in relativistic mechanics, Collection *Investigations of Transfer Processes. Problems of the Theory of Relativity*, Alma-Ata, Kazakhsk. Univ., 229 (1959).
341. N. V. Mitskevich, A non-linear vacuum effect in gravitation theory, *Zh. Eksp. Teor. Fiz.* **36,** 1207. English translation *Soviet. Phys. JETP.* **9,** 859 (1959).
342. R. Debever, Tenseur de super-énergie, tenseur de Riemann: Cas singuliers, *Compt. Rend. Acad. Sci.* (*(Paris)*, **249,** 1744 (1959).
343. R. Debever, Sur le tenseur de super-énergie, *Compt. Rend. Acad. Sci. (Paris)*, **249,** 1324 (1959).
344. H. Takeno, A note on the theory of gravitational waves, *Tensor* (New. Ser.), **9,** 73 (1959).
345. R. Arnowitt *et al.*, Energy and the criteria for radiation in general relativity, *Phys. Rev.* **118,** 1100 (1959).
346. T. L. Synge, A theory of elasticity in general relativity, *Math. Zeit.* **72,** 82 (1959).
347. V. I. Golikov, Thesis, Faculty of Gravitation, Kazan University (1959).

348. G. B. MAST and J. STRATHDEE, On the relativistic interpretation of astronomical observations, *Proc. Roy. Soc.* A, **252,** 476 (1959).
349. A. G. IOSIF'YAN, Problems of the unified theory of electromagnetic and gravitational fields, *Izd. Akad. Nauk. Arm. SSR*, Yerevan (1959).
350. H. TAKENO, On geometric properties of some plane wave solutions in General Relativity, I, *Tensor* (New Ser.), **9,** 76 (1959).
351. H. TAKENO, On geometric properties of some plane wave solutions in General Relativity, II, *Tensor* (New Ser.), **9,** 162 (1959).
352. M. IKEDA and Y. MIYACHI, Coordinate systems in general relativity and geometrical representation for physical quantities, *Progr. Theor. Phys.*, Supplement No. 9, 45 (1959).
353. H. BONDI *et al.*, I. Gravitational waves in general relativity, III, exact plane waves, Proc. Roy. Soc. A, **251,** 519, 533 (1959).
354. F. A. E. PIRANI, Gravitational waves in general relativity, IV, the gravitational field of a fast-moving particle, *Proc. Roy. Soc.* A, **252,** 96 (1959).
355. D. R. BRILL, On the positive definite mass of the Bondi–Weber–Wheeler time-symmetric gravitational waves, *Annals of Phys.* **7,** 466 (1959).
356. H. ARAKI, On weak time-symmetric gravitational waves, *Annals of Phys.* **7,** 456 (1959).
356a. A. PERES and N. ROSEN, On Cauchy's problem in general relativity I, *Il Nuovo Cimento* (Ser. 10), **13,** 430 (1959).
357. A. Z. PETROV *et al.*, Classification of general gravitational fields by groups of motion, II, *Izv. Vys. Uch. Zav. Mat. Kazan.*, No. 1, 175 (1960).
358. A. Z. PETROV *et al.*, Classification of general gravitational fields by groups of motion, III, *Izv. Vyz. Uch. Zav. Mat. Kazan*, No. 4, 158 (1960).
359. H. TAKENO, On the theory of gravitational waves in general relativity, *Tensor* (New Ser.). **10,** 34 (1960).
360. R. ARNOWITT *et al.*, Interior Schwarzschild solutions and interpretation of source terms, *Phys. Rev.* **120,** 321 (1960).
361. R. P. KERR, On the quasi-static approximation in general relativity, *Nuovo Cimento* (Ser X), **16,** 26 (1960).
362. R. PENROSE, A spinor approach to general relativity, *Annals of Phys.* **10,** 171 (1960).
363. P. JORDAN *et al.*, Strenge Lösungen der Feldgleichungen der Allgemeinen Relativitätstheorie, *Akad. Wiss. Mainz. Abh. Math.-Nat. Klasse*, 21 (1960).
364. Reference deleted.
365. V. HLAVATÝ, An almost complex space of order p and characteristic $(n_1, \ldots, n_p)$, *Tensor* (New Ser.), **10,** 90 (1960).
366. B. HOFFMAN, On Yilmaz' new approach to general relativity, *Jour. Math. and Mech.* **9,** 445 (1960).
367. V. HLAVATÝ, Rigid motion in a Riemannian space V_n, I, A reccurrent V_n. *Rend. Circ. Mat. Palermo* (Ser. II), **9,** 51 (1960).
368. V. HLAVATÝ, Rigid motion in a Riemannian space V_n, II, A regular V_n, *Rend. Circ. Mat. Palermo* (Ser. II), **9,** 125 (1960).
369. L. INFELD, A new form of the geodesic line equation, *Bull. Acad. Polon Sci.*, Ser. Math. Astr. Phys., **8,** 559 (1960).
370. L. INFELD, The EIH and the k-approximation methods, *Bull. Acad. Polon Sci.*, Ser. Math. Astr. Phys., **9,** 93 (1960).
371. M. IKEDA and S. ABE, On groups of motions in space-time with a non-symmetric fundamental tensor $g_{\mu\nu}$, I, *Tensor* (New Ser.), **10,** 26 (1960).
372. Reference deleted.

373. M. Lenoir, Sur les solutions à symétrie sphérique de la théorie du champ unifié, *Comp. Rend. Acad. Sci. (Paris)* **250,** 981 (1960).
374. A. Papapetrou and H. Treder, Zur Frage der Existenz von singularitätsfrein Lösungen der allgemein-relativistischen Feldgleichungen, die Teilchenmodelle darstellen könnten, I. *Ann. Physik* (Ser. 7), **3,** 360, (1959). II. *Ann. Physik* (Ser. 7), **6,** 311 (1960).
375. F. A. E. Pirani, A note on Fermi coordinates, *Proc. Roy. Irish. Acad.* **61,** *A*, 9 (1960).
376. Reference deleted.
377. J. L. Synge, Optical observations in general relativity, *Rend. Sem. Mat. Fis. di Milano*, **30,** 271 (1960).
378. T. Shibata, Geometrical definition of plane gravitational waves in general relativity, *Mem. Fac. Eng., Hiroshima Univ.* **1,** 165 (1960).
379. C. Soergel-Fabricius, Thirring-Effect im Einsteinkosmos, *Zeit. Phys.* **159,** 541 (1960).
380. I. Smorodinskij, Der gegenwärtige Stand der Theorie des β-Zerfalls, *Fortsch. Phys.* **8.** 426 (1960).
381. H. Treder, Über die allgemeine Form der Fortpflanzungsgesetze der Stosswellen des Gravitationsfeldes, *Ann. Physik* (Ser. 7), **6,** 307 (1960).
382. Reference deleted.
383. H. Takeno, On geometric properties of some plane wave solutions in general relativity, III, *Tensor* (New Ser.), **10,** 47 (1960).
384. Y. Takasu, Four transverse masses and a quaternary in relativistic physics and a ternary in classical physics, *Yokohama Math. Jour.* **8,** 69 (1960).
385. T. Takasu, Extended conformal geometry obtained by extending the group parameters to functions of coordinates, I, *Yokohama Math. Jour.* **8,** 85 (1960).
386. A. Das, Cellular space-time and quantum field theory, *Nuovo Cimento* (Ser. X), **18,** 482 (1960).
387. Le-Thanh-Phong, Vecteur de Poynting en relativité générale, *Comp. Rend. Acad. Sci. (Paris)* **250,** 987 (1960).
388. J. P. Terletsky, Le Principe de causalité et le second principe de thermodynamique, *Jour. de Phys. et Radium*, **21,** 681 (1960).
389. J. P. Terletsky, Sur la théorie statistique des champs non-linéaires, *Jour. de Phys. et Radium*, **21,** 711 (1960).
390. T. T. Vescan, Processes of microcosm and unified field theories, *Analele Ştiinţifice Univ.*, 'Al. I Cuza', Iaşi (New Ser.) Sec. I, **6,** 101 (1060).
391. G. Vranceanu, Reducerea le formâ canonicâ a unei forme pâtratice prin transformâri iperbolice, *Anal. Univ.* 'C. I. Parhon.' Bucharest, Ser. mat. fis., No. 25 (1960).
392. I. D. Novikov, The energy-momentum pseudo-tensor of the gravitational field, *Vestnik*, Moscow Univ., Ser. III, Fiz. Astr., No. 2, 59 (1960).
393. V. Hlavatý, Aperçu général de la Théorie du Champ unifié d'Einstein, *Colloque sur la Théorie de la Relativité*, 1959, 9, Centre Belge Rech. Math. (1960).
394. Reference deleted.
395. Reference deleted.
396. Reference deleted.
397. V. Hlavatý, Proper time, apparent time and formal time in the twin paradox, *Jour. Math. and Mech.* **9,** 733 (1960).
398. A. Z. Petrov, *Lenin on Space and Time and the General Theory of Relativity*, Report to the Jubilee Session at Leningrad University, April, 1960, pp. 119 (1960).

399. V. V. RYZHKOV, Tangentially degenerate surfaces, *Dokl. Akad. Nauk.* **135,** 20 (1960).
400. V. A. SKRIPKIN, Point explosion in a perfect incompressible fluid in the general theory of relativity, *Dokl. Akad. Nauk,* **135,** 1072 (1960). English Trans. *Soviet Phys. Dokl.* **5,** 1, 1183 (1960–1).
401. V. A. SKRIPKIN, Discontinuous centro-symmetric movement of an ultra-relativistic gas in the general theory of relativity, *Zh. Prikl. Mekh. Tekh. Fiz.* No. 4, 3 (1960).
402. A. A. SOKOLOV, The clock paradox in the motion of charged particles in a magnetic field, *Dokl. Akad Nauk,* **131,** 75 (1960). English Trans. *Soviet Phys. Dokl.* **5,** 287 (1960–1).
403. N. A. CHERNIKOV, Reduction of the relativistic collision integral to the Boltzmann form, *Dokl. Akad. Nauk,* **133,** 84 (1960). English trans. *Soviet Phys. Dokl.* **5,** 764 (1960–1).
404. N. A. CHERNIKOV, Relativistic kinetic equation and the equilibrium state of a gas in a static, spherically symmetric gravitational field, *Dokl. Akad. Nauk,* **133,** 333 (1960). English trans. *Soviet Phys. Dokl.* **5,** 786 (1960–1).
405. A. DAS, Birkhoff's theorem for electromagnetic fields in general relativity, *Prog. Theor. Phys.* **24,** 915 (1960).
406. R. DEBEVER, Tenseur de super-énergie et composantes irréductibles du tenseur de Riemann, *Comp. Rend. Acad. Sci. (Paris),* **250,** 64 (1960).
407. R. M. HEIDEL'MAN, Conformal theory of two-parameter families of spheres, *Dokl. Akad. Nauk,* **134,** 753 (1960).
408. L. A. DRUZHKIN, The problem of photon structure and the red shift, *Uch. Zap. Novgorod Gos. Ped. Inst.* **1,** No. 1 (1960).
409. A. E. LEVASHEV, The theory of the Lorentz connection in relativistic motion, Summary of Thesis for Doctorate (1960). Belo-Russia State University, Minsk.
410. E. M. LIFSHITS and I. M. KHALATNIKOV, Singularities of the cosmological solutions of the gravitational equations, I, *Zh. Eksp. Teor. Fiz.* **39,** 149 (1960). English Trans. *Soviet Phys. J. E. T. P.* **12,** 108 (1961).
411. E. M. LIFSHITS and I. M. KHALATNIKOV, Singularities of the cosmological solutions of the gravitational equations, II, *Zh. Eksp. Teor. Fiz.* **39,** 800 (1960). English Trans. *Soviet Phys. J. E. T. P.* **12,** 558 (1961).
412. F. A. E. PIRANI, *Introduction to the Concepts of Gravitational Radiation Theory,* Aeronautical Research Lab. Tech. Note, 60–143 (1960).
413. F. A. E. PIRANI, *Survey of Gravitational Radiation Theory. Recent Developments in General Relativity,* p. 89, Pergamon Press (1962).
414. A. PAPAPETROU, *Periodische Lösungen in der einheitlichen Feldtheorie mit nicht-symmetrischen* $g_{\mu\nu}$, Veb Deutscher Verlag der Wissenschaften (1960).
415. V. A. FOCK, Einstein statics in conformal space, *Zh. Eksp. Teor. Fiz.* **38,** 1476 (1960). English Trans. *Soviet Phys. J. E. T. P.* **11,** 1067.
416. J. WINOGRADZKI, Formalisme tensoriel et spinoriel incluant les parités, *Jour. Phys. et Radium,* **21,** 835 (1960).
417. A. PERES, Null electromagnetic fields in general relativity theory, *Phys. Rev.* **118,** 1105 (1960).
418. A. PERES, Gauge covariance of spinor geometry, *Nouvo Cimento* (Ser. X), **21,** 182 (1961).
419. A. PERES, On geometrodynamics and null fields, *Annals of Phys.* **14,** 419 (1961).
420. B. HOFFMAN, Noon–midnight red shift, *Phys. Rev.* **121,** 337 (1961).
421. V. HLAVATÝ, Einstein–Maxwell fields, *Jour. de Math.* (Ser. 9), **40,** 1 (1961).
422. V. HLAVATÝ, The holonomy group V. Weyl space W_4, First Part, *Jour. Math. and Mech.* **10,** 317 (1961).

423. L. INFELD, On the most Cartesian-like coordinate system, *Bull. Acad. Polon. Sci.* Ser. Math. Astr. Phys. **9,** 299 (1961).
424. P. JORDAN, The mathematical theory of quasi-order, semi-groups of idempotents and non-commutative lattices – a new field of modern algebra, European Office Aerospace Research (1961).
425. J. N. GOLDBERG and R. P. KERR, Some applications of the infinitesimal holonomy group to the Petrov classification of Einstein spaces, *Jour. Math. Phys.* **2,** 327 (1961).
426. R. P. KERR and J. N. GOLDBERG, Einstein spaces with four-parameter holonomy groups, *Jour. Math. Phys.* **2,** 332 (1961).
427. C. W. KILMISTER, *Newtonian Models for General Relativity*, Technical Note, Dept. of Maths. King's College, University of London, 5th July (1961).
428. A. Z. PETROV, *Space-Time and Matter (Prostranstvo-vremeni i materiya) An Elementary Account of the Modern Theory of Relativity (Elementarnyi ocherk sovremennoi teorii otnositel'nosti)*, Kazan' University (1961).
429. A. PAPAPETROU and H. TREDER, Das Sprungproblem nullter Ordnung in der allgemeinen Relativitätstheorie, *Math. Nachr.* **23,** 371 (1961).
430. F. A. E. PIRANI and A. SCHILD, Geometrical and physical interpretation of the Weyl conformal curvature tensor, *Bull. Acad. Polon. Sci.*, Ser. Math. Astr. Phys., **9,** 543 (1961).
431. J. L. SYNGE, How to pass to an exact solution of Einstein's vacuum field equations, starting from a linear approximation, *Proc. Roy. Irish Acad.* **61,** A, 37 (1961).
432. J. F. SCHELL, A classification of four-dimensional Riemannian spaces, *Jour Math. Phys.* **2,** 202 (1961).
433. H. TREDER, Zur Rolle der Planckschen Elementarlänge in Diracs Quantentheorie des Gravitationsfeldes, *Monatsber. Deutsch. Akad. Wiss. Berlin*, **3,** 241 (1961).
434. H. TREDER, Über eine einheitliche Feldtheorie mit symmetrischer Affinität, *Tensor* (New Ser.), **11,** 1 (1961).
435. H. TAKENO, Some plane wave-like solutions of non-symmetric unified field theory, *Tensor* (New Ser.), **11,** 263 (1961).
436. I. D. NOVIKOV, Curvature properties of the comoving space of some cosmological models considered in a quasi-Newtonian approximation, *Astr. Zh.*, **38,** 961 (1961). English Trans. *Soviet Astr.* **5,** 733 (1962).
437. I. D. NOVIKOV, The problem of the space-time metric inside the Schwarzschild singular sphere, *Astr. Zh.* **38,** 565 (1961). English Trans. *Soviet Astr.* **5,** 423 (1961).
438. A. Z. PETROV, The geodesic mapping of Einstein spaces, *Izv. Vys. Uch. Zav. Mat.* No. 2 (21), 130 (1961).
439. K. P. STANYUKOVICH, The radiation of gravitational waves by elementary particles, *Vestnik*, Moscow Univ. Ser. III, Fiz., Astr., No. 5, 71 (1961.)
440. A. S. SOLODOVNIKOV, *Riemannian Spaces with Common Geodesics*, Moscow State University (1961).
441. V. A. SKRIPKIN, Discontinuous centro-symmetric motions of an ultra-relativistic gas in the general theory of relativity, *Astr. Zh.* **38,** 192 (1961). English Trans. *Soviet Astr.* **5,** 143 (1961).
442. V. A. SKRIPKIN, An exact solution of the one-dimensional equations of relativistic hydrodynamics with a discontinuity in the transformation of the residual mass of the material, *Dokl. Akad. Nauk*, **138,** 81 (1961). English Trans. *Soviet Phys. Dokl.* **6,** 377 (1961).
443. V. A. SKRIPKIN, A class of progressive motions of an ultra-relativistic gas, *Dokl. Akad. Nauk*, **136,** 791 (1961). English Trans. *Soviet Physics Dokl.* **6,** 115 (1961).
444. V. A. SKRIPKIN, Problems with central symmetry in relativistic hydrodynamics, Summary of Doctorate Thesis, Moscow State University (1961).

445. P. G. BERGMANN, Observables in general relativity, *Rev. Mod. Phys.* **33,** 510 (1961).
446. P. G. BERGMANN, Gauge-invariant variables in general relativity, *Phys. Rev.* **124,** 274 (1961).
447. H. BONDI, On the reception of electromagnetic waves, *Proc. Roy. Soc.* A, **261,** (1961).
448. H. BONDI, Gravitational waves, *Endeavour*, **20,** 121 (1961).
449. Cospar, Abstracts of scientific papers by U.S. participants, Fourth Cospar Meeting, Florence (1961).
450. A. DAS *et al.*, Stationary weak gravitational fields to any order of approximation, *Proc. Roy. Soc.* A, **263,** 451 (1961).
451. V. F. KAGAN, *Subprojective Spaces (Subproektivnye prostranstva)*, Gos. Izdat, Fiz. Mat., Moscow (1961).
452. Reference deleted.
453. M. E. GERTSENSHTEIN, Conservation laws in general relativity, *Zh. Eksp. Teor. Fiz.* **40,** 114 (1961). English Trans. *Soviet Physics J. E. T. P.* **13,** 81 (1961).
454. YA. B. ZEL'DOVICH and YA. A. SMORODINSKII, An upper limit to the density of neutrinos, gravitons and barions in the universe, *Zh. Eksp. Teor. Fiz.* **41,** 907 (1961). English Trans. *Soviet Phys. J.E.T.P.* **14,** 647 (1962).
455. E. M. LIFSHITZ *et al.*, Singularities of cosmological solutions of the gravitational equations III, *Zh. Eksp. Teor. Fiz.* **40**, 1847 (1961). English Trans. *Soviet Phys.*, *J.E.T.P.* **13,** 1298 (1961).
456. N. V. MITSKEVICH, The transformation properties of some physical quantities the general theory of relativity, *Dokl. Bolg. Akad. Nauk* **14,** 439 (1961).
457. H. BONDI, *Research Program in Relativity Physics*, Aeronautical Research Lab. 16 (1961).
458. A. KOMAR, Positive-definite energy density and global consequences for general relativity, *Phys. Rev.* **129,** 1873 (1963).
459. P. S. FLORIDES, Applications of Møller's theory on energy and its localization in general relativity, *Proc. Camb. Phil. Soc.* **58,** 102 (1962).
460. J. L. SYNGE, On a certain non-linear differential equation, *Proc. Roy. Irish Acad.* **62,** A, 17 (1961).
461. A. PERES, Electromagnetic radiation from a freely gravitating charge, *Annals of Phys.* **12,** 86 (1961).
462. V. V. RADZIEVSKII, Two inclinometers for measuring the gravitational Effect, *Bull. Vsesoyuzn. astron. ob-va.* No. 31 (39) (1961).
463. V. V. RADZIEVSKII and B. N. GIMMEL'FARB, The apparent paradoxes of astronomical aberration, *Bull. Vsesoyuzn. astron. ob-va.* No. 31 (38) (1961).
464. P. S. FLORIDES, The electromagnetic energy and the gravitational mass of a charged particle in general relativity, *Proc. Camb. Phil. Soc.* **58,** 110 (1962).
465. I. ROBINSON, Some spherical gravitational waves in general relativity, *Proc. Roy. Soc.* A, **265,** 463 (1962).
466. A. PERES, Three-component spinors, *Jour. Math. and Mech.* **11,** 61 (1962).
467. A. LICHNEROWICZ, Propagateurs et quantification en relativité générale, *Proc. Conférence internationale sur les théories relativistes de la gravitation*, Warsaw (1962), p. 177, P.W.N., Warsaw (1964).
468. P. G. BERGMANN, Asymptotic properties of gravitating systems, *Proc. Conférence internationale sur les théories relativistes de la gravitation*, Warsaw (1962), p. 149, P.W.N., Warsaw (1964).
469. P.A.M. DIRAC, The motion of an extended particle in the gravitational field, *Proc. Conférence internationale sur les théories relativistes de la gravitation*, Warsaw 1962, p. 163, P.W.N., Warsaw (1964).

470. M. TRÜMPER, Zur Bewegung von Probekörpen in Einsteinschen Gravitations-Vakuumfeldern, *Zeit. Phys.* **168,** 55 (1962).
471. V. L. GINZBURG, Experimental verifications of the general theory of relativity, *Recent Developments in General Relativity*, p. 57, Pergamon Press (1962).
472. Reference deleted.
473. H. BONDI, Relativity and cosmology, *Recent Developments in General Relativity*, p. 47, Pergamon Press (1962).
474. J. L. SYNGE, Relativity based on chronometry, *Recent Developments in General Relativity*, p. 441, Pergamon Press (1962).
475. V. HLAVATÝ, The holonomy group VI, Weyl space W_4, Second Part, *Jour. Math. and Mech.* **11,** 35 (1962).
476. A. Z. PETROV, Invariant classification of gravitational fields, *Recent Developments in General Relativity*, p. 371, Pergamon Press (1962).
477. A. Z. PETROV, Gravitational field geometry as the geometry of automorphisms, *Recent Developments in General Relativity*, p. 379, Pergamon Press (1962).
478. A. Z. PETROV, *Modern Differential Geometry and the General Theory of Relativity*, Paper at the All-Union Conference on Geometry, Kiev (1962).
479. A. Z. PETROV *et al.*, Classification of general gravitational fields by groups of motion, IV, *Izv. Vys. Uch. Zav. Mat. Kazan*, No. 1, 130 (1962).
480. V. I. PUSTOVOIT and M. E. GERTSENSHTEIN, Gravitational radiation of a relativistic particle, *Zh. Eksp. Teor. Fiz.* **42,** 163 (1962). English Trans. *Soviet Phys., J. E. T. P.* **15,** 116 (1962).
481. S. SELEZNEV, The motion of test particles in a gravitational field (O dvizhenii probnykh chastits v pole tyagoteniya), Summary of a paper to the Student Scientific Conference at Kazan' State University (1962).
482. K. P. STANYUKOVICH, The interaction of bodies radiating gravitational waves, *Vestnik* (Moscow Univ.), Sec. III., Fiz. Astr., No. 1, 78 (1962).
483. P. L. STUSYAK, On the problem of the cause of the relation between the mass and velocity of a body. Conference of "Society for popularizing scientific knowledge", Odessa (1962).
484. E. A. TAGIROV, A scalar field admitting a group of motions, (o skalyarnom pole, dopuskayushchem gruppu dvizhenii), Summary of paper to the comprehensive scientific post-graduate students' conference for 1962, pp. 103–7.
485. F. I. FEDOROV, Combination of Lorentz group parameters, *Dokl. Akad. Nauk*, **143,** 56 (1962). English Trans. *Soviet Maths.* **3,** 357 (1962).
486. Bulletin on General Relativity and Gravitation, No. 1 Bern (1962).
487. E. J. BELINFANTE and J. C. GARRISON, On the uniqueness of Fock's harmonic coordinate systems in the presence of static, spherically symmetric sources, *Phys. Rev.* **125,** 1124 (1962).
488. P. F. BILYALOV, The generalization of a theorem on groups of motion to the case of groups of conformal transformations (Ob obobshchenii odnoi teoremy dlya grupp dvizhenii na sluchai grupp konformykh preobrazovanii), Summary of a paper to the comprehensive scientific post-graduate students' conference for 1962, pp. 96–100.
489. V. I. GOLIKOV, The motion of test particles in the general theory of relativity (o dvizhenii probnykh chastits v obshchei teorii otnositel'nosti), Summary of a paper to the comprehensive scientific post-graduate students' conference (1962), pp. 100–3.
490. I. I. GUTMAN, General covariant method of successive approximations in the general theory of relativity (Obshchekovariantny metod posledovatel'nykh prib-

lizhenii v obshchei teorii otnositel'nosti), Thesis for Doctorate, Uzbekistan Academy of Sciences (1962).
491. V. P. Kaigorodov, Motions in Einstein spaces (o dvizheniyakh v prostranstvakh Einshteina), Summary of paper to comprehensive scientific post-graduate students' conference (1962), Kazan' State University, pp. 89–92.
492. S. S. Moiseyev, The influence of relativistic effects on the kinetics of a charged plasma, Thesis for Doctorate, Siberian section of the U.S.S.R. Academy of Sciences (1962).
493. J. N. Goldberg, Dynamical variables and surface integrals, *Recent Developments in General Relativity*, p. 241, Pergamon Press (1962).
494. Reference deleted.
495. A. Peres, On Cauchy's problem in general relativity, II. *Il Nuovo Cimento* (Ser. 10) **26,** 53 (1962). III, *Il Nuovo Cimento*, **26,** 182, (1962).
496. N. Rosen, Flat-space metric in general relativity theory, *Annals of Phys.* **22,** 1 (1962).
497. E. Newman *et al.*, Empty-space generalization of the Schwartzschild metric, *Jour. Math. Phys.* **4,** 915 (1963).
498. P. Szymanski, Extraits de la théorie des intensives et de la théorie de matière continue, Warsaw, pre-print (1962).
499. Ya. T. Terletskii, Positive, negative and imaginary proper masses (Polozhitel'nye, otritsatel'nye i mnimye sobstvennye massy), Addition to syllabus of lectures on the theory of relativity for 1962. Moscow (1962).
500. A. Z. Petrov, Classification invariante des champs de gravitation, *Les théories relativistes de la gravitation*, p. 107, Centre National de la Recherche Scientifiques. Paris (1962).
501. R. P. Kerr, Scalar invariants and groups of motions in a V_n with positive-definite metric tensor, *Tensor* (New Ser.), **12,** 74 (1962).
502. J. L. Synge, Tensorial integral conservation laws in general relativity, *Les Théories relativistes de la Gravitation*, p. 75, Centre National de la Recherche Scientifique, Paris (1962).
503. J. L. Synge, Action at a distance and metaphysics, *Nature*, **194,** 47 (1962).
504. A. Das, A class of exact solutions of certain classical field equations in general relativity, *Proc. Roy. Soc.* A, **267,** (1962).
505. V. R. Kaigorodov, Einstein spaces of maximum mobility, *Dokl. Akad. Nauk.* **146,** 793 (1962). English Trans. *Soviet Phys. Dokl.* **7,** 893 (1963).
506. A. Z. Petrov, Centrally-symmetric gravitational fields, *Zh. eksp. teor. fiz.* **44,** 1525 (1963). English trans. *Sov. Phys. J.E.T.P.* **17,** 1026 (1963).
507. A. M. Anchikov, On conformally reducible gravitational fields with uniform distributions of matter, *Uch. Zap.* **123,** book 2, 96, Kazan State Univ. (1963).
508. A. M. Anchikov, Exact solutions of the equations of the general theory of relativity for metrics of conformally-reducible form. *Uch. Zap.* **123,** book 12, 44, Kazan State Univ. (1963).
509. A. Hamoui, Sur le théorème de Birkhoff et la solution "radiative" de Petrov, *C.R. Acad. Sci. (Paris)* **258,** 6085 (1964).

INDEX

PRINTED IN HUNGARY